INTRODUCTION TO
WILDLIFE AND FISHERIES

An Integrated Approach

INTRODUCTION TO
WILDLIFE AND FISHERIES

An Integrated Approach

Charles G. Scalet   Lester D. Flake   David W. Willis

South Dakota State University

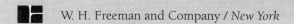 W. H. Freeman and Company / *New York*

Cover illustration: Ring-necked pheasant photograph courtesy of South Dakota Tourism; snail darter photograph courtesy of Richard T. Bryant.

Library of Congress Cataloging-in-Publication Data

Scalet, Charles G.

    Introduction to wildlife and fisheries: an integrated approach /
Charles G. Scalet, Lester D. Flake, David W. Willis.

      p.   cm.

    Includes bibliographical references and index.

    ISBN 0-7167-2816-8 (softcover)

    1. Wildlife conservation.   2. Fishery conservation.   3. Animal
ecology.   I. Flake, Lester D.   II. Willis, David W.   III. Title.

QL82.S3 1995

333.95'4—dc20

95-35545

CIP

Printed in the United States of America

First printing 1996, HP

To Ginger, Marcia, and Susan for their constant support, forbearance, and love.

CONTENTS

Part II THE BIOTA 59

Part IV THE HUMAN USERS   357

PREFACE

In writing this book we have attempted to fill a void that we have long believed should be addressed—the lack of integrated introductory information concerning both wildlife and fisheries. While there are numerous texts in each field, many of them are intended for upper-level courses, and none incorporate both fisheries and wildlife in an even manner.

We expect the primary audience for this book to be individuals taking their first course in wildlife and/or fisheries. In some cases, this first course will combine wildlife and fisheries; the utility of this book to those readers is obvious. In other cases, this initial course may include only wildlife or only fisheries; an integrated text should be useful to such readers by putting both of these similar fields into perspective. In most cases, such introductory courses will be in the first or second year of a college program; however, senior-level high school students in specialized courses and general interest audiences may also find this text useful. The book is designed as an introduction to the principles and practices involved in wildlife and fisheries and the ways in which those principles and practices relate to the organisms in question, their habitats, and their human users.

People new to wildlife and fisheries are generally unaware of the terminology specific to these fields, the biological concepts involved in fisheries and wildlife, the application of these concepts to practice, contemporary issues facing wildlife and fisheries, and numerous other building-block elements essential to basic understanding. By addressing such topics we hope to educate both those who take only an introductory course in this area and those who use an introductory course as a foundation for later, more in-depth courses.

Our intent is to cover breadth rather than depth, leaving detailed information to more advanced texts and courses. Content selection in any introductory book is a challenge. If a text is too long and detailed, basic ideas and concepts can be masked by more specific information that is better handled in advanced courses. If a text is too short and does not cover a sufficient array of topics, its utility is reduced, even for an introductory course.

We have tried to reach a balance between these two extremes.

The first acknowledgement we would like to make is to the thousands of students with whom we have been associated. Their need for an integrated text led a wildlife biologist and two fisheries biologists to work very closely together, a learning experience that caused the authors to recognize some of their own biases. When contentious issues arose during the writing of this text, they were handled in one of three ways. In some cases, because of the arguments provided, all reached agreement; in other cases, we compromised; in still others, we agreed to disagree.

There are many people who could be acknowledged for their contributions to this book; space precludes the addition of all of them, so only a few will be highlighted. We acknowledge Drs. Bruce Leopold, David Manuwal, Ray Oglesby, Harold Picton, and William Shelton for reviewing early drafts of the manuscript. We thank Dr. David Philipp for his excellent review of chapter 4. Dr. Larry Gigliotti provided much useful information and a critical review of chapter 16. Dr. George Mitchell, David Phillips, and Jack Kinnear are thanked for their assistance on Canadian material. Dan Plut is acknowledged for help on law enforcement information. Contributing in many ways were Drs. Richard O. Anderson, Charles Berry, Michael Brown, Walter Duffy, Kenneth Higgins, Daniel Hubbard, and Jonathan Jenks. We thank Beth Staehle for coordinating permission to use American Fisheries Society figures, and Dr. Harry Hogdon for coordinating permissions from The Wildlife Society. We acknowledge Betsy Sturm for providing artwork prior to actual publication of the new edition of *Fisheries Techniques* by the American Fisheries Society. We also thank Ginger and Sarah Scalet for their considerable efforts in manuscript preparation.

We would like to express our appreciation to the numerous people at W. H. Freeman who assisted with various phases of the publication of this book: Deborah Allen, acquisitions editor; Katherine Ahr and Kay Ueno, project editors, editorial assistants Ryan Southard and Amy Sheridan; Sheila Anderson, production coordinator; Blake Logan, designer; Bill Page, senior illustrator; and Sheridan Sellers, Electronic Publishing Center manager. We would also like to thank Norma Roche, who copyedited the manuscript, and our proofreader, Elizabeth Marraffino.

TO THE READER

The material appearing here would normally be found in a book preface. Unfortunately, few people read prefaces. Those using this text will need the following information if the full utility of the text is to be realized.

Each of the chapters begins with a numbered list of the chapter sections and the page on which each section opens. The numbered section headings are intended to allow quick access to chapter contents and also easy back reference to chapter sections referred to elsewhere in the text. Each chapter also begins with an overview that sets the stage for the chapter, describes why the material is important, and defines the chapter's objectives.

Within each chapter a variety of words and terms appear in boldface. These terms or words are generally defined or described within their context of use the first time they appear in the text. The alphabetized glossary contains all of the words or terms boldfaced in the text. This glossary should be useful because of its completeness. Many basic words and terms are included that are either not defined or insufficiently defined in most publications. It was difficult to determine which words and terms needed definition; our intent was to err on the side of defining too many, so some of the words and terms are probably already familiar to many readers.

At the end of each chapter is a list of selected readings that can be referred to for more in-depth information on the topics covered in the chapter. The selected readings section is not intended to be all-inclusive; it contains only a sample of appropriate readings. Also at the end of each chapter is a list of the literature cited in that chapter. Citations in the text have been kept to a minimum to ensure readability.

Also to ensure readability, scientific names are seldom used in the text. The appendix contains the current scientific names of all fish, bird, mammal, and other vertebrate species used as examples in the text. Most organisms used as examples in the text are North American birds, mammals, and freshwater fishes. These are the organisms with which we are most familiar and are also the ones that North American readers may recognize. A variety of species are used so that readers can become more familiar with the diversity of vertebrates present in North America.

Each chapter concludes with a brief summary and a short series of practice questions that readers can use to test their mastery of the chapter contents. It may be misleading to include practice questions because readers could interpret this to mean that the material covered in the questions is the most important material in the chapter. That is not our intent; they are simply examples.

The text is divided into four major parts. An introduction to wildlife systems and fisheries (part I) comprises chapters 1 and 2. The next three parts correspond to the three interactive components of any wildlife system or fishery: the biota, the habitat, and the human users. Part II, which includes chapters 3–11, addresses the biota component. Part III, which includes chapters 12–15, discusses habitats. Part IV, which includes chapters 16–20, discusses human users.

The authors welcome you to the study of wildlife and fisheries. We hope that you will find these important fields as fascinating as we do.

Part I

FISHERIES AND WILDLIFE SYSTEMS

1

DEFINING WILDLIFE AND FISHERIES

What is it about wildlife and fishery resources that interests and attracts people? Why or how are people motivated to pursue these areas as either a vocation or an avocation? The answers to these questions are probably as varied as the people who work in fisheries and wildlife or are interested in these fields. Experience tells us that interest in fisheries and wildlife is not necessarily associated with where one is raised. For example, one author of this text grew up on the Oregon coast, another came from the windswept plains of central North Dakota, and another is from the streets of Chicago. What all three have in common, and what most people who are oriented toward these resources have in common, is a desire to protect, conserve, and learn more about wild organisms.

This chapter is intended to introduce some of the basic terminology used in wildlife and fisheries, to discuss the components of a fishery or a wildlife system, to describe some of the major differences between wildlife and fisheries, to provide a brief history of the fields, and to provide additional introductory information.

1.1 Wildlife, Fish, Fishes, Fishery, and Fisheries: What Are They?

It would seem that defining which animals come under the purview of wildlife and which are in the fishery realm would be a simple matter. There are, however, differences of opinion. One of the authors of this text is a wildlife biologist, two are fishery biologists. It became readily apparent during the writing of this book that each perceives things somewhat differently; each sees through the lens of his own background, education, and experience. Whales are mammals, but they live their entire lives in water: are they a wildlife species or

do they belong in fisheries? How can a lobster, a crustacean, be the primary or target animal in a fishery? Issues such as these and others require further discussion.

What are **wild animals**? A dictionary definition of wild is "something living or growing in its original, natural state." Therefore, something wild cannot be domesticated or cultivated. Unfortunately, such a dictionary definition is not necessarily applicable from a wildlife or fishery perspective. Domesticated livestock such as cattle, sheep, or pigs are not wild. But what about domesticated pigs that escape human control? In a few generations they are wild pigs, like those found in North Carolina or Arkansas. Anyone who has been near a wild pig quickly realizes that it is anything but domesticated! A white-tailed deer in the Florida Everglades is wild. But what about a white-tailed deer that a game rancher owns? The captive deer surely is not in a natural state, nor is it publicly owned like its Everglades counterpart. In addition, in most states a wild white-tailed deer is under the control of the wildlife agency in that state, while a privately owned white-tailed deer is often under the control of a board that regulates domesticated livestock. Similarly, in fisheries, a channel catfish in a river in Mississippi is wild, but a channel catfish a few meters away in the pond of a fish farmer is not wild. Wild brook trout in California and wild gray partridges in Montana are species that have been introduced into those areas. They are not in their original natural state, but we consider them to be wild.

Because of these difficulties in determining whether or not an animal is wild, legal means are usually necessary to determine the state of wildness of particular animals in particular situations. In this text the term wild refers to those organisms that are usually legally defined as being wild.

Early in the development of these fields the term **wildlife** was considered to be inclusive of all wild animals, whether they were fishes, birds, mammals, insects, or other animals. Most biologists now take a more restrictive view, the view we will follow in this book, and consider wildlife essentially to be wild terrestrial and partly terrestrial vertebrate animals. Therefore, such organisms as mountain bluebirds, sage grouse, Dall sheep, thirteen-lined ground squirrels, smooth green snakes, and other totally terrestrial wild vertebrates are considered to be wildlife. There are also numerous aquatic mammals, such as walruses, muskrats, beavers, and hooded seals, that are categorized as wildlife even though a considerable portion of their lives may be spent in water. The same is true of numerous birds like **waterfowl** (ducks, geese, and swans), **shorebirds** (small wading birds, most of which feed along shores and mudflats), and **seabirds** (birds that are associated with oceans and, at times, their shorelines). Many amphibians and reptiles tend to be associated with both terrestrial and aquatic systems and can pose a problem with regard to whether they are considered to be in wildlife or fisheries. In addition, invertebrates such as butterflies and spiders are often categorized as wildlife. Most wildlife examples used in this book are birds or mammals. This should not be construed to mean that only birds and mammals may be considered to be wildlife; these are just the organisms that most people envision when they think of wildlife.

In addition, wildlife does not include only game or commercially important species. **Game (sport) animals** are species that are harvested for recreational purposes; **nongame animals** are the remaining majority of species that are not pursued for consumptive sport purposes. Species taken for commercial purposes are also at times placed in the game category, but there are many exceptions. In reality, whether an animal is considered to be a game or nongame species depends on legal definitions. For example, in some states where mourning dove hunting is legal, doves are game birds. In states where mourning dove hunting is illegal, they are nongame animals. The terms game and nongame will not be used often in this text; most of the topics covered have relevance to all wild animal populations.

The term **fish** is straightforward and refers to both jawless forms, such as silver lampreys and black hagfish, and jawed forms, such as thresher sharks, bluntnose minnows, and bull trout. The singular form, fish, is used when referring to a single species of fish or an individual fish. Thus, one alligator gar or 200 alligator gars are fish. If more than one species is being referred to, then the plural form, **fishes**, is used. One alligator gar and one spotted gar are fishes. The term **fishery** is more complicated. A fishery can either involve a single target or featured species, such as the chain pickerel fishery in New Jersey, or cover all the fishes in a particular area, such as the fishery of Beaver Lake, Arkansas. **Fisheries** is the plural form and refers to more than one fishery. Not only can a fish be the target or featured organism in a fishery, but so can a number of other aquatic organisms. The term "lobster fishery" may sound like an oxymoron, but lobsters, which are crustaceans, not fishes, can be the biotic focal point of a fishery. Fisheries include a wide range of living things, from lake trout to sturgeon chubs, from snapping turtles to bullfrogs, and from snails to crayfish. In addition, some aquatic mammals that essentially spend their entire lives in water, such as short-finned pilot whales and bottle-nosed dolphins, are often considered to be in the fishery realm. As with wildlife, some people think of only game or commercially important animals as the focal point of a fishery. This perception is outdated; a broad array of aquatic animals can be the focal point of a fishery.

1.2 What Constitutes a Fishery or a Wildlife System?

A fishery is composed of three interactive components: the **biota**, the **habitat**, and the **human users** (fig. 1.1). No comparable term is used in wildlife to describe the sum of these three components, but the components are still compatible

Biota

Habitat

Human Users

FIGURE 1.1 The three equally important, interactive components of a fishery or a wildlife system are the biota, the habitat, and the human users.

with wildlife. A comparable term "wildlifery" could be coined, but instead in this text we will use the term wildlife system as the counterpart term to fishery. Each of the three components of a fishery or a wildlife system influences the way it functions, and knowledge of each component is essential for understanding the entire system.

The biota includes the fishes, birds, mammals, reptiles, amphibians, other animals, and some of the plants in a fishery or a wildlife system. Biota can also be broadly defined to include all living organisms. Usually there will be a specific target or featured organism in a fishery or a wildlife system, as in the example of the chain pickerel fishery in New Jersey used earlier, but the biota component also includes all the animals and some of the

plants that interact with the chain pickerel. These animals and plants may function as its food, they may be parasites, or they may interact with the chain pickerel in a variety of other ways. Some consider all living organisms in a system to be part of the biota component (Nielsen and Lackey 1980). We believe that some living organisms, especially many plants, are more appropriately placed in the habitat component of a fishery or a wildlife system.

Habitat is where an organism lives. **Abiotic** (nonliving) components, such as rocks, water, wind, soil, humidity, salinity, oxygen level, stream gradient, water depth, and water **turbidity** (water opaqueness due to suspended particulate matter), constitute a portion of the habitat of an organism.

We also include some **biotic** (living) elements as habitat components. For most animals, organisms such as trees, grasses, and large aquatic plants represent important habitat components. It should also be remembered that habitats are not just natural settings—humans create habitats, as in urban settings, that affect organisms.

Any fishery or wildlife system also has a human user component. These human users represent what is often referred to as the **human dimension** of a system. Human users can be categorized as **direct users** or **indirect users**. Direct users are anglers, wildlife and fishery law enforcement officers, bait dealers, or any other humans who contribute to, use, or directly benefit from a fishery. They are also bird-watchers, binocular manufacturers, people who sell birdseed, or any others who contribute to, use, or directly benefit from a wildlife system. Humans can also be indirect users, such as the people who use or manage some aspect of the habitat for another purpose and in doing so affect a fishery or a wildlife system. Examples would be those who harvest trees, those who use water from streams, or those who graze domesticated livestock. These people may not actually use the fishery or wildlife resource in question, but their activities affect the system. Some people are **consumptive users**—those who harvest a deer, a tree, or a flower—while others are **nonconsumptive users**—those who canoe a river, climb rocks, photograph wildlife, or view autumn foliage. Even people who are nonconsumptive users are consumptive in the sense that they are using the biota or habitat in some way. Nonconsumptive users can affect consumptive users; the opposite is also true. Additionally, direct users can affect indirect users, and the converse also occurs.

The human dimension is now viewed as a more important component of a fishery or a wildlife system than has been the case in the past. All human users, whether consumptive or nonconsumptive, direct or indirect, have substantial effects on a fishery or a wildlife system and must therefore be viewed as portions of this component. Table 1.1 (on page 8) provides more in-depth examples of some of the potential direct, indirect, consumptive, and nonconsumptive users of a specific natural resource.

1.3 Fisheries and Wildlife: Similar, but Different

In many ways the study of fisheries and the study of wildlife are similar; that is why this text describes both. For example, the principles involving the use of fish scales or mammal teeth to determine age, the principles of the techniques used to capture aquatic and terrestrial animals, and the analysis of food habits are all similar for animals associated with wildlife systems and fisheries. A comprehensive list of similarities would be long; the similarities of the two fields far outnumber their differences. However, a fishery and a wildlife system can differ appreciably, and understanding some of the basic differences between them is important in understanding both kinds of systems.

Fishes generally exhibit **indeterminate growth**, while birds and mammals generally exhibit **determinate growth**—that is, most fishes continue to grow throughout their lives, while most bird and mammal species reach a certain adult size and essentially stop growing. A white perch increases in size as it increases in age. This growth will be determined by food supply, among other things, and as long as conditions for growth are satisfactory, the fish will continue to increase in length and weight throughout its life. If there are insufficient resources for growth, the fish will stop growing, but it will not necessarily die. If resources become sufficient at some time in the future, the fish can resume growing. Bird and mammal species tend to have a more uniform or specific adult size. Animals with determinate growth do not get taller and longer with age. This does not mean that a three-year-old badger or trumpeter swan may not be somewhat larger and heavier than a two-year-old badger or trumpeter swan, just that growth is

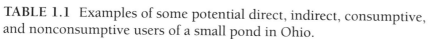

TABLE 1.1 Examples of some potential direct, indirect, consumptive, and nonconsumptive users of a small pond in Ohio.

Direct Users	Indirect Users	Consumptive Users	Nonconsumptive Users
Someone who fishes in the pond.	Someone who raises corn near the pond and applies fertilizer that runs into the pond.	Someone who removes fish from the pond.	Someone who canoes on the pond.
Someone who sells bait to the person who is fishing in the pond.	Someone who cuts and removes trees from near the pond, which results in increased wind action on the pond.	Someone who removes a beaver from the pond.	Someone who scuba dives in the pond.
Someone who traps muskrats from the pond.		Someone who collects cattails from the pond for floral arrangements.	Someone who picnics on the pond bank.
Someone who rents a motel room to someone using the pond.	An aerial sprayer who sprays crops near the pond, and has some of the spray drift into the pond.	Someone who removes ducks from the pond.	Someone who fishes on the pond but catches nothing.
Someone who sells gasoline used on the pond or to get to or from the pond.		Someone who uses water from the pond to irrigate a field.	Someone who photographs the pond.
Someone who views shorebirds on the pond.	Someone who drains a wetland near the pond, which affects pond water depth.	Someone who cuts and removes trees from near the pond.	Someone who uses the pond for swimming.
Someone who sells binoculars to the person viewing birds on the pond.	Someone who grazes domesticated livestock around the pond.	Someone who uses water from the pond for domesticated livestock watering.	
Someone who sells a license for someone to fish on the pond.	Someone who serves on a local board that decides zoning regulations near the pond.		
Someone who manages the fishery resource of the pond.			
Someone who enforces game and fish laws on the pond.			

not directly incremental with age and environmental conditions.

Mammals and birds are **endothermic (homeothermic)**, while fishes are **ectothermic (poikilothermic)**. That is, birds and mammals maintain their body temperatures at or near a specific level that is required for them to remain alive, regardless of the temperature of the environment in which they live. Their primary source of heat comes from within their bodies. Fishes and other ectotherms generally exist at or near the temperature of the surrounding environment, which is referred to as the **ambient temperature**. Their primary source of heat comes from outside their bodies. This does not mean that ectotherms have no control over their body temperatures. They can increase their body temperatures by moving to an area with a higher ambient temperature, and some species can even increase body temperature above ambient temperature by activity, such as rapid swimming. However, ectotherms do not maintain a specific body temperature under all environmental conditions. Endotherms must expend energy to maintain their body temperatures; ectotherms do not face this energy expenditure. This difference has direct ramifications for processes such as growth and metabolic rates of the two types of organisms. In the past the terms warm-blooded and cold-blooded were often used to differentiate between endotherms and ectotherms. Warm-blooded and cold-blooded are terms that are much too inaccurate to be useful. For example, a Red River pupfish living in 39°C water is anything but cold-blooded.

Fishes live in an aquatic environment, while most birds and mammals are primarily terrestrial. The two environments differ in many ways. The atmosphere of the earth contains approximately 19% oxygen; this amount is essentially unchanging. Well-oxygenated water contains only 10–15 milligrams of oxygen per liter (0.0010%–0.0015%), and this amount can vary by time of day, season, or location within a water body. This difference affects the amounts of oxygen available to terrestrial and most aquatic organisms, how they extract it, their respective metabolic rates, and other facets of their lives.

Gravity has a greater effect on terrestrial organisms than on aquatic ones. Most fishes have the ability to adjust their body density to that of the surrounding water by using the **gas bladder**. Fishes can increase or decrease the internal gas content of this organ and thus their body density. Even in fish species that lack a gas bladder, body density is similar to that of water. Thus, fishes can remain suspended in their environment with little expenditure of energy. Terrestrial organisms, in contrast, are much denser than air. Differences in characteristics such as the amount of bony structure needed for support, the specific density of the body, and mechanisms and energy expenditures for maintaining positions in the environment are a result of the differing effects of gravity on terrestrial and aquatic organisms. Fishes and other aquatic animals, however, have a problem that terrestrial organisms do not have: more energy is required to move water than air because of the greater density of water. This difference affects facets of life such as obtaining oxygen and movement.

Aquatic systems are generally more stable, in the short term, than terrestrial systems. Terrestrial organisms often must adapt to ambient temperature changes of 20°C or more during a 24-hour period. Aquatic ambient temperature changes during the same period are normally a couple of degrees at most. Terrestrial organisms, especially endotherms, can usually adjust more easily to rapid environmental changes than aquatic organisms.

The concept of day and night is quite different in the terrestrial and aquatic environments. Water quality characteristics, such as depth and turbidity, affect the ability of light to pass through water. The aquatic environment is very different from the terrestrial in terms of light intensity, duration, and color.

Terrestrial and aquatic organisms also differ in their mobility. How they are able to move from one place to another is affected by the medium in which they live. Freshwater fishes live in "islands" of water surrounded by land. Their movement is generally restricted to water areas. This restriction of movement can be compared with the situation of bird and mammal species found on islands in the middle of oceans. Most wildlife species are confronted by less formidable barriers to movement than the absence of the medium in which they live. For example, the overland movement of birds or mammals may be limited by barriers such as mountain ranges, forests, or deserts, but not by the lack of atmosphere. In general it is more difficult for freshwater fishes to move to new areas than it is for most terrestrial species. Marine fish mobility is more similar to that of terrestrial organisms than to that of freshwater fishes. While water may connect all oceans, barriers such as landmasses, wide expanses of deep water, water temperature and salinity differences, and other factors restrict the movement of many marine fishes.

Another substantial difference between wildlife and fisheries is that fishes tend to exist in a more three-dimensional world than do most wildlife species. Fishes live in a world of length, width, and depth, whereas most wildlife species live primarily in a length and width, or two-dimensional, environment. That is not to say that wildlife species are relegated only to the surface of the earth. Examples of wildlife species that are **fossorial** (burrow into the earth), **aerial** (have flight capabilities), or **arboreal** (live in trees) are numerous, but most wildlife species are fairly narrowly associated with the surface-to-atmosphere interface of the earth for most if not all of their lives. Even organisms such as birds using different horizontal layers of a forest are essentially associated with the surface of the earth.

Wildlife and fishes also differ greatly with regard to what organisms are generally managed for consumptive purposes as the target or featured species of a fishery or a wildlife system. Most of the fishes that are consumptively managed eat other animals, while most of the consumptively managed wildlife species eat plants. For example, fishes such as the muskellunge, smallmouth bass, and walleye generally eat other animals, while wildlife species such as the mule deer, Ross' goose, and scaled quail eat plants. There are obvious exceptions to this generalization, such as the plant-eating grass carp and the animal-eating mink, but overall, most past management activity associated with consumptively used animals has been directed toward animal-eating fish and plant-eating wildlife species.

An additional difference between fishes and most wildlife species involves the measurement or enumeration of their respective populations. Birds and mammals are usually counted in numbers per unit of area—for example, the number of pronghorns or American robins found per square kilometer. These numbers are usually referred to as the **population density** of a wildlife species. Fishes are sometimes counted in numbers per unit of area, but are most often measured in terms of **biomass**—for example, the number of kilograms of banded pygmy sunfish per hectare. One of the primary reasons for this difference in enumeration is the previously mentioned difference between determinate and indeterminate growth. For example, if a count involved the killdeers in an area, a count would most appropriately be made of all of the members of that species because they would all be the same approximate size. There might be a particular biomass of killdeers present, but killdeer numbers in one population would be about the same as the numbers in another killdeer population of similar biomass. If, however, a small pond had a bluegill population with a total biomass of 150 kilograms of this species, there could be a wide variety of bluegill numbers present. There could be 1,500 0.10-kilogram bluegills, 300 0.5-kilogram bluegills, 6 25-kilogram bluegills (which could be exciting for anglers and swimmers), or any combination of numbers that totaled 150 kilograms, depending on the sizes of fish present. Because of their three-dimensional aquatic environment, fish biomass can also be measured in numbers per unit of volume. Population number or biomass per unit of area or volume can be used as quantitative measures of a resource. The term **abundance** is often equated to these quantitative measures.

Most fishes and wildlife species also differ markedly in their reproductive potential. This difference affects their mortality and survival rates, behavior patterns, reproduction strategies, how rapidly they can repopulate an area, and how they are managed. Fishes generally produce many more offspring than bird or mammal species. Some fishes, such as the common carp, produce large numbers of eggs; a large female carp can produce in excess of 2 million eggs during one reproductive season. Even a fish such as the cutthroat trout, which is considered to have a moderately low rate of egg production, usually produces up to 4,500 eggs per mature female per year. When these fish egg production numbers are compared with the relatively small numbers of offspring that birds or mammals produce per year (usually fewer than a dozen), one readily sees that there are major differences between fishes and wildlife species with regard to reproductive potential. These differences are reflected in the use of different strategies for successful reproduction. Fishes tend to produce large numbers of offspring with low survival rates; mammals and birds tend to produce smaller numbers of offspring with higher survival rates.

Another difference between wildlife and fisheries is that human-created wildlife habitat is usually more fleeting than human-created fishery habitat. Particular types of vegetative communities developed or used as wildlife habitat tend to be more easily changed. The removal of trees and shrubs from around a building can be accomplished in a short time; these changes quickly affect wildlife species. Such rapid changes can also occur over larger areas. An example of such a change involves the Conservation Reserve Program (CRP), instituted in 1985, which shifts land use on erodible land from agricultural row crops to cover vegetation (see section 18.9). This CRP habitat, which covers many millions of hectares, could be radically changed very quickly with a plow. With such a change, large numbers and kinds of terrestrial wildlife species would be rapidly affected. On a short-term basis, most aquatic habitats tend to be more stable than terrestrial systems.

In addition, human-made water bodies are generally produced at the expense of wildlife habitat. The flooding of a river bottom for a reservoir or the building of a farm pond usually reduces some types of wildlife habitat while increasing aquatic habitat. This kind of change should not be thought of as a total shift because construction of water bodies also produces habitat for such wildlife species as shorebirds, waterfowl, and muskrats, but it does shift the area away from a totally terrestrial system. However, one should not think of aquatic habitat building as a one-way street. Some of the ponds and reservoirs we build today are the wetlands of tomorrow, and then potentially terrestrial areas again. While such a succession of changes may take hundreds of years, this is a relatively short period of time in a biological sense. Humans have a tendency to think of time in relation to the length of a human life span; this clouds our perception of time.

Finally, humans perceive most wildlife species differently than they perceive fishes and other species less similar to themselves. **Anthropomorphism** is the attributing of human shape or characteristics, to, among other things, other animals. Anthropomorphism is more often directed toward birds and mammals than toward fishes. Birds and mammals are endotherms and are therefore warm, often have eyelids and therefore blink, usually have hair or feathers and are therefore dry and smooth, have appendages such as legs that humans more closely identify with, and make sounds that humans can hear. They are more similar to humans in these ways than are ectothermic, nonblinking, scaled, wet (slimy), legless, and seemingly soundless fishes. In addition, humans can usually observe the activities of birds and mammals more readily than those of fishes because most wildlife lives in the same terrestrial environment as humans. We also perceive ourselves to be more like these other terrestrial organisms because we relate more readily to their sensory receptors and the way they perceive their environment. We generally think we know more of what a bird sees or what a mammal smells than what a fish sees or smells or how it uses a sensory structure such as its **lateral line system**, a mechanical sound reception system that humans do not possess. Humans also tend to be **anthropocentric**, that is, to believe that they are the center of the universe. If something seems more like us, it acquires added value in our eyes. Because we perceive other mammals and birds as being more like us than most fishes and other less advanced organisms, this influences the way we react and relate to them and how we perceive their relative importance. And yet there is a growing awareness of and subscription to the fundamental **biocentric** view that all creatures in an ecosystem are equally important. The human tendency to be anthropocentric is reflected in our division of wildlife systems and fisheries into three interactive components: biota, habitat, and human users. While we reserve for ourselves a special place in this triad, we should recognize that we are really a portion of the biota.

These are some of the most important differences between wildlife and fisheries. Realizing that differences exist allows one to understand the reasons why the study of fisheries and the study of wildlife systems may differ. There are other differences, as well as many similarities, between the two fields of study, many of which appear throughout this text.

1.4 A Brief History of Wildlife and Fisheries

Humans have always interacted closely with wildlife and fishery resources. Early humans living along the seashore or on lakes or rivers developed life rhythms that coincided with seasonal rhythms of fish movements and behavior. Nomadic hunter-gatherer societies were so closely intertwined with the wild organisms they required for life that they often viewed some of those organisms as deities. For example, North American Plains Indians were

dependent on bison for food, shelter, and many other products. It is no wonder that the bison is a focal point of their lives and important in their religious beliefs.

To survive, humans needed knowledge about the organisms on which they depended. Over time, some of the knowledge they gained allowed them to change or modify their relationships with these wild organisms. In some cases it led to domestication, while in others it led to human activities that changed habitats in favor of preferred wild organisms. The burning of grasslands to promote the return of wildlife to an area, the construction of shallow water areas in which to trap fishes, and the clearing of forest patches to attract animals are all examples of early attempts at modifying the relationship between humans and wild organisms.

While human knowledge about wildlife and fisheries spans many thousands of years, our knowledge concerning what early humans knew about wildlife and fisheries is primarily limited to the last few thousand years because only this time period is represented by a written record. Early written records are primarily Chinese, Greek, and Roman; most of these are oriented toward the natural history of organisms.

Early work (from the late eighteenth century to the late nineteenth century) on wildlife and fishery resources in the United States was primarily the result of efforts by two groups. One group is represented by people educated in the sciences whose work primarily involved classifying the organisms present, describing what they looked like, and determining where they were found. The second group is represented by writers, artists, politicians, philosophers, and others who championed the cause of saving wild organisms and unique natural areas.

At the end of the nineteenth century and the start of the twentieth century, the goals of protecting and preserving natural resources in the United States reached fruition and the pace of natural resource conservation activities quickened. Numerous U.S. government natural resource agencies were initially developed during this period, as was the scientific basis for many of our present-day activities directed toward natural resource conservation.

If the beginning of a comprehensive written account of the principles and practices involved in managing the wild animals of North America could be identified, one would have to point to the 1933 publication *Game Management* by Aldo Leopold. Leopold outlined current practices of wildlife management and also consolidated the theories and concepts used to support those practices. Using the Leopold wildlife book as a model, Carl Hubbs and Ralph Eschmeyer published a book in 1938 on fishery management. This publication, *The Improvement of Lakes for Fishing,* set the stage for fishery management as a field of study. These two publications laid the foundations for the variety of wildlife and fishery work that has continued to the present time. In addition to those who worked specifically in wildlife and fisheries, there has been a long succession of plant and animal ecologists, limnologists, geneticists, physiologists, ethologists, and people in other fields who have contributed to or forged foundations for the fields of wildlife and fisheries.

Not all of the seminal work and publications contributing to wildlife and fisheries were of a purely scientific nature. Few books have had the impact of Leopold's *A Sand County Almanac,* in which he philosophized on the land ethic, or *Silent Spring,* by Rachel Carson, which drew attention to the effects of humans on the environment. Because these books are basic to understanding the relationship of humans to their environment, each should be required reading for everyone, not just those who work in natural resource fields. Table 1.2 (pages 20–24) lists some of the individuals who have made important contributions, primarily to wildlife or fisheries in the United States, but also to other fields that have had direct implications for the development of wildlife and fisheries. The list is in general chronological order and contains brief descriptions of the contributions of each individual. A great number of past and present individuals could be added to such a list; most contemporary contributors are purposefully omitted.

1.5 Successes and Failures

The history of wildlife and fisheries is replete with examples of successes and failures with regard to increases or decreases in population levels. There have been successes in which native species that had been greatly reduced in numbers are now more numerous. Wood ducks were once thought to be on the road to extinction, but are now flourishing (fig. 1.2A). Atlantic salmon, once severely reduced, are now found in areas from which they disappeared decades ago (fig. 1.2B). White-tailed deer, after years of depletion and scarcity, currently have reached populations that probably exceed their numbers at the time of North American settlement by Europeans (fig. 1.2C). There are many other success stories, such as northern fur seals, wild turkeys, American alligators, and pronghorns. Depletion of these animals was usually a result of some human activity, such as uncontrolled harvest, pollution, or habitat degradation. However, their recoveries were also a result of human activities. These wildlife and fishery success stories are therefore a result of positive corrective actions by humans.

Not all success stories involve increased populations of native species. There have also been introductions of nonnative animals that have had positive results and are generally viewed as successes. The ring-necked pheasant, originally native to Asia, has populated new habitats in North America created by agricultural activity (fig. 1.3A). Native species that were historically present, such as the sharp-tailed grouse and greater prairie-chicken, could not withstand the agriculturally induced habitat changes in many areas, but the ring-necked pheasant was able to survive in these new landscapes. The nonnative brown trout, because it could withstand higher water temperatures, expanded into areas that could not support native trouts (fig. 1.3B). The brown trout introduction has had positive aspects, especially in relation to sport fisheries, but has also had some drawbacks

A

B

C

FIGURE 1.2 Three North American animals that were once greatly reduced in numbers but are now increasing. (*A*) The wood duck, once predicted to be on the road to extinction, is now a common bird in its native range. (*B*) The Atlantic salmon, a native of northeastern North America, the North Atlantic, and western Europe, while still low in numbers, has responded positively to pollution control and harvest regulations. (*C*) The population of white-tailed deer, once reduced to approximately 350,000 animals, now exceeds 12 million and continues to flourish. (*Part A photograph courtesy of South Dakota Tourism; part B photograph courtesy of D. Pugh; part C photograph courtesy of D. Naugle.*)

introductions that were unfortunately too successful. Many of these organisms are now found in large numbers, and their effects on native species, crops, and even buildings have been such that these introductions must be categorized as failures. Some of these were purposeful introductions; others were accidental. The common carp was a purposeful introduction into North America (fig. 1.4A). However, it has often displaced native fishes and has also had negative effects on habitats and on other organisms such as waterfowl. The sea lamprey was an accidental introduction into the Great Lakes (except Lake Ontario) that has reduced native fish populations (fig. 1.4B). House sparrows and European starlings (fig. 1.4C) were purposeful introductions that have displaced and reduced native bird species. Nutria were both accidental and purposeful introductions (fig. 1.4D). Originally brought into North America for fur purposes, many escaped or were released when they could not be sold. They were also purposefully released to control problem aquatic vegetation. They have displaced native animals and also cause damage to agricultural crops, trees, roadbeds, and other human-made structures. Any positive features of these nonnative introductions are far outweighed by their negative effects.

Potential success stories that are still in the formative stages are the attempts by humans to maintain species that have been reduced to low numbers. The ongoing efforts to save whooping cranes, Colorado squawfish, black-footed ferrets, and a

because of the negative effects of brown trout on the native species they can displace. The introduction of some species of **Pacific salmon** (pink, chum, coho, sockeye, and chinook salmon) into the Great Lakes of North America is also generally viewed as positive.

Unfortunately, for every nonnative introduction that is viewed as having primarily positive effects, there are multiple examples of nonnative

A   B

FIGURE 1.3 Some introductions of nonnative animals into North America have generally been viewed as having positive results. (A) The ring-necked pheasant has populated new habitats in North America created by agricultural activity. (B) The brown trout, because it can withstand warmer water temperatures, has expanded into areas that could not support native trouts. (*Part A photograph courtesy of South Dakota Tourism.*)

variety of other endangered and threatened species are potential successes (fig. 1.5); they are also potential failures (see chapter 11).

Efforts to save some species came either too late or not at all. These are failures in the truest sense because these species are gone and nothing can be done to ameliorate their loss. The heath hen, the Carolina parakeet, the blue pike (fig. 1.6A), the passenger pigeon (fig. 1.6B), and others are gone forever. Their loss is irreversible and unfortunate. How the loss of numerous plant and animal species, primarily as a result of human activities, will affect the integrity and functioning of the ecosystems of the world is currently unknown. Aldo Leopold compared the human experiment of modifying the earth to a tinkerer taking apart a machine (Leopold 1966). A good tinkerer does not throw away a part because its function or position in the machine is unknown. Humans continue to throw away parts of ecosystems at an ever-increasing rate.

There have been some magnificent successes in wildlife and fisheries, and there have also been many abysmal failures. Even with the successes taken into account, the future of many species is not bright. For every population that is flourishing, there are many others that are just holding their own or declining. All too often, humans are concerned only with large species, such as bison or California condors, or economically important species, such as canvasbacks or Pacific salmon, while a myriad of small and seemingly unimportant species, some known, some unknown, are relegated to inattention and benign neglect. All organisms, whether large or small, in great numbers or small numbers, economically important or unknown, represent a portion of the web of life—all are interconnected in some way. It would seem that the overall success or failure of the human species in the end will depend on the success or failure of many other species; their future and our future are intertwined.

FIGURE 1.4 Some nonnative introductions have been too successful and have had many negative effects. (A) Introduction of the common carp was too successful. It now ranges over almost all of North America in a wide array of habitats. (B) Sea lampreys entered the Great Lakes (except Lake Ontario) when the Welland Canal was built and allowed this fish a way around Niagara Falls. This species, along with overfishing and pollution, has resulted in a reduced Great Lakes fishery. (C) European starlings were purposefully introduced in small numbers. They are now found in most areas of North America. (D) The nutria, while having some positive attributes, is responsible for much damage to native organisms and to the environments in which it lives. (*Part B photograph courtesy of the U.S. Fish and Wildlife Service; part D photograph courtesy of R. Chabreck.*)

A

B

FIGURE 1.6 For some species any attempts to save them came too late; they are now extinct. (*A*) The blue pike, variously regarded as a color phase of the walleye, a walleye subspecies, or a species, was commercially important in Lake Erie up to the 1950s. It is believed to have become virtually extinct by the early 1970s. (*B*) The population of passenger pigeons at one time numbered approximately five billion birds and ranged from Canada to Mexico. The last individual died in 1914. (*Part A photograph courtesy of W. Carrick.*)

FIGURE 1.5 (*Bottom of opposite page*) Species that humans are attempting to save from extinction represent potential successes or failures. (*A*) Whooping cranes once had been reduced to 14 birds. They now number over 200. (*B*) Colorado squawfish have been diminished in numbers primarily as a result of dam construction and nonnative introductions, but their numbers are now increasing. (*C*) Because of their low numbers, all known black-footed ferrets were captured in an attempt to increase their numbers through captive breeding programs. (*Part A photograph courtesy of R. Drewien; part B photograph courtesy of the U.S. Fish and Wildlife Service; part C photograph courtesy of L. Parker, Wyoming Game and Fish.*)

1.6 Biology, Science, Management, Conservation, and Ecology: What Are They?

The terms fishery and wildlife are often followed by words such as biology, science, management, conservation, or ecology. For example, the phrases "wildlife conservation and management," "fishery science," and "fishery biology" are just a few of the commonly used combinations. Unfortunately, books and papers using these terms often define the same terms differently, or do not define them at all. This leads to unnecessary confusion. In an attempt to reduce this confusion (hopefully not add to it), we will give our definitions for various terms. Some people will disagree with our definitions, but these definitions are as correct as any.

Wildlife management and **fishery management** are the art and science of manipulating the biota, habitat, or human users to produce some desired end result. For example, habitat improvement efforts to increase Cape Fear shiner numbers and rules implemented to reduce the harvest of American black ducks fall into the category of management. **Fishery science** and **wildlife science** involve the process of obtaining knowledge about and studying fisheries or wildlife. For example, the development of stock recruitment curves in fisheries and the rationale and mechanics involved in the use of anesthetics in wildlife represent facets of science. In other words, science concerns the process of obtaining knowledge about something, while management represents the application of that knowledge.

Conservation is the wise management and use of natural resources. It thus follows that **fishery conservation** is the wise management and use of fishery resources and **wildlife conservation** is the wise management and use of wildlife resources. **Ecology** is the study of the interrelationships of organisms with other organisms and the environment. Therefore, **fishery ecology** is the study of the interrelationships of organisms that are considered to be a part of the fishery realm with other organisms and the environment; **wildlife ecology** is the study of the interrelationships of those organisms generally considered to be under the purview of wildlife with other organisms and the environment. **Biology** is the study of living things, thus **fishery biology** is the study of a fishery and **wildlife biology** is the study of a wildlife system. It should be apparent that the definitions of management and science are fairly specific and quite different from each other, while the definitions of the other terms are broad and overlap with one another as well as with the terms science and management.

These commonly used terms can also be combined in other ways. For instance, **conservation biology** is "the new, multidisciplinary science that has developed to deal with the crisis confronting biological diversity. Conservation biology has two goals: first, to investigate human impacts on biological diversity and, second, to develop practical approaches to prevent the extinction of species" (Primack 1993).

Definitions change with time. Leopold (1933) defined game management as "the art of making land produce sustained annual crops of wild game for recreational use." While it is not specified, many people assume that the "game" of which Leopold writes are animals that are harvested for sport purposes. Current perceptions of wildlife management would not be as restrictive; more than game animals would be involved, and few people would restrict wildlife management to only those animals used for consumptive recreational purposes. Interestingly, Leopold said nothing of science, but described management as an "art."

Rounsefell and Everhart (1953) began the preface of their book by defining fishery management as "the application of scientific knowledge concerning fish populations to the problems of obtaining the maximum production of fishery products, whether stated in tons of factory material or in hours of angling pressure." The defini-

tion is of fishery management, but their book title is *Fishery Science: Its Methods and Applications;* they provide no definition of fishery science. Their definition of management is directed only at consumptively used fishery resources, which is quite different than would be the case today. In addition, the term maximum production is used in their definition. In current definitions, optimum would replace maximum. There is a great deal of difference between maximum and optimum, as the reader will learn in section 3.6 and elsewhere in this book.

Changing terminology has led to the more common inclusion of terms such as ecology and conservation in the lexicon of fisheries and wildlife. However, the addition of such terms has not always been accompanied by rationales for their inclusion. For example, there are currently two books in print, both excellent, entitled *Wildlife Ecology and Management* (Caughley and Sinclair 1994; Bolen and Robinson 1995). The introductions to both books define and describe in depth the term wildlife management; neither specifically defines wildlife ecology in the same manner. It is obvious that ecology is added to the book titles to indicate a broad scope of content. Management is often described as applied ecology, so the inclusion of ecology is appropriate from this perspective also. Conservation might be added under similar circumstances because it implies a concern for wildlife resources. Biology might be added because of the positive connotations of that term; biology is typically perceived as being scientific. The reasons for using such terms are probably as numerous as the people who use them, with each having his or her own particular rationale.

What these additions of terms most likely represent is change and growth in language as it pertains to wildlife and fisheries. Whereas the terms management and science were once perceived to be sufficiently descriptive, that perception has now changed, and the addition of terms such as ecology, conservation, or biology is used to convey the intended message. These terms may be used to indicate a more modern or more all-encompassing approach to fisheries and wildlife than the terms management or science. Through time, word and term usage changes, and that has occurred and is occurring in wildlife and fisheries.

In closing, one additional item should be covered. Some confusion is produced by differences of opinion on the use of the terms fishery and fisheries as modifiers of nouns such as science or ecology. Some people contend that the term fishery should be used, while others contend that fisheries is just as appropriate. Eschmeyer (1990) stated that either fishery or fisheries is fully acceptable as a modifier in all usages. Thus, nouns such as management, science, ecology, conservation, or biology can be modified by either fishery or fisheries.

1.7 Use of Scientific Names

Only common names of organisms are generally used in this text, but this does not mean that scientific names are unimportant. Scientific names allow us to be certain what organism is being discussed regardless of the language being used. Whether a paper or a book is written in English, German, Russian, Japanese, or any other language, the currently accepted scientific name for an organism is the same. Common names often vary depending on location; this is not the case with scientific names. When standardized scientific names are used, there can be no confusion as to the organism under consideration. The appendix contains the scientific names of vertebrates used as examples in this text.

The scientific name of an organism generally consists of generic and specific names. For example, the current scientific name of the smallmouth bass is *Micropterus dolomieu*: *Micropterus* is

the genus and *dolomieu* is the species. People can sometimes be confused by the way the word species is commonly used. For example, consider the question, "What species is this?" If the organism is a smallmouth bass, the answer is *Micropterus dolomieu*. The smallmouth bass is considered to be a species, but both the generic and specific portions of the scientific name are required to identify the fish. The plural form of species is species, while the plural form of genus is genera. In a biological sense, there really is no singular form of species because a species name does not represent a single organism, but instead represents all of the individual organisms that constitute that species.

Scientific names can include more than just the genus and species. In situations in which subspecies have been described, the subspecies name follows the species. In addition, the person who named the organism and when it was described may be included. For example, the lake sturgeon has the name Rafinesque and the year 1817 following the species name. Its scientific name would appear as *Acipenser fulvescens* Rafinesque, 1817. This means that this species was described by Rafinesque in 1817. If the species was originally described in another genus, and thus the scientific name has been changed, the name of the person who originally described the organism and the date of that description are noted in parentheses. For example, the scientific name of the shovelnose sturgeon has been changed; therefore, it appears as *Scaphirhynchus platorynchus* (Rafinesque, 1820).

The genus, species, and, if present, subspecies names should be printed in italics, or if that is not possible, they should be underlined. The first letter of the generic name should be capitalized; no letters in the specific or subspecific portion of the name should be capitalized.

The term **taxon** is a general expression for a taxonomic group, whatever its rank. Therefore, a genus, species, family, or any of the other groupings used in the hierarchy of taxonomy is referred to as a taxon.

TABLE 1.2 A general chronological listing of some individuals, primarily from the United States, who have made important contributions to wildlife, fisheries, and related fields. Most contemporary contributors are purposefully omitted.

Fan Lee (Chinese)	circa 500 B.C.	Wrote a book on various aspects of oriental fish culture.
Aristotle (Greek)	384–322 B.C.	First zoologist of record who wrote on the structure and habits of animals in *Historia Animaculum.*
Pliny the Elder (Roman)	23–79 A.D.	Wrote a large number of books covering various aspects of the natural history of organisms.
Guillaume Rondelet (French)	Mid-1500s	Compiled much natural history information, especially concerning the animals of the Mediterranean Sea.
Pierre Belon (French)	Mid-1500s	Wrote zoological books concerning marine organisms that were among the first not written in Latin.
Robert Boyle (British)	Mid- to late 1600s	Primarily a chemist, but also worked in biological areas. He was the first to distinguish between chemical elements and compounds.
René Réaumur (French)	Early 1700s	Although his primary work was on insects, he did notable work on the natural history of organisms.
Carolus Linnaeus (Swedish)	Mid-1700s	Taxonomist who developed the binomial nomenclature system of scientific names.
Jean Lamarck (French)	Late 1700s	Contributed to the classification of both plants and animals, and was a pre-Darwinian evolutionist.
Thomas Malthus (British)	Late 1700s to early 1800s	Noted for his work on human population dynamics, which was the forerunner of work on the population dynamics of other organisms.
Alexander Wilson (U.S.)	Late 1700s to early 1800s	The father of U.S. ornithology.
John James Audubon (U.S.)	Early 1800s	Noted wildlife artist and naturalist.

(Table 1.2 continued on following page)

Constantine Rafinesque (U.S.)	Early 1800s	Early ichthyologist and natural historian who worked in the Ohio River area while it was still a frontier area.
George Catlin (U.S.)	Early to mid-1800s	Naturalist, artist, and writer who chronicled the early American West.
Ralph Waldo Emerson (U.S.)	Mid-1800s	Naturalist, writer, and philosopher who emphasized the natural environment.
George Perkins Marsh (U.S.)	Mid-1800s	Naturalist and writer who demonstrated how human activities affected the environment.
Louis Agassiz (Swiss, U.S.)	Mid-1800s	Biologist and many-faceted naturalist who gained fame in Europe before coming to the United States. He conducted studies in a variety of areas including ichthyology.
Theodatus Garlick (U.S.)	Mid-1800s	Father of American fish culture and also one of the founders of the American Fish Culturists' Association, which later became the American Fisheries Society.
Charles Darwin (British)	Mid-1800s	Developed theories on natural selection and evolution and wrote *On the Origin of Species*.
Alfred Russel Wallace (British)	Mid-1800s	Developed ideas similar to those of Darwin concerning evolution.
Gregor Mendel (Austrian)	Mid-1800s	Founder of modern genetics.
Spencer Baird (U.S.)	Mid-1800s	Commissioner of the first U.S. government fishery agency.
John Wesley Powell (U.S.)	Mid-1800s	Army officer and explorer who made natural history surveys primarily in the Rocky Mountain region.
Ulysses S. Grant (U.S.)	Mid-1800s	U.S. president who set aside Yellowstone National Park.
John Muir (U.S.)	Mid- to late 1800s	Noted naturalist and writer who supported the preservationist view of conservation. He also founded the Sierra Club.
Ernst Haekel (German)	Late 1800s	Early naturalist who formally defined the term ecology.

(Table 1.2 continued on following page)

François Forel (Swiss)	Late 1800s	Coined the term limnology and wrote early limnological books.
Theodore Roosevelt (U.S.)	Late 1800s to early 1900s	U.S. president noted for his conservation efforts. He initiated the development of the National Wildlife Refuge system, greatly expanded the National Forest system, and established a presidential mechanism for saving unique resource areas as National Monuments, one of which was the Grand Canyon.
Gifford Pinchot (U.S.)	Late 1800s to early 1900s	Forester who helped establish the American Society of Foresters, was director of the U.S. Department of Agriculture, Division of Forestry, and promoted wise use of resources on a sustained yield basis. He also first used the term conservation as it applies to natural resources.
William Hornaday (U.S.)	Late 1800s to early 1900s	Biologist and naturalist whose efforts helped to save the bison from extinction. He also promoted reforms of state wildlife conservation agencies.
C. Hart Merriam (U.S.)	Late 1800s to early 1900s	Headed the first U.S. government wildlife agency and also developed the life-zone concept of ecological classification.
Stephen A. Forbes (U.S.)	Late 1800s to early 1900s	Ecologist who described the lake community as an ecosystem.
George Bird Grinnell (U.S.)	Late 1800s to early 1900s	Naturalist and founder of the National Association of Audubon Societies.
Edward Birge (U.S.)	Late 1800s to early 1900s	Limnologist who collaborated with Juday and worked primarily on Wisconsin lakes. His work centered around zooplankton.
Ernest Seton (U.S.)	Late 1800s to early 1900s	Artist and mammalogist who is particularly noted for his natural history work on game animals.
David Starr Jordan (U.S.)	Late 1800s to early 1900s	Authored hundreds of ichthyological publications from 1874 to 1931, including the monumental work *Fishes of North and Middle America*.

(*Table 1.2 continued on following page*)

Barton Evermann (U.S.)	Late 1800s to early 1900s	Coworker with Jordan who contributed a great deal to ichthyological knowledge in North America.
Stephen Mather (U.S.)	Early 1900s	First director of the National Park Service who was especially adept at public relations.
Chancey Juday (U.S.)	Early 1900s	Limnologist who collaborated with Birge on the study of Wisconsin lakes.
Arthur Bent (U.S.)	Early 1900s	Authored a large series of books on the life histories of birds.
Victor Shelford (U.S.)	Early to mid-1900s	Plant ecologist who helped develop the study of communities.
Frederich Clements (U.S.)	Early to mid-1900s	Plant ecologist who collaborated with Shelford on the concept of biomes.
Charles Elton (British)	Early to mid-1900s	Oxford ecologist who helped establish the *Journal of Animal Ecology*.
Robert Marshall (U.S.)	Early to mid-1900s	Noted forester and founder of the Wilderness Society.
Konrad Lorenz (Austrian)	Mid-1900s	Founder of the science of animal behavior.
Aldo Leopold (U.S.)	Mid-1900s	Viewed as the founder of American wildlife management. Wrote the 1933 book *Game Management*. He was also a philosopher and developer of the land ethic concept, which he described in his 1949 book *A Sand County Almanac*.
J. N. "Ding" Darling (U.S.)	Mid-1900s	Writer, naturalist, and promoter of numerous wildlife causes. He helped persuade Congress to pass the Duck Stamp Act, was instrumental in the development of federal wildlife cooperative units at universities, helped establish the Wildlife Management Institute and the National Wildlife Federation, and was director of the Bureau of Biological Survey, the forerunner of the present-day U.S. Fish and Wildlife Service.

(Table 1.2 continued on following page)

Carl Hubbs (U.S.)	Mid-1900s	Worked in many areas of ichthyology and fishery management. Coauthored *The Improvement of Lakes for Fishing* in 1938, which inventoried contemporary theories on fishery management.
Ralph Eschmeyer (U.S.)	Mid-1900s	Early fishery management worker who coauthored the 1938 publication on fishery management with Hubbs.
Homer Swingle (U.S.)	Mid-1900s	Early fishery manager who demonstrated fish population dynamics in small ponds.
Adolph Murie (U.S.)	Mid-1900s	Biologist and naturalist noted for his predator-prey studies on Dall sheep and wolves in Alaska.
H. Albert Hochbaum (U.S., Canadian)	Mid-1900s	Noted for his work on waterfowl ecology and was also the longtime director of the Delta Waterfowl Research Station in Canada.
David Lack (British)	Mid-1900s	The author of numerous books on animal population dynamics, particularly noted for his work on birds.
E. S. Russell (British)	Mid-1900s	Worked on fish population dynamics and most effectively presented the concept of maximum sustained yield.
Paul Errington (U.S.)	Mid-1900s	Both a biologist and writer, he was particularly noted for his work on the population dynamics of furbearers.
E. Raymond Hall (U.S.)	Mid-1900s	Taxonomist whose primary contributions concerned the distribution of mammals.
Durward Allen (U.S.)	Mid-to late 1900s	Biologist, conservationist, and author noted for his wolf-moose predator-prey relationship work in Isle Royal National Park.
William Ricker (Canadian)	Mid- to late 1900s	Developed many concepts of fish population dynamics, such as stock-recruitment relationships, and also worked on computational techniques for fishery data.
Raymond J. Beverton (British)	Mid- to late 1900s	With S. J. Holt, developed modeling systems to predict fish yield.

SUMMARY

Wild animals are organisms that are legally defined as being wild. Terrestrial and partly terrestrial vertebrate animals are generally categorized as wildlife. Fishes are considered to be in the fishery realm, but a variety of other organisms are also usually included. A fishery or a wildlife system is composed of three interacting components; the biota, the habitat, and the human users. The biota consists of the target or featured organism of a fishery or a wildlife system and most other living organisms in the system. Habitat consists of abiotic components and some biotic components, primarily large plants. Human users can be direct users, indirect users, consumptive users, or nonconsumptive users.

There are some basic differences between the target animals of a fishery and those of a wildlife system that make the study of the two fields different. Aspects of these animals such as growth patterns, body temperature regulation, the environment in which they live, their mobility, the stability of their habitats, how their abundance is measured, their reproductive potential, and the way humans view them are all quite different.

The history of wildlife and fisheries involves a variety of naturalists, philosophers, artists, politicians, and biologists who have made contributions. The past successes and failures of wildlife and fisheries revolve around (1) successes involving native species that were reduced in numbers but are now flourishing, (2) successes of some nonnative introduced organisms that have had positive effects, (3) failures resulting from nonnative introduced organisms that have had negative effects, (4) potential successes or failures involving species currently low in numbers, and (5) failures resulting from species that have become extinct.

Wildlife science and fishery science involve the process of obtaining knowledge about and studying wildlife or fisheries. Wildlife management and fishery management involve the manipulation of biota, habitat, and human users. The terms ecology, conservation, and biology are also commonly used in reference to these fields. Scientific names allow us to communicate with people around the world with no confusion as to the organism in question.

PRACTICE QUESTIONS

1. Why is it difficult to define which animals are wild and which are not?

2. What are the three components of a fishery or a wildlife system? Select a fishery or wildlife resource in your region and provide examples of each component.

3. Choose a specific fishery or wildlife system in your region, then develop a list of six of each of its direct, indirect, consumptive, and nonconsumptive users.

4. What is the primary difference between population density and biomass?

5. Why do humans relate more closely to birds and mammals than to fishes?

6. With regard to their ability to move from one area to another, how are marine fishes and land mammals and birds similar?

7. Describe two major written contributions of Aldo Leopold.

8. List a number of possible benefits of introducing nonnative organisms, then list a number of negative consequences associated with nonnative introductions.

9. What is science, and what is management?

SELECTED READINGS

Allen, D. L. 1962. *Our wildlife legacy.* Rev. ed. Funk and Wagnalls, New York.

Anderson, S. H. 1991. *Managing our wildlife resources.* 2d ed. Prentice Hall, Englewood Cliffs, N.J.

Benson, N. G., ed. 1970. *A century of fisheries in North America.* Special Publication No. 7. American Fisheries Society, Washington, D.C.

Bolen, E. G., and W. L. Robinson. 1995. *Wildlife ecology and management,* 3d ed. Macmillan, New York.

Carson, R. 1962. *Silent spring.* Houghton Mifflin, Boston.

Kallman, H., chief ed. 1987. *Restoring America's wildlife 1937–1987.* U.S. Department of the Interior, Fish and Wildlife Service, Washington, D.C.

Leopold, A. 1949. *A Sand County almanac and sketches here and there.* Oxford University Press, London.

Nielsen, L. A. 1993. History of inland fisheries management in North America. Pages 3–31 *in* C. C. Kohler and W. A. Hubert, eds. *Inland fisheries management in North America.* American Fisheries Society, Bethesda, Md.

Reiger, J. F. 1975. *American sportsmen and the origins of conservation.* Winchester Press, New York.

Trefethen, J. B. 1975. *An American crusade for wildlife.* Winchester Press, New York.

LITERATURE CITED

Bolen, E. G., and W. L. Robinson. 1995. *Wildlife ecology and management,* 3d ed. Macmillan, New York.

Caughley, G., and A. R. E. Sinclair. 1994. *Wildlife ecology and management.* Blackwell Scientific Publications, Boston.

Eschmeyer, P. H. 1990. Usage and style in fishery manuscripts. Pages 1–25 *in* J. Hunter, ed. *Writing for fishery journals.* American Fisheries Society, Bethesda, Md.

Leopold, A. 1933. *Game management.* Charles Scribner's Sons, New York.

Leopold, A. 1966. *A Sand County almanac with other essays on conservation from Round River.* Oxford University Press, New York.

Nielsen, L. A., and R. T. Lackey. 1980. Introduction. Pages 3–14 *in* R. T. Lackey and L. A. Nielsen, eds. *Fisheries management.* John Wiley & Sons, New York.

Primack, R. B. 1993. *Essentials of conservation biology.* Sinauer Associates, Sunderland, Mass.

Rounsefell, G. A., and W. H. Everhart. 1953. *Fishery science: Its methods and applications.* John Wiley & Sons, New York.

2

ECOLOGICAL
CONCEPTS

An important activity in wildlife and fisheries is working with populations of living organisms. To understand these organisms it is necessary to understand the ecological principles that govern various facets of their lives, such as where they live, what roles they play, and how they interact with the abiotic and biotic components of their environment. Much of the work conducted by wildlife and fishery biologists involves the application of this knowledge of ecological principles.

The decisions of wildlife and fishery biologists often have far-reaching implications for an ecosystem, several ecosystems, or even the entire biosphere. For example, a wildlife biologist may be involved in evaluating and prescribing timber harvest practices on private or public lands in North America. These practices may directly influence nesting sites, available food sources, and other needs of several migrant bird species, some of which winter in tropical portions of South America. Think of the myriad effects that a fishery biologist could have while working with a large and important group such as migratory **salmonids** (members of the fish family Salmonidae). These fishes can have ecological importance on a global scale. A total ecosystem approach to conservation is being increasingly emphasized, particularly on public lands.

The purposes of this chapter are to present a variety of ecological concepts, to illustrate the importance of these concepts with examples, and to introduce some basic ecological terminology. Many of these concepts will be referred to in later chapters.

2.1   Ecosystems, Biotic Communities, and Populations

An **ecosystem** is an interacting system of biotic and abiotic components in a particular area or place. Many different types of ecosystems can be identified. Ecosystem type classification is generally based on some major characteristic, such as the dominant plant in terrestrial ecosystems or the water body type (pond, stream, or lake) in aquatic ecosystems. Ecosystem classification can be applied at broad levels, such as grasslands or mountain lakes, or applied more specifically to areas with the same species of dominant trees (oak-hickory forest), animals (coral reef), or with other recognizable characteristics. Even a pile of cow manure in a pasture or a pool of rainwater in a natural rock bowl could be considered small temporary ecosystems.

A water body such as a pond or lake is an aquatic ecosystem with more or less clearly defined boundaries and with aquatic organisms in large part tied to those boundaries. However, other more mobile species that are part of this aquatic ecosystem, such as little blue herons, southern water snakes, and raccoons, regularly traverse the boundaries of the ecosystem. These animal species remove energy and nutrients from the system in excess of what they may replace. Thus the influence of such an ecosystem, particularly because it is highly productive, reaches far beyond its shoreline.

Ecosystems in general have influences reaching beyond their geographic boundaries. For instance, the northern regions of Canada and Alaska provide nesting and rearing areas for many species of shorebirds that spend the remainder of the year in southern temperate or tropical climates. The lesser golden-plover, for example, breeds in the Arctic and winters in southern South America. As these birds move between and through ecosystems, nutrients are transferred great distances, and the animal life of two or more widely separated ecosystems becomes interdependent. Likewise, the nutrients obtained by many Pacific salmon species in the marine ecosystem are important to the **productivity** (the rate at which energy is captured and stored in an ecosystem) of the freshwater and terrestrial ecosystems where they migrate to spawn and die during their life cycle. Oxygen produced in marine ecosystems and tropical forests is important to the atmosphere of the earth and to the survival of living organisms. Thus, even though we identify specific ecosystems, the influences of each ecosystem extend to other ecosystems and to the entire **biosphere** (the layer around the earth in which all living things are found).

Sometimes extensive regions of similar vegetation and animal life, such as Arctic tundra or tropical rain forest, are identified as **biomes**. Biomes are major terrestrial communities or major ecosystems occurring on the various continents and are discussed further in section 12.1.

The living portion of an ecosystem is called a **biotic community**; it is usually just referred to as a **community**. This interacting assemblage of living organisms is sometimes subdivided into plant and animal communities. At times the terms community and ecosystem are incorrectly used as though they were synonymous. Ecosystem is a broader term because it includes the interaction of the biotic and abiotic components of the system, while a community includes only the biotic component. In Europe the term **biogeocenosis** is a synonym for ecosystem, as is **biocenosis** for biotic community.

Ecosystems are often studied in terms of their structure and function. **Ecosystem structure** refers to the composition of the system with regard to species of plants, animals, and microbes, the nutrients available, the availability of water, and so forth, or generally a "grocery list" of what is there. **Ecosystem function** refers to how the ecosystem works in its many interacting ways, particularly how energy and nutrients move through and within the system.

In studies of ecosystems or their plant and animal communities, the number of species in a particular community is often called **species richness**. Species richness is sometimes used as a rough measure of **species diversity**. Species diversity, however, includes both the number of species present and some measure of **species equity** or **evenness**. Species equity or evenness refers to the relative abundance of the various species in a community. For example, two forests (A and B) with 20 tree species occurring in each would have the same tree species richness. However, if the most common tree species in forest A accounted for only 10% of all the trees present, while the most common tree in forest B accounted for 90% of all the trees present, forest A would have a higher species diversity than forest B. The greater the chance that a random tree selected from one of the forests would be of the same species as a previously selected tree, the lower the species diversity.

More recently, terms similar to species richness and species diversity have become widely used by both biologists and the general public. These terms, **biodiversity** and **biological diversity**, are synonymous, and they are often used in situations in which other terms, such as species diversity or species richness, would be more appropriate. The U.S. Congress Office of Technology Assessment (1987) defined biological diversity as the variety and variability among living organisms and the ecological complexes in which they occur. Primack (1993) divided biological diversity into three components: (1) genetic diversity within populations, (2) species diversity within ecosystems, and (3) community and ecosystem diversity across the landscapes of entire regions.

Species diversity is being eroded by increasing human populations and their effects on natural resources and ecosystems (fig. 2.1). Losses of species and species diversity can have many negative influences on humans. For example, insect populations have many checks and balances, which in part depend on their interrelationships with other organisms within an ecosystem. Reductions or losses of other species may cause some insects, once kept under control by natural processes, to increase in abundance, causing economic damage to crops or becoming a nuisance to humans. By losing species diversity, ecosystems become inherently less stable. There are many other reasons for maintaining species diversity; some of these are discussed in chapter 11. Biologists, realizing the importance of species diversity, attempt to maintain populations of species and to restore those that are endangered, threatened, or declining.

A **population** is defined as all the individuals of a species within a specified area at a given time. The area specified defines the scope of the

A   B

FIGURE 2.1 Human encroachment can reduce species diversity and cause substantial changes in ecosystems. *(A)* Urban development of montane and foothill areas about 30 kilometers east of Salt Lake City. *(B)* Plowing of native prairie for small grain production in western South Dakota. This native prairie site is in the initial phase of destruction by tillage and will soon be void of native plants.

population in question. A population can include all the individuals of a species in the world, for example, all the skipjack herrings. A population can also include only certain segments of a species, for example, the rock sandpipers in western Washington or the lake chubs in a specific lake.

At times the term population is used loosely to refer to more than one species, such as all the fish species in a pond or all the small mammals in a given area. This use of the term population is incorrect; multispecies assemblages are communities.

A population existing as a set of geographically separate subpopulations, but with some genetic interchange among them, is sometimes referred to as a **metapopulation**. Greater prairie-chickens, for example, currently exist in several subpopulations throughout the Great Plains that have minimal genetic interchange; these subpopulations together could be considered a metapopulation. Likewise, the subpopulations of mountain goats associated with different mountains or mountain ranges in western Canada could be considered a metapopulation. Coho salmon have many subpopulations

associated with different rivers; these subpopulations compose a metapopulation.

2.2 Biogeochemical Cycles and Energy Flow

In any ecosystem, energy and the essential elements of living cells must be available to support the system. Living organisms obtain essential elements from their environment, including the soil, water, air, and other organisms. In this way these elements are cycled through the ecosystem, thus the term **biogeochemical cycles**. Cycles of elements essential to the growth of living organisms may also be called **nutrient cycles**.

The elements essential as nutrients to most plants include carbon (C), hydrogen (H), oxygen (O), phosphorus (P), potassium (K), iodine (I), nitrogen (N), sulfur (S), calcium (Ca), iron (Fe), magnesium (Mg), chlorine (Cl), copper (Cu), boron (B), manganese (Mn), zinc (Zn), and molybdenum (Mo). Essential elements for animals include those used by plants, with the exception of boron. Additional elements required by animals include selenium (Se), cobalt (Co), fluoride (F), chromium (Cr), and silicon (Si). Essential elements for animals are further discussed in section 5.1. Understanding how these elements cycle

through ecosystems is important to understanding ecosystem function. Let us consider nitrogen, carbon dioxide, and sulfur as examples of element cycling.

The Nitrogen Cycle

Nitrogen (N_2) occurs as an inert gas in the atmosphere, but is unusable by most plants in this

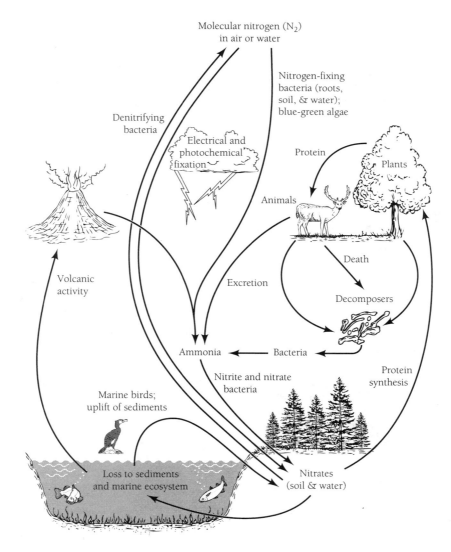

FIGURE 2.2 The nitrogen cycle. Nitrogen-fixing bacteria are the primary mechanism by which inert nitrogen is made available to plants and, through plants, to other living organisms.

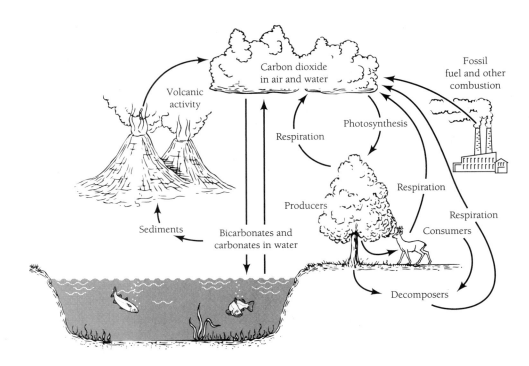

FIGURE 2.3 The carbon cycle. Carbon dioxide is required by plants for photosynthesis and thus is closely tied to energy flow in ecosystems.

form. The conversion of inert, gaseous nitrogen to ammonia (NH_3), nitrites (NO_2^-), or nitrates (NO_3^-) is termed **nitrogen fixation**. The nitrogen cycle (fig. 2.2) is especially interesting because certain plants, particularly legumes, in association with **symbiotic** bacteria in their root nodules (a symbiotic relationship is one in which interacting species are closely and permanently dependent on each other), can fix nitrogen in the form of ammonia (NH_3), making it available to plants and animals. Some free-living **aerobic** (living in the presence of oxygen) bacteria in the soil are also able to make nitrogen available to the biota, as can blue-green algae in aquatic systems. Biological nitrogen fixation produces ammonia by splitting the nitrogen molecule and combining it with hydrogen; this process requires energy. Most fixation of nitrogen is biological, although nonbiological fixation, the formation of nitrates from sources such as lightning and volcanoes, also occurs. Other parts of the cycle involve additional types of bacteria

that convert ammonia to nitrites (nitrite bacteria) and nitrites to nitrates (nitrate bacteria), as well as denitrifying bacteria living under **anaerobic** conditions (in the absence of oxygen) that release inert nitrogen back into the atmosphere. The amount of nitrogen fixed daily in the biosphere and the amount returned to the atmosphere is roughly equal. Collapse of the nitrogen cycle would quickly place most of the organisms in the world, including ourselves, in jeopardy of a very short existence.

The Carbon Cycle

The carbon (C) cycle (fig. 2.3) is currently receiving much attention because of its possible role in global warming (Stern et al. 1992; Ricklefs 1993) (see section 14.1). Carbon dioxide (CO_2) is required by plants for photosynthesis (see below). It is produced by plants and animals through normal processes of cellular respiration. Relatively small concentrations of CO_2 occur in the atmosphere and water, yet these concentrations are critical to ecosystems, and changes in these concentrations are dangerous in terms of possible perturbations to the climate of the planet. Carbon is in a dynamic equilibrium among the forms present in the oceans and other waters (bicarbonates and carbonates), in sediments and fossil fuels, and in the atmosphere, plants, and soils.

The Sulfur Cycle

Sulfur (S) is essential to living organisms, primarily for the formation of certain amino acids and enzymes. Sulfur occurs in **organic** (carbon-containing compounds including living or dead organisms) and **inorganic** (nonorganic minerals) deposits and is generally present in adequate supply for plants and animals. Sulfur at the surface of the earth is released into the air, water, and soil by decomposition of organic matter, including that in stored deposits such as peat, and by surface weathering, erosion, and volcanic eruptions (fig. 2.4). Sulfur is also released in gaseous form and in salt spray from the surface of the ocean. Sulfur occurs in various forms, including hydrogen sulfide (H_2S) and sulfur dioxide (SO_2), in the atmosphere and water, and as sulfate (SO_4) in water and soil. Sulfur in the atmosphere first appears as toxic hydrogen sulfide, but is quickly converted to sulfur dioxide (Smith 1990). Sulfur dioxide is soluble in water and forms weak sulfuric acid (H_2SO_4) in precipitation, which, upon reaching the surface of the earth, again becomes available to plants. Bacteria and fungi that act as **decomposers** (organisms that obtain energy from dead organisms and waste products) release sulfur from organic material. Certain bacteria are also able to use inorganic sulfur in the soil and make it available to the ecosystem. Plants obtain sulfur primarily as sulfate from soil and water, although some plants are able to use sulfur dioxide directly. Animals obtain adequate sulfur by eating plants or other animals.

Exposure of sulfur in mining deposits, particularly coal deposits, has in some cases led to acidification of water by sulfuric acid. The release of sulfur into the atmosphere through industrial processes, particularly the burning of high-sulfur coal or associated high-sulfur sediments, results in hydrogen sulfide formation in the atmosphere. The volatile gases containing sulfur combine with water to increase the level of acidity in rain, snow, dew, and fog, causing acid deposition, which can alter the acidity of streams, lakes, and soils and damage ecosystems (see section 14.1).

The Water Cycle

Water also cycles through ecosystems (fig. 2.5). The oceans provide moisture to the continents in this **hydrologic cycle** (water cycle). Most precipitation that falls on land evaporates, is stored as surface water or groundwater, or is returned to the oceans via river flow. Water composes only a small proportion of the atmosphere, and the turnover rate is high. Human-induced or natural changes in the hydrologic cycle can affect ecosystems in a broad way. Human activities that influence runoff or evaporation rates, use surface water or groundwater, or reduce natural recharge of **aquifers** (underground water in permeable formations of sand, gravel, or rock) can influence this important cycle.

Energy Flow

A major aspect of any ecosystem is **energy flow**. The word flow is used because energy comes into the system, usually as sunlight, is fixed by plants through photosynthesis, is passed through the system, and is eventually lost as heat. Most of the energy captured by plants is bound in chemical bonds through photosynthesis, and part of this energy is harvested by other organisms and passed through the system. Because energy captured by the ecosystem through photosynthesis eventually leaves the system, the process is called energy flow rather than cycling.

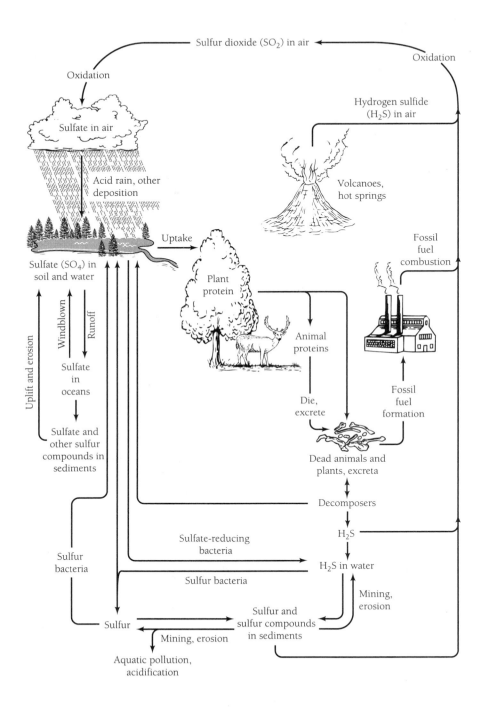

FIGURE 2.4 The sulfur cycle. Sulfur is essential for living organisms, but excess amounts released by the burning of fossil fuels and from acidic drainage of coal mines have caused environmental damage.

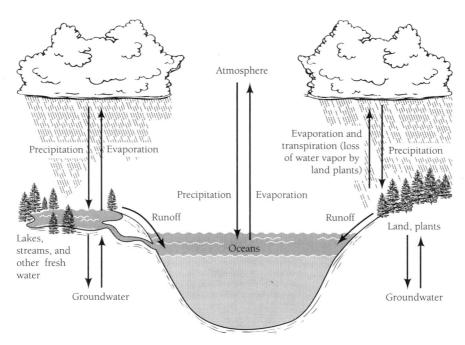

FIGURE 2.5 The hydrologic cycle involves the movement of water between the oceans, land, and atmosphere.

Photosynthesis involves the use of carbon dioxide and water to form complex organic molecules. The reaction requires light energy, which is captured by chlorophyll molecules in the leaves or sometimes in the stems or other parts of plants. The primary products of photosynthesis are sugars and oxygen, but other organic products, such as fatty acids, fats, vitamins, and proteins, are also produced by plants. The products of photosynthesis may vary by plant species and even with the age or part of the plant; for example, young leaves on a shrub may produce much more protein than older leaves. The basic chemical reaction of photosynthesis is:

$$6\ CO_2\ +\ 6\ H_2O\ \xrightarrow{\text{Sunlight}}\ C_6H_{12}O_6\ +\ 6\ O_2$$

Carbon dioxide + Water → Glucose + Oxygen

This formula shows the basic products and raw materials of photosynthesis. The photosynthetic process involves a light phase, in which energy is captured and water split into oxygen and hydrogen ions; oxygen is produced in this reaction (Brewer 1994). It also involves a dark phase, in which carbon dioxide and hydrogen are combined to form a variety of organic compounds, including glucose. Photosynthetic pathways and mechanisms for the capture of energy and formation of sugars are presented in greater detail in Jones (1983) and Smith (1990).

While photosynthesis uses energy from the sun to build complex organic molecules and produces oxygen, decomposition breaks down these organic compounds, releases carbon dioxide and water, and returns nutrients to an inorganic state. Decomposition of plant and animal matter is accomplished primarily by **microflora** such as fungi and bacteria, along with invertebrates and other organisms that feed directly on **detritus** (dead organic matter). Invertebrates that feed on litter and break it into fragments are usually called **detritivores**, while bacteria and fungi involved in decomposition are called decomposers. However, microfloral decomposers and invertebrates that serve as detritivores are sometimes grouped together as decomposers because they are all important in the decomposition process. Fungal and bacterial decomposers and detritivores obtain nutrients and energy in the process of breaking down dead organic matter. At the same time, these decomposers and detritivores release the energy bound in these materials and return nutrients to the inorganic state. Other larger organisms also break down detritus by their physical activities or by ingestion. Even the effects of precipitation in leaching nutrients from detritus and of weather in breaking down detritus, primarily plant matter, into smaller fragments are important in decomposition.

The primary decomposers of animal matter are bacteria, while fungi are more important in the decomposition of plant litter. Microflora secrete enzymes into plant detritus to assist in the decomposition process. Many species of bacteria can function in both aerobic and anaerobic environments. Bacteria that function in anaerobic environments are particularly important in the decomposition of organic matter that collects at the bottoms of lakes or slow-moving streams.

Mites, protozoans, nematodes, caddisfly larvae, stonefly nymphs, earthworms, and millipedes are all examples of detritivores. Detritivores play an important role in the decomposition process by fragmenting organic material and inoculating it with bacteria and fungi. The composition of the microflora and detritivores changes with the stage in decomposition. At one stage, organisms that live on simple sugars may be successful, while at a later stage, the remaining decomposers and detritivores may be those that specialize in the use of more complex carbohydrates, such as cellulose and lignins in the woody cell walls of plants. Another group of organisms, the **microbivores**, feed on the nutrients and energy in bacteria and fungi. Examples of microbivores include larval beetles, flies, mites, and amoebas.

While decomposers may use nutrients and keep them out of circulation in their own body tissues (**nutrient immobilization**), their role in decomposing detritus and releasing inorganic nutrients (**mineralization**) is invaluable to the ecosystem. Microflora and detritivores are essential to the release of nutrients and energy bound in the ecosystem and thus to maintaining ecosystem productivity.

2.3 Trophic Levels, Food Chains and Webs, and Pyramids of Energy

Energy is transferred from plants to animals when the animals consume the plants. Subsequently, this energy is passed through additional feeding or **trophic levels** as animals are eaten by other animals. **Autotrophs** are organisms, such as plants, that can produce their own food; **heterotrophs** must utilize autotrophs or other heterotrophs to obtain food. Energy flows through the ecosystem as organisms eat and are eaten. This path of energy flow can be diagrammed as a **food chain**, or more accurately, as a **food web** (fig. 2.6). Food chains are simple examples often used to introduce this topic, but nature really functions in food webs. A chain implies linearity, but that is not the way in which these trophic interactions generally occur.

The trophic levels, in the order of energy flow, include **primary producers** (autotrophs or plants), **primary consumers** (**herbivores** that eat primarily autotrophs), and **carnivores** (animals that feed primarily on other heterotrophs). The term **secondary consumer** is often used for carnivores such as the mountain lion that feed primarily on primary consumers. The term **tertiary consumer** can be used for carnivores such as the white shark that feed primarily on secondary

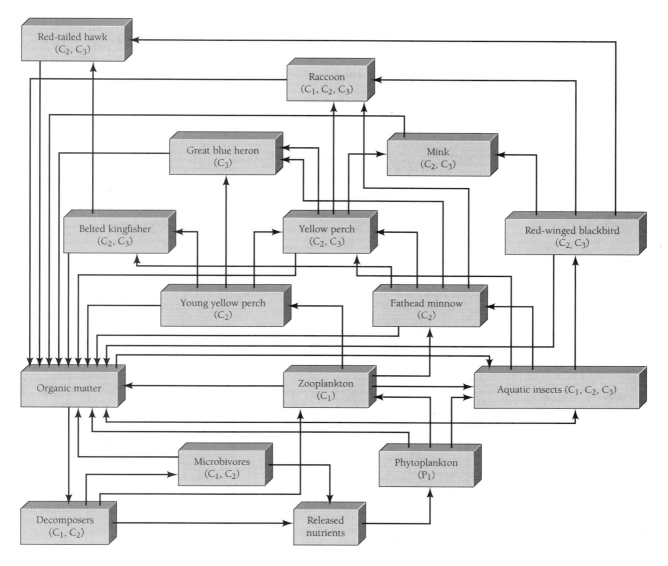

FIGURE 2.6 A simplified food web for a small farm pond in the Midwest. The arrows show the paths of energy flow and nutrient cycling among the various trophic levels. P_1, primary producer; C_1, primary consumer; C_2, secondary consumer; C_3, tertiary consumer. Many other organisms could be added to such a food web.

consumers. The term **predator** is also used for organisms that attack, kill, and eat other animals, which are known as their **prey**.

Other feeding levels are also used to group organisms. Animals such as black bears, channel catfish, and raccoons, like humans, readily eat both autotrophs and heterotrophs and are called

omnivores. Omnivorous species fill a mixed role as both primary consumers and carnivores (secondary and/or tertiary consumers). Also contributing to the food web are the decomposers, detritivores, and microbivores discussed in section 2.2. Depending on what they feed on (plant or animal matter), these organisms can also be grouped as primary consumers or carnivores. Larger organisms such as black vultures that obtain much of their food from dead animal matter are often termed **scavengers**; they are also carnivores. **Saprophytes** are nonanimals, primarily fungi, that feed mainly on dead plant material; they are also appropriately called decomposers. **Parasites**, such as fleas, sea lampreys, and intestinal nematodes, obtain nutrients and energy by living inside of or on other organisms.

Plants, whether they are trees, grasses, shrubs, or **phytoplankton** (tiny aquatic plants that float or drift in the water), are primary producers. They are the living base of food webs. Organisms such as ring-necked ducks, grass carp, mule deer, muskrats, and **zooplankton** (tiny aquatic animals that primarily float or drift in the water) are, for the most part, primary consumers. A qualifying statement is needed here because an animal such as the muskrat will eat animal material, including other muskrats, if the need and opportunity arises, but such material constitutes only a small portion of its diet. Numerous other species alternate as primary consumers and secondary consumers depending on their nutrient needs. For instance, female ducks, such as northern pintails and mallards, are largely herbivorous much of the year, but feed heavily on invertebrates such as aquatic insects, fairy shrimp, snails, and earthworms during late winter, spring, and early summer when preparing to reproduce and during egg laying. Ducklings also depend on the high protein content of invertebrates, especially in the first several weeks after hatching. The young of **gallinaceous** birds (members of the order Galliformes, such as the sharp-tailed grouse, California quail, and gray partridge) exhibit a similar age dependence on invertebrate foods. These examples illustrate why food webs are a more appropriate metaphor than food chains for the flow of energy through an ecosystem.

Numerous organisms also shift between being secondary consumers and tertiary consumers. Predatory great barracuda feed primarily on invertebrates as secondary consumers in the early portion of their lives, but shift to a fish diet and a tertiary consumer role as they grow. Thus, even though trophic levels are often used to categorize species, these levels may be dynamic within a species in relation to age, reproductive status, and opportunity to exploit new food sources. For species that switch trophic levels seasonally or with age, it works best to think of them at seasonal or age-specific trophic levels instead of assigning them one overall trophic level designation.

Energy is transferred through the various trophic levels. Although energy cannot be created or destroyed (**first law of thermodynamics**), a considerable loss of energy occurs with each conversion in the food web. Producers fix a given amount of energy from the sun in various chemical bonds. Primary consumers use this energy through herbivory and digestion, but much of it is dispersed in unusable forms such as heat. A reasonable rule of thumb is a 90% loss of energy at each conversion. In other words, if 1,000 kilocalories of plant material (primary producers) were consumed by primary consumers, only about 100 kilocalories would be converted to primary consumer biomass. This loss of energy from one trophic level to the next follows the **second law of thermodynamics**. According to this law, when energy is transferred or transformed, much is lost or dispersed in unusable forms, primarily heat. Thus, species at higher trophic levels, such as muskellunge or bobcats, are limited in population biomass by the amount of energy they can receive from lower trophic levels. By the laws of energy flow within ecosystems, a population of a **piscivorous** (fish-eating) fish species such as the northern pike is incapable of reaching the biomass that shorthead redhorses, bluegills, or other lower trophic level fishes can reach. Pyramids of energy illustrate the manner in which energy is transferred and lost as it flows through trophic levels (fig. 2.7). Other similar **ecological pyramids** include those for numbers and biomass.

In fisheries, the primary game and food fishes are secondary and tertiary consumers, while almost all game wildlife species are at lower trophic

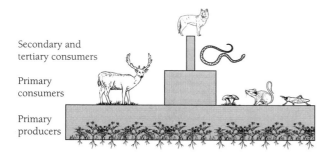

Secondary and
tertiary consumers

Primary
consumers

Primary
producers

FIGURE 2.7 A generalized pyramid of energy for a terrestrial ecosystem, showing the transfer of energy through different trophic levels. Primary producers include not only aboveground but also belowground plant matter (roots) as well as dead plant materials. Organisms involved in decomposing dead plant or animal material are grouped with the appropriate consumers.

levels, filling the role of primary consumers. Some predators managed as **furbearers** (mammals commonly harvested for their hides, such as mink and red foxes) fill roles at higher trophic levels and, as in the fishery example above, cannot by the laws of energy flow reach the population biomass of a primary consumer species such as the muskrat.

2.4 Ecological Succession

Ecological succession involves the processes that occur when an ecosystem changes from an early, immature, transient stage, such as that found on a newly disturbed site, to a more mature, stable final stage termed a **climax community**. Examples of climax communities include old-growth tropical rain forests in Brazil, old-growth oak-hickory forests in the southeastern United States, and deserts dominated by giant saguaro and palo verde in southern Arizona. Succession involves changes in the structure and function of the ecosystem and is

reasonably predictable in terms of the dominant plants and other associated plants and animals that will occur in the various successional stages. In succession, the biotic community modifies the environment and exerts much control over the direction of succession. The physical environment of the general area also exerts a strong influence. The transient but recognizable stages in succession are called **seral stages**, while the entire sequence of seral stages from early succession to climax is termed a **sere**. Succession occurring on sites unchanged by living organisms is called **primary succession**. An example of primary succession would be the development of a biotic community on a recently cooled lava flow. Succession occurring on sites that were previously occupied by other organisms, as in an abandoned agricultural field or burned forest, is **secondary succession**.

A ponderosa pine forest burned by a wildfire in which all the trees and many shrubs were killed provides an example of succession. The earliest phase of succession in this sequence might be a grass-forb seral stage (a **forb** is a nonwoody plant such as poison ivy that is not a grass or grasslike), followed by grass-forb-shrub, shrub-pine sapling, and other intermediate seral stages, until the climax ponderosa pine forest with its associated **understory** (plants growing on forest floors under trees) and animal life is eventually reached in 200 or 300 years (fig. 2.8). The ponderosa pines form an **overstory** or **canopy** above the understory. During this sequence the species composition and energy flow within the system changes, and the ecosystem itself, along with the physical environment, dictates the nature of the system.

In general, the structure and function of an ecosystem becomes increasingly more complex as succession progresses; this complexity imparts greater stability to the ecosystem. A mature tropical rain forest is a relatively stable ecosystem due to its great complexity in terms of species diversity, energy flow, and nutrient cycling. Its many checks and balances resist perturbations to the system. The rain forest system would probably be very resistant, for example, to attacks of harmful insects

A   B

FIGURE 2.8 Secondary succession from a burned-over forest (grass-shrub-sapling seral stage) (A) to a climax forest dominated by large ponderosa pines (B) will take from 200 to 300 years. The areas shown in these photographs are in the northern Black Hills in Wyoming; climax forests here feature unusually lush understory vegetation compared with most climax ponderosa pine forests.

due to checks and balances in the form of a broad array of **insectivores** (insect eaters). The diversity of plants in such an ecosystem also adds to its stability. The loss or reduction of a single plant species due to insects or a disease, for example, would tend to have less of an effect on an ecosystem with a great diversity of plant species than on a simpler ecosystem with few plant species.

An ecosystem maintained at an earlier seral stage by continued disturbance such as fire or agriculture is called a **disclimax**. For example, portions of the native tall-grass prairie in North America were maintained by periodic fires that favored grassland over forest species. Specific types of forests can also be maintained by fire. Lodgepole and jack pines are fire-tolerant species that are maintained by periodic fires that remove competing species. In the southeastern United States, longleaf pine forest is maintained by periodic fires that retard invasion by deciduous trees.

Succession is not confined to terrestrial ecosystems; it also occurs in aquatic systems. For example, over time a river progresses through various successional stages that can be illustrated by

looking at river sections of different ages. The headwaters of rivers typify an early successional stage. Headwaters are characterized by V-shaped valleys, colder and clearer water, lower fertility (concentrations of soluble minerals that are important as nutrients in the ecosystem), rock and gravel bottoms, and lower species diversity. Downstream successional stages in older sections of the river are characterized by having U-shaped valleys, water that is warmer and more turbid, increased fertility, more silted bottoms, and increased species diversity. Standing water bodies such as lakes also move through successional stages. Early successional stage lakes are deeper and have more oxygen, more inorganic bottoms, better water transparency, lower fertility, and generally less vegetation than do older lakes. As these characteristics change in water bodies, so do the fish and other animal and plant populations.

One of the reasons biologists are interested in succession is the direct effect it has on the occurrence and abundance of organisms. Post-fire succession in forested regions from grass or grass-shrub communities to climax communities is accompanied by predictable changes in animal communities. For example, early forest clearing by means of plow, axe, and fire in central Wisconsin transformed once-forested regions into excellent sharp-tailed grouse and greater prairie-chicken habitat. These grouse species thrived temporarily until the grass, early shrub, and agricultural communities were replaced by young quaking aspens and eventually by coniferous forest. Many other wildlife species also fluctuated in abundance and even occurrence as succession progressed from grassland-shrub-agriculture to quaking aspen and eventually back to the climax forest.

Biologists are aware of the effects of succession and use these principles to manage for species associated with various seral or climax stages. A biologist attempting to maintain summer breeding habitat for the endangered Kirtland's warbler in northern Michigan may use burning to maintain an early successional stand of jack pines. Ruffed grouse are managed by maintaining different-aged stands of quaking aspen, an early successional stage plant species. Other animals, such as northern spotted owls in Oregon or Sitka black-tailed deer in southeastern Alaska, directly benefit from maintenance of old-growth coniferous forest. These are examples of management for individual species; however, management for an individual species through manipulation of seral stages influences the entire animal community, not just the target species. Thus, the emphasis on the northern spotted owl in relation to old-growth forests in the Pacific Northwest should be broadened to include the variety of other animal species and subspecies that depend on old-growth forests, such as the marten, pileated woodpecker, marbled murrelet, northern goshawk, and others. Plants dependent on the environment created by these old-growth forests should also be considered.

It may be best to provide a variety of seral stages to increase the diversity of wildlife in an area. Additionally, many species require several widely divergent seral stages. White-tailed deer and elk (also called wapiti), for example, readily use clearings for feeding, but seek nearby forest cover for escape and protection from the elements.

Even within woodland areas, thick stands of small trees may provide hiding cover, while more mature forests with extensive overstory can be important as thermal protection in both winter and summer.

Newly constructed reservoirs provide an excellent example of aquatic succession and its effects on fish populations. When Lake Oahe, a Missouri River reservoir in North and South Dakota, was being filled, northern pike were abundant. As the reservoir filled with water and for a few years after filling, the flooded soils and terrestrial vegetation released abundant nutrients that led to high levels of productivity; the vegetation also provided reproductive habitat for the northern pike. After the reservoir had been filled for some time, productivity declined, and flooded terrestrial vegetation was no longer available for northern pike reproduction. The northern pike population declined, and this species was replaced by other fishes such as walleyes that were better adapted to the reservoir at this later successional stage. Continued succession in Lake Oahe can be expected in the future.

2.5 Niche, Habitat, and Environment

The role or function of an organism in the biotic community, its "occupation," has been defined as its **niche** (Elton 1927). This definition emphasizes the way an organism functions in an ecosystem in terms of obtaining energy. For example, the Mexican long-nosed bat is a primary consumer that specializes in feeding on nectar and pollen. The term **multidimensional niche**, proposed by Hutchinson (1957), can be defined as the variety of niche components that constitute the total niche of an organism. This term is often used because it better explains the breadth and complexity of the niche concept. The multidimensional niche concept was first developed in relation to studies of

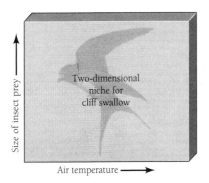

FIGURE 2.9 Air temperature and size of flying insects are two potential dimensions of the niche of a cliff swallow.

competition between organisms that occupy similar niches. Hutchinson (1978) viewed the niche of an organism as comprising a variety of components, including trophic level (functional role), where it seeks food (for example, open forest understory or the rocky bottom of a stream), when it feeds, what size food it eats, and where it reproduces. The multidimensional niche concept emphasizes the role the organism plays in an ecosystem and is broader than, but inclusive of, the definition provided by Elton (1927). For example, the niche of a cliff swallow, an aerial feeder, might be viewed as two-dimensional, comprised of air temperature and size of insect prey (fig. 2.9); distance of feeding sites from nesting or roosting sites would add a third dimension to this multidimensional niche.

The niche of any species involves many dimensions and can be an abstract and difficult concept. However, most studies of niches target from one to three dimensions. For example, a study comparing the niches of mule deer and elk on winter ranges could target overlap in food habits and spatial distribution in relation to escape cover such as dense forest or shrub cover. A study examining possible niche overlap in nesting red-tailed and Swainson's hawks might target prey size as one dimension and nest substrate as a second dimension. Researchers examining possible niche overlap in largemouth, smallmouth, and spotted

bass might focus on habitat used, diversity in prey items, and the sizes of prey items.

In section 1.2 we defined habitat as the place where an organism lives. That term needs to be further discussed. Brewer (1994) defined habitat as the specific set of environmental conditions under which an individual, species, or community exists. Odum (1971) defined habitat as the place where an organism lives; this definition implies longer-term use of an area. Perhaps the term "lives" in the definition provided by Odum should be replaced by "is found" so that areas used regularly during migration would also be included. When animals migrate between areas of regular seasonal use, they may spend only a few hours, days, or weeks at any particular site. For example, caribou are a highly mobile species. All the areas used seasonally by caribou herds, such as calving grounds, windy hilltops and cooler slopes used to escape insects in the summer, wintering areas, and areas they migrate through, are needed to support caribou and constitute habitat for that species. Likewise, on its route to Arctic nesting grounds, the lesser golden-plover migrates from southern South America through the middle of North America. The shallow flooded areas and mudflats where lesser golden-plovers stop to feed and rest during migration qualify as migration habitat, even though the birds may remain at such sites for only a few hours or days. Habitats used by migrating animals to rest and replenish food supplies are sometimes referred to as staging areas, particularly in birds. An organism moving rapidly through an area without using the resources present would not necessarily be in habitat suitable for that organism. For example, a coyote that is just crossing an extensive stretch of frozen lake is not in coyote habitat. The airspace occupied by migrating Arctic terns over the Atlantic Ocean is not tern habitat. However, the stream through which a chinook salmon migrates provides resting and escape habitat even in areas where reproduction may not occur, and is considered a type of migrational habitat.

Even on a localized basis, year-round resident species may shift habitats seasonally. During winter in the northern Great Plains, white-tailed deer and ring-necked pheasants move into areas with dense winter cover, such as cattail or willow stands, to seek protection from winds and extreme cold. Fish species such as lake trout shift from deep water habitat to shallower, warmer surface water habitat as temperatures cool during the fall.

The habitats of organisms are usually described by the general characteristics of the area being used. For example, the breeding habitat of pied-billed grebes is a shallow body of water that has an interspersion of emergent vegetation and open water. **Emergent vegetation** refers to rooted aquatic plants that can grow through the water column and above its surface; cattails are an example. The water and emergent vegetation are emphasized as the dominant characteristics of the habitat, but other aspects, including prey availability, size of the water body, escape cover, and humidity, may also be important parts of the habitat.

The term **environment** includes the habitat of the organism but is much broader in scope. We define environment as all the surroundings of an organism, including various conditions or influences that may occur irregularly. Aspects such as acid deposition, unusual flooding, hurricanes, volcanoes, and extreme temperatures are not generally used to describe habitat but are part of the environment of an organism.

Microhabitat is defined in a variety of ways, but is generally viewed as a small area of intensive use within the overall habitat of an organism. For example, a wild turkey usually selects a nest site microhabitat that provides some type of visual obstruction, such as shrubs or dead tree limbs, around or over the nest (fig. 2.10).

The term **guild** is sometimes used to refer to a group of organisms that use resources, such as food sources, in similar patterns and thus often have considerable niche overlap. For example, fishes that feed on organisms on the bottom of a stream, such as spotted suckers and suckermouth minnows, could be considered part of the guild of "bottom feeders." Birds that catch insects primarily in the air, such as common nighthawks and cliff swallows, could be considered a guild. Small mammals, such as the little pocket mouse and Ord's kangaroo rat, would belong to a seed-eating guild.

2.6 Interspecific and Intraspecific Competition

The niche concept is closely associated with interspecific competition. **Interspecific** means between or among different species; **intraspecific** means within a species. **Competition** occurs when two or more organisms living in the same area have overlapping niche requirements for a resource that is in limited supply. Note that for competition to occur, the resource must be in limited supply. For example, there could be competition

FIGURE 2.10 Wild turkeys usually place their nests in microhabitats that provide overhead and lateral concealment.

for nesting sites such as **secondary tree cavities** (cavities that have already been constructed), or for a specific food resource. In North America the eastern bluebird has declined, in part because of interspecific competition for secondary nesting cavities with two bird species introduced from Europe, the European starling and the house sparrow. However, two or more species may overlap almost completely on several dimensions of a niche and still not be in competition. For example, great horned owls and red-tailed hawks are found in the same geographic areas and habitats and have some of the same niche requirements. However, the owl generally hunts at night, while the hawk is active during the day. This difference in temporal (time) activity patterns, a single niche dimension, reduces the competition between these two species. The tendency for species that live in the same area and require similar resources to have niche requirements that differ in one or more dimensions is called **niche segregation**.

The **competitive exclusion principle** (**Gause's principle**) addresses the matter by stating that no two species can simultaneously and completely occupy the same niche for an indefinite period of time. This principle has also been succinctly stated as "complete competitors cannot coexist." The principle has many applications even though it is in large part untested (and untestable) scientifically.

A good example of the flexibility of organisms with regard to interspecific competition occurs in the **sunfishes** (members of the fish family Centrarchidae). The pumpkinseed eats a broad variety of invertebrates when it is the only sunfish species in a pond. However, if bluegills and green sunfish live in the same pond, pumpkinseeds will depend heavily on snails. Likewise, where yellow-headed and red-winged blackbirds occur in the same area, the two species compete for overwater nesting sites in emergent vegetation. When nesting, the yellow-headed blackbirds will displace the red-winged blackbirds from the emergent vegetation over the deeper parts of the water body; the red-winged blackbirds will be forced to set up territories in the shallower peripheral areas of the water

body or on moist upland habitats. Red-winged blackbirds will establish territories and nests over the deeper parts of water bodies where yellow-headed blackbirds are absent. Mule and white-tailed deer in southeastern Arizona were found to have considerable overlap in their food habits but to have relatively little overlap in local distribution due to differences in habitat selection (Anthony and Smith 1977). In all of these examples, some niche overlap occurs, but not to the extent of elimination of the other species. Some niche overlap between species that live in the same area is common. The **fundamental niche** of an organism is the niche occupied by that organism when there is no competition from other species. The **realized niche** of an organism is the niche occupied by that organism when some competition with other species is occurring. Most organisms occupy a realized niche instead of a fundamental niche because competition with other organisms is usually occurring to some degree.

Interspecific competition can also be human induced. Introductions of trout into areas and systems lacking native trout species, such as the Black Hills in South Dakota and Wyoming and the streams of New Zealand, have provided considerable potential for competition with native fishes. Brown, rainbow, and brook trout were all introduced into the troutless streams of the Black Hills. Native fish species in those streams, such as the mountain sucker and creek chub, although not endangered, must now compete for prey with the introduced trout species. The brown trout introduced into New Zealand are well known for their rapid growth rates, but probably compete with native piscivorous fishes. If competition from introduced species is strong enough, native species may be reduced in numbers, eliminated in some areas, or possibly brought to extinction.

Few questions were asked about possible interspecific competition with native species when the common carp, brown trout, house sparrow, burro, horse, and European starling were intentionally brought to North America from foreign lands. Biologists should be especially concerned about possible competition between organisms that are native and those that are introduced into an area. Animals found on islands are particularly susceptible to competition and possible elimination after the introduction of new species. For example, the mallard, introduced from Europe and North America, appears to be competing strongly with and increasing in relation to the native gray duck in New Zealand (Caithness et al. 1991); the two species also hybridize.

Intraspecific competition is competition for limited resources among individuals of the same species. The strongest competitor of any individual organism is another organism of the same species, because these organisms have essentially the same niche requirements. In some cases niche requirements vary by sex or age categories, and this variation can reduce direct intraspecific competition. For example, males of most hawk species are generally markedly smaller than females and may exploit different-sized prey, reducing intraspecific competition and expanding the ability of a pair to feed their young. Intraspecific competition for space in territorial species is a mechanism that assures the availability of certain resources exploited by that species.

Intraspecific competition is sometimes divided into **exploitative competition** and **contest competition** (**interference competition**). In exploitative competition the competing individuals divide the resource somewhat equally, and all suffer similar effects, such as death or reduced growth or reproductive rates, from resource shortages. For example, in a stream with a high density of adult yellow bullheads, the fish are competing for a limited food supply. The result of competition is that all the yellow bullheads may have reduced growth rates, but all are basically influenced in the same manner by the lack of food. Contest or interference competition occurs when some individuals compete more successfully for the resource at the expense of other competing individuals. Nesting common barn-owls often hatch several young, and because they initiate incubation as soon as they lay their first egg, these nestlings are of different ages and sizes. If the food supply during the rearing stage is limited, the

adults may not be able to obtain adequate food for all of the nestlings. In this case the older, larger nestlings will dominate the "contest" for food brought to the nest and thus be more likely to survive.

The consequences of intraspecific competition can also be observed in reduced reproductive rates in some species. For example, when a northern bobwhite population is high, the number of young produced per adult is generally lower than in years when the population is comparatively low. Elk on overcrowded ranges will similarly reflect competition with lowered survival rates of young and reduced birth rates. In some species, younger individuals delay entry into the breeding population by a year or more where competition for resources, including space, is severe.

2.7 Home Range and Territory

Some animals, particularly vertebrates, tend to establish a normal daily area of use as individuals or groups. The area within which an individual animal normally travels in its daily activities is termed its **home range**. Home range should not be confused with the **geographic range** of a species, which defines the geographic boundaries within which the species occurs (see section 2.9). Geographic range is often just referred to as the **range** of a species.

Within an annual period, the home range of an animal normally changes in response to factors such as reproductive condition, weather and habitat conditions, and food availability. For year-round resident species, shorter-term home ranges, such as those used in the nesting season, can be identified, as can longer-term or annual home ranges. Measuring home ranges during different periods of the year can give more realistic information on animal needs in terms of habitat than a single annual measurement. For example, the annual home range for female wild turkeys would include the general area used during nesting, brood rearing, and wintering. A biologist would want to know specific habitat needs during each of these periods to provide the most useful information. Likewise, large northern pike may have home ranges in the warmer near-shore zone of a lake during spring and move to deeper, cooler portions of the lake as water temperature increases. Putting these seasonal home ranges together would mask much important information.

All organisms have space requirements, even if just for a place to attach for life, as with a barnacle. However, many organisms, especially vertebrates, reflect their spatial needs through territorial behavior. **Territory** can be defined as a portion of the home range defended against others of the same or sometimes closely related species. Thus, a large rainbow trout will defend part of a stream against other rainbow trout that would compete for that space as a resting site, feeding site, or for other uses. Likewise, one would expect a snapping turtle or black bear to defend certain areas against others of the same species. During the breeding season, birds often proclaim their nesting territories through songs, calls, and displays. Other types of territories, such as feeding territories and wintering territories, have been identified for various animals. Even a long-domesticated animal such as the dog retains the territorial traits exhibited by its wild kin. The smallest dog can have a big bark and a hard bite in its own territory; its aggressiveness wanes rapidly when it is removed from that territory.

Gray wolves defend pack territories and act as a group. Reproduction within the pack is restricted to the dominant male and female, or at the least to a few dominant individuals, and the entire pack helps in rearing the young. The pack territories are marked by scent and extend over a large area that often exceeds 100 square kilometers. The reason for such large territories is clear when the extensive prey needs of gray wolves are noted. A large territory assures a food supply throughout

the year. Gray wolves can suffer severe population losses due to starvation during winters when their primary food sources are scarce. Ignoring pack boundaries can lead to injury or even death for trespassing individuals should the pack not accept them.

Male largemouth bass are highly defensive toward other largemouth bass, as well other fish species, around their nest sites. Thus, largemouth bass nests are widely spaced. In contrast, bluegills nest in colonies, with each male bluegill defending only a small area around his individual nest site.

Some species group together at certain times of the year, or even for most of the year, and appear to show little evidence of territoriality. However, many of these species have an **individual distance** requirement even when found in groups. Observe red-winged blackbirds perched together in the early spring or fall or migrating birds such as cliff or bank swallows on a power line (fig. 2.11). Each individual maintains a fixed, although minimal, distance between itself and

FIGURE 2.11 Birds in flocks, such as these red-winged blackbirds, still maintain a space, or individual distance, between themselves and the nearest bird or birds.

the next bird. Birds crossing that distance do what is apparently socially unacceptable to the next bird. Many fish and mammal species behave in a similar fashion.

2.8 Limiting Factors and the Law of Tolerance

Organisms, both plant and animal, have a variety of needs that must be obtained from the environment. These needs can include food, water, escape cover, habitat for reproduction or nesting, habitat for rearing young, and many others. Populations may be limited by the lack of sufficient amounts of these items. A single component that limits population growth is considered to be a **limiting factor** (**limiting resource**). Populations may also be limited by a combination or interaction of factors.

Biologists attempt to identify the most important limiting factors that affect a population. A smallmouth bass or spottail shiner population normally remains at a certain biomass level in a specific water body. That level is often determined by a limiting factor that could, for example, be prey availability in one area or lack of reproductive habitat in another. Bald eagle breeding populations may be limited by the availability of large nesting trees near water, human disturbance, or an adequate prey base. If human populations and activity in an area are too great, nesting bald eagles may be unable to use potentially valuable habitat. If food sources are lacking, bald eagles will not be able to use large trees isolated from human activity that otherwise would provide suitable nest sites. Determination of limiting factors allows biologists to develop management plans to increase populations of target organisms. If food is the limiting factor, for example, and it is increased, the population can expand, but at some point a new limiting factor will restrict population growth. This discussion

could be about mayflies, white-footed mice, western hog-nosed snakes, or humans—they are all influenced by limiting factors (see section 3.5).

Available space can be a limiting factor for a territorial species. Usually biologists must work with the given space requirements of the species, realizing that the upper limit is set by the behavior of the organism and ultimately, by limited resources. However, for some species, territory size is at least partly related to the availability of other resources or even to the degree to which territorial individuals are visible to each other; in such cases, management to increase populations may be possible. For example, some fish species are highly territorial around their nests during the breeding season. If that territoriality were dependent on visual contact with other breeding fish at other nest sites, the use of visual obstructions could enhance breeding densities of that species of fish.

For some pest species, the biologist's goal is to manipulate limiting factors to reduce populations. Rock doves (pigeons) and house sparrows are introduced species that are immensely successful in North America. Most city officials would gladly accept no or fewer rock doves and house sparrows instead of the abundant populations that often exist, but some buildings provide perfectly engineered habitat for these troublesome species. Engineers could manipulate limiting factors on these species by designing buildings with few sites for rock dove roosts or house sparrow nests. Conversely, barn and cliff swallows, which are desirable in most situations, are limited by the availability of "clifflike" nesting sites protected from excessive rain. Humans have accidentally influenced a limiting factor for these species with the construction of bridges and other structures that provide protected sites for attachment of their mud nests (fig. 2.12).

Within the geographic range of a species, one finds regions where the species thrives and other areas where it may occur only sporadically. Often, conditions near the edge of the geographic range of a species approach the limits of the species' tolerance in regard to one or more environmental

FIGURE 2.12 The construction of bridges has removed a limiting factor by providing nesting habitat for cliff swallows in areas of North America once unsuitable for nesting populations of these birds.

no longer occur, or will occur only sporadically. For black bullheads, the range of tolerance for embryonic fish has been clearly defined: black bullhead embryos can survive only in water between 18° and 28°C, with an optimal temperature between 20° and 22°C (Stuber 1982) (fig. 2.13).

Northern bobwhites are found across most of the eastern, midwestern, and southern United States. To the north, they reach into Minnesota, Wisconsin, and South Dakota, but fluctuate greatly there with severity of winters and snow cover. It appears that northern bobwhites can withstand a limited duration and depth of snow cover, but cannot withstand continuous snow along with extreme cold for more than a few days. Thus, they reach the limits of tolerance for snow cover and temperature in the northern parts of their geographic range. The variation in tolerance for minimum temperatures among small birds is well illustrated by the ability of the horned lark to winter in the harsh cold of the northern Great Plains while

conditions; this is the **law of tolerance**, a concept attributed to the ecologist V. E. Shelford (1913). Odum (1971) stated the law of tolerance in the following manner: "Absence or failure of an organism can be controlled by the qualitative or quantitative deficiency or excess with respect to any one of several factors that may approach the limits of tolerance for that organism." Thus, species tolerate ecological minimums and maximums in regard to various conditions. For instance, greenback cutthroat trout living in streams on the eastern slope of the Rocky Mountains in Colorado can exist only under environmental conditions amenable to that species. If one moves downstream along a hypothetical Rocky Mountain river with a greenback cutthroat trout population, summer water temperatures will at some point exceed the limits of tolerance for that species, and it will

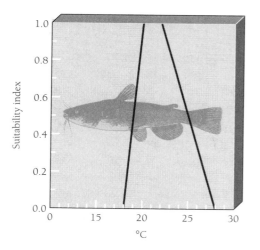

FIGURE 2.13 The suitability of water for embryonic development of black bullheads is dependent on water temperature. The suitability index is directly proportional to population abundance. Minimum and maximum temperatures provide an example of the range of tolerance during embryonic development for this species. Optimal temperatures appear to fall between 20°C and 22°C. (*Adapted from Stuber 1982.*)

many similar-sized species retire to tropical or subtropical areas to winter.

2.9 Dispersion Patterns and Dispersal

Each wildlife and fish species has a geographic range over which it is distributed. For example, the geographic range of the Nuttall's cottontail covers most of the sagebrush deserts found in the intermountain region of western North America. Geographic ranges vary from very small areas to almost worldwide. A species' geographic range reflects factors such as the mobility of the organism, the distribution of habitat conditions appropriate for the species, its ability to tolerate ranges of conditions, the distribution of other species that might exclude the organism, and other physical or behavioral barriers to movement.

Individuals within populations of plants and animals are dispersed over the surface of the earth in various patterns. **Dispersion** refers to the spatial pattern or distribution pattern of individuals in a population. These patterns are sometimes categorized as **random**, **uniform** (**spaced**), or **clumped dispersion** patterns (fig. 2.14). Random dispersion of individuals in animal populations may occur in species that lack social tendencies toward clumping and where there is little variation in the environment. In a random dispersion, the position of an organism is independent of the positions of other like organisms. Uniform dispersion of animals is often related to territoriality or other strong intraspecific competition. The dispersion of most animals can be described as clumped and reflects the distribution of suitable habitats as well as social tendencies in various species. A school of bleeding shiners or a herd of muskoxen represent clumped dispersion patterns. The dispersion patterns of animal species often reflect seasonal tendencies toward formation of large groups, pairs, family groups, or other social conglomerations. Dispersion of pairs or other groupings of animals, such as coveys of quail or pods of whales, can also be categorized as random, uniform, or clumped. For example, packs of gray wolves or territorial pairs of passerine birds would probably be uniformly dispersed if the habitat were somewhat uniform. **Passerine birds** are members of the largest order of birds, the Passeriformes. Many families of passerine birds have well-developed vocal abilities and are called **songbirds**.

Dispersal is defined as outward movements of individuals away from their established areas of activity. For example, when a moose is occasionally found in an Iowa cornfield, it is likely to be a dispersing individual, usually a young male

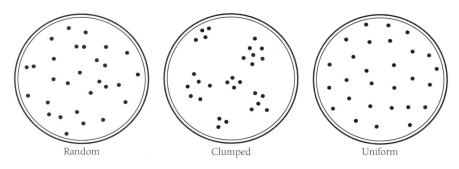

FIGURE 2.14 Individuals within populations may be dispersed across their geographic range in various patterns.

who has left his **natal area** (place of birth or hatching). The terms dispersal and dispersion are often confused.

Migration is defined as a two-way movement to and from an area that occurs with characteristic regularity or with changes in life history stage; it is often related to reproduction, feeding, or movement between summer and winter ranges. Migration out of an area is **emigration**; migration into an area is **immigration**. Sometimes, as in Pacific salmon, migration involves a return to the natal area to reproduce, with death of the adult occurring after reproduction.

Both dispersal and migration reflect the mobility of an organism. Birds tend to be highly mobile, but their degree of mobility varies greatly even if flightless birds are excluded from comparison. A northern pintail will sometimes cross the Pacific Ocean to reach the Hawaiian Islands, while ruffed grouse will seldom cross even a few hundred meters of water to reach an adjacent shoreline. Birds are the best dispersers in the animal kingdom, although fishes, mammals, and insects can and do make extensive dispersal movements.

Various barriers inhibit movements of animal species, depending on the species' mobility. The movement and dispersal of birds and mammals are greatly influenced by physical barriers such as mountain ranges, deserts, and large bodies of water. A waterfall can limit the movements and distribution of Atlantic salmon during migration. Fishes may be limited in their dispersal by water temperature or salinity beyond the range of tolerance of a species. Animals may be limited in their distribution by biological barriers such as lack of prey species, changes in vegetation, and competition with other closely related species. A forested area may be a major barrier to a species adapted to grasslands; the converse is also true.

Some species are highly adaptable and by nature are distributed across broad geographic regions. The white-breasted nuthatch breeds and winters from the southern United States to the northern forests of Canada. Deer mice are found over most of North America. Likewise, the range of the channel catfish extends from southern Canada to Mexico. Other species are much less adaptable and are more limited in their geographic ranges. Seaside sparrows are limited to specific coastal wetlands of the eastern and southern United States. The Colorado squawfish, an endangered species, is limited to the turbid river habitat of free-flowing portions of the Colorado River system and is not adaptable to other water types. Most endangered species lack the ability to adapt to changes in their habitats, including human-induced changes. Species with a narrow range of tolerance for various environmental conditions, if those conditions are limited as to the geographic areas in which they occur, are often the species that reach threatened or endangered status.

2.10 Exotics, Transplants, and Restorations

In order to define the terms introduced in this section, we must first discuss the zoogeographic regions of the world. The distributions of animals are often described in terms of zoogeographic regions first defined by Alfred Russel Wallace (1876). These zoogeographic regions, or realms, represent continents or portions of continents that are relatively isolated from one another in terms of the dispersal of animals (fig. 2.15). One zoogeographic region, the **Australian** (Australia, New Zealand, and New Guinea) has long been separated from other regions by extensive areas of ocean and has a unique assemblage of organisms. The **Neotropical region** (South America, Central America, and the West Indies) was also separated from other regions until the development of a land bridge in the last few million years; this isolation plus the variety of habitats and the stability of the climate have contributed to a diverse and

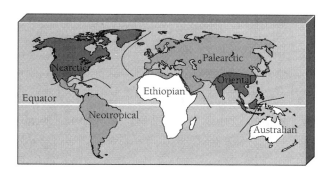

FIGURE 2.15 The major zoogeographic regions of the world.

unique assemblage of organisms in this region. Both the **Oriental** (India, Burma, Indochina, Malaysia, Sumatra, Java, Borneo, and the Philippines), and **Ethiopian** (Africa south of the Sahara, southern Arabia, and Madagascar) regions, because of their variety of habitats, suitable climates, and long period of stability compared with glaciated regions, also have considerable species diversity. The northern zoogeographic regions, the **Nearctic** (North America and Greenland) and **Palearctic** (Europe and Asia north of the Himalayas), have less species diversity than the southern regions; this lack of species diversity may be due to recent glaciation and colder climates, particularly in the northern portions of these two zoogeographic regions.

Because of the activities of humans, many organisms, both plants and animals, have been moved to areas where they did not naturally occur. A variety of terms have arisen to describe these organisms. These terms should be standardized for legislative, enforcement, and scientific purposes, but unfortunately they are not. We define an **exotic** as an organism introduced from another zoogeographic region. Exotics are sometimes defined as organisms introduced from a foreign land; however, this definition seems inadequate in light of changing national boundaries and the lack of relationship between zoogeography and national boundaries. The European starling is an example of an exotic species in North America (Nearctic region)

because it was introduced from another zoogeographic region, the Palearctic.

A **transplant** is an organism moved outside its native range but within the same zoogeographic region. The **native range** of an organism is the geographic area in which the species is **indigenous** (originally found in the area). The rainbow trout, a native of the Pacific coast areas of the Nearctic region, is a transplant when introduced into areas of North America outside its native range. For example, rainbow trout in New York or Manitoba are transplants. Rainbow trout introduced into New Zealand or Europe are exotics. Both exotics and transplants can be referred to as **introduced organisms (introductions)**.

Once animal species have been introduced into an area, they may or may not establish a naturally reproducing population there. Many introduced species, such as rainbow trout and ring-necked pheasants, reproduce in their new areas after initial introduction. Naturally occurring offspring produced by exotics or transplants are called **naturalized organisms**.

There is another type of organism classification that should be added: **feral animals**. These are animals that have reverted to the wild state after having been domesticated. Examples of such organisms are feral horses and feral pigs in the Nearctic region.

Species eliminated from portions of their native geographic range and then successfully released and reestablished on these vacated portions of their native range are **restored species**; the act of reestablishing a native population is called a **restoration**. The reestablishment of breeding populations of giant Canada geese is an example of a successful restoration. Restoration biology is a topic of growing importance because of its relationship to species diversity and richness.

The various terminologies used for organisms moved by humans are a subject of debate. Some of these terms are used interchangeably, and this leads to unnecessary confusion. Throughout this text the definitions provided above will be used in reference to "moved" species.

2.11 Landscape Ecology, Patch Size, Corridors, and Ecotones

Landscape ecology is an area of study that spans several disciplines and is tied directly to ecosystems (Forman and Godron 1986). Landscape ecology is defined as the study of structure, function, and change in landscapes. A **landscape** is a heterogeneous land area composed of interacting ecosystems that are repeated in similar patterns throughout an area or region. For our purposes, land area should be interpreted to include aquatic ecosystems such as streams and lakes as well as terrestrial ecosystems. For instance, an agricultural landscape in eastern Kansas might include various crops (winter wheat, milo, soybeans) interspersed with hedgerows, shelterbelts, stream bottoms, and constructed ponds, each of which might be described as a type of ecosystem. Such a system is highly disturbed from its original state, but is now a new landscape with characteristic habitat patterns, plants, and animals. This agricultural landscape pattern would repeat itself in similar form throughout a region with similar agricultural potential.

When the needs of individual organisms and populations of organisms are viewed in terms of the minimum-sized living spaces required to meet those needs, one begins to see the importance of landscapes. For instance, forest **patch size**, a landscape characteristic, as well as other characteristics of the forest, such as age, tree diameter, and understory, are related to the success of various plant and animal species. A **patch** is a nonlinear (not long and narrow) surface area within a landscape that differs from surrounding areas. For example, a continuous area of ponderosa pine covering 10 hectares (patch size) surrounded by agricultural fields is a patch, as would be a similar-sized area of grassland surrounded by urban habitat on all sides.

The decline in the number of northern spotted owls in the Pacific Northwest may be related to the concepts of landscapes and minimal patch size. Populations of the northern spotted owl are thought to be partially dependent on the patch size of old-growth or climax forest. Remaining patches of old-growth forest are also prime logging habitat and are thus susceptible to **habitat fragmentation** (the breaking of larger patches of habitat into smaller patches), or elimination. Logging and associated lumber industries are the major sources of employment in many areas of the Pacific Northwest. Logging restrictions to protect remaining old-growth forest are extremely unpopular with segments of the public dependent on logging for their income. Many people have already lost their logging-related jobs, and local economies have in some cases been seriously impaired. Often those dependent on logging place the blame for their economic losses on the northern spotted owl; this is by no means the only, nor even the most important, reason for economic losses in that region. Understandably, northern spotted owls have become a politically sensitive issue because they were used as a primary reason to reduce logging on old-growth timber. Other segments of the public want the remaining old-growth forest on public lands preserved for aesthetic reasons, for tourism, and as habitat for the northern spotted owl and other species. The debate concerning the northern spotted owl, its abundance and population status, and its dependence on old-growth forest is still continuing.

Patch size is an important concept in the management of many other species as well. The occurrence of western meadowlarks and a host of other grassland birds is very much tied to grassland patch size. The success of the reintroduction of the endangered black-footed ferret, whose primary prey is the black-tailed prairie dog, may be closely tied to the availability of patches of prairie dog towns of some suitable size. Reduction of patch size will result at some point in the area no longer being able to support a viable black-footed ferret population, regardless of how the ferrets are otherwise protected.

The concept of patch size has seen little use in fisheries, but the principles are applicable. Areas of submergent vegetation provide escape cover and may determine the survival rates of some prey fishes. **Submergent vegetation** is aquatic vegetation that generally does not extend above the water surface; examples would be kelp and sago pondweed. Perhaps the total area of submergent vegetation as well as size of submergent vegetation blocks (patch size) would relate well to survival rates of prey species such as fathead minnows, golden shiners, or brook sticklebacks. Coral reefs nicely tie into the idea of patch size with regard to the types of organisms that find a particular-sized reef to be suitable habitat.

The concept of **corridors** has application to both wildlife and fisheries. A terrestrial corridor is a narrow strip of land that differs, usually in terms of dominant vegetation form (such as forest or grassland), from surrounding areas. The same concept can be applied to rivers, chains of lakes, or other linear arrangements of aquatic systems. A narrow strip of trees extending across a large meadow and connecting extensive patches of forest on either side of that meadow is a corridor (fig. 2.16). Similarly, a narrow stretch of grassland running through a forested region, or a series of periodically flooded shallow water habitats in close proximity to one another, are also types of corridors. Disruption of these corridors may have substantial effects on the movement of some animals. In prairie or agricultural regions, the wooded corridors next to rivers may be important for migration and dispersal of small woodland birds and mammals. The importance of corridors is being questioned and needs further study.

Various fish species use streams and rivers as corridors of movement. Humans have in many instances altered fish corridors with dams, dredging, timber operations, or other activities that cause changes such as siltation and temperature increases. A good example of corridor disruption with resultant negative effects on fish populations is the series of dams on the Columbia River and its tributaries, which have negatively affected Pacific salmon migrations. In addition, sections of rivers

FIGURE 2.16 The North Platte River in Nebraska provides an example of a wooded corridor bordered by agriculture and grasslands. *(Photograph courtesy of C. Johnson.)*

may become too polluted for some aquatic species, thus interrupting a corridor.

The extensive reduction of shallow water habitats and their associated vegetation (such as bulrushes, spike rushes, and cattails) in North America and many other areas of the world can be viewed as a corridor and patch problem; it is also a form of habitat fragmentation. As shallow water habitats are drained, the distance between them increases and the number of such habitats per unit of area is reduced. How are organisms such as tiger salamanders, southern leopard frogs, and other seldom-studied species affected by the densities and linkages of these shallow water habitats? The decline of frogs in many areas may be related to loss of wet areas and thus the landscapes in which the frogs once thrived. Other factors, such as pesticides, that could affect populations of such organisms should not be ignored. Nevertheless, losses of shallow water habitats to urbanization, agricultural drainage, and other causes are closely related to patch and corridor concepts and the welfare of a great variety of organisms. In some states over 90% of the original shallow water habitats have been drained.

Some corridors, if too narrow, can be viewed as predator traps for species using them. For example, ground-nesting birds are more vulnerable to predation by red foxes, raccoons, and other mammalian predators if nesting areas are reduced to narrow corridors that can be easily traversed and searched by these predators. A fenceline habitat with cropland on either side can be a dangerous nesting area for a blue-winged teal. However, a small increase in corridor width could reduce predation rates. Unfortunately, in areas where most of the land is tilled for crops, the only nesting areas remaining for ground-nesting birds are often narrow corridors or strips of grass along field or fenceline edges. In most agricultural landscapes, the tendency has been to reduce the width of these corridors. Species adapted to extensive and continuous stands of forest may be at greater risk of predation if the forest is reduced to patches connected by thin strips of forest. The relationship of corridor width to predation has received little study and deserves more consideration.

The transition area between two different ecosystems is called an **ecotone**, and is also known as **edge**. For instance, the area where a mixed-grass prairie and a burr oak woodland come together is an ecotone. Such transition areas are often associated with greater species diversity because the two habitats in juxtaposition provide an increased array of trophic, spatial, and habitat possibilities.

In fisheries, the word ecotone is not commonly used, but there are transition zones in aquatic environments to which it could be applied. For example, the transition zone of a lake bed between a shallow water area and a deep water area is an ecotone. The structural change at the edge of a coral reef is an ecotone. Even the interface between beds of emergent or submergent vegetation and open water fits this description. A variety of species, both predators and prey, find transition areas attractive.

Some ecosystems are rich in species in part because they tend to be narrow and transitional in nature. A **riparian zone** is the habitat associated with the edges of rivers and streams. The riparian zone is really a transition from the stream or river to the adjacent upland shrubland, grassland, or other community. Because of the varieties of habitat at these edges, riparian zones have high species diversity and are important to the reproduction, movements and other activities of a disproportionately large number of species.

2.12 Predator-Prey Interactions

As already described, anything that hunts, kills, and eats other animals to obtain most of its energy and nutrients is a predator; the organism eaten is the prey. Animals such as the red-tailed hawk, mountain lion, or muskellunge normally come to mind when the word predator is used. One does not normally think of eastern kingbirds or blacktail shiners as predators, but they fit this description because they hunt and kill other animals.

Predator-prey interactions are not as simple as they may seem. The lines between predators and prey are often blurred. The interaction between largemouth bass and bluegills is a good example. The largemouth bass is usually viewed as the predator, with the bluegill assuming the prey role. However, bluegills can eat largemouth bass eggs or young. In addition, a juvenile largemouth bass may be a predator on small bluegills, but can be a prey item for a larger predator such as a walleye, bowfin, or northern pike.

Gray wolves are predators that specialize in taking large **ungulates** (hoofed animals), though they may also seek smaller prey, such as beavers. The predator-prey interactions of gray wolves have been studied intensively. When gray wolves feed on deer, moose, or caribou, they are often seen as competitors with humans. The effects of gray wolves on large ungulate populations have received and continue to receive much attention from researchers, managers, and the public. When packs of gray wolves roamed the prairies of North America, both wolves and ungulates seemed in ample supply. The ungulates thrived on the abundant grasses and forbs, and the gray wolves, limited by their territorial interactions, likely attained relatively stable population levels. The gray wolves probably had little influence on the abundant ungulate populations in these rich, undisturbed ecosystems.

Work in northern Minnesota has shown that gray wolves depend on white-tailed deer as a primary food source (Fuller 1989). In areas with abundant white-tailed deer, the effect of wolves on the deer population may be negligible, but if an area within the home range of a gray wolf pack has a low deer density, the possibility of an effect becomes greater. Some researchers have indicated that the likelihood of gray wolf effects on a prey population is closely tied to the predator-to-prey ratio in terms of biomass. Where white-tailed deer or other ungulate populations are reduced due to severe winter weather or other reasons, the gray wolf population may depress prey populations even more. However, reproduction and thus population levels in northern Minnesota white-tailed deer are also greatly influenced by vegetation changes related to forest management and succession. Thus, gray wolves are just one factor in a complex relationship between white-tailed deer and their environment.

It has been suggested that gray wolves maintain caribou populations at low levels in areas where there is a low number of prey per predator. It is argued that under such conditions gray wolf populations could be reduced by human intervention to allow caribou recovery, and that wolf control would eventually allow both prey and predator to thrive once the ratio of prey to predator increased. This argument was used in Alaska, where some caribou and moose populations were at low levels and experiencing poor calf survival due, in part, to predation from gray wolves and grizzly bears (Gasaway et al. 1983). Other factors, such as long-term changes in habitat, weather patterns, and normal cycles or fluctuations in ungulate populations, also need to be considered in attempting to understand changes in abundance of caribou and moose. The idea that predation is the major limiting factor on some populations of moose and caribou remains controversial (Boutin 1992).

Fishery biologists look at predator-prey relations in different ways, depending on the type of fish community being managed. For example, largemouth bass can be managed to create high-quality panfisheries in small impoundments (Gabelhouse 1984). **Panfish** are small game fishes such as bluegills, black crappies, white perch, and yellow perch. A high density of small (20–30 centimeter) largemouth bass can reduce overabundant young of some panfish species so that the surviving panfishes experience decreased competition. These panfishes grow quickly and reach sizes of interest to anglers. However, northern pike and yellow perch can have a very different predator-prey relationship. When northern pike density in Horseshoe Lake, Minnesota, became too high, the number and sizes of yellow perch in the lake declined (Anderson and Schupp 1986). Larger northern pike preferred to feed on large prey, represented by the larger yellow perch. Thus, northern pike actually competed with humans for large yellow perch.

Certain outbreaks of insects in forest areas demonstrate an additional economic aspect of predator-prey interactions. For instance, some insectivorous warblers are highly mobile and seek areas with abundant insects. In areas experiencing an outbreak of an insect causing damage to timber, this relationship of predator and prey is of potential economic interest to humans. Biologists may have an additional interest in what happens to the insectivorous birds in terms of the effect of abundant prey on reproduction and survival of young.

FIGURE 2.17 The abundance of prey can influence reproductive success in species such as these ferruginous hawks in southeastern Idaho, which feed heavily on black-tailed jack rabbits. *(Photograph courtesy of R. Hansen, Environmental Science and Research Foundation, Inc.)*

For instance, the increased availability of prey in an area with an insect outbreak may remove food as a limiting factor for a warbler species.

Thus, in predator-prey interactions it is important to understand what happens to the predator as well as to the prey. Both prey and predator are part of the interaction, and prey effects on the predator population can be as substantial, or more so, than predator effects on the prey. When prey is scarce, animals such as snowy owls, lynx, and ferruginous hawks (fig. 2.17) reduce their reproductive efforts and may suffer increased losses of young. Prey shortages commonly influence reproductive success in predators.

The effect of prey shortages on a predator also depends on the degree to which the predator is a **generalist** or a **specialist**. Specialists, which require a more specific prey base, are more at risk with fluctuations of their primary prey; generalists can more readily switch to other prey sources. For example, a northern grasshopper mouse will eat a great variety of insects, other invertebrates, and even other mice—it is a generalist. In comparison,

the black-footed ferret is a species strongly dependent on black-tailed prairie dogs for its prey—it is a specialist.

▼ SUMMARY

Ecosystems represent interacting systems of organisms and their abiotic environment and are characterized by dominant vegetation or some other dominant feature. Ecosystems have a structure in terms of the types of organisms and abiotic materials present. Ecosystem function relates to nutrient (mineral) cycling and energy flow. Nutrients and energy move through ecosystems in a complex manner that can be represented as a food web. Energy is captured by photosynthesis by primary producers, but much is lost in unusable forms at each trophic level as animals consume primary producers or other animals. Energy pyramids illustrate this loss of energy.

Ecological succession relates closely to what organisms are present and to the energy available to those organisms. Wildlife and, to a lesser extent, fishery management practices directed at improving habitat are often designed to manipulate succession to create seral stages favorable for the target organism or organisms.

The role of an organism in an ecosystem is termed its niche; this complex concept can be illustrated using the multidimensional niche concept. Habitat must often be described on a seasonal or functional basis for animals that migrate or otherwise make seasonal changes in habitat use. Competition can occur when resources are in short supply and some aspects of the niches of two or more organisms living in the same geographic area overlap. Territories are established to defend an area of habitat, usually against others of the same species. The strongest competitor of an organism is usually another organism of the same species.

Animal populations can be limited by a single factor (limiting factor) or by one or more factors acting in concert. The conditions near the edge of the geographic range of a species usually approach the limits of tolerance of that species for one or more environmental conditions.

Animals may be dispersed in random, uniform (spaced), or clumped patterns. Animals moving away from established home ranges are dispersing, while migration is a two-way movement to and from an area occurring on an annual or periodic basis. Animals introduced into a new area from a different zoogeographic realm are exotics, while those introduced into new areas within the same zoogeographic realm are transplants.

Landscape ecology is a field of study that examines larger patterns that exist within a heterogeneous land area. Animal populations are often influenced by landscape characteristics such as patch size and habitat fragmentation. Animal movements may be dependent on corridors of certain types of habitat. The transition zone between two ecosystems is an ecotone and may be particularly rich in plant and animal species.

Predator-prey interactions influence both prey and predator populations. The effects of predators on prey are complicated by other factors, such as habitat quality, that influence reproduction and survival of the prey.

PRACTICE QUESTIONS

1. Select two ecosystems with which you are familiar and describe their dominant features.

2. Select two landscapes with which you are familiar and describe those landscapes in terms of interacting ecosystems.

3. Give two specific examples in which widely separated ecosystems (terrestrial, marine, freshwater, or any combination) are interrelated.

4. How is species diversity different from species richness? Give an example in conceptual terms using a taxon and ecosystem familiar to you.

5. Give two examples, other than those provided in the chapter, of populations and metapopulations.

6. Diagram the primary aspects of the carbon, sulfur, and nitrogen cycles.

7. Develop a food web example of your own. Show routes of energy flow and categorize species by trophic level.

8. Develop an energy pyramid for the food web in question 7.

9. Give an example of a climax community (other than the examples provided in this chapter). Why does your example fit the definition of a climax community?

Name two animal species primarily associated with the climax stage. Describe an earlier seral stage of this community. Give two animal species characteristic of that seral stage.

10. Give three dimensions of a niche for an organism with which you are familiar. Relate each dimension to the niche of the animal.

11. Describe the habitat of two animal species. Indicate whether the habitat is seasonal.

12. Name three exotic species found in the area where you live. How might one of these exotic species be competing with a native species?

13. Give your own examples of contest competition and exploitative competition.

14. What are some possible advantages and disadvantages of territorial behavior? What are some of the ways that animals mark or otherwise advertise territory boundaries?

15. Give four of your own examples of possible limiting factors for different aquatic or terrestrial animal populations.

SELECTED READINGS

Bormann, F. H., and G. E. Likens. 1979. *Pattern and process in a forested ecosystem.* Springer-Verlag, New York.

Brewer, R. 1994. *The science of ecology.* 2d ed. Saunders College Publishing, Fort Worth, Tex.

Errington, P. L. 1967. *Of predation and life.* Iowa State University Press, Ames.

Laycock, G. 1966. *The alien animals.* Natural History Press, Garden City, N.Y.

Mech, L. D. 1970. *The wolf: The ecology and behavior of an endangered species.* University of Minnesota Press, Minneapolis.

Ricklefs, R. E. 1993. *The economy of nature.* 3d ed. W. H. Freeman, New York.

Rodiek, J. E., and E. G. Bolen. 1991. *Wildlife and habitats in managed landscapes.* Island Press, Washington, D.C.

Smith, R. L. 1990. *Ecology and field biology.* 4th ed. HarperCollins, New York.

Sprent, J. I. 1987. *The ecology of the nitrogen cycle.* Cambridge University Press, New York.

Stroud, R. H., and H. Clepper, eds. 1979. *Predator-prey systems in fisheries management.* Sport Fishing Institute, Washington, D.C.

Wilson, E. O., ed. 1988. *Biodiversity.* National Academy Press, Washington, D.C.

Wilson, E. O. 1992. *The diversity of life.* The Belknap Press of Harvard University Press, Cambridge, Mass.

Wootton, R. J. 1990. *Ecology of teleost fishes.* Chapman and Hall, London.

LITERATURE CITED

Anderson, D. W., and D. H. Schupp. 1986. *Fish community responses to northern pike stocking in Horseshoe Lake, Minnesota.* Investigational Report no. 387. Minnesota Department of Natural Resources, St. Paul.

Anthony, R. G., and N. S. Smith. 1977. Ecological relationships between mule deer and white-tailed deer in southeastern Arizona. *Ecological Monographs* 47:255–277.

Boutin, S. 1992. Predation and moose population dynamics: A critique. *The Journal of Wildlife Management* 56:116–127.

Brewer, R. 1994. *The science of ecology.* 2d ed. Saunders College Publishing, Fort Worth, Tex.

Caithness, T., M. Williams, and J. D. Nichols. 1991. Survival and band recovery rates of sympatric grey ducks and mallards in New Zealand. *The Journal of Wildlife Management* 55:111–118.

Elton, C. 1927. *Animal ecology.* MacMillan, New York.

Forman, R. T. T., and M. Godron. 1986. *Landscape ecology.* John Wiley & Sons, New York.

Fuller, T. K. 1989. *Population dynamics of wolves in north-central Minnesota.* Wildlife Monographs 105.

Gabelhouse, D. W. Jr. 1984. An assessment of crappie stocks in small midwestern private impoundments. *North American Journal of Fisheries Management* 4:371–384.

Gasaway, W. C., R. O. Stephenson, J. L. Davis, P. E. K. Shepherd, and O. E. Burris. 1983. *Interrelationships of wolves, prey, and man in interior Alaska.* Wildlife Monographs 84.

Hutchinson, G. E. 1957. *A treatise on limnology.* Vol. 2. *Introduction to lake biology and the limnoplankton.* John Wiley & Sons, New York.

Hutchinson, G. E. 1978. *An introduction to population ecology.* Yale University Press, New Haven, Conn.

Jones, H. G. 1983. *Plants and microclimate: A quantitative approach to environmental plant physiology.* Cambridge University Press, Cambridge.

Odum, E. P. 1971. *Fundamentals of ecology.* 3d ed. W. B. Saunders, Philadelphia, Pa.

Primack, R. B. 1993. *Essentials of conservation biology.* Sinauer Associates, Sunderland, Mass.

Ricklefs, R. E. 1993. *The economy of nature.* 3d ed. W. H. Freeman, New York.

Shelford, V. E. 1913. *Animal communities in temperate America.* University of Chicago Press, Chicago.

Smith, R. L. 1990. *Ecology and field biology.* 4th ed. HarperCollins, New York.

Stern, P. C., O. R. Young, and D. Druckman, eds. 1992. *Global environmental change: Understanding the human dimensions.* National Academy Press, Washington, D.C.

Stuber, R. J. 1982. *Habitat suitability index models: Black bullhead.* U.S. Fish and Wildlife Service, FWS/OBS-82/10.14, Washington, D.C.

U.S. Congress, Office of Technology Assessment. 1987. *Technologies to maintain biological diversity.* OTA-F-330, Washington, D.C.

Wallace, A. R. 1876. *The geographical distribution of animals.* 2 vols. Harper and Brothers, New York.

Part II

THE
BIOTA

3

POPULATION DYNAMICS AND STRUCTURE

Before one can understand how communities and ecosystems function, an understanding of how populations function is necessary. When a biologist considers the dynamics and structure of a population, specific concepts come to mind. **Population dynamics** is the study of changes in numbers or weight (biomass) of organisms in populations and of the factors influencing those changes.

Populations are affected by three dynamic rate functions: natality or recruitment, growth, and mortality. Natality and recruitment add new individuals to a population, growth adds biomass to a population, and mortality causes individuals to be lost from a population. These three dynamic rate functions interact to influence a population and determine **population structure**. Population structure generally is assessed as size structure, age structure, or sex ratio. Population structure can also be called **population form**, and population dynamics is sometimes referred to as **population function**. This chapter introduces the basic concepts of population dynamics, and describes how the interaction of the three dynamic rate functions influences population structure, population regulation, production, and yield.

▼
3.1   Natality/Recruitment

Natality refers to birth or hatching, and is commonly expressed as the number of young individuals born or hatched within a specified period of time. An example would be the number of big brown bats added to a population through birth in one year. Most often, wildlife biologists express

natality rate as the number of offspring per 1,000 females. Natality can also be used in reference to fishes, although most fishes are hatched, not born. Fishery biologists consider **recruitment** more often than natality. Recruitment can be defined in numerous ways; however, fishery biologists often define recruitment as the number of fish hatched or born in any year that survive to reproductive size. Thus, fishery biologists are not so concerned with the number of fish that hatch; they are more concerned with the number that survive to reach adulthood and enter the reproductive pool.

Why are fishery biologists more concerned with recruitment than with natality? The answer has to do with differences in reproductive strategies (see sections 1.3 and 6.10). Many wildlife species have low numbers of offspring. For example, white-tailed deer usually have only one or two fawns per year. Species that produce few young generally provide them with substantial parental care, and a large percentage of the young survive. Conversely, most fishes tend to produce many eggs and invest much less energy in parental care. For example, a large female common carp can produce one or two million eggs per year. These adhesive eggs are placed on flooded vegetation, and the parents swim away and let the eggs and newly hatched offspring fend for themselves. For most fishes, as egg numbers increase, parental care decreases; the converse is also true.

Only a fraction of a percent of the fish eggs that are produced will actually hatch, and only a fraction of the newly hatched fish will be recruited into the adult population. Following is a hypothetical, but realistic, example using sockeye salmon. These fish attain most of their growth as adult fish in the ocean, but return to fresh water to **spawn** (release their eggs and sperm). Sockeye salmon use their body reserves to get to the spawning stream, and die after spawning. A female sockeye salmon might contain 3,000 eggs. Of those 3,000 eggs:

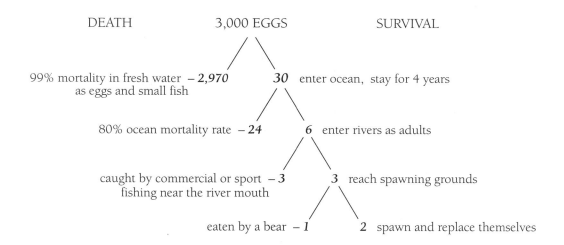

DEATH   3,000 EGGS   SURVIVAL

99% mortality in fresh water − *2,970*   *30* enter ocean, stay for 4 years
as eggs and small fish

80% ocean mortality rate − *24*   *6* enter rivers as adults

caught by commercial or sport − *3*   *3* reach spawning grounds
fishing near the river mouth

eaten by a bear − *1*   *2* spawn and replace themselves

In a balanced system, only two offspring from a pair of adults need to return to the stream and reproduce to continue this age-old cycle.

Although it is not as strongly stressed as in fishery work, wildlife biologists are also concerned with recruitment. Monitoring of fawn to doe ratios in white-tailed deer and counts of the young of songbirds, waterfowl, and upland game birds are common examples. **Upland game** is an artificial category describing the many birds and small mammals, such as ruffed grouse and fox squirrels, that are found in terrestrial systems and used for sport hunting.

Natality and recruitment are affected by both **density-dependent** and **density-independent factors**. A density-dependent factor is an environmental factor that affects a population based on its density. For example, when the size of an elk herd reaches the point at which the animals are competing for a limited food supply, a lower percentage of cow elk will produce calves. A density-independent factor is an environmental factor that acts independently of population density. An example in a fishery could be the influence of a major flood on a stream. Flood conditions could result in no or reduced survival of young creek chubs, regardless of the population size. In other words, the survival of one young fish would not be dependent on how many others were present; the environmental factor (the flood) would have killed many of the young creek chubs, regardless of population density.

Density-dependent influences on natality and recruitment are commonly discussed by biologists. Wildlife biologists often refer to the concept of **inversity**. When the density of adult white-tailed deer is low, the proportion of yearling does that bear fawns increases, and adult does often have twins or even triplets. At such times there tends to be less competition for resources such as food, and well-nourished white-tailed deer are more likely to breed at a younger age and produce larger numbers of offspring. When the density of adult white-tailed deer is high, food may be a limiting factor, and the nutritional status of the deer may be poor. As a result, few yearling white-tailed does bear fawns, and adult does usually have only a single fawn (fig. 3.1).

Fishery biologists typically are more concerned with changes in recruitment than with changes in reproductive capability. **Stock-recruitment curves** demonstrate the influence of parental abundance on subsequent recruitment (fig. 3.2). A **stock** is a group of organisms with a common ancestry or parentage that is adapted to a particular environment; stock is primarily a fishery term (see section 4.2). Wildlife

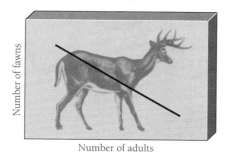

FIGURE 3.1 The concept of inversity is illustrated by the relationship of natality in a white-tailed deer population to adult population density. As the number of adult white-tailed deer in a population increases, the number of fawns produced decreases.

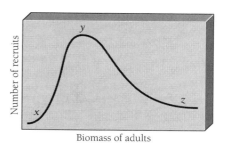

FIGURE 3.2 The greatest number of young fish are recruited to a fish population when the biomass of adults is intermediate (*y*). At extremely low adult biomass, there are too few adults to quickly rebuild the population (*x*). When biomass of adults is high, fewer young fish will be able to recruit to the adult population because of density-dependent factors such as competition for food resources (*z*).

biologists generally use the term population, but at times use stock. Chinook salmon that are spawned in different Alaskan and Canadian rivers will intermingle in the ocean. However, when they return to the rivers to spawn, each stock will return to the place where it was spawned.

When the density of an adult stock is high, there is room for little recruitment because adults are abundant (point *z*, fig. 3.2). As adult density decreases, recruitment increases until a point is reached at which recruitment is at its maximum (point *y*, fig. 3.2). If adult density decreases below this point, recruitment decreases because there are too few adults left to replace themselves (point *x*, fig. 3.2); the stock then "collapses." Stock-recruitment curves have most often been developed for commercially harvested fishes in the ocean or large lakes, but stock-recruitment relationships have been noted for wildlife species such as white-tailed deer (McCullough 1979).

3.2 Growth

Most birds and mammals exhibit determinate growth, while fishes exhibit indeterminate growth

(see section 1.3). Organisms with determinate growth tend to grow quickly and reach their approximate adult size in a year or two. For organisms with indeterminate growth, growth patterns are more variable.

Fish growth is very much density dependent. Indeterminate growth allows the development of high-density, slow-growing fish populations that are often referred to as **stunted populations**. Yellow perch growth can be used as an example. In a high-density, slow-growing population, yellow perch may reach 125 millimeters after four years (a weight of 20 grams). In low-density, fast-growing populations, yellow perch can reach 260 millimeters after four years (approximately 290 grams). When density is high, there is much competition for a limiting factor such as food, and growth is slow. When density is low, food is more abundant, and growth rates are substantially faster.

The growth of individual wildlife organisms can be affected by population density, but not to the same extent as that of fishes. Because most birds and mammals have determinate growth, stunting to the extent possible in fish populations cannot occur. While there may be some density-dependent slowing of growth in birds and mammals, such a problem usually changes to density-dependent mortality (see section 3.3).

3.3 Mortality

Mortality is the death of an individual, but is often expressed as the percentage of a population that dies in one year. Mortality can be categorized as density dependent or density independent. An example of mortality that is primarily density independent occurs in ring-necked pheasant populations in the northern Great Plains. An extensive winter blizzard can result in high mortality in these birds; that is, a certain percentage of the ring-necked pheasants will be killed due to the blizzard, regardless of the density of the birds. This does not mean that all storm-related mortality is necessarily density independent; a density-dependent factor such as competition for winter cover might also influence mortality from the blizzard.

Density-dependent mortality caused by factors such as predation or limited food supplies can occur in animal populations. When muskrat populations are high, they may outstrip the supply of available food. An increase in mortality may occur that reduces the population to a level that the available habitat can support. Another example of density-dependent mortality involves walleye predation on yellow perch. In a year when young yellow perch are abundant in a lake, walleye predation may account for a substantial portion of total mortality for the perch. Conversely, in years when young yellow perch density is low, the walleyes are likely to switch to other, more abundant fishes as primary prey items, and walleye predation will account for a smaller portion of the mortality of young perch.

Wildlife biologists monitor both natural and hunting mortality for game species, while fishery biologists monitor both natural and fishing mortality for game and commercial fishes. **Natural mortality** is mortality caused by predation, starvation, diseases, accidents, or other natural causes. **Harvest mortality** results from human activities directed at taking organisms for food or sport pur-

poses, and includes both hunting and fishing mortality. Harvest mortality is an anthropocentric term; this mortality is actually one more type of predation, and could be included in the definition of natural mortality. **Total mortality** is the sum of natural and harvest mortality. Natural mortality is closely tied to the growth rates and longevity of animals. A longer-lived species such as the walleye may have natural mortality rates near 30% per year, while a shorter-lived species such as the bluegill may have natural mortality rates closer to 50%. Similarly, shorter-lived wildlife species such as eastern cottontail rabbits have higher natural mortality rates than longer-lived species such as black bears.

Because fishes are ectotherms, growing season length affects fish growth rates and ultimately also influences natural mortality rates. In general, slower-growing individuals of a fish species tend to live longer than faster-growing individuals of that species. There certainly are exceptions, but this general rule often holds. For example, a 45-centimeter walleye in a Kansas reservoir is likely to be 3 years old. Such populations may have natural mortality rates near 40% per year. A 45-centimeter walleye in a northern Canada lake may be as old as 15 years; natural mortality rates for such populations are likely to be less than 10% per year.

The term natural in natural mortality can be misleading. Natural mortality can be influenced by humans, especially in situations in which human activities affect habitat. If human disturbances affect the quantity or quality of the habitat an organism needs to reproduce or survive, then natural mortality rates can be increased.

A population will experience a certain level of natural mortality whether or not harvest mortality occurs. For example, approximately 70% of a northern bobwhite population will die in a normal year, regardless of whether hunting mortality occurs. Due in part to density-dependent mortality factors, removal or addition of mortality factors will likely have little influence on total annual mortality rates. Therefore, within limits, a **harvestable surplus** of animals can be removed without adversely affecting a population. The harvestable surplus is that portion of the population that can be taken by humans without affecting subsequent populations of that organism. For example, if hunting accounted for 35% of the annual mortality of northern bobwhites in a given area, then 35% mortality would occur from natural causes, and total mortality would equal approximately 70%. If hunting mortality accounted for 15% of the mortality, then natural mortality would account for the remaining 55%, and total mortality would remain at 70%. For the animal population, what caused the total mortality would be essentially irrelevant. In such a situation, hunting mortality is said to be **compensatory mortality**. That is, the 35% or the 15% hunting mortality replaces part of the natural mortality that would have occurred even without hunting mortality. Compensatory mortality is not just a phenomenon associated with harvest mortality. If a particular species had an annual total mortality of 50%, with 25% of that mortality being contributed by predation and 25% by disease, compensatory mortality could occur. For example, if disease mortality increased to 40% and predation mortality decreased to 10%, compensatory mortality would have occurred.

The question of when harvest mortality is compensatory mortality and when it is **additive mortality** has stirred controversy. Harvest mortality is additive when it results in a total mortality that exceeds what would have occurred due to natural mortality alone. If a portion of harvest mortality is additive, then population density may be reduced. In the northern bobwhite population discussed above, harvest mortality would be compensatory only up to a certain level of harvest. If 60% of the northern bobwhites were harvested by hunters, total mortality would very likely exceed 70%; harvest mortality would obviously result in additive mortality if it exceeded 70%. The amount by which total mortality exceeded 70% (the expected total mortality in a nonhunted population) would be additive mortality.

Even when additive mortality due to harvest is possible in a population, that level of mortality may not occur. There are a number of reasons why additive mortality may not occur. Human interest in harvest may simply decline if animal population density decreases. With fewer organisms present, hunting success may be reduced to the point at which harvest ceases simply because the organisms are too difficult to find. Also, human interest can decline if the animals become elusive. For example, sharp-tailed grouse become increasingly wary after being hunted, and they soon begin to flush (rise from the ground) out of shooting range. In either case, the law of diminishing returns enters the picture, and hunters stay home because success becomes increasingly unlikely.

There also may be no additive mortality because of harvest regulation. For example, pronghorns are quite vulnerable to hunting because of their tendency to use open habitat, their highly visible coloration, and the accuracy of rifles with telescopic sights. However, additive mortality sufficient to cause population declines in this species can be avoided by proper harvest regulation. That is, the state, provincial, or territorial conservation agency issues only the number of hunting permits that ensures that harvest mortality is not additive.

Conversely, it is possible for harvest mortality to become additive if harvest regulations are inappropriate (most often this means too liberal) or if additive mortality is a management goal. Additive mortality can be a purposeful management strategy. For example, white-tailed deer populations have been increasing in many areas during the last two decades. These increased populations have caused increased damage to crops, urban area problems, and car-deer accidents. Thus, many management agencies have increased the number of white-tailed deer permits issued in an attempt to reduce the deer population density. This reduction would not harm the ability of the species to maintain a population; therefore, some additive mortality would not necessarily be negative.

Other factors complicate an assessment of whether harvest mortality is additive or compensatory in a particular situation. Some species respond to increased mortality by increasing natality or recruitment (remember the discussion of inversity in section 3.1). For example, the reproductive rate of coyotes may increase as mortality rates increase. Also, movements of animals such as ruffed grouse and northern bobwhites from lightly hunted surrounding areas into a heavily hunted area can complicate an assessment of mortality. Estimates of total mortality for northern bobwhites can exceed 100% in some hunting areas. Obviously, mortality in excess of 100% is not possible. Movement of birds from surrounding lands onto the hunted lands can result in a harvest greater than the original number of birds present.

North American duck populations have been declining through the late 1980s and into the 1990s. The causes of these declines are complex, and they have been attributed to various factors, including habitat loss and potential additive hunting mortality. Most studies have indicated that waterfowl harvest at current levels is compensatory; however, the compensatory versus additive mortality question in waterfowl management will continue to remain a point of controversy.

The issue of when and whether harvest mortality is additive or compensatory is further complicated because mortality rates are typically assessed based on "expected" or "normal" annual mortality rates. Mortality rates are not necessarily stable or normal. As an example, consider the previously described population of northern bobwhites. Through regulations, these birds may be managed in an area to maintain a certain population level. Great care may be taken to ensure that harvest mortality is not additive. All of these efforts may be to no avail if a density-independent factor, such as a severe blizzard or a substantial loss of reproductive habitat, reduces the northern bobwhite population to some exceedingly low level. All of the management efforts over a number of years to ensure the maintenance of the population at some level would be negated by that single storm or habitat loss event.

In natural systems, surplus organisms in a population die from natural causes such as predation. For example, mountain lions were a common predator on mule deer in the southwestern United States before humans reduced lion numbers. In perturbed systems, there may be no predators other than humans. Once mountain lion populations were reduced by human activities, regulated hunting became the primary source of predation for the mule deer. In highly perturbed systems, such as urban areas, this becomes a special problem. With no or reduced predation and the unavailability of traditional human harvest methods (hunting), increasing numbers of some wildlife species have become a major concern. In many areas, species such as Canada geese and white-tailed deer often reach problem population levels.

The topic of compensatory and additive mortality for harvested species will remain a point of contention. It is an important aspect of the human dimension in wildlife and fisheries, a subject addressed later in this text.

3.4 Interactions of the Dynamic Rate Functions

Natality or recruitment, growth, and mortality interact to determine the nature of a population (fig. 3.3). Consider a population of white catfish in

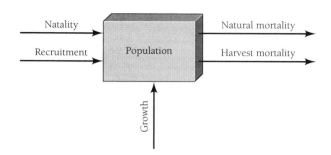

FIGURE 3.3 The three dynamic rate functions (natality/recruitment, growth, and mortality) interact to determine the nature of a animal population. Natality and recruitment add new individuals to the population, growth adds biomass to the population, and mortality results in a loss of both individuals and biomass.

a South Carolina river; recruitment adds new individuals, growth adds biomass, and mortality reduces the number of fish in the stock. In a stable system, the inputs (recruitment and growth) are balanced by outputs (harvest and natural mortality).

Few systems are stable, however, especially those that have been perturbed by humans. If mortality exceeds natality or recruitment, then the population density will decrease. The converse is also true: if natality or recruitment exceeds mortality, then the population density will increase. However, only a certain biomass can be supported by any particular environment, and that biomass cannot be exceeded for any extended time period.

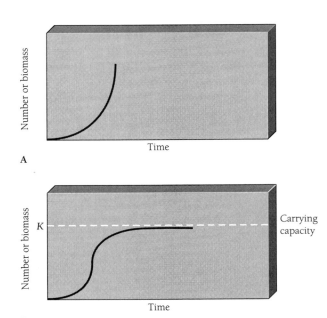

FIGURE 3.4 (*A*) When a species is first introduced into a new and favorable environment, the resulting increase in population resembles a J-curve. (*B*) Population growth cannot continue unchecked, however, because the population will reach carrying capacity (*K*), often defined as the maximum biomass for that population that can be supported over an entire year. Thus, population growth slows, and the J-curve becomes an S-curve (also known as the logistic growth curve).

3.5 Population Growth and Regulation

A species in a new and favorable environment often increases at an exponential rate. This type of population growth curve is termed a **J-curve** (fig. 3.4A). However, this exponential increase in numbers can continue only as long as no limiting factors are reached. For example, as numbers increase, there could be increasing competition for food. At some point, there would be insufficient food for new individuals to be added to the population. Thus, the J-curve flattens at the upper end and becomes an **S-curve**, also known as a **logistic growth curve** (fig. 3.4B). Think back to the white-tailed deer inversity example in section 3.1. When white-tailed deer density is low, the reproductive rate is high. As adult density increases, the reproductive rate (offspring per adult) decreases.

The level at which the logistic growth curve flattens is generally close to **carrying capacity** (*K*) (fig. 3.4B). Carrying capacity is the maximum biomass of a population that can be sustained within a defined area throughout a specified period of time. Very often, the period of time used

by biologists is an entire year. Thus, a small Iowa pond might have a carrying capacity of 40 kilograms per hectare for largemouth bass. The same pond might have a carrying capacity of 150 kilograms per hectare for bluegills, which are at a lower trophic level (see section 2.3). In an Alabama pond, which is less fertile than the Iowa pond because the soils there are highly leached by abundant rainfall and the waters tend to be less productive, carrying capacity may only be 15 kilograms per hectare for largemouth bass and 60 kilograms per hectare for bluegills. Thus, the trophic level of a particular fish species and the fertility of a water body both influence carrying capacity. In wildlife, carrying capacity is generally given in numbers of animals rather than biomass. In addition, wildlife biologists generally consider carrying capacity to be the maximum population number that can be supported without causing habitat damage.

Management practices can be directed at either increasing or decreasing carrying capacity for a species, depending on management objectives for that particular area. For example, the carrying capacity for Baird's sparrows on a section of grassland can be quickly changed by the manner in which the grassland habitat is managed: mow or heavily graze the area and the carrying capacity for the species declines; enhance the grass species on which the Baird's sparrows depend and the carrying capacity increases. Management practices to increase or decrease the carrying capacity for a target population affect other nontarget species in that area as well, some favorably and others unfavorably. That is, nontarget species may increase or decrease in numbers depending on the extent to which the management practice affects them and their habitat.

Unfortunately, carrying capacity is not measurable; it represents a potential. In fact, biologists do not even agree on a standard definition for carrying capacity (MacNab 1985). Although carrying capacity is neither measurable nor well defined, it is a necessary concept for understanding population dynamics.

Standing stock (**standing crop**), unlike carrying capacity, is measurable. Standing stock is the abundance of organisms present at a given time in a given area and does not include the year-round portion of the carrying capacity definition. It represents the amount, usually by weight for fishes and number for birds and mammals, of a given species or complex of species present in a specific habitat or area at a specific moment. In reality, the term is commonly used in fisheries but seldom by wildlife biologists. Fishery biologists commonly refer to standing stock for a single fish species or to the total standing stock for the entire fish community. Wildlife biologists usually refer to population density rather than standing stock, but the concept is certainly the same.

Assume that the standing stock for redear sunfish in a body of water is 50 kilograms per hectare. The population biomass can be the same whether the population contains 500 or 1,000 individuals. In the 1,000-fish population, mean weight per individual is one-half of that in the 500-fish population. Stable biomass with changing numbers of individuals is possible in fish populations because of the indeterminate growth characteristics of fishes. For most wildlife populations, a decline in population number results in a decline in biomass because of the determinate growth characteristics of these animals.

The number or biomass of organisms in a population is ultimately a result of the combined effects of one internal factor, **biotic potential**, and one external factor, **environmental resistance**. Biotic potential is synonymous with reproductive potential, and is the maximum rate of population increase under ideal conditions. As shown in figure 3.5, biotic potential can be thought of as a "force" that pushes up on the population growth curve.

In the natural world, a series of environmental factors (both biotic and abiotic) interact to limit the abundance of a population; these factors constitute environmental resistance. Environmental resistance acts to hold the rate of population increase below the level theoretically possible as a result of biotic potential. As shown in figure 3.5, this environmental resistance is the "force" that pushes down on the population growth curve. Environmental resistance might include predation, limited sites for nesting or reproduction, climatic conditions, or competition for a limited supply of food. Carrying capacity is the theoretical balance between biotic potential and environmental resistance. Standing stock is the actual balance, in numbers or biomass, between the two "forces" at a particular moment in time.

Figure 3.5 is also useful as an illustration of population dynamics. Few populations in the natural world remain stable for very long. Changes in environmental resistance allow changes in population density. For example, if predation is reduced, population density may increase. If food supplies decline, population density may decrease.

Populations may increase and decrease on either a regular or an irregular basis. A **cyclic**

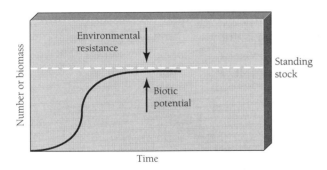

FIGURE 3.5 Biotic potential, or reproductive potential, can be thought of as a "force" that pushes upward on the population growth curve. Environmental resistance can be thought of as a "force" composed of factors such as predation and competition that limits abundance. The balance between these two forces results in standing stock, which is the population abundance at any given moment.

population is one in which abundance increases or decreases on a regular basis. Cycles may be short, like those of brown lemmings, which have a 3- to 5-year cycle of abundance, or longer, as for snowshoe hares, which seem to have a 9- to 11-year cycle. Black-tailed jack rabbits in the shrub deserts of western North America have populations that are **irruptive**, here defined as populations that dramatically increase and decrease at irregular intervals. Some biologists consider black-tailed jack rabbits to have cyclic populations and argue that they increase and decrease at somewhat regular intervals in approximately a 7-year cycle. Whether their population dynamics are cyclic or irruptive, studies of black-tailed jack rabbit populations indicate that predation, primarily by coyotes, is not what causes the initial population downturn that occurs when jack rabbit populations reach a peak. At the peak of black-tailed jack rabbit abundance, jack rabbits per predator are well beyond the level that the predators can control. At the time when the black-tailed jack rabbit population starts to decline, the coyotes are abundant because of the abundant food that has been available to them over the past few years. As the black-tailed jack rabbit population continues to decline, the prey per predator ratio becomes so low that predation depresses the population of jack rabbits further. Possible delays in recovery of

the black-tailed jack rabbit population may be related to predators continuing to depress the prey population. At some time in the future, the vegetation recovers, and the prey species again begins to increase. By this time, coyote populations have declined to low levels along with those of the black-tailed jack rabbits. The coyotes, because of their dependence on black-tailed jack rabbits, have adjusted their reproduction to prey availability. Thus, whether predator or prey is regulating the cycle is unclear. In reality, the regulating agent changes depending on the time in the cycle. In fact, weather, vegetation, predators, and prey can all have an influence at some point.

Black and white crappies are often referred to as cyclic species. This terminology is incorrect because changes in their abundance are not regular. Crappie populations in many waters commonly experience high variation in recruitment among years. Thus, their abundance is irregular rather than cyclic; irruptive would probably be a better descriptor.

3.6 Production and Yield

Production is defined as the biomass accumulated by a population during a year or some other specific period, and includes both living organisms and those that died during that period. **Surplus production** is that portion of production that can be removed from a population by natural causes or human harvest without adversely affecting future population levels. Surplus production results in the harvestable surplus discussed in section 3.3. In commercial fisheries, surplus production is synonymous with sustainable yield (Ricker 1975). That portion of the population taken by humans is referred to as **yield** and is just one of the several fates of production.

Wildlife biologists generally consider surplus production in terms of numbers of animals (see section 1.3). An example might be a harvest of a certain number of giant Canada geese. Fishery biologists, however, generally consider surplus production in terms of weight, not number. A Georgia farm pond will support a certain biomass of largemouth bass, but that biomass may be many small fish, fewer large fish, or moderate numbers of small, medium, and large fish. Fishery biologists often attempt to influence the size structure of fish populations through harvest regulations (see chapter 17).

In fisheries, **maximum sustained yield (MSY)** is a traditional management philosophy with a goal of achieving maximum harvest from a population over a series of years; that is, the maximum biomass that can be harvested without affecting future harvest. As a simple example, commercial ocean fisheries might allow harvest of 60% of the adult fish and leave 40% of the adults for reproduction. These percentages can vary tremendously depending on species and location. The problem with MSY is that it does not take into account the quality (size) of the organism being harvested. Perhaps the maximum harvest of walleyes from a Minnesota lake, in terms of biomass, might result from a 70% exploitation rate, with the mean length of walleyes harvested being only 30 centimeters. Few anglers would be satisfied with that type of fishery. Thus, biologists now tend to manage for **optimum sustained yield (OSY)**, which can be defined as a management philosophy with objectives that consider ecological and socioeconomic factors. Biologists consider such factors as user desires and expectations and cost-benefit relationships—the human dimension of a fishery. Such a philosophy pertains to both commercial and sport activities. In contrast to the above MSY example using walleyes, one management objective under an OSY strategy might be to produce a mean length of 40–45 centimeters for walleyes in the angler harvest.

If a wildlife population, such as elk, were managed under MSY, the biologists would attempt to keep the population at the inflection point on the S-curve (fig. 3.6). However, MSY is a dangerous management strategy because biologists seldom know exactly where this point lies on the S-curve, and because an event such as an extremely harsh winter or an outbreak of disease could

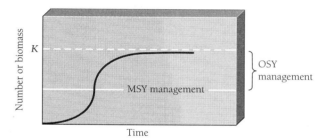

FIGURE 3.6 Maximum sustained yield (MSY) is a management strategy that typically seeks to produce the maximum harvest without endangering future reproduction and recruitment. Optimum sustained yield (OSY) represents a strategy under which maximum harvest is not the goal; rather, biologists also consider factors such as size of the animals harvested. *K* represents carrying capacity.

easily push the population below this point, causing a substantial population decline. Thus, wildlife biologists also usually manage at OSY (somewhere between the inflection point on the S-curve and *K* in fig. 3.6) to avoid population catastrophes. If a population was managed for trophy hunting, it would be managed under OSY rather than MSY.

While the MSY and OSY philosophies may seem to be simple, they are not. It is difficult to determine sustainable levels for populations. To be assured of sustainability, much more information is generally needed than is available. Therefore, a determination of the OSY or MSY for a particular population is at most a best guess based on currently available data.

3.7 Population Structure

Population structure is, as previously mentioned, the result of the three interacting rate functions. Biologists are concerned with both **age structure** and **size structure**, the age and size distributions of a population.

Consider the age structure of harvested buck white-tailed deer in two different states (fig. 3.7).

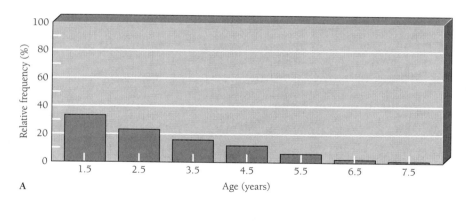

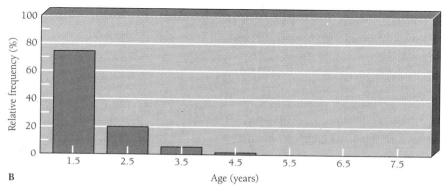

FIGURE 3.7   Age structures for buck white-tailed deer harvested from two populations that are under different management strategies. *(A)* This population is managed with limited buck licenses so that at least 20% of the harvest will be composed of bucks that are 4.5 years old and older. *(B)* This population is managed with unlimited buck license sales to allow maximum hunting opportunity.

Which of these harvests likely occurred where the state conservation agency operated under a management philosophy seeking to maximize hunting opportunity, and thus did not limit the number of licenses for antlered white-tailed deer? Which of these two age structures likely represents the harvest in a state that limits the sale of buck licenses to maintain a certain level of quality in the buck white-tailed deer harvested—perhaps 20% of harvested bucks being at least 4.5 years old? High mortality of white-tailed deer bucks caused by unlimited buck harvest results in an age structure that is dominated by young deer, with few bucks older than 2.5 years being harvested (fig. 3.7B).

Conversely, lower mortality of bucks under limited harvest results in more older bucks being harvested (fig. 3.7A). Either management strategy can be defended, but the age structures of the two populations are certainly different.

Fishery biologists also consider age structure. Which of the two black crappie populations represented in figure 3.8 has consistent recruitment and which has inconsistent recruitment? Black crappies in Lake Mitchell, South Dakota, successfully recruited every year for a six-year period (fig. 3.8A). In contrast, the black crappies in Roy Lake, South Dakota, successfully recruited only three times in a similar six-year period (fig. 3.8B).

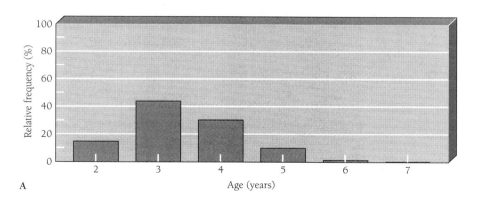

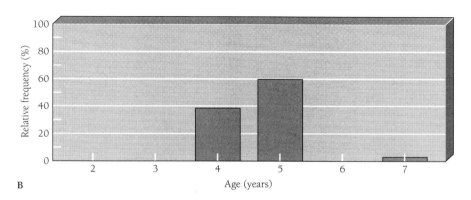

FIGURE 3.8 Age structure of black crappies sampled from *(A)* Lake Mitchell, South Dakota and *(B)* Roy Lake, South Dakota. Recruitment is inconsistent in the Roy Lake population; no individuals from age groups 2, 3, and 6 were collected in this sample.

Anglers are more likely to consistently catch black crappies in a population such as that in Lake Mitchell than in one like Roy Lake's. Black crappie recruitment in South Dakota was found to be more consistent in artificial impoundments than in natural lakes, apparently because of physical differences in the amount of shoreline irregularity and the degree of water level fluctuation (Guy and Willis, 1995). Lake Mitchell is an impoundment, while Roy Lake is a natural lake.

Fishery biologists rely heavily on assessments of fish population size structure, while wildlife managers seldom are concerned with size structure. Bluegills are popular sport fish in much of the United States, but have the tendency to overpopulate and stunt. Two different bluegill populations are depicted in figure 3.9. Population B is at or near carrying capacity, and growth is slow. This high-density, slow-growing population is a classic example of a stunted bluegill population. Population A is below the theoretical carrying capacity for that water body that might be expected given the available nutrients. As a result, growth is faster and there are larger individuals in the population. Given a choice, most anglers would prefer to fish for bluegills such as those represented in figure 3.9A rather than those in figure 3.9B.

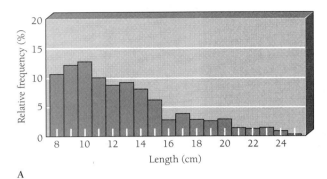

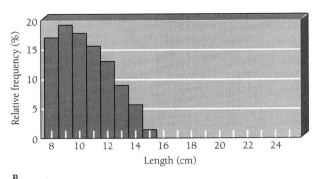

A   B

FIGURE 3.9 Size structures of bluegill populations in two different water bodies. (A) Bluegills in this population are faster growing because their standing stock is below the theoretical carrying capacity that could be supported by the nutrients available in that water body, and they reach 17 centimeters (cm) by age 4. (B) Bluegills in this population are slower growing because of high population density, and reach only 12 centimeters by age 4.

▼ SUMMARY

Three dynamic rate functions (natality or recruitment, growth, and mortality) form the basis for the population dynamics of any species. Natality and recruitment add new individuals to a population, growth adds biomass to a population, and mortality results in a loss of both individuals and biomass. Density-dependent and density-independent factors affect all three rate functions. While stable populations would exhibit a balance among these three rate functions, most wild populations are not stable, but instead fluctuate in response to factors such as variations in environmental conditions. If recruitment or natality exceed mortality, the population will increase; if mortality exceeds recruitment or natality, the population will decline.

When a species is introduced into a new and favorable environment, the population will increase in number and biomass. At first, this increase in abundance can be exponential. However, such rapid increases do not continue because the carrying capacity of the system is reached. Populations cannot be sustained beyond the capacity of a system to provide for their needs. Biotic potential, or reproductive potential, can be thought of as a "force" that is driving the population to increase. Environmental resistance is the sum of the biotic and abiotic environmental factors that interact to limit populations.

Production is the biomass that is accumulated by a population during a specific time period, most often one year. Surplus production results in a harvestable surplus of many sport and commercial fish and wildlife species, and the portion of the population taken by humans is termed yield. The maximum sustained yield philosophy has as its goal obtaining maximum yield without affecting future yield, but does not consider the quality of the harvested organisms. Today, most biologists operate under a philosophy of optimum sustained yield, whereby ecological and socioeconomic factors, rather than simply number or biomass to be harvested, are considered.

The interactions of the dynamic rate functions are expressed in population structure. Biologists assess both age and size structure for populations. Assessment of age structure can provide information concerning irregular recruitment or excessive mortality. Fishery biologists in particular assess size structure, which provides information on the quality of the fish population to anglers.

PRACTICE QUESTIONS

1. What is the primary reason why wildlife biologists use natality and fishery biologists use recruitment to adult size when considering how many individuals are added to a population?

2. What is inversity?

3. Consider two brook trout populations in beaver ponds in Wyoming. Neither has been fished by human anglers. In one pond, the brook trout density is high and few individuals exceed 20 centimeters. In the other, density is low and the majority of adult brook trout are longer than 20 centimeters. In which pond would you expect faster growth rates? How would intraspecific competition relate to expected growth rates in the two ponds?

4. Consider two mammal species that reside in your geographic area. Choose one species that is short-lived and in which most individuals do not live more than one or two years. Examples of the shorter-lived species might include Arizona pocket mice and pygmy rabbits. Choose a second species at the other end of the longevity spectrum, in which individuals have the capacity to live for five, ten, or more years. Longer-lived species might include Dall sheep, brown bears, moose, or one of the deer species. Contrast the rates of reproduction/recruitment and mortality for the two species you have chosen.

5. Choose an animal species with which you are familiar. Make a list of the various biotic and abiotic factors that are part of the environmental resistance for a population of this species. Now, decide which of these factors are density-dependent, which are density-independent, and which might be both.

6. Why is carrying capacity not a measurable number?

7. Wildlife and fishery biologists typically consider surplus production as being a harvestable surplus for game or commercial species. Think back to the lessons from ecology in chapter 2. In general, will the harvestable surplus be greater for a species at a low trophic level or a high trophic level? In other words, is the biomass available for harvest generally greater for an herbivore or a carnivore? Why?

8. How do determinate and indeterminate growth affect surplus production?

9. Describe how total mortality, natural mortality, harvest mortality, compensatory mortality, and additive mortality may interact in an animal population.

SELECTED READINGS

Begon, M., and M. Mortimer. 1986. *Population ecology: A unified study of animals and plants.* 2d ed. Blackwell Scientific Publications, Oxford, England.

Busacker, G. P., I. R. Adelman, and E. M. Goolish. 1990. Growth. Pages 363–387 in C. B. Schreck and P. B. Moyle, eds. *Methods for fish biology.* American Fisheries Society, Bethesda, Md.

Caughley, G. 1977. *Analysis of vertebrate populations.* John Wiley & Sons, New York.

Getz, W. M., and R. G. Haight. 1989. *Population harvesting: Demographic models of fish, forest, and animal resources.* Monographs in Population Biology 27. Princeton University Press, Princeton, N.J.

Hilborn, R., and C. J. Walters. 1992. *Quantitative fisheries stock assessment: Choice, dynamics, and uncertainty.* Chapman and Hall, New York.

Van Den Avyle, M. J. 1993. Dynamics of exploited fish populations. Pages 105–135 in C. C. Kohler and W. A. Hubert, eds. *Inland fisheries management in North America.* American Fisheries Society, Bethesda, Md.

LITERATURE CITED

Guy, C. S., and D. W. Willis. 1995. Population characteristics of black crappies in South Dakota waters: A case for ecosystem-specific management. *North American Journal of Fisheries Management* 15:754–765.

MacNab, J. 1985. Carrying capacity and related slippery shibboleths. *Wildlife Society Bulletin* 13:403–410.

McCullough, D. R. 1979. *The George Reserve deer herd: Population ecology of a K-selected species.* University of Michigan Press, Ann Arbor.

Ricker, W. E. 1975. *Computation and interpretation of biological statistics of fish populations.* Fisheries Research Board of Canada, Bulletin 191, Ottawa, Canada.

4

GENETICS IN WILDLIFE AND FISHERIES

Any discussion of fisheries and wildlife requires some coverage of genetics because understanding the inheritance of traits is basic to understanding organisms and their populations. Genetic variability occurs at two different levels: there are genetic differences among individuals within populations, and there are genetic differences among different populations of a species. Understanding how genetic variability is transformed into the differences within and among populations is fundamental to the study of evolution.

While it has always played an important role in fisheries and wildlife, genetics is coming more to the forefront in these fields as greater attention is given to the implications of genetics. Genetic engineering, the effects of stocking on native and introduced populations, and captive breeding programs are examples of recent areas of heightened interest. This chapter will address some basic genetic concepts and briefly describe how they relate to speciation, natural selection and adaptation, artificial selection, hybridization, and other genetically related topics.

4.1 Basic Genetics

As in all organisms, traits in species associated with wildlife and fishery resources are transferred from one generation to the next by **genes**. Genes are segments of hereditary material that are positioned on **chromosomes** within cells. Chromosomes are structures that carry genes and are found in the nuclei of cells. Genes are heritable and determine the characteristics of an organism. Some traits are transmitted by single genes. For example, in rainbow trout, normal skin color versus colorless skin (albino) is controlled by a single gene. Other traits are controlled by multiple genes and are called **polygenic traits**. For example, in

rainbow trout, growth is controlled by multiple genes present at a variety of locations on chromosomes. Genes that encode amino acid sequences or the structure of proteins are called **structural genes**, while those that control the expression of the structural genes during development or among different tissues are called **regulatory genes**.

Most vertebrates are **diploid** organisms; that is, they have two complete sets of chromosomes. Each individual organism has two copies of each gene, usually one received from its mother and one from its father. These copies are called **alleles** and may be alternative forms of the same gene. Alleles for the same gene occur at the same **locus**. A locus is the location of a particular gene on a chromosome. The two chromosomes that each carry the same set of genes are called **homologous chromosomes**. The specific genetic information possessed by an individual is called its **genotype**; the actual physical expression of traits is referred to as its **phenotype**.

The phenotype of an individual is a product of its genotype interacting with its environment. Two organisms can have the same genotype for a particular trait but have different phenotypes. For example, two individuals of a species may have the same genotype for growth, but be of different sizes because one receives more or less food than the other. The converse can also be true: two individuals of a species can have different genotypes for a particular trait but, because of environmental conditions, have the same phenotype. For example, one individual may have a genotype for faster growth than another individual, but the two may grow at the same rate if the individual with the genotype for faster growth receives less food than its counterpart with a genotype for slower growth.

Individuals with two copies of the same allele at a particular locus are **homozygous**; individuals with two different alleles at a particular locus are **heterozygous**. An individual may have two **dominant alleles**, two **recessive alleles**, or one of each at a particular locus. Dominant alleles are typically expressed over recessive alleles in a heterozygous individual. With other genes, both alleles are expressed phenotypically but to different degrees; this is called **incomplete dominance**. Both alleles at a locus may also be expressed to the same degree; this situation is called **codominance**.

Genes and their alleles determine not only what characteristics are present in an individual, but how well an individual or population is adapted to its environment. The **gene pool** of a population is composed of all the alleles present within all the individuals constituting that population. If little genetic variability is present in the gene pool, a population will be less able to adjust to changes in environmental conditions. The opposite is also true; the greater the genetic variability, the greater the ability to adapt or change.

The terms **heterozygosity** and **homozygosity** are commonly used in referring to both individuals and populations. Heterozygosity is the condition of having one or more pairs of dissimilar alleles at one or more loci; homozygosity is the condition of having identical alleles at one or more loci. The use of these terms differs somewhat depending on whether individuals or populations are being discussed. With regard to an individual organism, heterozygosity and homozygosity can refer to the average of all the genes in the individual, but most often it refers to the probability of heterozygosity or homozygosity for any one gene. When used in reference to populations, heterozygosity or homozygosity usually refers to the percentage or proportion of the population that is homozygous or heterozygous for a particular trait or traits.

As a very simplified example of inheritance, we can use eye size in a hypothetical fish species. Assume that eye size is controlled by a gene or genes that determine whether a fish has large or small eyes. If a fish population had only alleles for large eyes, then only large-eyed fish would occur in that population. If only small-eye alleles were present in another population, then only individuals with small eyes would occur in that population. Neither of these populations would have any heterozygosity for eye size. Therefore, no within-population variation would be present for this trait, but a difference between the two popu-

lations would exist. If, however, a third population had both large-eye and small-eye alleles in its gene pool, then some within-population variation would exist, and this population would have some level of heterozygosity. Various levels of heterozygosity can be present among different populations depending on the frequency of the alleles in question. When eye size is combined with the myriad of other genetic traits an organism can possess, one can readily see the endless variety of combinations that can be available to produce genetic similarities and differences.

4.2   Species, Subspecies, and Strains

Evolution is the theory that pertains to the process of continuous and gradual transformation of lines of descent from a common ancestor. It involves changes in the gene pool of a population that over time lead to adaptation to the environment in which the population lives. The products of evolution are genetically new organisms. A **species** is a naturally occurring group of organisms that can successfully interbreed and produce fertile offspring; they usually are unable to successfully interbreed with organisms outside that group, or if they do so they tend to produce less viable offspring. **Subspecies** (**races**) are formally recognized subdivisions of a species whose members resemble one another to some extent and can interbreed to produce fertile offspring, but differ from other subdivisions in some easily recognizable way. **Strains** (stocks) are groups within a species with common ancestry or parentage that are adapted to similar environments.

Subspecies and strains differ from other members of the species in genetic composition and in the phenotypic expression of that composition. For example, there are a number of subspecies of cutthroat trout in the Rocky Mountain region. These subspecies are separated by barriers that have not allowed them to intermingle for some time. Although they are all the same species, the differing conditions in which they live have resulted in allelic frequencies in each population's gene pool that are somewhat different from that of the next. Another example would be the dozen or so subspecies of Canada geese. Canada geese mate for life; they also winter at traditional sites with their young from the previous summer, and return to the same nesting area with the young birds in the spring. These behavioral characteristics have led to the evolution of different subspecies of Canada geese that interbreed only to a limited extent with subspecies adapted to other geographic areas.

On the surface, such small genetic differences may seem to have little practical importance, but this is not the case. An example of the potential importance of such differences is the case of the Audubon bighorn sheep. This bighorn sheep subspecies was historically found in the badlands and Missouri River breaks of North and South Dakota, but became extinct by the early twentieth century. Since then, the introduction of other bighorn sheep subspecies into that area has met with limited success. It is likely that the Audubon subspecies was specifically suited to the local environmental conditions, and if that subspecies were still in existence today, restoration efforts would be more successful. Another example of the importance of subspecies revolves around the issue of whether subspecies should or should not be considered in the assessment of endangered or threatened status. This issue has many economic, ethical, legal, and other implications (see section 11.2).

Formally named fish strains are often developed by humans for aquaculture purposes. **Aquaculture** is the culture and husbandry of aquatic organisms. Rainbow trout have been cultured for a long period, and sufficient records of parentage exist to allow development of named strains with specific characteristics. For example, a rainbow trout strain that was initially developed at the fish hatchery in Manchester, Iowa is called the Manchester strain. Although different strains of rainbow trout are all the same species, they can differ markedly in spawning time, growth rate, vulnerability to angling, susceptibility to diseases, and a wide range of other traits. It is also quite possible that with the advent of **game ranching** and **game farming**, specific strains of various wildlife species will soon be or have already been developed. Game ranching is the husbandry of native animals, usually large ungulates, for the production of meat and other products. It is a variation of range management that in North America involves species such as elk and mule deer rather than domesticated sheep or cattle. At times game ranching is also used to refer to pay-for-hunting enterprises in which animals, primarily exotic ungulates, are harvested for sport purposes. Game farming usually involves the commercial production of smaller organisms such as ring-necked pheasants and northern bobwhites. These animals can be used for food or sport purposes.

Similarly, exposure of natural populations to varying environmental conditions has given rise to different stocks, each with its own specific traits. For example, a variety of stocks of Pacific salmon have developed naturally on the west coast of North America.

4.3 Isolating Mechanisms

Populations can have their gene pools segregated by a variety of **isolating mechanisms**. Isolating mechanisms are mechanisms that tend to keep organisms from interbreeding. **Speciation** (the process of species formation) without some form of isolation among groups is almost impossible. If populations are **allopatric** (having nonoverlapping geographic ranges), they normally do not interbreed unless changes occur that mix them in a common environment (fig. 4.1). When organisms

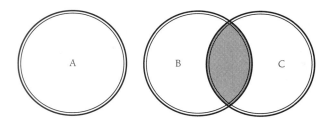

FIGURE 4.1 A diagram of the ranges of allopatric and sympatric populations. Population A is allopatric with populations B and C; it is effectively isolated from the others, at least by space. Populations B and C have overlapping ranges (shaded area) and are, therefore, sympatric. Isolating mechanisms must separate them if they are to become different species.

are separated in this way, each group exists in a somewhat different habitat. These differences in habitat often result in the organisms in each group becoming different genetically. If they become different enough, they may no longer have the capacity to interbreed. Thus, the spatial separation of populations allows for genetic differentiation and species formation.

Even if populations of closely related species are currently **sympatric** (their geographic ranges coincide or overlap), previously evolved isolating mechanisms usually prevent them from interbreeding (fig. 4.1). There are a variety of mechanisms that can prevent interbreeding between sympatric species, and often several of these mechanisms will work in concert. Premating isolating mechanisms, such as sexual or other behavioral differences, physiological differences, anatomical differences, and mating time differences, can work to reduce interbreeding. Even if interbreeding occurs, several postmating isolating mechanisms, such as natural developmental abortion and hybrid sterility, can prevent the successful establishment of hybrid populations.

The ultimate test of whether speciation has occurred happens when two previously allopatric populations come back into contact. For example, the red-shafted and yellow-shafted races of the northern flicker have been brought back into contact through forest plantings by humans in the Great Plains. Where their ranges now overlap, the two groups interbreed freely, producing fertile offspring, indicating that they have not been isolated long enough to form separate species.

4.4 Natural Selection and Adaptation

Natural selection and **adaptation** are closely interrelated concepts. The theory of natural selection is an attempt to account for the adaptation of organisms to their environment. The major tenet of this theory is that organisms of different genotypes in a population contribute differently to the gene pool of succeeding generations. Those organisms most successful in an environment will survive better and produce more offspring with the characteristics that led to their success. Differences between individuals with regard to survival and subsequent reproduction result in genotypic differences between generations. In this way, natural selection allows for adaptation in organisms. Adaptation, the change in structure, physiology, behavior, or mode of life, allows species to adjust to their environment.

Think back to the simplified, hypothetical, large-eyed/small-eyed fish species example in section 4.1. Assume that a population of these fish lives in a body of water where water clarity keeps improving each year. Also assume that these fish use sight as a primary sense during feeding, mate selection, or for some other life function, and that alleles are present in the population for both eye sizes. Also assume that larger eyes are beneficial to this population under these improving water clarity conditions. Large eyes might not necessarily be beneficial under these circumstances, it could be just the opposite, but assume this to be the case. Under these circumstances, large-eyed fish would have a competitive advantage over small-eyed fish and would be able to produce more offspring because they would survive better and grow faster. Their offspring would receive large-eye genes from their parents, and in succeeding generations larger numbers of fish in the population would have large eyes. Each succeeding generation would have increasingly larger numbers of large-eyed fish, and it is probable that in time, few small-eyed fish would be present. Thus, through natural selection, the population would have adapted to the changing environmental conditions.

This evolution within the population would not necessarily mean that the alleles for small eyes would be completely lost from the population; the phenotypic or genotypic incidence of small eyes likely would just be greatly reduced. Continued heterozygosity for eye size would be beneficial because it would allow for future adaptation if environmental conditions were to change again. Try to envision what would happen to this population if some heterozygosity for eye size were retained and at some time in the future water clarity began decreasing.

An example such as this does not have to portray an either/or situation. Incomplete dominance or codominance could result in a range of eye sizes. In addition, multiple genes controlling a trait could also result in a range of phenotypic expression of eye size.

The above large-eye/small-eye example is hypothetical, but a real-life example of such adaptation is exhibited by the Mexican tetra. This fish species has cave-dwelling populations that are eyeless, normal populations living outside of caves that have well-developed eyes, and populations that live near the mouths of caves that have intermediate-sized eyes (Mitchell et al. 1977). These fish were able to adapt to the conditions of their environment because of the genetic composition of their populations.

Some species have a limited ability to adapt to changing conditions. Some of the species that are currently endangered are ones that are very specialized with regard to the environmental conditions they require. If environmental conditions change over all or part of the range of these species, their populations may be either lost or further reduced. Often, environmental changes happen too rapidly for adaptation to occur. How rapidly a population can adapt depends on the genotypes of the individuals it comprises. Examples of organisms in this type of danger include the Kirtland's warbler, which can reproduce only in early successional stages of jack pine forests in upper Michigan, and the Devils Hole pupfish, a species naturally found only in one spring area of Death Valley. Changes in the amount of new-growth forest in upper Michigan, as determined by fire frequency, or in the water regime of Devils Hole would threaten to reduce or eliminate these species.

On a long-term basis, some species can display striking adaptive change or evolution. The different types of forelimbs exhibited by mammals are an example. Adaptation is reflected in the shape and purpose of these limbs: in chimpanzees the forelimbs are adapted for climbing, in pallid bats for flying, in eastern moles for digging, in killer whales for swimming, in coyotes for walking, and in humans for handling objects.

The striking effects of natural selection and adaptation are also seen in examples of **convergent evolution**. Such evolution occurs when species with different origins and histories develop similar traits and characteristics that allow them to occupy similar niches. Often these species are widely separated by geographic location and by their positions in the evolutionary sequence. For example, several bird species have evolved into flightless, running organisms. This characteristic is exhibited by birds such as dwarf cassowaries, greater rheas, ostriches, and emus. There are some advantages to flightlessness. Flight is an energetically expensive activity. If a bird can avoid

A   B

FIGURE 4.2   The Nile tilapia, an African cichlid *(A)*, and the bluegill, a North American sunfish *(B)*, are good examples of fishes from families that exhibit convergent evolution. Note the similarity of features such as general body shape, fin size and location, and mouth size and location. These characteristics and others are indicative of the fact that the two families to which these species belong exhibit convergent evolution. *(Part A photograph courtesy of T. Batterson; part B photograph courtesy of D. Garling.)*

predation by nocturnal or secretive activity patterns, keen vision, rapid running, or aggressiveness rather than by flight, it can be evolutionarily successful. The point is that flightless running birds have developed in widely separated areas of the world; this convergent evolution is a result of natural selection and adaptation to similar ecological niches.

Another excellent example of convergent evolution involves the sunfishes of the Nearctic region and the **cichlids** (members of the fish family Cichlidae) of the Ethiopian region (fig. 4.2). In their respective regions each family has evolved to fill a wide variety of freshwater ecological niches. Both families have species that are prey organisms, while others are predators; some eat primarily snails, while others eat small soft-bodied bottom invertebrates, hatching insects, or a number of other food organisms; some live in rocky areas, while others live in vegetated areas. In many cases a cichlid and a sunfish that have adapted to similar ecological niches closely resemble each other in characteristics such as body shape, fin location, and mouth size. They have evolved to fill similar niches, but on two different continents.

4.5   Artificial Selection

Just as natural factors can affect selection, so can artificial or human-induced factors. In **artificial selection** humans choose the parents of each generation based upon the phenotypic traits they want to produce. Such selective breeding has produced many different kinds of domesticated organisms that differ markedly in physical, physiological, and behavioral characteristics from their wild ancestors. The variety of different shapes, sizes, and colors of dogs, cattle, and chickens that exists today is the result of artificial selection.

Artificial selection, like natural selection, is a **systematic process**, that is, a genetic process that changes allelic frequencies in a population in some predictable, nonrandom fashion. While artificial selection has benefits in domestication, such selection is usually not beneficial to wild populations. A bulldog may do quite well living in your house, and a Holstein cow may survive well on a dairy farm, but such organisms would not likely survive or maintain their current forms in a wild situation.

The rainbow trout is cultured throughout the world for both commercial and sport purposes. Through selective breeding, a population of rainbow trout can be developed to have some trait desired by humans, such as rapid growth or a high protein content. Such selection may be beneficial only for those individuals raised in an artificial situation while being prepared for market. The same fish would be unlikely to compete and grow well in a wild situation. In fact, it would probably be at a distinct disadvantage because of its lack of a variety of useful genetic traits. Its introduction might also have negative effects on local wild populations if it survived and reproduced because genetic material unsuited for that environment might be added to the gene pool.

Artificial selection is not practiced only in fisheries. As game ranching and game farming become more prevalent, so does artificial selection in farmed and ranched "wild" animals. For example, white-tailed deer are being selectively bred to produce qualities such as large antlers. Records of which males have genotypes for large antler production are kept, and those animals are used for breeding purposes. The offspring of such animals may lose other beneficial genetic traits needed for survival in a wild situation.

Even when the intent is to raise a "wild" organism for stocking, unintentional artificial selec-tion can often occur. By rearing or holding organisms in a hatchery or some other artificial situation, humans inadvertently select for eggs, young, or adults that survive well in that particular artificial situation. Individuals that have the genetic composition to survive such an experience will live to pass on their genetic information; those that do not do well in such an environment may not. This differential survival and reproductive success changes the genetic composition of future generations.

Biologists need answers to a variety of genetic questions before better conservation of natural resources can be accomplished. The problem of artificial selection, whether intentional or unintentional, faces the profession whenever an organism is placed in a controlled or artificial environment, and then it or its offspring are expected to perform like their wild counterparts. Such organisms may lack what is called **fitness**, which is a measure of the reproductive success of individuals or populations. Fitness is a trait that is the product of many genes and the environment interacting throughout the lifetime of an individual organism. When we raise organisms in artificial conditions, we do not necessarily end up with the same organism (genetically) with which we started (fig. 4.3). What is the effect, for example, on the genetic composition and fitness of mallards, elk, or razorback suckers

A

B

FIGURE 4.3 A trout hatchery (*A*) and a natural stream (*B*) are dissimilar habitats that will produce dissimilar organisms, each adapted to its particular environment.

when they are raised in captivity? Should these captive-reared organisms be released into natural environments? If so, under what conditions and in what situations? Domesticated turkeys have lost much of their potential to survive in the wild due to genotypic change; if they breed with wild turkeys, the inherent wildness and survival ability of the wild birds could be reduced. Some pen-reared ring-necked pheasants are bred for large size and are poor flyers, yet many of them are released into the wild. How do they affect the wild ring-necked pheasant population?

Whenever mixed stock populations result from management activities, this question of **genetic integrity** arises. Genetic integrity refers to the genetic characteristics of a group of organisms that distinguish it from other groups or populations of the species. Little is known concerning the effects of the loss of genetic integrity. Unfortunately, there are also many populations, both native and introduced, in which genetic integrity is really no longer a relevant issue because human activities have already altered them so much.

In some situations the effects humans have on a population due to captive holding or rearing are unavoidable. Black-footed ferrets and California condors were so low in numbers that all of the known individuals were captured and held for captive breeding. Obviously the intent of doing this was to increase their numbers and then later move them back to the wild. However, we cannot know what genetic consequences will result from their time in captivity.

4.6 Inbreeding and Outbreeding

When populations are small, either naturally or in captive situations, the probability that individuals in the population will mate with closely related individuals increases. Such mating is called **inbreeding**. Both the individual and the population can be affected by the loss of genetic variability that results. Inbreeding is usually associated with small populations; it is most often a problem in fish culture and other artificial situations and in small wild populations. **Inbreeding depression**, a reduction in fitness or vigor due to increased homozygosity, is the result. Inbreeding depression is difficult to demonstrate in natural populations, but is a common phenomenon in domesticated animals and laboratory populations that are extremely inbred.

In natural populations, the **founder effect** occurs when a few individuals leave a large population to establish a new population and inbreeding results. Rare alleles found in the large population will be lost if no individuals in the new population carry that genetic material. The smaller the initial number of individuals that found a population, the greater the risk of inbreeding depression. It is important to remember that once inbreeding has occurred, its effects remain, even if the population increases to a much greater number; the individuals in the population are still the offspring of relatively few parents and thus have the genetic diversity of only those few.

Inbreeding is a **dispersive process**, that is, a process that causes random change in allelic frequencies and a loss of genetic diversity in populations over time. What ramifications might this have for species such as the previously mentioned black-footed ferrets and California condors, which have been reduced to extremely small numbers and are now essentially maintained only as captive populations? Will the combination of small population size and rearing in captivity result in black-footed ferrets or California condors that cannot survive well in the wild? While inbreeding is a problem in these situations, there are really few alternatives to captive breeding programs if these particular species are to survive.

There is much utility in knowing the repercussions of inbreeding and small population size. Is the stocking of a lake with walleyes as effective as it might have been if fish with greater genetic diversity had been introduced? Would using parental walleyes from a nearby, similar lake have produced a better fishery? Could the introduction of the exotic gray partridge into North America have produced a better wildlife system if greater attention had been paid to genetic diversity and if larger numbers of partridge had been originally stocked? It is difficult to incorporate genetic diversity into a large population once the population has been established; for example, adding "new" gray partridge genetic material to an existing large population would be difficult at best. Such new genetic information is usually swamped by that of the population already present. It might be possible in this situation to have an effect, but it might take many generations. What will be the genetic diversity of gray wolf populations resulting from the current restoration efforts in the western United States? Gray wolf restorations are usually made with only a few animals, often captured from the same location. Will these restored populations be less successful than hoped for—or unsuccessful—because of the specific animals used in the restoration effort? Inbreeding and small population size must be a consideration in such efforts.

It might be assumed that if inbreeding leads to decreased genetic variability (homozygosity) and is not desirable, then the best situation would be for organisms to mate with their most distant relatives. This is not necessarily the case. Such distant mating is termed **outbreeding**. Populations are often adapted to local environments; while outbreeding increases heterozygosity, it can also result in the loss of genetic fitness for specific conditions. Outbreeding is usually not a problem in natural situations because members of distantly related populations do not normally meet for mating.

The maintenance of gene pools and genetic heterozygosity is important if we are to have long-term success with populations that are managed or maintained by adding organisms to the population or by reestablishing populations. If gene pools and genetic heterozygosity are not maintained, these populations, no matter how large they become, will always be less adaptable to changing conditions than most naturally occurring populations.

The loss of within-population variability through inbreeding and the loss of among-population variability through outbreeding both must be considered, especially when small populations are involved.

4.7 Hybridization

Hybrids are the offspring resulting from the mating of parents that are genetically unlike (Piper et al. 1982). In most cases hybrids represent crosses between different species, subspecies, or strains. An increase in some phenotypic value (growth rate, for example) of a hybrid relative to its parents is called **hybrid vigor** (**heterosis**). Hybrid vigor is usually attributed to increased heterozygosity. In some situations hybrids are unable to reproduce, while in other cases they can produce offspring. In some situations hybridization is a totally natural occurrence, while in others it is human induced.

Some management practices have led to negative population effects due to hybridization. The cutthroat trout has been eliminated or reduced in parts of its native range because of hybridization with introduced rainbow trout. After several generations of interbreeding, the rainbow trout phenotype often becomes dominant. Such a result is called **introgression**. This shift to rainbow trout is obviously not beneficial to the native cutthroat trout because they are reduced in numbers or eliminated. Various subspecies of wild turkeys have been introduced into new regions in North America. In some cases, introductions of wild turkeys of mixed-subspecific ancestry or introductions of two subspecies into the same area have occurred. The possible loss of identifiable subspecies and their gene pools resulting from such practices should be of long-term concern.

Hybrids constructed and used in a controlled fashion can have positive effects. The decision to use hybrids depends on the management objectives and whether those objectives can be accomplished with little or acceptable risk to other populations. A commonly used hybrid in fishery work is the striped bass–white bass hybrid. This hybrid apparently does not require a coolwater refuge during the summer, something that the striped bass needs. However, the hybrid retains some preferred attributes of the striped bass, especially with regard to size. The risk of using this hybrid is its potential adverse genetic and ecological effects on populations of white bass and other species.

▼

4.8 Mutation, Chromosome Aberrations, Genetic Drift, and Genetic Engineering as Modes of Genotypic Change

Selection, inbreeding, and outbreeding are all ways in which genotypic change can occur. There are other ways that such change can occur. **Mutation** of genes, a change in their **DNA (deoxyribonucleic acid)**, is one of those other ways. DNA is the chemical used to store genetic information in most organisms. Gene mutation occurs when the DNA of a gene permanently and spontaneously changes, resulting in a new allele. An allele produced by mutation may be entirely new to a population. If the mutation in some way makes the organism more successful, the trait may be passed to the next generation and have an increasing frequency in the population. Therefore, the mutation can lead to different genotypic and phenotypic frequencies in future generations; it can also lead to a totally new species, especially if isolating mechanisms separate populations. It is actually much more likely that the mutation will have a negative effect or a neutral effect.

Mutation is a systematic process that increases genetic diversity in populations and counteracts the loss of genetic diversity caused by dispersive processes. It is a normal phenomenon in nature. In general, mutation occurs at random or in an undirected manner. The genetic materials supplied by mutation provide the building blocks that allow for natural selection and, therefore, evolution.

Another mechanism of genetic alteration is chromosomal change resulting from **chromosome aberrations**. Such changes can occur through **deletion**, **duplication**, **inversion**, or **translocation**. Deletion occurs when a chromosome breaks, then reunites, but loses a segment in the process. Duplication occurs when a segment of a chromosome is duplicated; this results in that segment appearing twice in the genetic composition of the organism. Inversion occurs when the sequence of genes on a chromosome becomes inverted. Translocation occurs when a segment from one chromosome becomes attached to a **nonhomologous chromosome**. Nonhomologous chromosomes are chromosomes that carry genes for different characters. Each of these modes of genetic alteration can cause genotypic change in an organism and can, therefore, affect populations.

Genetic drift is another mode of genotypic change. Genetic drift occurs due to random changes in allelic frequencies caused by sampling errors that occur in each generation. In this situation, sampling refers to the small sample of parental eggs and sperm that actually develops into offspring. For example, if something causes the death of organisms in a population and the cause has nothing to do with the fitness of those organisms, genetic drift could occur. Assume that there is a northern oriole population with only eight males, all with a different genotype for a trait such as bill color. Also assume that the five males with the darkest bills are killed, and that the genetic composition of those five has nothing to do with their dying. The three remaining males would be the only ones that could mate, and therefore, gene frequency in the population would change in the direction of their genotypes. Genetic drift removes genes from the population, but cannot add them. The chance of genetic drift increases as population size decreases, and it is usually associated with small populations. Genetic drift, like inbreeding, is a dispersive process, so it tends to increase homozygosity.

There is some recent evidence that genetic drift may not always be an overriding factor in genetic change, even in small populations. For example, Karl and Avise (1992) and Vrijenhoek et al. (1992) reported that **balancing selection** was probable in the animal populations they studied. Balancing selection occurs when heterozygous individuals have superior fitness and the two kinds of homozygous individuals have equal fitness. Because of processes such as balancing selection, genetic drift does not have to occur, but it remains an important mode of genotypic change in organisms, especially when small populations are involved.

Dispersal can also cause shifts in genetic composition. The movement of an organism to a new population that is genetically different from its original population can cause genetic change. Dispersal can have a much more immediate effect if small populations are involved (see section 4.6).

Mutation and dispersal are processes that increase the genetic diversity of populations. Inbreeding and genetic drift decrease genetic diversity. Under natural conditions the combined effects of inbreeding, genetic drift, selection, mutation, and dispersal determine the genetic composition of a population (Allendorf and Ferguson 1990; Kapuscinski and Jacobson 1987).

There is another source of genetic change that should be addressed: the relatively new field of study called genetic engineering. This field involves **gene transfer**, induced **polyploidy**, and other artificial methods of producing "new" organisms. Gene transfer involves the insertion by humans of genetic information from one species into the DNA of another. The result is the construction of a novel genotype. The new gene can be obtained from any other species and may confer some specific beneficial trait to improve the recipient organism, such as disease resistance, faster growth, or greater tolerance for some environmental condition.

Polyploidy occurs when an organism has more than two sets of chromosomes. While some organ-

FIGURE 4.4 This triploid grass carp looks the same as its diploid counterpart, but genetically, it is not really the same organism. Producing triploid fish reduces the potential of such an exotic to establish reproducing populations. (*Photograph courtesy of B. Blackwell.*)

isms are naturally occurring polyploids, this condition can be artificially induced in organisms in numerous ways. Polyploidy, especially **triploidy** (having three sets of chromosomes) is induced in some species, such as grass carp, to produce fish that cannot reproduce (fig. 4.4). This practice reduces the potential of this exotic species to become established where it is stocked. In other cases, for example, in rainbow trout, triploidy is induced in an attempt to have the fish grow more rapidly. Because triploid fish are generally sterile, they can direct energy intake into body growth rather than gonadal development and reproductive behavior.

Other genetic engineering methods are currently being used in aquaculture. Such efforts include the production of **monosex populations**, **gynogenesis**, and **androgenesis**. Monosex populations are populations that consist of only one sex. These may be useful because one sex may be more valuable than the other; for example, females of the species may grow more rapidly than males. Another potential benefit of monosex populations is that reproduction can be controlled. Monosex populations are usually produced by hybridizing two closely related species, by using hormones to artificially reverse the sex of organisms, or by induced gynogenesis. Gynogenesis is the production of viable young that have only the genetic material from their mother; androgenesis is the production of viable young that have only the genetic material from their father. Both gyno-genesis and androgenesis, which involve reproduction without fertilization, can be used to produce diploid monosex or polyploid fishes.

The potential hazards of genetic engineering and the creation of "new" organisms are current topics of both scientific and public debate, but are beyond the scope of this text. This debate is, however, not just a biological one. Genetic engineering relates to economics, ethics, and a variety of other social issues.

4.9 Results of Evolution

Evolution results from genetic change in populations. Because genetic modifications provide the building blocks for evolution, understanding the way in which inherited variations occur is basic to our understanding of the mechanisms of evolution. Much of what has been discussed in this chapter relates directly to the evolution of organisms. As the genetic composition of a species changes, the phenotypes of the individuals change and evolve. A combination of the individual's genetic composition and the pressure of environmental change acts to determine how successful it is in surviving.

The time scales over which genetic change occurs are highly variable. Some organisms are able to speciate or change in only a few generations, while others remain apparently unchanged or little differentiated over hundreds of thousands or millions of years. The genetic composition of organisms such as the American alligator, the duckbilled platypus, the brown kiwi, and the paddlefish has allowed these organisms to survive in relatively unchanged form for long periods of time and through many environmental changes. The genetic composition of groups such as minnows, warblers, and mice has allowed them to develop

multiple species in many new habitats as these habitats became available and changes occurred. Many contemporaries of these species or groups did not have the genetic composition necessary to adapt to or confront change; they have become extinct and are just fossil memories.

SUMMARY

Genes transfer traits from one generation to the next. Some traits are transmitted by single genes, while others are transmitted by multiple genes. Most vertebrates are diploid organisms that have two complete sets of chromosomes, one from their mother and one from their father. The specific genetic information possessed by an organism is called its genotype; the actual physical expression of traits is called its phenotype. The phenotype is a product of the genotype interacting with the environment. Alleles are alternate forms of a gene. An organism with two of the same alleles is homozygous for that trait. If the alleles are different, the organism is heterozygous for that trait. The amount of heterozygosity or homozygosity present in a population determines its ability to change or adapt.

Species are naturally occurring groups of organisms that can successfully interbreed and produce fertile offspring. Subspecies are formally recognized subdivisions of species whose members can interbreed but which differ from other subdivisions in some recognizable way. Strains are groups of organisms of a species with common parentage.

Isolating mechanisms tend to keep organisms from interbreeding. Without some form of isolation, speciation is almost impossible. Natural selection and adaptation are closely intertwined. The theory of natural selection is an attempt to account for the adaptation of organisms to their environments.

Artificial selection is a process in which humans select for some desired phenotypic trait. Artificial selection, whether intentional or unintentional, has far-reaching genetic ramifications for organisms that are kept for a part or all of their lives in an artificial environment. The fitness of the offspring of these organisms is usually reduced in natural settings.

Inbreeding is a dispersive process that is usually associated with small populations and often results in increased homozygosity. Outbreeding, while promoting heterozygosity, may be detrimental if it reduces the capacity of an organism to survive in a particular environment. Hybrids usually represent crosses between different species, subspecies, or strains. These crosses often exhibit hybrid vigor because of their heterozygosity.

Mutation is a systematic process that increases genetic diversity. Genetic drift, usually a problem only in small populations, is a result of chance fluctuations in allelic frequencies. It is a dispersive process that tends to increase homozygosity.

Gene transfer and induced polyploidy are among the methods of a new field of study called genetic engineering. The potential hazards of genetic engineering and the creation of "new" organisms have economic, ethical, and other social implications. Genetic modifications provide the building blocks for evolution.

PRACTICE QUESTIONS

1. How can two organisms with the same genotype for a trait have different phenotypes for that trait?

2. How can two organisms with different genotypes for a trait have the same phenotype for that trait?

3. Define a species.

4. Select two closely related sympatric organisms and describe some of the isolating mechanisms that might keep them from interbreeding.

5. Excluding the large-eyed/small-eyed fish example used in this text, develop a hypothetical organism and a set of environmental conditions that would lead to some specific scenario of natural selection in that organism.

6. Think of the conditions that a particular organism would experience in a human-controlled artificial environment, then list a dozen ways in which that artificial environment could artificially select for particular genetic traits.

7. Why is inbreeding a problem even after a population founded by a small number of individuals grows large?

8. What is introgression?

9. Describe a hypothetical mutation that would be beneficial to an organism, then describe how that characteristic could lead to the formation of a new species.

10. Why and for what reasons could a variety of different people be supportive of or negative toward genetic engineering?

SELECTED READINGS

Allendorf, F. W., and M. M. Ferguson. 1990. Genetics. Pages 35–63 *in* C. B. Schreck and P. B. Moyle, eds. *Methods for fish biology.* American Fisheries Society, Bethesda, Md.

Gill, F. B. 1995. *Ornithology.* 2d ed. W. H. Freeman, New York.

Kapuscinski, A. R., and L. D. Jacobson. 1987. *Genetic guidelines for fisheries management.* Minnesota Sea Grant, University of Minnesota, St. Paul.

Lacy, R. C. 1987. Loss of genetic diversity from managed populations: Interacting effects of drift, mutation, immigration, selection, and population subdivision. *Conservation Biology* 1:143–158.

Ryman, N., and F. Utter, eds. 1987. *Population genetics and fishery management.* Washington Sea Grant Program, University of Washington Press, Seattle.

Watson, J. D., J. Witkowski, M. Gilman, and M. Zoller. 1992. *Recombinant DNA.* 2d ed. W. H. Freeman, New York.

LITERATURE CITED

Allendorf, F. W., and M. M. Ferguson. 1990. Genetics. Pages 35–63 *in* C. B. Schreck and P. B. Moyle, eds. *Methods for fish biology.* American Fisheries Society, Bethesda, Md.

Karl, S. A., and J. C. Avise. 1992. Balancing selection at allozyme loci in oysters: Implications from nuclear RFLPs. *Science* 256:100–102.

Kapuscinski, A. R., and L. D. Jacobson. 1987. *Genetic guidelines for fisheries management.* Minnesota Sea Grant, University of Minnesota, St. Paul.

Mitchell, R. W., W. H. Russell, and W. R. Elliott. 1977. *Mexican eyeless characin fishes, genus Astyanax: Environment, distribution, and evolution.* Special Publication 12. Texas Tech Museum, Lubbock.

Piper, R. G., I. B. McElwain, L. E. Orme, J. P. McCraren, L. G. Fowler, and J. R. Leonard. 1982. *Fish hatchery management.* U.S. Department of the Interior, Fish and Wildlife Service, Washington, D.C.

Vrijenhoek, R. C., E. Pfeiler, and J. D. Wetherington. 1992. Balancing selection in a desert stream-dwelling fish, *Poeciliopsis monacha. Evolution* 46: 1642–1657.

5

NUTRITION AND ENVIRONMENTAL PHYSIOLOGY

The ways in which organisms obtain nutrients and the ways in which they are adapted to their particular environments are important aspects of animal and plant populations and communities. The nutritional needs of wild animals are met primarily through feeding patterns that have evolved to promote the survival and reproduction of each species. Animals must obtain an array of mineral elements, vitamins, proteins, fats, and carbohydrates in their diets. While broad similarities in nutritional needs exist among animals in the same guild, there are interesting differences among species; differences also often exist within species by season, sex, and age. In exploiting available food sources, vertebrate species have evolved various structural and functional differences in their digestive tracts that allow the processing of widely differing food materials. Many animals, such as ungulates, even enlist the help of an array of bacteria, protozoans, and other microbes in extracting needed nutrients from their diets.

The physiology of a species determines the environment in which it can survive and reproduce. Endotherms and ectotherms live in environments that are both static and changeable. Organisms must be able to adapt to a variety of conditions if they are to be successful. For example, aquatic ectotherms are adapted to an environment with low oxygen concentrations. Aquatic ectotherms and some endotherms must be able to adapt to changing salinities. All organisms must be able to adapt to changing temperatures, but such adaptations are most developed in endotherms because they must maintain their body temperatures within a very narrow range.

A single chapter on topics as extensive as nutrition and environmental physiology cannot be all-inclusive or in-depth. Nevertheless, because of their importance to understanding how organisms function, nutritional and physiological comparisons can and should be made between and among different animals. The objectives of this chapter are to provide an overview of the nutritional needs of fish and wildlife species and to discuss some selected physiological adaptations of these species to their environment.

5.1 Minerals and Other Elements

Both fish and wildlife species need a number of inorganic elements in large amounts (**macroelements**), as well as several that are needed in only trace amounts (**trace elements** or **microelements**). Several additional elements are required by captive animals in sterile environments and on purified diets; these **ultratrace elements** are not known to be lacking under natural conditions (Robbins 1993). Macroelements, microelements, and ultratrace elements occur in soil and water as minerals (table 5.1). Other elements that are needed in large quantities but are not listed with the macroelements occurring as minerals include oxygen, hydrogen, carbon, and nitrogen. The seven macroelements—calcium, phosphorus, sodium, potassium, magnesium, chloride, and sulfur—constitute approximately 4% of the body weight of vertebrates, while the microelements contribute less than 1%. Oxygen, carbon, hydrogen, and nitrogen constitute about 96% of the body weight of vertebrates.

Calcium is involved with activities such as muscle contraction, nerve impulse transmission, bone formation, eggshell formation in birds, blood clotting, and acid-base balance. As much as 99% of the calcium in the body may be found in the **hydroxyapatite matrix** (mineral matrix of calcium phosphate) of which vertebrate bones and teeth are primarily composed (Kincaid 1988; Schmidt-Nielsen 1990). The growing young of fish and wildlife species need high levels of calcium. During egg laying in birds, the intestine increases its absorption of calcium; this is reflected by increased levels of calcium in the blood and increased calcium deposition in the bones. The female uses this additional calcium from her diet and some of her stored bone calcium for eggshell formation. The eggshell is about 94% calcium carbonate. Calcium is also mobilized from bones during **lactation** (milk secretion) in mammals.

TABLE 5.1 Macroelements, trace elements (microelements), and ultra-trace elements that occur as inorganic minerals are needed by animals.

Macroelements	Trace Elements	Ultratrace Elements[a]
Calcium (Ca)	Iron (Fe)	Silicon (Si)
Phosphorus (P)	Zinc (Zn)	Tin (Sn)
Sodium (Na)	Manganese (Mn)	Boron (B)
Potassium (K)	Copper (Cu)	Bromine (Br)
Magnesium (Mg)	Molybdenum (Mo)	Cadmium (Cd)
Chloride (Cl)	Iodine (I)	Lead (Pb)
Sulfur (S)	Selenium (Se)	Lithium (Li)
	Cobalt (Co)	Vanadium (V)
	Fluorine (F)	Nickel (Ni)
	Chromium (Cr)	Arsenic (As)

Note: Oxygen (O), hydrogen (H), carbon (C), and nitrogen (N) are usually grouped separately as nonmineral elements or organic constituents and, although they are required in large quantities by animals, are not listed with as macroelements.

[a] Captive animals in sterile environments and on diets completely lacking these elements may show deficiencies. For practical purposes, these elements are not deficient in the diets of wild animals.

Antlers contain about 22% calcium and create a substantial calcium demand in many ungulates during the late spring and early summer when new antler growth occurs (Robbins 1993).

Calcium is obtained through the diet, but wild animals sometimes encounter deficiencies. Young carnivores must consume adequate amounts of bone in the diet because animal flesh generally does not provide adequate amounts of calcium. Various bird species, from ring-necked pheasants to shorebirds, ingest tidbits of calcium-containing items such as snails, snail shells, and small pieces of bone. **Raptors** (predatory birds such as hawks and owls) and their young may at times ingest rather large bones, and rodents may chew on bones or shed antlers. Fishes often obtain calcium and several other elements through the gills as well as obtaining them in food.

Phosphorus is involved in a wide array of metabolic activities, including muscle contraction, nerve metabolism, and metabolism of fats, carbohydrates, and amino acids. Phosphorus is extremely important as a mineral element in combination with calcium in the hydroxyapatite matrix of bone (Hays and Swenson 1993). Bone includes approximately 360 grams of calcium, 170 grams of phosphorus, and 10 grams of magnesium per kilogram (McDonald et al. 1981). Phosphorus is also used in small amounts in eggshell formation and composes about 11% of the weight of antlers (Robbins 1993). Like calcium, additional phosphorus can be obtained in the diet through ingestion of bone.

Sodium is the primary **cation** (positively charged ion) needed for maintaining the cells of vertebrates in a saline fluid environment. It is also important in nerve impulse transmission and for general growth and reproduction. Carnivorous aquatic and terrestrial vertebrates generally obtain adequate sodium in the tissues of their prey. Herbivorous vertebrates may have difficulty

FIGURE 5.1 White-tailed deer and mule deer may seek soils high in sodium where they develop mineral licks by ingesting or licking the soil, as at this site in the Black Hills in eastern Wyoming. (*Photograph courtesy of J. Kennedy.*)

obtaining sodium because it is not essential to plants. Plants that do contain high sodium levels, such as submergent vegetation, may be sought as sodium sources by herbivores such as the moose. Ungulates such as white-tailed deer and elk may seek sites with high sodium levels and ingest sodium-rich soil to satisfy their needs (fig. 5.1). Ingestion of soil by animals to obtain minerals is termed **geophagia**. Such mineral sites may be used over long periods of time by various ungulate species. African elephants in Wankie National Park in Zimbabwe excavate scrapes and pits in areas with unusually high concentrations of soluble sodium (Weir 1969). Termite mounds contain elevated concentrations of soluble sodium and may also be used as licks by African elephants in areas where sodium concentrations cannot be located near the ground surface. Salt, usually in block form, can be placed on the ground for use by domesticated livestock and may also be used by wild ungulates. Some natural licks may actually represent remnant salts in the soils around old salt block sites.

Both potassium and magnesium are readily available in plant and animal food sources. Potassium is critical for processes such as nerve impulse transmission, muscle contraction, maintenance of cell osmotic pressures, body metabolism, and enzyme activation. Potassium is normally available in adequate amounts in vegetation or in ingested animal tissue; deficiencies are rare in wild animals. High levels of potassium in ingested vegetation can reduce the retention of other elements such as sodium or magnesium and thus contribute

to the deficiency of another mineral (Robbins 1993).

Magnesium is critical to functions such as enzyme activation and energy metabolism and is also needed for bone and tooth formation; about 70% of the magnesium found in animals occurs in bone (Hays and Swenson 1993). Magnesium is present in ample amounts in plant and animal tissue, and deficiencies are rare in wild animals.

Sulfur is needed for the production of sulfur-containing amino acids, while chlorine functions as the primary **anion** (negatively charged ion) in the blood and body fluids. Chlorine is also important in maintaining body pH and in hydrochloric acid formation in the stomach. Chlorine is obtained by fishes, in large part, through the gills. All of these macroelements must be obtained in relatively large amounts; in some cases feeding preferences have been related to animals seeking needed macroelements. Table 5.2 describes some of the macroelement requirements in prepared diets of captive animals.

Trace elements are used in lesser amounts than macroelements, but are still necessary to normal body functions in animals. Geophagia observed in wildlife may be related to trace element requirements. Though these elements are needed in only small amounts by wildlife and fish species, they do not represent elements of only minor importance. For instance, iron and manganese are both trace elements, but they are essential to animals, the former in red blood cells and the latter for bone formation and energy metabolism as well as other functions. Iodine, as in humans, must be present

TABLE 5.2 Nutrient requirements for selected macroelements in captive animals, expressed as a percentage of dry matter in the diet.

Macroelement	Species	Life Stage	% of Dry Diet
Calcium	Ring-necked pheasants; northern bobwhites	Growth	0.5–1.0
	Ring-necked pheasants; northern bobwhites	Breeding female	2.5
	Mallards	Growth	0.6
	Mallards	Breeding	2.8
	Mink; red foxes		0.3–0.6
	Common carp		0.3
	Atlantic salmon		None required in seawater
Phosphorus	Ring-necked pheasants; northern bobwhites; mallards		0.3–0.7
	Mink; red foxes	All stages, but needs highest at lactation	0.3–0.6
	Rainbow trout		0.5–0.8
Sodium	Ring-necked pheasants; northern bobwhites		0.15
Salt (NaCl)	Mink; red foxes		0.5
Potassium	Japanese quail		0.4
	Chinook salmon	Juvenile in fresh water	0.6–0.8
Magnesium	Ring-necked pheasants; northern bobwhites; mallards		0.03–0.05
	Guinea pigs	Growth	0.06
	Rainbow trout; common carp; channel catfish		0.04–0.06

Sources: National Research Council 1982, 1993, 1994; Robbins 1993.
Note: Mineral requirements in fishes are difficult to study because in addition to those obtained in the diet, several elements can be taken in through the gills, through the skin, or by drinking water.

for normal thyroid gland function and growth. A lack of selenium, though it is toxic in high amounts, will cause muscle degeneration and labored breathing. Even the lack of fluoride and chromium can cause reduced growth and other problems. Most of the trace elements needed by fish and wildlife species are found in sufficient quantities in natural diets, and deficiencies are uncommon except in captive situations (Robbins 1993).

5.2 Vitamins

Vitamins are organic substances required by animals for normal body functions. Vitamins are of two types, water soluble and fat soluble. Fat-soluble vitamins are often stored in large quantities in the bodies of animals; these stores can be used when dietary amounts are inadequate. Water-soluble vitamins, with some exceptions, are generally not stored in large amounts in the body and must be consistently available in the diet. The fat-soluble vitamins include vitamins A, D, E, and K, while the water-soluble vitamins include thiamin (B_1), riboflavin (B_2), nicotinic acid or niacin, pyridoxine (B_6), pantothenic acid, folic acid, cyanocobalamin (B_{12}), choline, biotin, and ascorbic acid (C). The vitamin needs of fishes have been largely discovered through feeding experiments with various species of salmon and trout, channel catfish, common carp, and other cultured fishes. Likewise, the vitamin requirements of wildlife species have largely been discovered from studies and observations on a variety of captive animals.

In general, the small quantities of vitamins needed are found in adequate proportions in the diets of wild animals. Problems occur when fishes or wildlife are reared in captivity if needed vitamins are inadequate in prepared diets. Vitamins can also be lost from prepared diets during feed

production or if the feed is stored too long. Some fat-soluble vitamins, such as A and D, can be highly toxic when excessive amounts are included in the diet. For example, the livers of some carnivores, such as polar bears and various seal species, accumulate large quantities of vitamin A and can be toxic if fed to other animals. At times some animals eat feces (**coprophagia**). In the case of some species of rabbits, this is done to obtain and conserve vitamins produced by bacteria in the rabbits' digestive tract.

Vitamins perform a variety of functions in animal species (table 5.3). Lack of various vitamins can cause serious consequences, such as disturbances in the normal calcification of growing bones, a lack of appetite and growth, nervous disorders, hair loss, difficulty in blood clotting, gill disease, anemia, hemorrhagic kidneys, or even sudden death (Robbins 1993). Many of the effects of vitamin deficiencies in hatchery-reared fishes and captive-reared wildlife were thought to be caused by parasites, diseases, or other factors until the role of vitamins in diets was demonstrated.

5.3 Amino Acids

Amino acids are the building blocks of proteins. They can be obtained by animals in two ways. Amino acids that an animal can manufacture within its own body are called **nonessential amino acids**. Others, called **essential amino acids**, must be obtained in the diet because the animal either cannot manufacture them, or the quantities it can manufacture are insufficient to provide for its needs. Do not be confused by these terms: all of these amino acids can be "essential" to an animal. The essential versus nonessential designation refers to how they are generally obtained.

Most fishes and terrestrial vertebrates with simple stomachs have similar lists of essential amino

TABLE 5.3 Vitamins and the major functions of each.

Vitamin	Known Functions
FAT SOLUBLE	
A	A constituent of visual pigment (rhodopsin); maintenance and growth of epithelial tissue; synthesis of glycoprotein; bone and tooth development
D	Calcium absorption from intestine and resorption from bone; calcium metabolism; promotion of bone growth
E	Maintains functional integrity of cell membranes; prevents breakdown of vitamins A and C in gut
K	Electron transport role in body energetics; needed in formation of blood clots
WATER SOLUBLE	
Thiamin (B_1)	Formation of connective tissue; needed in use of iron; coenzyme in carbohydrate metabolism
Riboflavin (B_2)	Functions with coenzymes flavin adenine dinucleotide (FAD) and flavin mononucleotide (FMN) in breakdown of glucose and other carbohydrates
Niacin (nicotinic acid)	Functions with coenzyme nicotinamide adenine dinucleotide (NAD^+) and nicotinamide adenine dinucleotide phosphate ($NADP^+$) in breakdown of glucose and other carbohydrates
Pyridoxine (B_6)	Coenzyme role in protein metabolism
Pantothenic acid	Coenzyme role in carbohydrate, fat, and amino acid metabolism
Folic acid	Coenzyme role in amino acid and nucleic acid metabolism
Cyanocobalamin (B_{12})	Coenzyme role in carbohydrate metabolism; also important in nucleic acid metabolism
Biotin	Coenzyme role in fat and glycogen formation, and in amino acid metabolism
Choline	Needed in nervous system (component of acetylcholine); component of phospholipids in cells and tissues
C (ascorbic acid)	Needed in bone, teeth, and collagen formation and carbohydrate metabolism; may reduce infections

Sources: Adapted from Starr and Taggart 1992; Robbins 1993.

acids, which include arginine, histidine, isoleucine, leucine, lysine, methionine, phenylalanine, threonine, tryptophan, and valine. Nonessential amino acids for most fishes and other vertebrates include alanine, aspartic acid, citrulline, cystine, glutamic acid, glycine, norleucine, proline, serine, and tyrosine. Essential amino acid requirements are reduced or eliminated in **ruminants**, animals such as mountain goats and white-tailed deer that have four-chambered stomachs in which microbial populations break down plant matter (see section 5.5). Essential amino acid requirements are also reduced or eliminated in blue grouse, spruce grouse, African elephants, and other birds and mammals with well-developed **gastric caeca** (see fig. 5.11). These gastric caeca are outpocketings of the intestine in which microbial populations break down complex plant matter and improve digestive efficiency. The bacteria involved in this process in ruminants and animals with gastric caeca are themselves digested, providing an important source of amino acids.

5.4 Proteins, Fats, and Carbohydrates

Body metabolism can be divided into **anabolic processes**, which build up tissue, maintain body tissue, and store energy, and **catabolic processes**, which break down substances and release energy. Fish and wildlife species get their energy from proteins, fats, and carbohydrates in the diet. These food sources undergo chemical breakdown (are **catabolized**) through the various processes in the tricarboxylic acid cycle (Krebs or citric acid cycle) to produce the energy needed for growth, body heat, and muscular contraction (Frandson and Spurgeon 1992). Carbohydrates obtained in the diet include simple sugars such as glucose (a **monosaccharide**) as well as more complex molecules such as **oligosaccharides** (a short chain of two or more simple sugars) or **polysaccharides** (a straight or branched chain of hundreds or thousands of sugar molecules). Glucose is the primary carbohydrate found in the blood and other tissue fluids, and can be readily obtained from plants or from the breakdown of other carbohydrates.

When blood glucose and other carbohydrates are in short supply, as during fasting, glucose can be synthesized from noncarbohydrate sources, such as fats and proteins, in a process occurring primarily in the liver that is called **gluconeogenesis**. Use of proteins in gluconeogenesis requires catabolism of amino acids and **deamination** (removal of an amino group [NH_2] from the amino acid); deaminized amino acids can also be used directly for energy without being converted to glucose. Fats represent a stored form of potential energy and can be broken down by hydrolysis (a catabolic process known as **ketogenesis**) to enter the tricarboxylic acid cycle

Protein is more important for building body tissues, particularly muscle, than as an energy source. The protein component of bone, including antlers, is substantial at all stages of life, but particularly during growth. Vertebrates catabolize ingested proteins into amino acids and **peptide groups** (combinations of two or more amino acids) and later **anabolize** (construct) proteins as needed by the animal. In many cases, the most efficient manner of obtaining needed proteins would be to eat other animals of the same species, something many fish species do on a regular basis. Even a rogue muskrat, normally an herbivore, obtains added amino acid benefits from eating its own kind.

Protein can be obtained from vertebrate muscle tissue, invertebrates, and new vegetation growth. Fats and carbohydrates are normally ingested in sufficient amounts by carnivores, which tend to eat the entire prey organism, including fat deposits. It should be noted, however, that animals contain a low proportion of carbohydrates compared with plants. Herbivores and omnivores may seek plant material particularly high in fat or carbohydrates, such as seeds and fruits.

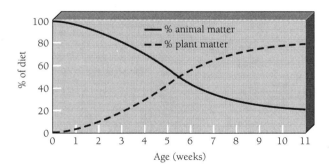

FIGURE 5.2 Percentages of animal matter (primarily invertebrates) and plant matter in the diets of the young of gallinaceous birds and ducks at different ages.

Protein requirements vary among vertebrates by species and within species by age, sex, and season of the year. Vertebrates feeding on the flesh of other animals consume high levels of protein during all adult life stages. In terrestrial carnivores, protein requirements for maintenance in adults often exceed 25% of the diet. Carnivores, unlike herbivores and most omnivores, are adapted to obtain most of their energy requirements from proteins. Carnivores have enzymes in their livers that are used to deaminate proteins and that have higher activity levels than similar enzymes found in herbivores and omnivores. In comparison, ruminants have protein requirements of 5%–9% of the diet; their protein requirements during growth are commonly 13%–20%, but are still less than those of adult or juvenile carnivores (Robbins 1993). As noted earlier, the ruminants, unlike carnivores, supplement their amino acid intake with proteins from the bodies of microorganisms in their complex stomachs.

Fishes exhibit great species differences in the amounts of proteins, fats, and carbohydrates that they need or can tolerate in the diet. Carbohydrate needs in most fishes are generally quite low. If trout or salmon in captivity are fed a high-fat diet, they develop fat-infiltrated livers, while common carp thrive on the same diet (Lagler et al. 1977). In general, protein levels below 6% of the diet are too low to sustain even herbivorous fishes. Many piscivorous fishes ingest protein levels exceeding 40% of the diet, depending on water temperature and the physiological needs of the species.

Young birds and mammals can be expected to require higher percentages of protein in the diet than at most times later in the life cycle (fig. 5.2). American wigeon ducklings require large amounts of invertebrates in the diet, particularly in the first few weeks of life. The young of greater white-fronted geese graze on new-growth sedges and other Arctic plants which are better sources of protein than mature plants. Young mammals such as nursing Ord's kangaroo rats or desert cottontails obtain their protein needs from the lactating female. In fishes that are herbivorous as adults, such as grass carp, the young initially feed on zooplankton, then switch to herbivory as they grow. The protein needs of young white-tailed deer are higher than those of adults, even after weaning, because they are growing rapidly.

Reproduction in fishes and terrestrial vertebrates is often accompanied by increased protein demands in the females. In mammals, **gestation** (growth of the embryo in the uterus) and lactation require extra dietary protein. Likewise, the production of bird eggs requires considerable parental investment in terms of proteins and fats in the yolk. Several species of waterfowl and gallinaceous birds will attempt to renest if the original **clutch** (group of eggs) is destroyed or otherwise terminated. Each clutch may weigh an appreciable amount, in some cases more than the female, so the costs to the female in proteins, fats, and calcium (for the eggshell) are enormous (fig. 5.3). Additionally, the hen must stay on the nest for prolonged periods of time during incubation, which limits her opportunities to feed.

In ducks such as northern pintails, blue-winged teals, and gadwalls, the females include an unusually high percentage of invertebrates in the diet while preparing to nest and during egg laying (Krapu and Reinecke 1992). Protein and energy demands are met in large part by feeding intensively during egg development. During mid- to late incubation, the hen remains on the nest most

FIGURE 5.3 Egg production can be energetically demanding, particularly in species that produce large clutches like this one, laid by a mallard.

of the time and has few opportunities to feed; therefore some body protein may be needed to supplement the diet. Female ducks such as northern pintails, blue-winged teal, and gadwalls, also use stored body fat and carbohydrates during incubation. However, the resulting weight loss in these female ducks not as severe as in most Arctic-nesting geese.

Arctic-nesting geese, such as lesser snow geese, forage extensively on corn and other waste grains as well as winter wheat during their slow spring migration. Prior to the extensive development of agriculture, migrating lesser snow geese probably sought roots and other nutritious parts of native plants in wetlands and associated uplands. During this northward migration, lesser snow geese store large supplies of fats, carbohydrates, and proteins for use during the nesting season. On the Arctic nesting grounds, nutrient-rich food in the form of new-growth grasses and sedges is especially limited early in the year, and the reproductive season is short. Nesting must be initiated soon after arrival by the already mated females to allow time for incubation and rearing of the young to flight stage before fall migration.

If the year is colder than normal, snow may cover potential nesting sites, and many lesser snow goose pairs may not attempt to nest that year. The short season does not allow adequate time for renesting if initial nesting fails. Because avian predators are quick to take advantage of un-protected clutches of lesser snow goose eggs, successful females seldom leave the nest, so most of their protein, as well as energy, must come from body stores (fig. 5.4). The female uses protein from her breast and leg muscles as well as from internal organs such as the gizzard and the once enlarged oviduct. After an incubation period of about 23 days, the female lesser snow goose has lost over 25% of her body weight, and may approach a point (about 40% weight loss) at which further weight loss could mean death (Ankney and MacInnes 1978). A small number actually die on

FIGURE 5.4 Female lesser snow geese remain on the nest throughout incubation with only brief periods of feeding. During this period, they generally lose 25% or more of their body weight. *(Photograph courtesy of D. Ankney.)*

the nest, so precipitous is this contest between time and body condition.

Even the male lesser snow goose, while guarding the female from other males, is required to use more protein and energy than he can replace by feeding during the nesting period (Ankney 1977). However, weight loss by paired males is not nearly as substantial as that of nesting females, and the males are able to accept increased parenting duties after the young hatch. Females are in an emaciated condition by that time and enter into a period of **hyperphagy** (rapid and lengthy feeding) to replenish the lost supply of proteins, fats, and carbohydrates.

The migration of Pacific salmon could be viewed as being similar to the ordeal of some of the Arctic-nesting geese, except that these parents die soon after reproduction. In the months before they enter spawning streams, Pacific salmon undergo a period of hyperphagy. While still in the ocean, the female needs protein and energy to form the large egg masses. These fishes undertake a long, arduous upstream journey that requires a large amount of energy. In sockeye salmon making spawning migrations of 596–1,023 kilometers, females depleted over 90% of their fat reserves and 55%–61% of their body protein by the time of death; male loss of fat and protein was also severe, but not as great as that in females (Burgner 1991). The change from a saltwater to a freshwater environment is an added stress to migrating salmon. Because Pacific salmon species, while in the ocean, feed on fishes or invertebrates, they naturally have a high-protein diet in comparison to herbivores.

Death of the parents immediately after reproduction is a successful reproductive strategy in Pacific salmon because the eggs do not require post-spawning care; it would not work in lesser snow geese. In a sense, however, the dead Pacific salmon continue to feed their young. By dying in the home stream, they provide a nutrient source that enriches the stream and eventually feeds invertebrates and other stream organisms, which then in turn may be eaten by the young salmon.

FIGURE 5.5 Pacific salmon migrate up freshwater rivers and streams, then spawn and die. The nutrients from adult salmon enrich the stream and thus indirectly serve to nourish the young fish. *(Photograph courtesy of R. Labisky.)*

Even those Pacific salmon eaten by brown bears have carried nutrients upstream that will eventually enter the stream and nourish young salmon (fig. 5.5).

Nutritional needs and body condition can vary greatly through the seasons of the year in white-tailed deer. Forage with adequate protein and energy content is available to the deer from mid-spring or early summer through early fall in most areas. White-tailed deer feed heavily during summer and fall to increase their body fat and nutrient reserves. Energy demands due to factors such as lactation, growth, and **rut** (the period associated with male breeding activity in some ungulates) delay or reduce the accumulation of fat and nutrient reserves. These reserves are essential because of the poor quality (low protein content) of the food usually available during the winter months. For white-tailed deer does, the winter season is also the time when protein and energy is needed for gestation. Individuals that cannot accumulate adequate reserves are faced with a greater risk of death during the winter (fig. 5.6) (Mautz 1978).

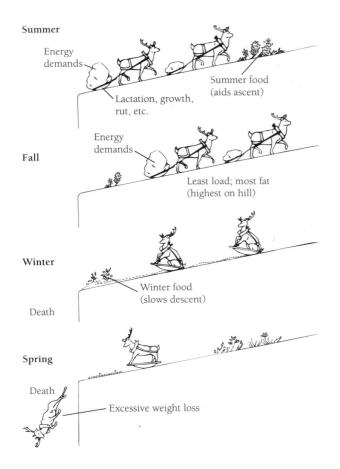

Summer

Energy demands

Lactation, growth, rut, etc.

Summer food (aids ascent)

Fall

Energy demands

Least load; most fat (highest on hill)

Winter

Death

Winter food (slows descent)

Spring

Death

Excessive weight loss

FIGURE 5.6 An analogy of the yearly body fat cycle in deer. The uphill climb represents fat accumulation; the downhill slide represents fat catabolism or weight loss. *(Adapted from Mautz 1978.)*

tween body condition and reproduction is also quite clear. Ungulates that are in good condition are more likely to become pregnant at a younger age and to have multiple births. White-tailed deer with diets high in protein and energy, such as those in the agricultural regions of the upper Midwest, reproduce at a younger age and have more twins and triplets than do populations in nutrient-poor environments; for example, in nutrient-rich portions of the northern Great Plains approximately 70% of the white-tailed deer does born in late spring become pregnant in the following fall and have fawns the next spring, and 33% of those does may carry two fetuses (Petersen 1984) (fig. 5.7). Carnivores, such as the lynx, have larger litters in years when prey is abundant than during years of food scarcity. A number of bird species respond to improved diets with more eggs, multiple clutches, and more young **fledged** (reaching flight stage).

As a general principle, nutritionally fit animals are much more likely to be reproductively successful. In fishes, nutrient-poor environments produce smaller individuals and lower egg numbers and quality, thus decreasing reproductive success and offspring survival. In terrestrial vertebrates, especially mammals, the relationship be-

FIGURE 5.7 White-tailed deer does in good nutritional condition often have twins or triplets. *(Photograph courtesy of D. Linde.)*

5.5   The Digestive Tract

The digestive tract of fishes provides the basic design found throughout the vertebrates. The system includes a mouth, oral cavity (mouth cavity), **pharynx** (between the oral cavity and the esophagus), **esophagus** (connecting the pharynx to the stomach), stomach, small intestine, large intestine, and anus.

In fishes, the mouth cavity, pharynx, esophagus, and stomach are lined with a mucous membrane containing glands that secrete mucus that lubricates food items being swallowed. Fishes lack salivary glands, with a few exceptions such as the parasitic lampreys (Lagler et al. 1977). The digestive tract in general is highly elastic and allows the passage of large food items. If a prey item can fit into the mouth of a fish, it will rarely be too large for the digestive tract. Occasionally, however, a fish with spines, such as the black bullhead, can become lodged in the anterior part of the digestive tract of a predatory fish.

Birds, like fishes, with the exception of many aquatic bird species, have glands that secrete mucus in the mouth cavity, pharyngeal region, and esophagus. As in fishes, the mucus aids the passage of food items to the stomach. Birds generally have salivary glands in the pharynx; these glands are particularly abundant in seed-eating birds and assist in initiating the breakdown of starches into sugars. Some birds, such as chimney swifts, also use salivary secretions to cement nesting material together.

Like fishes, many birds have a highly elastic pharynx and esophagus that allow the passage of large food items or the storage of food items in the esophagus. Piscivorous fishes and birds are sometimes found with large prey items extending from the mouth, but these seem to present no problem. In birds, the swallowing of large food items is made possible by a lower jaw that hinges with movable bones on the skull (quadrates), giving the bird wide gaping abilities. Reptiles, particularly snakes, are especially adept at gaping and have a mobile articulation of the jaw to the skull similar to that of birds. An eastern diamondback rattlesnake, for example, can capture and swallow an eastern cottontail that appears much too large for the snake to swallow. After the rabbit is swallowed, a lump that is larger than the normal diameter of the snake can be observed as the prey is slowly digested.

In many birds that eat seeds, such as mourning doves, the esophagus has an expanded area known as the **crop** where food can be stored until space is available in the stomach. In the case of the mourning dove and other species of doves and pigeons, the lining of the crop contains specialized cells that produce a rich milklike material (**pigeon milk**) that is fed to the young; both males and females have this capability.

Most mammals have abundant salivary and mucous glands in the mouth cavity and pharyngeal region, but lack mucous glands in the esophagus. Unlike that of fishes, reptiles, and birds, the jaw articulation of mammals does not allow wide gaping, so mammals generally do not swallow large prey whole. The mammalian esophagus also is not as distensible as that of fishes, reptiles, and birds.

The stomach in fishes is single chambered but can vary in shape from elongate to small or even be absent; in the latter case the esophagus connects directly to the intestine. In some fishes, such as the gizzard shad, the stomach is modified into a muscular **gizzard**. The lining of the gizzard is greatly thickened and muscularized for grinding food material. The gizzard lacks digestive glands. Stomach glands are common in other fishes, particularly those that are carnivorous. As in higher vertebrates such as birds and mammals, these glands secrete hydrochloric acid and pepsinogen, which help to catabolize protein molecules for digestion. Hydrochloric acid also is important in dissolving bones.

Birds have an anterior **glandular stomach (proventriculus)** and a posterior muscular stomach

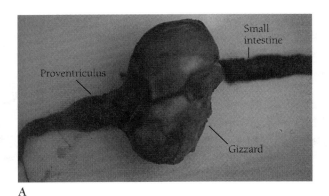

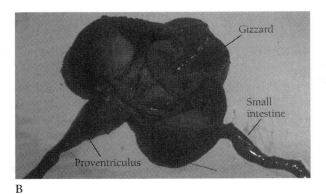

A

B

FIGURE 5.8 *(A)* The digestive tract of a lesser snow goose, showing the proventriculus, the muscular gizzard, and the exiting small intestine. *(B)* The hardened ridges and grinding surface inside the gizzard.

or gizzard that generally lacks glands (fig. 5.8). The birds with the best-developed glandular stomachs include predatory species such as American white pelicans, sharp-shinned hawks, and snowy owls. In these species the gizzard is weakly developed. Birds that eat foods that require crushing, such as seeds, snails, or small clams, have well-developed muscular gizzards. The lining of the gizzard in these birds has hardened ridges that assist in the grinding of food. Birds also eat grit (sand and small stones), which is deposited in the gizzard to assist in the grinding of seeds and other hard food objects. Some reptiles also have gizzards.

Although the enzymes associated with digestion are similar among vertebrates, mammalian stomachs vary with food type and can be more complex than those of other vertebrates. Many mammals, such as humans and northern grasshopper mice, have simple stomachs consisting of one chamber. Other mammals, particularly the ruminants, have complex stomachs consisting of multiple compartments. Ruminants are ungulates with an even number of toes (such as the two toes on each foot of a giraffe or mule deer) that regurgitate and chew their food repeatedly (chew their cud). Ruminants have stomachs that generally consist of four chambers, the **rumen**, **reticulum**,

omasum, and **abomasum** (fig. 5.9). Ingested plant material generally flows through the chambers in the order given above.

Plant material eaten by the animal enters the rumen, where it is moistened and mixed into a mass. It then passes into the reticulum, where the coarser material is formed into small cuds or masses that are regurgitated and rechewed while the animal rests. The animal reswallows the regurgitated material after rechewing, and it reenters the rumen for reprocessing. The finer plant materials and the digested materials from the rumen pass into the omasum. In the omasum the plant material is subjected to further churning before it is released into the glandular stomach, the abomasum, for digestion. Small fingerlike projections on the walls of the rumen, reticulum, and omasum increase the surface area for absorption of water and nutrients. The reticulum, as its name implies, has a reticulated (netlike) pattern of absorptive tissue on its inner surface.

The rumen is a small ecosystem that contains abundant bacteria, protozoans, fungi, and other microorganisms capable of living in an anaerobic environment. These microorganisms break down plant material through a fermentation process. The rumen is inoculated with microorganisms from sources such as ingested plant material, animal-

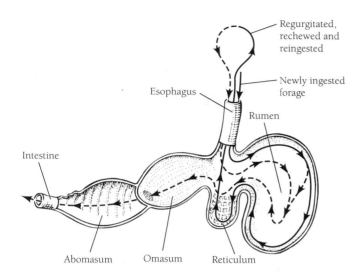

Esophagus

Intestine

Abomasum · Omasum · Reticulum

Regurgitated, rechewed and reingested

Newly ingested forage

Rumen

FIGURE 5.9 The stomach of a ruminant, illustrating the arrangement of the three nonglandular stomachs (rumen, reticulum, and omasum) and the glandular abomasum. The solid line indicates the flow of newly ingested material; the dashed line, the flow of most regurgitated, rechewed, and reingested material. Some forage may be regurgitated more than once.

to-animal contact, feces, and milk. In the young, inoculation of the rumen with microorganisms from the parent or through other animal-to-animal contact may be essential; this inoculation can occur through mechanisms such as exchange of cud or salivary transfer (Yokoyama and Johnson 1988). The microorganisms living in the rumen are involved in a symbiotic relationship with the ruminant. The ruminant provides the physiological conditions necessary for the microorganisms to survive and reproduce. In return, the microorganisms break down fibrous plant matter that the ruminant otherwise could not digest. The bodies of these microorganisms also serve as a protein source for the host ruminant.

Although the ruminant stomach is designed for the digestion of coarse food materials, the capacity of ruminants for digesting coarse or fibrous material varies considerably by species. Species that are less selective feeders, such as bighorn sheep and bison, must have highly advanced ruminant stomachs; these species are sometimes grouped as **grass/roughage eaters**. Other species, the **concentrate selectors**, have a simple rumen capable of handling only vegetation with low fiber concentrations; species in this group, such as the white-tailed deer, mule deer, and moose, are more selective in their food habits

(Hofmann 1988). Many other ruminants, such as pronghorns, caribou, and elk, are grouped as intermediate-type feeders and have a ruminant stomach intermediate between the simple and highly advanced forms (fig. 5.10). Mammals with ruminant stomachs are sometimes referred to as **foregut fermenters**.

Many vertebrate species have caeca, or outpocketings of the digestive system, that are also involved in digestion. Most fishes have **pyloric caeca** that occur at the junction of the stomach and intestine; these may number from one to more than one hundred. They probably function in both digestion and absorption. Many reptiles have a caecum arising at the junction of the small and large intestine. Birds with gastric caeca have only two, which are located at the junction of the large and small intestine. The length and development of the caeca are highly variable, with the most highly developed caeca rivaling the length of the intestine. Gastric caeca are well developed in birds that eat coarse plant material, such as lesser snow geese, spruce grouse, and blue grouse (fig. 5.11). Most mammals have a single caecum at the junction of the large and small intestine. Even ruminants have a caecum. The caecum is highly developed in some of the large herbivores that lack ruminant stomachs. For example, the white

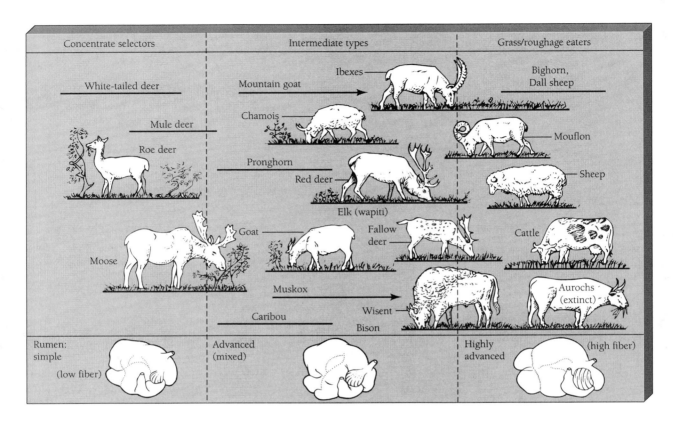

FIGURE 5.10   Positions of European and North American ruminant species within the system of feeding types. The farther the baseline for a species extends to the right, the greater its ability to digest fiber in the rumen. An arrow on the baseline indicates that it extends farther to the right. (*Adapted from Hofmann 1988.*)

rhinoceros, Asiatic elephant, and warthog lack ruminant stomachs; microbial fermentation of fibrous plant material in these species occurs in the greatly enlarged caecum and in the large intestine (Owen-Smith 1988). Species in which fermentation occurs primarily in the gastric caecum and intestine are sometimes grouped as **hindgut fermenters**.

Among endotherms and ectotherms, primary consumers tend to ingest a higher percentage of their body weight daily than do organisms at higher trophic levels. This is particularly true of organisms that require a diet of vegetable matter other than high-energy seeds. Therefore, a grass carp or a sage grouse, both herbivores, could be expected to ingest a greater percentage of their body weight

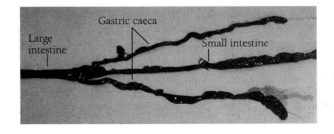

FIGURE 5.11   The intestine and gastric caeca of a lesser snow goose. The caeca branch anteriorly from the intestine at the junction of the small intestine and large intestine.

daily than a white shark or a bobcat. Vertebrate species that feed on vegetation, particularly green or leafy matter, tend to have much longer small intestines than do predators; this allows for more efficient processing of ingested plant material. Therefore, one would expect a predatory Pacific barracuda to have a shorter intestine in relation to its body length than a grass carp. In general, vegetable matter is much more difficult to digest than animal matter.

Some generalizations can be made concerning the digestive secretions and hormones released into the small intestine in vertebrates. Digestive secretions from the pancreas and liver are released into the anterior portion of the small intestine. The pancreas releases enzymes important in catabolizing carbohydrates, fats, and proteins in the intestine. The pancreas also releases a bicarbonate that helps to buffer the hydrochloric acid reaching the intestine from the stomach. Specialized cells in the pancreas (**alpha cells**) also produce **glucagon**, a hormone that causes stored polysaccharides and amino acids in the liver to be converted to glucose, thus increasing blood glucose levels. When blood glucose levels are high, **beta cells** in the pancreas produce **insulin**, a hormone that stimulates the liver, muscles, and fat tissues to take up glucose, and thus reduces blood glucose levels. The liver releases bile into the small intestine through the bile duct that helps to **emulsify** fat (to finely divide globules of fat), to break down the fat with enzymes, and to encourage absorption of fat into the blood.

5.6 Body Temperature, Metabolic Rate, and Body Size

Most ectotherms, including fishes, have body temperatures that are close to the ambient temperature, and their temperatures change with the ambient temperature. Endotherms, in contrast, maintain nearly constant body temperature levels. Small changes in internal temperatures that would cause severe stress or death to most bird and mammal species can be easily tolerated by ectotherms. Most ectotherms, however, are not as well adapted to withstand wide ranges of or rapid changes in ambient temperature as are endotherms.

The relatively slower and generally narrower temperature changes in aquatic environments as compared with terrestrial ones ideally fit the needs of aquatic ectotherms (see section 1.3). The geographic ranges of ectotherms are often limited by high or low temperatures that are lethal to particular species. For example, threadfin shad cannot exist in waters that become cold enough to allow ice formation; conversely, Arctic char cannot survive in the warm waters found in an Oklahoma reservoir during midsummer. Freshwater fish species are commonly categorized as **warmwater**, **coolwater**, or **coldwater species** based on the general optimal water temperatures at which they survive, grow, and reproduce. Common warmwater species include largemouth bass, bluegills, striped bass, channel catfish, red shiners, and brook silversides. Common coolwater fishes include walleyes, yellow perch, northern pike, and brook sticklebacks. Common coldwater fishes include lake trout, brook trout, coho salmon, Arctic graylings, mountain whitefish, and spoonhead sculpins.

Ectotherms sometimes use behavioral methods to warm their bodies above the ambient temperature. A western fence lizard, for instance, can warm itself by using a warm rock and the heat-absorptive capacity of its body. Some fishes, such as the albacore, can maintain body temperatures considerably above the ambient temperature by physiological mechanisms (Carey 1982). Vigorous swimming activity generates metabolic heat, and these fishes' systems of blood vessels keep much of that heat from reaching the body surface, where it would be lost. These fishes also have a special hemoglobin that, unlike those of most other fishes, increases its oxygen-holding capacity with warming (Cech et al. 1984). Fishes that maintain their body temperatures above the ambient

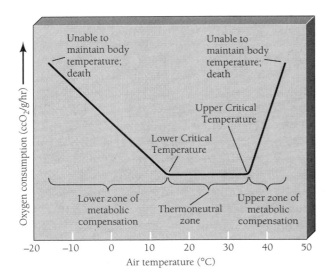

FIGURE 5.12 The relationship between energy used (as measured by oxygen consumption) by a hypothetical bird in maintaining body temperature homeostasis and the ambient air temperature. The thermoneutral zone is the temperature range within which no extra work is required to maintain the body temperature; the upper and lower critical temperatures are the points at which oxygen consumption begins to increase.

temperature in this manner are sometimes grouped as **heterotherms**.

Endotherms can survive great variations in ambient temperature, but must maintain their body temperatures within a narrow range. **Homeostasis** is the tendency to maintain relatively stable conditions within the body. To maintain body temperature homeostasis, endotherms must expend energy. This energy expenditure is often accompanied by increased oxygen consumption. Within a certain range of ambient temperatures, called the **thermoneutral zone**, no additional oxygen beyond that needed for normal body metabolism is required to maintain body temperature (fig. 5.12). At ambient temperatures above or below the thermoneutral zone, below the **lower critical temperature** or above the **upper critical temperature**, the animal must increase its oxygen consumption to maintain temperature homeostasis. The temperature zones within which the animal must expend additional energy to maintain homeostasis are termed the **upper** and **lower zones of metabolic compensation**. The lower zone of metabolic compensation represents a wider range of ambient temperatures than does the narrower upper zone of metabolic compensation for almost all species. At some point on the temperature scale, at both low and high tempera-

tures, the animal is no longer able to maintain its body temperature, and it dies.

The breadth of the thermoneutral zone is highly variable among species and reflects their adaptation to a particular environment. A snowy owl in winter has a broad thermoneutral zone with the lower critical temperature set at a considerably colder temperature than that for the migratory ring-billed gull; the upper critical temperature is probably also lower in the snowy owl, but not much lower than that of the ring-billed gull. A similar comparison could be made between the cold-adapted muskox of Alaska and the pronghorn of the North American Great Plains.

When endotherms are faced with ambient temperatures above the upper critical temperature, they can use several strategies to maintain homeostasis. They may move to sites out of direct sunlight or otherwise choose more favorable locations to avoid high ambient temperatures. Many desert species remain hidden during daylight hours and become active only at night. Panting in birds and mammals, fluttering of the **gular sac** (a distended membrane in the throat area) in cormorants and pelicans, sweating in horses and some other species that have sweat glands, and exposure of the skin to the surrounding air to allow heat loss by convection (transfer of heat by movement of heated liquid or gas) in birds are all adaptations to maintain homeostasis when ambient temperatures are above the upper critical level. Additional adaptations for cooling the body include loss of water directly through the skin in animals lacking sweat glands, dilation of blood vessels in the lower legs and feet of wading birds to increase circulation to the extremities and thus dispersal of heat into the environment, and seasonal losses in the coat of insulative fur or feathers.

When endotherms are faced with temperatures below the lower critical level, the strategies they use include shivering, fluffing of the fur or feathers to increase insulating layers of stable air around the body, increases in insulating fur, feathers, or layers of fat during the colder months, shunting of blood away from the extremities, and exchange of heat between warmer blood traveling to the extremities and cooler blood returning from the extremities. Many behaviors, such as sleeping in aggregations by small birds, use of roosts in snow tunnels by ruffed and other grouse, curling up to reduce exposed surface area by most mammals, use of tree cavities for roosting, seeking habitats protected from the wind, seasonal migration, and seeking warm background areas and direct sunlight, are used as means of reducing heat loss in endotherms.

Endotherms use increased amounts of energy to warm the body during the colder periods of the year and must either increase their food intake as ambient temperatures decline or depend on stored body energy accumulated during summer and fall. In contrast, the body temperatures of ectotherms are reduced as ambient temperatures decrease. Reduction of body temperature in ectotherms automatically reduces their metabolic activity and thus reduces their energy expenditures.

As a generalization, birds and mammals must consume considerable quantities of food during winter to support endothermy and maintain their body weight. In cold weather, it is not unusual for small birds or mammals to ingest 25% of their body weight or more in food daily; at the same temperatures, larger birds or mammals eating energetically similar foods ingest a markedly lower percentage. Actual intake of food depends greatly on the energy and protein content of the food; it also varies within species with factors such as age and activity, and it varies greatly among species. The **metabolic rate** (the rate at which chemical reactions occur in an organism) of fishes, amphibians, and reptiles is reduced during cold periods, and their energy demands are also

FIGURE 5.13 Large subspecies of Canada geese are able to winter farther north than the smaller subspecies because of their larger body size, lower surface area-to-body mass ratio, and lower energy use per unit of body weight.

low. For example, during the summer, bluegills may consume about 5% of their body weight daily, while at colder winter temperatures they consume 1% or less of their weight in food.

Small endotherms have special problems in cold weather as they attempt to maintain body temperature homeostasis. Small endotherms have a high surface area-to-body mass ratio, and they lose body heat proportionally faster than do larger endotherms. Because of their higher surface-to-mass ratio, smaller animals also tend to be cooled more rapidly by wind through convection (fig. 5.13). Thus, wind combined with cold environmental temperatures would be much more stressful to a seven-month-old mule deer than to a mature mule deer with twice the body mass. Small endotherms must also consume proportionally more energy in relation to their body weight than larger endotherms because of the higher metabolic rates and greater energy demands of small bodies (Schmidt-Nielsen 1990).

These relationships between body size and heat loss are encapsulated in **Bergmann's rule**.

FIGURE 5.14 Giraffes are megaherbivores with ruminant stomachs, and are able to obtain relatively nutritious vegetation by moving between clumped food sources. *(Photograph courtesy of A. Emmrich.)*

Bergmann's rule states that species living in colder climates often have larger body sizes than subspecies or closely related species living in warmer regions. The range in size of white-tailed deer from the large northern subspecies to the smaller southern subspecies well illustrates Bergmann's rule. Geist (1986), however, questions the validity of Bergman's rule and the conclusion that reduction in relative surface area in mammals is an adaptation to conserve heat. **Allen's rule** states that ears, bills, tails, and other extremities tend to be shorter in the colder portions of the range of a species; these adaptations would also serve to reduce the surface area-to-body mass ratio.

Even though larger endotherms require less energy per kilogram of body mass than smaller ones, they still require more total food per individual. Thus, an African elephant must take in more total energy in food than a blue wildebeest, even though the wildebeest requires more energy per unit of body mass. However, because of the lower metabolic rates of larger animals, less energy would be required to support 100,000 kilograms of African elephants than 100,000 kilograms of blue wildebeests. Likewise, 10,000 kilograms of herbivorous mice would use more energy than an equal mass of moose (Peters 1983).

Large herbivores, particularly **megaherbivores** (those that weigh over 1,000 kilograms), must take in large amounts of forage to satisfy their energy needs. Among the ruminants, the proportion of crude protein in the diet generally declines with body size, indicating that larger animals are better adapted to processing poorer-quality, more abundant forage (Owen-Smith 1988). However, some megaherbivores, such as giraffes, are able to obtain a diet higher in protein than would be expected based on their body size (Owen-Smith 1988). Giraffes, which are ruminants, are able to move rapidly between clumped food sources and have relatively high crude protein levels in their diet (fig. 5.14). If an herbivore is highly selective, it is generally dependent on foods that are in more limited supply in the environment.

Because of their lower metabolic rates, large endotherms can survive on an equivalent proportion of body energy stores for longer than small endotherms. Thus, a wild turkey could survive longer on stored energy than a horned lark, and a bison longer than a mule deer. Mobility also tends to be less costly in terms of energy used (oxygen used per unit of body weight) in larger mammals (Schmidt-Nielsen 1990). Body size and mobility can also be related to the habitat requirements of an animal. For example, mule deer may not adapt as easily to logging debris or deep snow as elk because locomotion per unit of body weight is more costly in the smaller animal. In reality, differences in the costs of locomotion between mule deer and elk would interact with additional factors such as average bite size, food consumption rate, and forage access in influencing energetics and habitat needs in these species (Wickstrom et al. 1984).

The heart in endotherms, particularly birds, is a larger, more powerful organ than that in ectotherms. Unlike the two-chambered heart of fishes and the three-chambered hearts of amphibians and most reptiles, the endotherm heart has four chambers with completely separated left and right ventricles. The separation of the heart into four distinct chambers segregates newly oxygenated blood returning from the lungs from oxygen-depleted blood returning from the rest of the body. Thus, the endotherm is able to pump a blood supply higher in oxygen content through the body. Endotherms require higher oxygen levels in the blood than do ectotherms to support their higher level of metabolic activity.

The benefits of endothermy, such as the capacity for increased mobility, especially flight, and the capacity to remain active in the cold, have proved successful for a large number of bird and mammal species. Flight in birds and bats, for example, is an energy-intensive activity that requires the high metabolic rates found in endotherms. Even the flying dinosaurs, the pterosaurs, like other dinosaurs, may have been endothermic or partially endothermic, although controversy exists on this matter (Weishampel et al. 1990). Many species of birds migrate to the Arctic tundra for breeding each summer, a feat that would not be possible in their ectothermic cousins, the reptiles. The low species richness of reptiles and amphibians in Arctic regions compared to bird and mammal species demonstrates the advantage of endothermy.

Among the endotherms, some have seemingly reached back in time and borrowed certain ectothermic strategies. Hummingbirds are too small to maintain their high body temperatures throughout cool nights; to conserve energy, these birds go into a type of daily **torpor** (lethargy or sluggishness) or **hypothermy**. Hypothermy is defined here as the lowering of body temperature in endotherms to save energy during periods of stress. Most physiologists define **hibernation** as a type of winter hypothermy, in which the body temperature drops to approximately that of the surroundings of the animal (Schmidt-Nielsen 1990). Periods of torpor during the summer are referred to as **aestivation**. Because of a lack of precise definitions, hibernation, aestivation, and torpor are all referred to as hypothermy in this text. The common poorwill in the southwestern United States

goes into a winter hypothermy and may remain in this state for a month or more until insect prey populations warrant its return to an active lifestyle. Mammals such as the thirteen-lined ground squirrel go into a summer hypothermy when food resources become scarce in mid- to late summer, then, usually after a brief emergence from this inactive period, go into a winter hypothermy. Numerous species of small mice go into daily or seasonal hypothermy to conserve energy (Vaughan 1986). In many small mammals exhibiting seasonal hypothermy, periodic arousal or awakening occurs; this explains why thirteen-lined ground squirrels may at times be seen on warm late fall or winter days. Thus, the benefits of ectothermy in saving energy, especially in cold periods or periods without food, can be partially matched by these endothermic species.

5.7 Obtaining Oxygen in Terrestrial and Aquatic Environments

Fishes differ greatly from endotherms and terrestrial ectotherms in their capabilities for obtaining and circulating oxygen. In section 1.3, it was noted that the terrestrial atmosphere consists of approximately 19% oxygen, while well-oxygenated water may have only 10–15 milligrams per liter (0.0010%–0.0015%) available to organisms. Oxygen levels in aquatic environments also fluctuate much more than those in terrestrial systems. For example, oxygen in aquatic habitats can fluctuate greatly on a seasonal basis, and may also fluctuate on a 24-hour basis. Terrestrial organisms obviously have a more readily accessible and constant supply of oxygen available than do totally aquatic organisms. When one thinks of limiting factors, oxygen can be very important to aquatic species, but is seldom mentioned as a limiting factor for terrestrial species.

Aquatic ectotherms, such as fishes, have gills or other specialized structures that they use to extract oxygen from water. Terrestrial vertebrates typically use lungs for respiration. All of the aquatic endotherms, such as whales and seals, use lungs and breathe atmospheric oxygen. Given the small amounts of oxygen available in water, it is unlikely that an endothermic animal that had to extract oxygen from water could obtain sufficient oxygen to satisfy the needs of endothermy. In fact, no such endotherms exist.

To exacerbate problems for aquatic ectotherms, the capacity of water to hold dissolved oxygen decreases as water temperature increases. Aquatic ectotherms, therefore, often face decreased oxygen availability as well as increased oxygen consumption as water temperature increases.

Another difference between endotherms or other terrestrial vertebrates and aquatic ectotherms is the energy expenditure needed to pass the oxygen-bearing medium (air or water) over their respiratory surfaces. Water is 800 times denser than air; therefore it requires much more energy to pass water over gills than air over lungs. Add this to the different levels of oxygen available in the two media, and one can readily see that aquatic ectotherms must work harder to extract oxygen than endotherms and other terrestrial vertebrates.

Birds and mammals have lungs that differ in their basic structure and function. In mammalian lungs, the air enters a dead-end sac, and gas exchange with the blood occurs in tiny pockets called **alveoli**. Because each lung is basically a single sac, the used air is never completely exhaled, and newly inhaled air mixes with the air from previous inhalations. This mixing reduces oxygen uptake efficiency. In bird lungs, the air flows in a one-way direction, avoiding the problem of stale air in the lungs. The respiratory systems of birds feature several air sacs, usually one single and four paired, which lack blood vessels and are not involved in gas exchange; air flow occurs between the lungs and air sacs. A volume of air first enters the more posterior air sacs on inhalation, then on the following exhalation progresses into the lungs for gas exchange in the **parabronchi**, then flows into the more anterior air sacs on the next inhalation, and out of the body on the following exhalation (fig. 5.15). Bird lungs are highly efficient in oxygen uptake.

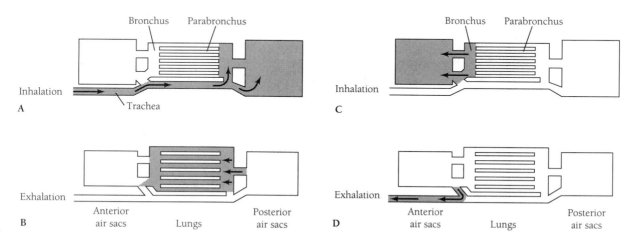

FIGURE 5.15 An illustration of the flow of a single volume of air (shaded area) through the air sacs and lungs of a bird. (*A*) An inhaled volume of air travels through the trachea and bronchus, expanding the posterior air sacs with fresh air. (*B*) On the following exhalation, the volume of air enters the lungs and passes through the parabronchi for gas exchange. (*C*) On the next inhalation, the volume of air passes into the anterior air sacs. (*D*) On the following exhalation, the volume of air is expired from the body. (*Adapted from Schmidt-Nielsen 1971.*)

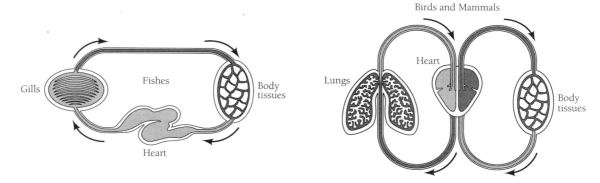

FIGURE 5.16 In fishes, blood is pumped to the gills, then flows directly to the body tissues. Delivery of blood through the blood vessels in fishes is therefore slower, at lower pressure, and less efficient than that in other vertebrates in which oxygenated blood returns to the heart before being pumped to the body tissues.

As we have seen, fishes have to work harder than endotherms to obtain oxygen, but they do have some strategies for obtaining oxygen more efficiently and reducing oxygen demand. Fishes make use of a countercurrent system of gas exchange between the water and the blood in their gills to increase their efficiency in obtaining oxygen and ridding the body of excess carbon dioxide (Schmidt-Nielsen 1990). Most fishes are also able to save energy by maintaining a neutral buoyancy using the gas bladder. Gas can be moved into the gas bladder in two ways: through swallowing as is sometimes the case in fishes in which the gas bladder opens to the digestive tract (**physostomous** fishes), or through a complex system of blood vessels and gas exchange as is usually the case in fishes in which the gas bladder lacks a connection to the digestive tract (**physoclistous** fishes). Neutral buoyancy creates an effect similar to weightlessness and clearly results in much energy savings. The gas bladder is one of the reasons why swimming fishes use less energy than flying birds or running mammals per unit of distance traveled (per kilogram of body weight) despite the high density of water (Schmidt-Nielsen 1990).

Differences between endotherms and aquatic ectotherms also exist in the delivery of oxygenated blood to the body tissues. As mentioned in section 5.6, birds, mammals, and a few reptiles have four-chambered hearts, while fishes have two-chambered hearts. In organisms with four-chambered hearts, deoxygenated blood is pumped to the lungs, where it undergoes oxygenation. To be oxygenated, the blood must be passed through a fine capillary network where oxygen uptake (and carbon dioxide elimination) can be accomplished. This passage of blood through the lungs reduces blood pressure and is the reason why blood must go back to the heart for further pumping. The added pumping stage ensures that the oxygenated blood is rapidly distributed to the body. In animals with three-chambered hearts, such as amphibians and most reptiles, part or most of the blood is also pumped through the lungs and returned to the heart before being pumped to the body. In fishes, blood is pumped from the heart to the gills. The blood is oxygenated in the gills and flows directly to the body. No secondary pump is used; therefore blood pressure is low and oxygen distribution is less efficient (fig. 5.16).

Some birds and mammals are able to dive under water and remain completely submerged for considerable periods of time. The sperm whale can dive to over 900 meters and remain under water for over an hour; other whales and seals are also highly adapted for diving and staying submerged for long periods. Among birds, common murres may reach depths in excess of 150 meters, razorbills 120 meters (Piatt and Nettleship 1985), and emperor penguins 265 meters (Kooyman et al. 1971). In terms of respiration, these endotherms are adapted to diving in several ways. Deep-diving birds and mammals can withstand much higher levels of carbon dioxide in the blood than can their terrestrial relatives. Diving birds can store considerably more oxygen in the blood and muscle than can nondiving birds, and can recirculate air in the air sacs through the lungs. The heart rate slows greatly in diving birds and mammals when they are submerged, and blood is shunted to critical areas such as the nervous system and heart. In addition, their muscles can operate anaerobically and incur an oxygen debt that is repaid after the animal resurfaces.

▼

5.8 Adapting to Fresh Water and Salt Water

Fishes live in aquatic habitats, and as such face very challenging physiological problems with regard to water and mineral balance (**osmotic balance**) in the body fluids and cells. Some fishes live in marine (saltwater) habitats, others live in freshwater systems, and some move between both; in each case fishes face different osmotic problems because of the differing levels of salinity they encounter. In addition, salinity does not remain constant. As water evaporates from water bodies, the ions that cause salinity remain. Thus, salinity can increase with evaporation or decrease with precipitation. To further add to the problem, some inland waters are very saline, and waters in coastal areas where fresh water and salt water meet can vary daily in their salt content.

In freshwater bony fishes, the blood and other body fluids contain higher concentrations of minerals and organic materials than does the surrounding water. Bony fishes are fishes with skeletons primarily composed of bone, in contrast to cartilaginous fishes, such as sharks and rays, which have skeletons made of cartilage. The higher **osmotic pressure** of the blood and body fluids in freshwater bony fishes leads to constant movement of water from the environment into the body. These fishes are said to be **hyperosmotic**. The rate at which water moves into the blood and body tissues is reduced by a partially impervious layer of skin and sometimes scales, but water easily crosses selectively permeable membranes such as the gills, mouth lining, and intestinal surface. Sharks and rays are also hyperosmotic.

Freshwater bony fishes use several strategies in their constant battle against water intrusion from the surrounding medium. The kidneys of freshwater bony fishes are specially adapted to remove water from their bodies. Freshwater bony fishes excrete copious amounts of dilute urine in order to maintain osmotic balance. In fact, although the main function of the kidneys in most vertebrates is the removal of nitrogenous wastes, the kidneys of most fishes are primarily used for maintaining osmotic balance; most nitrogenous wastes are eliminated through the gills. In general, freshwater bony fishes have many more and larger kidney **glomeruli** (conglomerations of blood vessels associated with the functional units of a kidney) than marine bony fishes; this reflects their increased ability and need to excrete water. These freshwater species excrete little salt because they are in a constant struggle to maintain body salts against the osmotic gradient. They compensate for salt losses to the environment by **active transport** of salts from the surrounding medium across the gills (fig. 5.17A). Active transport involves the use of body energy to transport salts across selectively permeable membranes and hold them against the osmotic gradient.

Marine bony fishes face the problem of an environment with a higher osmotic pressure than their bodies, which results in a constant loss of

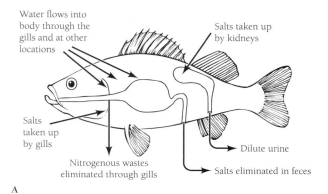

Water flows into body through the gills and at other locations

Salts taken up by kidneys

Salts taken up by gills

Dilute urine

Nitrogenous wastes eliminated through gills

Salts eliminated in feces

A

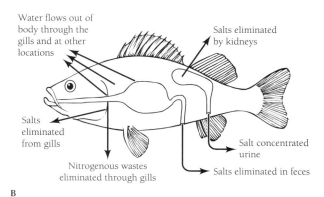

Water flows out of body through the gills and at other locations

Salts eliminated by kidneys

Salts eliminated from gills

Salt concentrated urine

Nitrogenous wastes eliminated through gills

Salts eliminated in feces

B

FIGURE 5.17   Net inflow and outflow of salts and water in (*A*) freshwater bony fishes that are hyperosmotic to the surrounding medium and (*B*) marine bony fishes that are hyposmotic to the surrounding medium. In freshwater bony fishes, active transport in the gills through chloride cells transports salts into the body. In marine bony fishes, the chloride cells in the gills actively transport salts out of the body.

ic balance differently than marine bony fishes by retaining large quantities of organic salts. The osmotic balance system of marine sharks and rays is therefore more like that of most freshwater bony fishes than that of marine bony fishes.

Some fishes are able to migrate between marine and freshwater systems and are able to adjust kidney and gill function to these extremes of changing osmotic pressure. Considering the adaptations needed to survive in either of these water types, the ability to switch from salt water to fresh water, or vice versa, is truly remarkable. These changes in the structure and function of the glomeruli of the kidneys and the gills involve the active transport system. Apparently, the chloride cells in the gills of some fishes, such as the chum salmon, can change their function from transferring salts into the body in fresh water to removing salts in marine environments. The adaptability of some of these fishes has been surprising to biologists. Chinook and pink salmon were once thought to need both freshwater and saltwater life phases, but have been able to adapt to totally freshwater environments, as in the Great Lakes. In a similar case, the Santee-Cooper Dam in South Carolina blocked the return routes of a striped bass population to the ocean, but biologists soon found that these fish could complete their life cycle in a large freshwater lake.

Birds living in the ocean or using highly saline lakes do not have a serious problem of water intrusion into the body fluids because of their relatively impermeable skin. However, many bird species commonly drink ocean water. Ocean water has a concentration of approximately 3% salt, compared with 1% in the blood; the urine of birds contains

body fluids and gain of salts. These marine bony fishes are said to be **hyposmotic**. Salts are taken in continuously with food and water. Most of these salts, along with 90% of nitrogenous waste, are excreted through the gills. The excretion of salts through the gills against an osmotic gradient requires active transport, which is in part associated with cells called **chloride cells** (fig. 5.17B). Because osmotic pressure favors loss of fluids in marine bony fishes, their kidneys are adapted to reabsorb water.

Primitive fishes, such as the Pacific hagfish, have body fluids similar in osmotic pressure to the surrounding water (they are **isosmotic**); they do not have substantial problems with osmotic balance. Marine sharks and rays regulate their osmot-

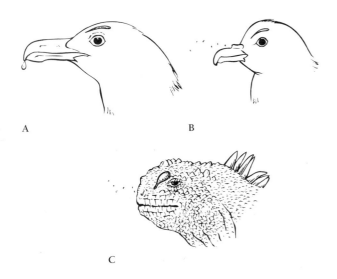

A   B

C

FIGURE 5.18 The salt gland in marine birds and reptiles is generally located above or near the eye and often opens into the nasal cavity. (*A*) The salt secretion emerges from nostrils, but drips from the tip of the beak in gulls. (*B*) In various petrel species and other seabirds with a tubular nostril, the secretion is forcibly expelled through the nostril. (*C*) In marine iguanas, the secretion of the salt gland emerges from the nasal cavity. (*Adapted from Schmidt-Nielsen 1959.*)

only about 0.3% salt. Their solution to this problem involves the active transport of salts out of the bloodstream through a gland above the eye known as the **salt gland**. In birds with salt glands, the size of the gland varies with the degree to which the birds are drinking marine or other saline waters or feeding on highly saline foods. The salty exudate (approximately 5% salt) secreted from the salt gland drips from the bill or nasal area of marine birds and the marine iguana (fig. 5.18). A similar salt removal system is found in marine turtles, which secrete salt from an opening near the back corner of the eye.

5.9 Standing and Swimming in Cold Water

Many fishes are adapted to extremely cold water; however, having a body temperature at or near the surrounding ambient temperature is normal for ectotherms. Some mammals, such as certain seals, dolphins, and whales, and some birds, such as king penguins, also can live in extremely cold water, but they must maintain a body temperature very different from that of their surroundings. Heavy layers of fur or feathers and underlying fat layers are critical to protecting these marine mammals and birds from heat loss. Their appendages project into an extremely cold environment; these appendages can allow a desirable loss of heat during periods of high metabolic activity or an undesirable loss of heat at other times.

In marine mammals, wading birds, and many other endotherms, particularly those that live in cold climates, the arterial blood going to the extremities (legs, wings, flippers, ears) comes in close contact with returning venous blood. In many cases the veins are specifically arranged to facilitate the flow of heat from the adjacent arteries into these veins. Thus, there is a **countercurrent exchange** that results in the warming of blood returning to the core of the body and the cooling of blood going to the extremities. This countercurrent exchange reduces heat loss in the extremities as well as the energetic costs of heating venous blood returning from the extremities. Similar systems are widely used by humans to reduce energy costs associated with heating and cooling in various industrial processes. In the flippers of seals and other marine mammals and in the legs (in the tibiotarsus or drumstick area) of some wading birds, there is a special netlike assemblage of blood vessels called the **rete mirabile** (meaning "wonderful net") that brings large numbers of small arteries and veins into close contact to increase countercurrent exchange.

Animals with countercurrent exchange systems are also able to shunt blood away from the

Air   −16°C

38°C
24°C
15°C
8°C
7°C
5−0°C

FIGURE 5.19 Glaucous, ivory, Iceland, and other northern gulls can stand for long periods on ice or in extremely cold water because of their ability to maintain cold temperatures and reduced blood flow in their lower legs and feet. The temperature of this gulls' feet is below 5°C while the internal body temperature is approximately 38°C. Waterfowl such as Canada geese and mallards are similarly adapted. *(From Irving 1966.)*

extremities when required to conserve heat or to dilate the blood vessels to increase blood flow to the extremities to dissipate heat. A Canada goose, mallard, or glaucous gull, for instance, can stand on ice for hours in extremely cold temperatures. The temperature of the feet may be at 0°–5°C while the body remains at approximately 38°C; the flow of blood to the feet is reduced during this period (fig. 5.19). The noses, ears, and legs of terrestrial mammals may also show a temperature gradient, with the extremities often close to ambient temperature; freezing of body extremities is unusual but occasionally occurs in nature. These adaptations for heat conservation provide substantial energy savings, without which these organisms could not survive in a cold environment.

At other times, increased blood flow to the extremities is necessary to dissipate metabolic heat. Even marine mammals in cold waters must dissipate excess metabolic heat through their flippers during active periods. Because they have thick insulating layers of fat (blubber) under the skin, seals, sea lions, whales, and some other marine mammals generate excess metabolic heat during active periods or when the water is warm. Part of that heat can be released by dilation of the blood vessels going to the appendages. In addition, these animals also have blood vessels in the skin that can carry heat to the body surface for release into the environment. These marine mammals differ from terrestrial mammals because the primary insulator (blubber) is located below the surface from which heat is dissipated. In most terrestrial endotherms the primary insulator (hair or feathers) is external to the skin surface, so heat dissipation must occur through other methods (Schmidt-Nielsen 1990).

 SUMMARY

Animal species require an array of inorganic elements in their diets. The elements needed in the greatest amounts are referred to as macroelements, while those needed in trace amounts are the trace elements or microelements. Terrestrial vertebrates obtain essential inorganic elements primarily from food, although some animals locate and ingest soil with concentrations of elements such as sodium. Fishes obtain most of the essential elements through food or directly from the water.

Vitamins are grouped as fat soluble and water soluble. They serve a variety of functions in animals. Most vitamins are obtained in adequate quantities in the diet. Lack of vitamins may be manifested in a variety of symptoms, depending on the vitamin.

Amino acids are grouped into those that animals can manufacture within their own bodies (nonessential amino acids) and those that must be obtained, at least in part, from the diet (essential amino acids). The need for essential amino acids may be reduced or eliminated in animals with multiple-chambered stomachs or gastric caeca in which microbial populations contribute to the digestive process.

Animals obtain their energy from ingested proteins, fats, and carbohydrates. Energy is normally obtained by breaking down carbohydrates into simple sugars such as glucose. When blood glucose and other carbohydrates are in short supply, glucose can be synthesized from fats and proteins. Protein, however, is more important for building muscle tissue than as an emergency source of energy. Protein requirements vary by species, with carnivores having the highest requirements; protein requirements within species may also vary by season, sex, and age. Nutritionally fit animals are much more likely to reproduce successfully than those with deficiencies.

Broad similarities exist in the digestive enzymes and morphology of the digestive tracts of vertebrates. The stomachs of most birds are divided into a glandular stomach and a separate muscular stomach (gizzard) for grinding. Ruminant mammals have complex stomachs in which microorganisms break down plant materials through a fermentation process. Many vertebrate species have gastric caeca in which digestion of coarse materials occurs through fermentation.

High metabolic rates are characteristic of endotherms. Within the thermoneutral zone, little energy expenditure beyond that required for normal body metabolism is needed to maintain body temperature. Above or below the thermoneutral zone, endotherms use a variety of behavioral, physiological, and morphological adaptations to help them maintain temperature homeostasis. Smaller endotherms have a larger surface area-to-body mass ratio and higher metabolic rates than do endotherms with greater mass. Because of their lower metabolic rates, larger endotherms can generally survive on less energy per unit of body mass than can smaller endotherms. Larger endotherms can also normally walk or fly more efficiently than can smaller endotherms.

Oxygen is in much more abundant and constant supply in terrestrial ecosystems than in aquatic ecosystems. Strictly aquatic animals must extract oxygen from a medium with extremely low and fluctuating oxygen concentrations. Endotherms could not obtain sufficient oxygen from water to survive. Because of their neutral buoyancy, fishes can move more efficiently through water than can birds through air or mammals overland. The lungs of birds are more efficient than those of mammals because of their one-way air flow pattern and the resulting reduction of stale air remaining in the lungs after exhalation.

Fishes face challenges in maintaining the osmotic balance of their body fluids. Freshwater bony fishes are hyperosmotic to their environment and must contend with the loss of salts from the body and the gain of water. Freshwater bony fishes have kidneys especially adapted for the secretion of water from the body; the gills use active transport to pump salts into the body against the osmotic gradient. Marine bony fishes are hyposmotic to their environment and use active transport to pump salts out of the body (through the gills); their kidneys are adapted to reabsorb water. Birds that drink salt water and eat saline foods often have salt glands that secrete salt through the nostrils and bill.

Endotherms living in cold aquatic environments must have mechanisms for heat conservation. They must also be able to dissipate metabolic heat during periods of activity. Loss of heat through the extremities is controlled by a countercurrent system that cools blood going to the extremities and warms blood returning to the body. Dilation or constriction of blood vessels leading to the extremities or body surface can also control heat loss. Special networks of blood vessels in the flippers or legs of some species enhance heat exchange and assist in reducing heat loss and blood flow to the extremities.

PRACTICE QUESTIONS

1. List four macroelements and a function of each in animals.

2. List four trace elements and a function of each in animals.

3. What is geophagia?

4. Provide examples of three fat-soluble and three water-soluble vitamins and list the important functions of each.

5. What is the definition of a nonessential amino acid? What is the definition of an essential amino acid?

6. What is gluconeogenesis?

7. Why are protein requirements in some species related to season, age, and sex?

8. Compare the reproductive energetics of northern pintails and snow geese.

9. How do Pacific salmon, in a sense, nourish their young even after the adults are dead?

10. How is nutritional condition related to reproductive success in fish and wildlife species?

11. Describe the form and function of a foregut fermenter and of a hindgut fermenter.

12. Compare the digestive tract of a California quail with that of a northern grasshopper mouse.

13. How is intestine length related to diet?

14. What is temperature homeostasis in endotherms? How do endotherms maintain temperature homeostasis when ambient temperatures are below the lower critical temperature or above the upper critical temperature?

15. Compare the changes in energy demands of house sparrows and bluegills from summer to winter.

16. Relate metabolic rates and energy requirements to body mass in endotherms.

17. What is hypothermy, and what advantages does it offer an animal species? Give an example.

18. Discuss oxygen availability and the difficulties aquatic ectotherms face in obtaining oxygen from the water.

19. How does the heart and basic blood circulation plan differ between fishes and mammals?

20. Compare the ways in which fishes in salt water and fresh water are adapted to these different osmotic environments.

21. How are some birds adapted for drinking salt water and eating highly saline foods?

22. What are some methods that seals in the Arctic Ocean might use to reduce or promote the loss of metabolic heat?

SELECTED READINGS

Bond, C. E. 1979. *Biology of fishes.* W. B. Saunders, Philadelphia.

Caughley, G., and A. R. E. Sinclair. 1994. *Wildlife ecology and management.* Blackwell Scientific Publications, Cambridge, Mass.

Eckert, R. 1988. *Animal physiology: Mechanisms and adaptations.* W. H. Freeman, New York.

Georgievskii, V. I., B. N. Annenkov, and V. T. Samokhin. 1982. *Mineral nutrition of animals.* Butterworths, London.

Gill, F. B. 1995. *Ornithology.* 2d ed. W. H. Freeman, New York.

Halver, J. E. 1972. *Fish nutrition.* Academic Press, New York.

Moyle, P. B., and J. J. Cech Jr. 1988. *Fishes: An introduction to ichthyology.* 2d ed. Prentice-Hall, Englewood Cliffs, N.J.

Owen-Smith, R. N. 1988. *Megaherbivores: The influence of very large body size on ecology.* Cambridge University Press, Cambridge.

Robbins, C. T. 1993. *Wildlife feeding and nutrition.* 2d ed. Academic Press, San Diego, Calif.

Schmidt-Nielsen, K. 1971. How birds breathe. *Scientific American* 225:72–79.

Schmidt-Nielsen, K. 1990. *Animal physiology: Adaptation and environment.* 4th ed. Cambridge University Press, Cambridge.

Scholander, P. F. 1957. The wonderful net. *Scientific American* 196:96–107.

Welty, J. C., and L. Baptista. 1988. *The life of birds.* 4th ed. Saunders College Publishing, Fort Worth, Tex.

LITERATURE CITED

Ankney, C. D. 1977. The use of nutrient reserves by breeding male lesser snow geese *Chen caerulescens caerulescens. Canadian Journal of Zoology* 55: 1984–1987.

Ankney, C. D., and C. D. MacInnes. 1978. Nutrient reserves and reproductive performance of female lesser snow geese. *Auk* 95:459–571.

Burgner, R. L. 1991. Life history of sockeye salmon (*Oncorhynchus nerka*). Pages 3–117 in C. Groot and L. Margolis, eds. *Pacific salmon life histories.* University of British Columbia Press, Vancouver.

Carey, F. G. 1982. Warm fish. Pages 216–233 in C. R. Taylor, K. Johansen, and L. Bolis, eds. *A companion to animal physiology.* Cambridge University Press, Cambridge.

Cech, J. J. Jr., R. M. Laurs, and J. B. Graham. 1984. Temperature-induced changes in blood gas equilibria in the albacore, *Thunnus alalunga,* a warm-bodied tuna. *Journal of Experimental Biology* 109: 21–34.

Frandson, R. D., and T. L. Spurgeon. 1992. *Anatomy and physiology of farm animals.* 5th ed. Lea and Febiger, Philadelphia.

Geist, V. 1986. Bergmann's rule is invalid. *Canadian Journal of Zoology* 65: 1035–1038.

Hays, V. W., and M. J. Swenson. 1993. Minerals and bones. Pages 517–535 in M. J. Swenson and W. O. Reece, eds. *Dukes' physiology of domestic animals.* Cornell University Press, Ithaca, N.Y.

Hofmann, R. R. 1988. Anatomy of the gastro-intestinal tract. Pages 14–43 in D. C. Church, ed. *The ruminant animal: Digestive physiology and nutrition.* Prentice-Hall, Englewood Cliffs, N.J.

Irving, L. 1966. Adaptations to cold. *Scientific American* 214 (1): 94–101.

Kincaid, R. 1988. Macro elements for ruminants. Pages 326–341 in D. C. Church, ed. *The ruminant animal: Digestive physiology and nutrition.* Prentice-Hall, Englewood Cliffs, N.J.

Kooyman, G. L., C. M. Drabek, R. Elsner, and W. B. Campbell. 1971. Diving behavior of the emperor penguin, *Aptenodytes forsteri. Auk* 88:775–795.

Krapu, G. L., and K. J. Reinecke. 1992. Foraging ecology and nutrition. Pages 1–29 in B. D. Batt, A. D. Afton, M. G. Anderson, C. D. Ankney, D. H. Johnson, J. A. Kadlek, and G. L. Krapu, eds. *Ecology and management of breeding waterfowl.* University of Minnesota Press, Minneapolis.

Lagler, K. F., J. E. Bardach, R. R. Miller, and D. R. M. Passino. 1977. *Ichthyology.* 2d ed. John Wiley & Sons, New York.

McDonald, P., R. A. Edwards, and J. F. D. Greenhalgh. 1981. *Animal nutrition.* 3d ed. Longman, London.

Mautz, W. W. 1978. Sledding on a bushy hillside: The fat cycle in deer. *Wildlife Society Bulletin* 6:88–90.

National Research Council. 1982. *Nutrient requirements of mink and foxes.* 2d rev. ed. National Academy Press, Washington, D.C.

National Research Council. 1993. *Nutrient requirements of fish.* National Academy Press, Washington, D.C.

National Research Council. 1994. *Nutrient requirements of poultry.* 9th rev. ed. National Academy Press, Washington, D.C.

Owen-Smith, R. N. 1988. *Megaherbivores: The influence of very large body size on ecology.* Cambridge University Press, Cambridge.

Peters, R. H. 1983. *The ecological implications of body size.* Cambridge University Press, Cambridge.

Petersen, L. E. 1984. Northern plains. Pages 441–448 in L. K. Halls, ed. *White-tailed deer ecology and management.* Stackpole Books, Harrisburg, Pa.

Piatt, J. F., and D. N. Nettleship. 1985. Diving depths of four alcids. *Auk* 102:293–297.

Robbins, C. T. 1993. *Wildlife feeding and nutrition.* 2d ed. Academic Press, San Diego, Calif.

Schmidt-Nielsen, K. 1959. Salt glands. *Scientific American* 200 (1): 109–116.

Schmidt-Nielsen, K. 1990. *Animal physiology: Adaptation and environment.* 4th ed. Cambridge University Press, Cambridge.

Starr, C., and R. Raggart. 1992. *Biology: The unity and diversity of life.* Wadsworth, Belmont, Calif.

Vaughan, T. A. 1986. *Mammalogy.* 3d ed. Saunders College Publishing, Fort Worth, Tex.

Weir, J. S. 1969. Chemical properties and occurrence on Kalahari sand of salt licks created by elephants. *Journal of Zoology* 158:293–310.

Weishampel, D. B., P. Dodson, and H. Osmolska. 1990. *The dinosauria.* University of California Press, Berkeley.

Wickstrom, M. L., C. T. Robbins, T. A. Hanley, D. E. Spalinger, and S. M. Parish. 1984. Food intake and foraging energetics of elk and mule deer. *The Journal of Wildlife Management* 48:1285–1301.

Yokoyama, M. T., and K. A. Johnson. 1988. Microbiology of the rumen and intestine. Pages 125–144 *in* D. C. Church, ed. *The ruminant animal: Digestive physiology and nutrition.* Prentice-Hall, Englewood Cliffs, N.J.

6

BEHAVIOR

When an organism responds to its environment, it often does so by performing specific behaviors. The behavior of individuals or groups of animals is closely tied to sensory perception; therefore, an understanding of behavior requires some general knowledge of sensory structures. In addition, because animals often respond to their environment through behavior, any modification of that environment can affect their behavior. Thus, animal behavior is influenced by natural processes that affect the environment as well as by human modifications of the environment, both intentional and unintentional.

Succession is an excellent example of the way naturally occurring environmental change can affect an animal's behavioral patterns. Terrestrial animals whose behaviors are adapted to open areas are affected when trees or shrubs appear at a later seral stage. Fishes that are dependent on clear water are behaviorally affected by succession that results in reduced water clarity.

Human-induced modification of the environment can also affect animal behavior patterns. Plowing or urbanization of a grassland may not be done with the intent to modify the habitats of an organism, but its effects may be substantial. For example, human effects on the grassland habitat of the lesser prairie-chicken can eliminate its reproductive habitat. Habitat, and therefore behavior, can also be intentionally affected. For example, if there are no suitable locations for hunting perches or nesting sites for ferruginous hawks in a particular area, such structures can be provided. Thus, an understanding of behavior may be a necessary prerequisite for programs that are intended to increase or decrease the abundance of a particular species or groups of species. Knowledge of animal behavior is also important with regard to activities such as animal sampling. A biologist cannot sample effectively without knowing when and where a species is vulnerable to a sampling design or method.

This chapter first provides an overview of the senses used by organisms in their behavioral activities. Then, a variety of individual and group behaviors are addressed.

6.1 Senses

When humans think of senses, we typically think of sight, hearing, touch, smell, balance, and taste. We interact with our surroundings through the use of these senses. Terrestrial and aquatic animals use these same senses, but often differ from humans in the level of their capabilities. For example, the range of sounds heard by a mammal such as the red fox is broader than the range of sounds heard by humans. Night vision is better developed in a bird such as the common barn-owl than it is in humans. Color vision is well developed in humans, but poorly developed in many mammals. In addition, some animals have sensory structures that humans do not possess, such as the lateral line system in fishes. This sensory structure is used for detecting sound, but it is quite different from the "typical" ears of birds, mammals, or even fishes themselves.

The eyes of different fish species vary in effectiveness depending on the habitat in which they live and how they have incorporated sight into their sensory repertoire. Some species, such as the predatory chain pickerel, have acute vision. Others, such as the southern cavefish, have no eyes; they rely on cues from their other senses to compensate for their inability to see. A good deal of what a fish sees is influenced by the medium in which it lives. The intensity of light is reduced as the light passes through water. The rate of reduction depends on water transparency as well as depth. Light reaching the bottom of a clear mountain stream is of a totally different intensity than is the light reaching the same depth in a muddy lake. In addition, light passing through water loses different portions of the color spectrum depending on water transparency. For example, red light penetrates deeper than blue light when water transparency is low, while the opposite is true when water transparency is high (Horne and Goldman 1994).

Birds generally gain more information from their surroundings by sight than from any of their other senses. Depending on the species, they can detect direction, distance, size, shape, brightness, color, intensity, and motion. In comparison to birds, most mammals have poorer sight but a better sense of smell (**olfactory** capabilities), probably because mammals initially evolved as **nocturnal** organisms (active during periods of darkness). Mammals are said to possess "smell brains" because their brains have large olfactory lobes, while birds are said to possess "eye brains" because of their large optic lobes. In mammals, color vision is primarily found in the primates, a group that has relatively well developed vision. For many mammal species, the ability to detect motion appears to be more important than the ability to distinguish colors. Another indication that color vision is less important to most mammals is their own rather drab coloration compared with that of birds, fishes, and many other animals.

Most birds are sight feeders, although some, such as turkey vultures and brown kiwis, are quite sensitive to smell. Air can be a difficult medium through which to follow a scent, depending on the amount of air movement and turbulence. In contrast, the terrestrial surface environment where most mammals live usually retains odors for a longer time than air. Although mammals, like other organisms, can smell odors that have **volatilized** (vaporized at a relatively low temperature), the source of the odor often remains in place and produces odor for an extended time period.

Most terrestrial animals use their sense of taste for short-range perception of chemical stimuli and their sense of smell for long-range perception. Aquatic animals, on the other hand, can use both senses for long-range perception because of the different medium in which they live. Chemical cues for both taste and smell must pass through a moist membrane to be perceived. For a terrestrial animal to taste something, the material usually must be present in its mouth; this is not the case for an organism such as a fish. The external body surface of terrestrial organisms is not moist; the external body surface of fishes is.

In the aquatic environment, odors can remain in solution for a long time. Fishes perceive primar-

ily organic odors, which can be either smelled or tasted. They use both an olfactory apparatus contained in a nasal sac and their taste buds to detect these odors. Some fishes, such as Pacific salmon and many sharks, have developed an acute sense of smell. In fishes, the sense of smell is usually more acute than the taste sense.

Most fishes have taste buds in the mouth, lip, and throat regions, but some, like the channel catfish, have taste buds over all of their bodies. The common carp, like mammals, can distinguish salty, sweet, bitter, and acid tastes. Birds also have an acute sense of taste, but have few taste buds compared with other vertebrates. For example, a mallard has only 375 taste buds, while a human has 9,000, and a rabbit has 17,000 (Welty and Baptista 1988). A black bullhead can have 100,000 taste buds located over all of its body (Lagler et al. 1977).

Hearing is the sense that allows animals to detect sound waves. The hearing apparatus is relatively simple in fishes, and becomes progressively more refined in reptiles, birds, and mammals.

The inner ear portion of the sound receptor in fishes has no opening to the outside of the fish. Because the density of fish flesh and the density of water are similar, sound passes through the body of the fish, strikes structures with differing densities such as bones or the gas bladder, and is transmitted to the inner ear.

The lateral line system is an additional system for sensing mechanical sound that is unique to fishes and some amphibians. The most obvious portion of the lateral line system found in most fish species is located in a canal that runs along the side of the fish and has numerous openings to the outside of the fish (fig. 6.1). In scaled fishes, pores through the scales on the lateral line allow water to enter the canal; in scaleless fishes, the pores do not need to pass through scales. Within the canal are structures called **neuromasts** that can detect displacement of water and are thus sensitive to mechanical sound waves and motion around the fish. In some fishes that have particularly sensitive lateral line reception, no canal is present and the neuromasts are on the exterior of the fish; Ozark

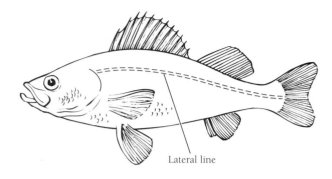

Lateral line

FIGURE 6.1 A portion of the lateral line system of many fishes is located along the side of the body. The line is actually a series of pores that allow sensory cells inside the lateral line canal to respond to the displacement of water.

cavefish are an example of a species that has many exposed neuromasts on the skin surface. Finally, the lateral line system is not limited to the side of a fish. Most fishes also have a portion of the lateral line system on the head.

The lateral line system has numerous uses. For example, a predatory fish can use the lateral line system to detect prey, schooling fish can use it to detect one another, and fishes can also detect non-moving objects when water currents "bounce" off the objects. This "distant touch" organ therefore allows location of both moving and nonmoving objects.

Mammals and birds have ears that are divided into three parts: the external ear, middle ear, and inner ear. In most birds, the external ear is simply an opening, often shaped to enhance sound gathering, that allows sound waves to enter. Some birds, such as owls, have feathers associated with ear openings that may direct sound into the ear. Most mammals have an external ear (**pinna**) that helps to intercept and direct sound waves into the ear. Different species or groups of species can sense different frequencies of sound. Some birds are capable of hearing notes in their songs that

humans cannot hear because they occur too rapidly. Bats have developed a sound production and reception system for very high frequencies that helps them locate prey. They produce a high-pitched sound wave, then use their large pinnae to monitor return signals (echoes) from prey organisms such as flying insects. This same system allows them to fly in areas such as caves without striking the walls.

The middle ear of birds and mammals transmits sound waves from outside the animal to the inner ear. A single elongated bone comprises the middle ear of birds, while mammals have three connected bones. The inner ear is the location where sound is converted from mechanical energy to electrical energy (the form in which it travels through the nervous system). In fishes, **otoliths** (earstones) located in fluid-filled sacs in the inner ear stimulate sensory cells lining the sacs that convert mechanical to electrical energy. In birds and mammals, vibrations passing through the middle ear are transmitted to the viscous fluid in the membrane-surrounded chambers of the middle ear. Vibrations in the fluid of the inner ear are then received by tiny sensory cells and transmitted as electrical signals to the brain.

In vertebrates, the sense of hearing is closely tied to the sense of equilibrium. The inner ear arose in the earliest vertebrates as a mechanism for maintaining equilibrium. In mammals, birds, and fishes, the **semicircular canals** of the inner ear and otoliths both contribute to the sense of equilibrium.

Birds and mammals both have touch receptors in the skin, but these receptors are much better developed in mammals. There is little evidence of such receptors in most fish species. The skin is also the location where many organisms can sense temperature. The skin of birds and mammals contains nerve endings that can detect heat and cold. Higher bony fishes also have nerve endings in the skin and can detect temperature changes as small as 0.03°C (Lagler et al. 1977).

Various fishes have electroreceptors that can sense electrical energy (Bleckmann 1993). In addition, some fishes have specialized organs that can produce electrical fields. Some catfish species have electroreceptors termed **ampullary organs** that are also sensitive to pressure, touch, temperature, and chemical stimuli. While many amphibians also have this electrical sense, the duck-billed platypus and spiny anteater are the only mammals known to possess electroreceptors. There is much that humans do not know or understand about the ability of animals to produce or detect electrical stimuli.

There are numerous other sensory capabilities that animals possess. Humans again do not always have a good understanding of these capabilities because such senses often have no externally obvious sensory reception area, such as an eye or ear. We are also less aware of some of these senses because the inputs they provide are not as distinct as those provided by senses such as sight or hearing. For example, many birds apparently can detect magnetic fields. In fact, many of the birds that migrate long distances probably use the magnetic field of the earth for at least part of their navigation. Determination of direction using the sun and stars also appears to be important in animal migration, as do other senses involving lunar and tidal perception. Additionally, little is known concerning the function of the **pineal gland** (**pineal organ**). This structure, which is located in the brain, is sensitive to light, and may function in many animals as an ultrasensitive light receptor.

6.2 Innate and Learned Behavior

Behavior displayed by an organism can be **innate** (**inborn**) **behavior** or **learned behavior**. These are not always easy to differentiate; biologists commonly debate whether various behaviors are innate or learned. For example, does a scarlet tanager instinctively know how to build a nest, or is part or all of that behavior learned by observation? In truth, many behavior patterns are likely to be a blend of innate and learned behaviors.

At the simplest level, innate behavior can consist of **autonomic responses** that act independently of volition. Common examples of autonomic responses include breathing and the response of the pupil of the eye to changes in environmental brightness. Another simple type of innate behavior is a **reflex**, which is an autonomic response to a stimulus. The knee-jerk reflex of a human during a medical examination is a familiar example. If a house mouse touches a hot surface, quick reflex movement away from the surface is its immediate response. A **taxis**, another innate behavior, is the oriented movement of an organism toward (positive taxis) or away from (negative taxis) a stimulus. A taxis may occur in response to such environmental stimuli as light, air or water currents, or soil or water salinity. For example, many moth species are positively phototactic, meaning that they move toward light, while a brown bullhead is negatively phototactic, meaning that it moves away from light. An **instinct** is an innate, stereotyped behavior that is characteristic of a given species. A newly hatched northern pike has a holdfast structure on its upper snout. It instinctively "knows" that it can use the holdfast to attach to a piece of vegetation while it absorbs its yolk sac. Young birds respond to touch or shaking of the nest, or even shadows, by gaping for food. Adult male mallards and green-winged teal perform stereotyped and species-specific courtship activities toward hens in the late winter and spring. All of these behaviors are examples of instincts.

Learned behaviors are those that have been shaped as a result of experience. American goldfinches learn to feed at thistle feeders in the backyards of bird-watchers. Some largemouth bass that have been caught and released by anglers apparently learn from this experience; they may be less vulnerable to angling after an initial capture. Similarly, coyotes that have escaped from a leghold trap are much more wary with regard to traps. Conversely, some animals may actually become "trap-happy." Waterfowl biologists banding ducks during the summer on the Canadian prairie and using grain as bait often capture the same duck day after day. Has this bird learned that the grain is available and that it will be released unharmed, or has it failed to learn to avoid the trap? Questions such as these are difficult to answer. Learning capacity is higher for some bird species, such as American crows and blue jays, than for others. Fish and mammal species also vary considerably in ability to learn.

The simplest form of learning is **habituation learning**, which means learning not to respond to meaningless stimuli. A bird nesting near a construction site, for example, must learn not to respond to frequent, familiar loud noises. **Trial-and-error learning** involves sorting through several nonrewarding actions to find an action that will provide a reward or avoid a punishment. A western kingbird may feed on many types of insects. Perhaps some insects taste better than others, or perhaps some might even make the bird sick. Trial-and-error learning allows the bird to learn which insects should be avoided and which should be sought. **Insight learning** is a much higher form of learning, and requires that an animal have the ability to perceive relationships. It needs to have the capacity to relate a similar experience to the one at hand. Some biologists consider the use of sticks to remove ants from burrows by chimpanzees to be an example of insight learning, although this is arguable. Humans obviously use insight learning.

As mentioned earlier, most behaviors are likely to be a blend of innate and learned behavior. For example, the basic template of a bird song is innate, but each individual needs learning and practice to perfect its song. A young male western meadowlark hears adult male meadowlarks and perfects its own song. A northern mockingbird innately knows how to sing, but it increases its song complexity as it hears the songs of other birds.

6.3 Mobility in Early Life

At hatching or birth, young organisms can be classified as either **precocial** or **altricial**, depending on their particular behavioral pattern and state of development. A precocial bird, such as the lesser golden-plover or sharp-tailed grouse, hatches with down, is alert and well developed, and is mobile soon after hatching; in most cases, precocial young are soon able to fend for themselves with respect to activities such as feeding or avoiding predators. Mammals that bear precocial young include black-tailed jack rabbits and pronghorns. Although this term is seldom used in fishery work, it is still applicable. A newly hatched walleye is free-swimming, and can be called precocial.

Altricial birds, such as mourning doves, are typically naked, blind, and weak at the time of hatching. They require parental care to survive. The same can be said for the altricial young of eastern cottontails and deer mice, which are naked and blind at birth. Although they are not born naked, the young of coyotes and tigers are also altricial, relying on parental care for their survival. Newly hatched northern pike can also be termed altricial. After hatching, they struggle to attach themselves to stems of vegetation. They remain attached for several days while they absorb their yolk sac. The energy provided by the yolk sac allows them to develop further so that they will be free-swimming when they detach from the vegetation.

Whether its offspring are precocial or altricial, each species has evolved a particular reproductive strategy. Parental investment is needed to enhance the likelihood of offspring survival, and these two strategies simply vary the time of parental investment. At first, it might seem that parents of altricial offspring have to expend more energy because they have to feed and care for their young. However, they typically do not invest as much energy in the actual production of an egg or young animal as do parents of precocial offspring. Even though parents of precocial offspring do not need to invest as much time in feeding and caring for their young, they generally invest more energy in the development of young before hatching or birth.

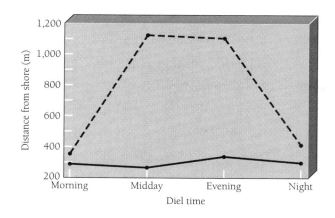

FIGURE 6.3 Diel activity patterns for northern pike in Lake Thompson, South Dakota, during August. Small northern pike (solid line; mean total length = 53 centimeters) remained approximately the same distance from shore throughout the day, while large pike (dotted line; mean total length = 84 centimeters) moved farther offshore during the warmer midday and evening periods. (*Adapted from Neumann 1994.*)

6.4 Rhythmic Behavior

The activity patterns of individual organisms often involve **rhythmic behavior** associated with daily or seasonal changes. Commonly recognized examples include **diel (circadian) behaviors** that change throughout the day (diel refers to the 24 hour period) and **circannual behaviors** that change throughout the year.

Mourning doves have a typical diel pattern of activity. For example, Howe and Flake (1989) found that mourning doves in Idaho sought water by visiting a pond most frequently at midmorning and late evening during early September (fig. 6.2).

Diel behavior can vary among species. For example, although many bats are nocturnal, there is substantial variation in activity patterns at various times of the night among bat species (Erkert 1982). Many insectivorous bats are most active

near sunset, while many bats that feed on fruit or nectar are most active just prior to sunrise.

Diel behavior can also vary within a species. For example, Neumann (1994) found that the activity patterns of small and large northern pike differed in Lake Thompson, South Dakota, during August (fig. 6.3). Small northern pike remained about the same distance from shore throughout the day, while large northern pike were closer to

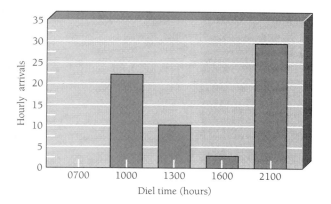

FIGURE 6.2 Arrivals of mourning doves at various times of the day during early September at a pond located at the Idaho National Engineering Laboratory. (*Adapted from Howe and Flake 1989.*)

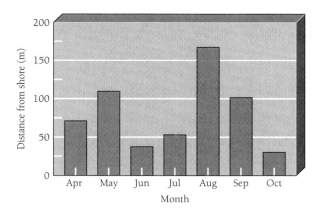

FIGURE 6.4 Average distance from shore for adult white crappies in Lake Goldsmith, South Dakota, at different times of the year. Fish were closest to shore during spawning in June and again in the fall, and farthest from shore during the warm weather period in August. (*Adapted from Guy et al. 1994.*)

shore at night and in the morning but farther off-shore during midday and evening.

Adult white crappies exhibited circannual changes in their distribution pattern in Lake Goldsmith, South Dakota (Guy et al. 1994) (fig. 6.4). They were located near the shoreline during the spawning period in June and during the fall. They were furthest offshore during August, the warmest time of the year.

Elk also exhibit a circannual behavior pattern. Summers tend to be spent at high elevations, even above the **timberline** (the highest altitude at which trees can grow in a particular area). Winters are spent at lower elevations that are warmer and provide more access to food. Other examples of circannual behavior are the various migrations exhibited by some organisms (see section 6.8).

Most biologists believe that many circannual behaviors are related to **photoperiod**, the proportion of light and dark in a 24-hour period. Increasing day length during the spring may be an important cue for species such as northern hog suckers and white-breasted nuthatches that breed at that time of year. Similarly, decreasing day length in the fall probably cues reproductive behavior in species such as brown trout and elk.

6.5 Early Social Behavior

As soon as most animals are born or hatched, they must participate in a variety of social behaviors. These can be intraspecific or interspecific interactions, and can take numerous forms. Immediately upon hatching, a blue-winged teal duckling must interact with its siblings and parents. It will have to communicate in order to survive. A young red fox will need to interact with others of its species, but will also need to learn to interact with prey species such as rodents and waterfowl, and with competitors and potential predators such as the coyote.

Social behavior may begin just prior to hatching in the young of many precocial birds. Some researchers believe that the calls of the parent during **pipping** (breaking of the eggshell during hatching) in precocial species are important in early bonding between the parent and its young. During the first 10–20 hours after hatching, the young become strongly attached to the parent in a process known as **filial imprinting**. Imprinting is critical to the survival of precocial young that must follow their parent soon after hatching. Interestingly, if they are exposed to a moving object (e.g., a wooden decoy, an orange toy truck, or even a human) during this imprinting period, precocial birds will often become imprinted to the substitute parent. Clucking sounds similar to those made by the natural parent will enhance the imprinting process. Once the brief imprinting period passes, the young will no longer accept a substitute parent.

Filial imprinting does not occur to the same extent for altricial young as it does for precocial young. Apparently, the most likely imprinting of altricial young is habitat imprinting or sexual imprinting, which occur somewhat later in life.

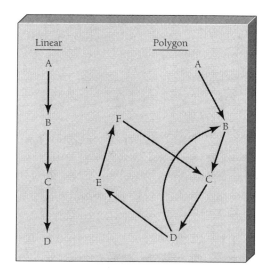

FIGURE 6.5 Dominance hierarchies may have a simple linear pattern *(left)*, but are typically more complicated, and often resemble complex polygons *(right)*. In the complex polygon shown here, individual A is dominant over B, B over C, C over D, D over B, D over E, E over F, and F over C.

6.6 Social Organization

Sociality is the tendency to gather in organized groups. Social organization is highly developed in many animals. Even invertebrates such as some bees and ants have developed high levels of social organization. Many social animals establish a **hierarchial order** (order of dominance) to help them deal with social gatherings. This hierarchy is often referred to as a **pecking order**. A flock of domesticated chickens will establish an order of dominance, in which dominant individuals will dominate subordinate individuals, often by pecking them. Such hierarchies allow animals to avoid conflict that requires energy expenditure. Rather than having continual, energy-consuming fights over day-to-day matters, the social hierarchy determines which individual has dominance, and there is little need for fighting. The position of an animal in a hierarchy is generally established through fighting or threat, and is influenced by such things as breeding status, age, size, and sex. The hierarchy of dominance may be linear, but is often more complicated, with dominance relationships resembling a polygon (fig. 6.5).

Examples of social hierarchies are plentiful. A wildlife biologist may want to capture wild turkeys during winter for relocation. Often, this is accomplished by placing a bait under a large net. The net is dropped when a sufficient number of birds are under the net. However, the dominant wild turkeys, generally older males, will keep other turkeys from feeding on the bait, complicating the task of the watching biologist, who would usually prefer to capture more than just the older males. Gray wolves and mountain gorillas also operate under social hierarchy systems. South American cichlid fishes, such as the firemouth cichlid and Jack Dempsey, will quickly establish social hierarchies when moved into an aquarium. Wild rainbow trout in a stream maintain specific feeding locations, with the dominant fish maintaining the best locations.

6.7 Communication

Most communication in the animal world occurs within or between species by sound, sight, smell, and taste. Electrical communication is rare, but does occur in some organisms (see section 6.1). Touch can also play a role in animal communication.

Sound has several advantages over visual displays or scents for communicating: a wide range of vocal frequencies and intensities is possible; little energy expenditure is needed; sound can penetrate visual barriers; and communication is possi-

FIGURE 6.6 The song of a male red-winged blackbird announces his establishment of a territory during the breeding season. *(Photograph courtesy of South Dakota Tourism.)*

Numerous mammal species use sound for similar functions. Some mammals use little sound communication, while others, such as coyotes, are great vocalizers. The sounds produced by mammals vary from the bugling of an elk (accomplished with specialized teeth) to the distress screams of an injured white-tailed jack rabbit. However, most mammals are poor vocalizers compared with birds.

Sound also plays an important role in aquatic ecosystems. Humans hear poorly in the underwater environment, so we label it the "silent world." Nothing could be further from the truth; we are just not physically adapted to hearing underwater sound. Sound waves actually travel farther and faster in water than in air. Many fish species use sound as a form of communication. An example is the freshwater drum, which produces a drumming sound using its gas bladder. The songs of the humpback whale are well known, and also serve as a means of communication. With devices designed to detect underwater sounds, humans

ble in the dark. Sound can be used for spacing individuals of the same species in a given environment, keeping a group together when visibility is limited, conveying information about predators, or indicating the presence of food. Sound is also important for reproductive functions, such as the proclamation of sex, development of pair bonds, and mate identification.

Birds use sound as a communication tool in their calls and songs. The "oak-a-lee" song of a male red-winged blackbird in a cattail marsh during the spring is a clear message to other males and potential mates that the bird has chosen a territory (fig. 6.6). Similarly, a northern bobwhite covey that has been separated can regroup through calls. Not all sound communication in birds is by vocalization. The male ruffed grouse, for example, uses rapid wingbeats to produce its characteristic "drumming" sound during the mating season (fig. 6.7).

FIGURE 6.7 During the breeding season, the male ruffed grouse (here on a drumming log, but not drumming) produces a drumming sound by means of rapid wing movements, not by vocalization. *(Photograph courtesy of South Dakota Tourism.)*

FIGURE 6.8 Male bluegills have a large, dark tab on the gill flap. Female bluegills may use the tab as one visual cue to help ensure that they breed with a male bluegill and not a male of another sunfish species. *(Photograph courtesy of Nebraska Game and Parks Commission.)*

find that the underwater world is far from a silent place. In reality, it is a rather noisy place.

Visual forms of communication, some relating to color and others to posture or structure, are common in animals. Visual forms of communication have the advantage of being visible over long distances in open habitat. Color, which is a function of vision, is commonly used by aquatic and terrestrial animals for activities such as courtship and species recognition. Removal of the moustache (black or red in color, depending on the race) from a male northern flicker will confuse sex recognition by females. The display of colorful body parts is a major aspect of courtship for many species. The males of most ducks, such as wood ducks, mallards, and ruddy ducks, display bright breeding plumage, while the females have drab coloration. The colorful buttocks of hamadryas (sacred baboons) provide information on female reproductive status and also on dominance status. Even in humans, who have the most highly developed verbal communication skills, visual cues are very important during interpersonal communication.

Many North American sunfish species are reproductively isolated by visual cues associated with color. In the wild, a female green sunfish will generally not breed with a male bluegill. However, bluegill × green sunfish hybrids can be produced in a fish culture pond. If the fish culturist removes the dark tab on the gill flap of a male bluegill (fig. 6.8), a female green sunfish will be more likely to accept him as a mate. The tab is probably a visual cue that serves as one mechanism that aids in reproductive isolation. Hybrid sunfish in the wild are often found where submerged aquatic vegetation is very dense and visual cues are obscured.

Another type of visual signal common to many animals is a signal displayed as a sign of danger. Examples include the white rump of a pronghorn (fig. 6.9) and the white underside of the tail in white-tailed deer. Many prairie birds, such as western meadowlarks or vesper sparrows, that are vulnerable to predators have light undersides on their tails. These light undersides flash during escape maneuvers, and may serve as a distraction to predators, a danger signal to **conspecifics** (others of the same species), or both.

FIGURE 6.9 Pronghorns have a highly visible white rump patch that can provide a danger signal to other pronghorns across substantial distances in the open habitat that this species typically inhabits. *(Photograph courtesy of South Dakota Tourism.)*

Visual communication can also involve the position or posture an animal assumes or the use of specially designed structures. For example, when a rock bass is disturbed, it erects the front portions of its dorsal and anal fins (the spined parts); this gives the fish the appearance of being larger. The **dewlap** (fold of skin on the neck) of some lizards can be enlarged for use in visual communication during reproduction. The tails of male sage grouse and wild turkeys are opened or fanned during courtship behavior. Many such structures are conspicuously colored when displayed. In addition, particular types of movements may be used. During reproduction, brook trout males display in a sideways fashion directly in front of females with whom they are trying to mate. A lizard that extends its dewlap may also perform a series of "push-ups" or "bobs." Male and female western grebes run along the water surface during their courtship displays.

Visual communication does not necessarily mean drawing attention to the organism; in fact, it can work in the opposite manner. Cryptic or camouflage coloration patterns can make an organism difficult to see. Female ducks, such as wood ducks, mallards, and ruddy ducks, are drab in color, probably because it makes them less conspicuous while nesting. This form of communication conveys the message, "I am not here." Coloration patterns may also be used to send false messages. For example, some fishes have large black spots on their tails. This makes the tail look like the head of a large fish (the large spots resemble the eyes). In this case, two false messages are conveyed, "this is my head," and "I am big." In some cases, coloration is combined with body shape to hide an organism. This form of communication sends the message, "I am a tree leaf," "I am a twig," or other similar messages.

In many animals, agonistic displays involve both sight and sound. **Agonistic behavior** is behavior involving aggression or threat displays. A gray wolf pack has a social hierarchy, and visual displays as well as sounds (growls) are a part of keeping this order with little actual fighting. The curled-lip aggressive stance that displays the teeth

of a gray wolf is well recognized within the pack. Lowering of the tail and exposure of the vulnerable head or neck regions are submissive responses to aggressive displays. Such submissive responses usually serve to repress any escalation of aggressive behavior, such as attack, by the dominant animal.

Animals can also use chemical forms of communication by means of their olfaction and taste senses. Chemical messages are sometimes lumped together into a group of compounds called **pheromones** (Grier 1984). Although some biologists use a more specific definition of pheromones, we will use the term in this more general sense. One common form of chemical communication involving pheromones is the territorial marking or signposting of an area by male mammals. Anyone who has ever watched a male dog proceed around its territory (which often corresponds to the property boundaries of its owner) while periodically stopping to sniff and urinate is viewing chemical communication between dogs. Territorial marking is also commonly used by animals such as deer, rabbits, and wolves. In most mammal species, males ascertain the reproductive status and receptiveness of females by scent. Chemical communication is also important for some fish species. For example, female blue and channel catfish use odors (sexual pheromones) to be certain of the species of a potential mate (Todd 1971). In addition, chemical substances produced by some fishes, especially minnows, can warn nearby conspecifics of danger.

6.8 Movements and Migrations

Migration is normally defined as a two-way movement to and from an area with characteristic regularity or with changes in life history stage (see section 2.9). **Movements** can be defined as

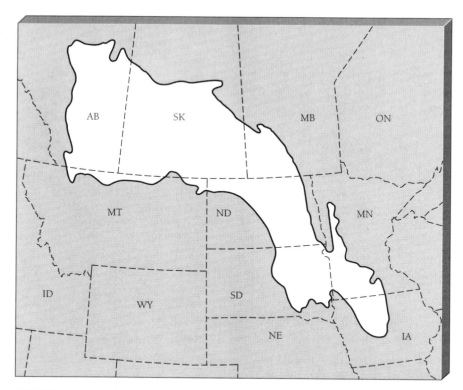

FIGURE 6.10 The prairie pothole region of North America extends from northwestern Iowa to southeastern Alberta.

changes in position or posture, and are not necessarily regular in nature. Most migrations are associated with food supply, reproduction, or changes in ambient temperature. They are often cued by photoperiod, and can be examples of rhythmic group behavior (see section 6.4).

Some fishes are **diadromous**; that is, they migrate between salt and fresh water. **Anadromous** fishes spawn in fresh water, but live most of their lives in the ocean. The various species of Pacific salmon spawn in freshwater lakes and rivers in northwestern North America, and their offspring migrate to the Pacific Ocean, where they grow to adult size. The striped bass, native to the Atlantic coast of North America, is another example of an anadromous fish. **Catadromous** fishes spawn in the ocean and return to fresh water to accomplish much of their growth. For example, American eels that live and grow in the Mississippi River system migrate to the Sargasso Sea area of the Atlantic Ocean (east of southern Florida) to reproduce.

Fish migrations can also occur strictly within fresh or salt water. For example, a strain of rainbow trout living in Lake McConaughy in western Nebraska migrates hundreds of kilometers up the North Platte River and actually spawns in Wyoming. Other fish migrations may be shorter in distance. For example, black crappies spend much of the year in the open water of a lake, but migrate to the shoreline at spawning time.

Both elk and mule deer in the Rocky Mountains migrate from summer feeding areas at high elevations to wintering areas at lower elevations. Caribou migrate over tremendous distances in Alaska and northern Canada. They feed far to the north in the summer, then migrate south for wintering.

Birds are some of the most famous of all animal migrants. For example, lesser golden-plovers breed along the northern coast of Alaska and across northern Canada, then winter in the southeastern part of South America, migrating a distance of 12,000 kilometers or more. Most North American waterfowl species migrate north during the spring and summer to areas with abundant and productive shallow water. Such areas include the extensive prairie pothole region (fig. 6.10) as well as northern river deltas and floodplains. In this way waterfowl can take advantage of the high productivity of these water bodies during the summer while avoiding the harsh winter. Many snow geese nest north of the Arctic Circle, then migrate south during the winter, and many spend the winter on the Gulf of Mexico coast. Birds may rely on a variety of cues for navigation during these long migrations, such as the positions of celestial bodies, the magnetic field of the earth, and landscape features.

Homing is the ability of an animal to return to a familiar site that is outside the range of the direct senses, such as vision and hearing. Homing is useful for finding past successful breeding sites, wintering areas, feeding sites, and so forth. Pacific salmon home to the site where they hatched, probably with the aid of their sense of smell (**chemical imprinting**). Little brown bats in New England home as far as 300 kilometers to a wintering cave (Davis and Hitchcock 1965). Giant Canada geese frequently return to breed at the location where they fledged, as do many other bird species.

▼

6.9 Flocking, Herding, and Schooling

Terrestrial and aquatic animals may gather for a variety of reasons. Social groups may demonstrate flocking, herding, and schooling or other grouping behavior. A fish school is defined here as a group of fish that is together because of social attraction for one another and displays coordinated movement. Bird flocks also are social assemblages, as are mammal herds.

Many social assemblages are the result of predator-prey interactions. For fishes that are prey species, schooling can provide protection from predators. A closely packed school may appear to be a larger organism and thus deter predation. In addition, a predator is more likely to be observed by at least one of the many individuals in a school than by a lone fish. Some predator species are known to take more prey items when the items are presented individually rather than in a group. It is also possible that there is less likelihood of a compact mass being found than widely scattered individuals. A predator may simply be confused by the high level of motion in a school, and not be able to strike as effectively. Some biologists have suggested that schooling provides a greater chance of group survival through the sacrifice of some individuals in the group. The more individuals there are in the group, the greater the chance that some individuals will escape predation. A school of newly hatched smallmouth bass is typically composed of siblings; therefore the sacrifice of some individuals in the school may increase the chances of survival for related individuals. Most fish species school as juveniles, but only about 20% school as adults (Bond 1979).

Birds may or may not flock with parents and siblings. Canada geese tend to flock and migrate in parental/sibling groups. Snow geese and other Arctic-nesting geese also remain in family groups and form large flocks during migration. The chances of survival of the young are probably increased when they flock with adults. Increased survival may result from direct protection by the adults or the experience of the older birds may lead to higher survival rates of the young during migration.

Large ungulates of the open plains, such as pronghorns, commonly form herds as a defense against predators. Muskoxen form a circle when threatened, with young animals in the center (fig. 6.11).

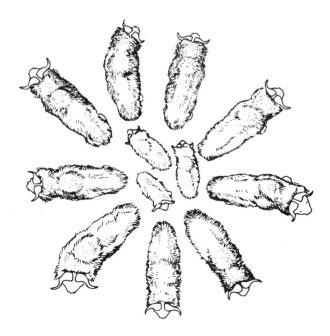

FIGURE 6.11 When threatened, muskoxen form a circle with the young herd members in the center.

ing food. Fishes may also simply be in aggregations. A group of adult bluegills may aggregate for feeding purposes, or may concentrate near submerged brush when seeking cover.

Breeding seasons can also result in social grouping by some animals. Dominant male elk will guard a group of females. Wild turkey gobblers will similarly try to maintain a group of hens.

Some birds form colonies during breeding seasons. Both double-crested cormorants and great blue herons form breeding colonies, often in large, dead trees or in live trees near water (fig. 6.12). The trees provide nest sites that are relatively safe from predators, and the adults use the trees as a nighttime roosting site. Nesting in a colony also helps to synchronize breeding, and information about food sources can apparently also be shared.

Some birds are intolerant of conspecifics during the breeding season, but may flock together for the remainder of the year. The red-winged

The predators themselves may also operate in social assemblages. Gray wolves tend to group in packs of genetically related individuals. Packs apparently offer a feeding advantage for this species; the gray wolves can capture prey more effectively in groups than as individuals. Striped bass are effective open water predators that feed in schools. Killer whales also cooperate in a group for more effective predation.

Schooling fishes and flocking birds may also gain hydrodynamic or aerodynamic advantages. Fish schools pass through water more efficiently than do individuals, as the lead fish allows for easier passage by the following individuals. Similarly, many birds flock together during migration and thus gain an aerodynamic advantage. In snow goose flocks, birds take turns flying in the lead position. It takes less energy to follow the leader than to be in the lead position.

Birds may flock, or they may simply be in short-term aggregations. Songbirds gathering at a bird feeder are in an aggregation, primarily seek-

FIGURE 6.12 Great blue herons nest in colonies in either live or dead trees over water.

blackbird is a good example of such a bird. During the breeding season, a red-winged blackbird is highly territorial. During the remainder of the year, it may join other red-winged blackbirds, yellow-headed blackbirds, European starlings, common grackles, and other birds in mixed-species flocks. Such mixed flocks typically join together for feeding and roosting; the flock may also be a good defense against predation.

6.10 Reproductive Behavior

Mating systems in higher animals are generally categorized as **promiscuous**, **monogamous**, or **polygamous**. Promiscuous species generally form no pair bonds between males and females. Rainbow smelt in the Great Lakes are promiscuous breeders. They spawn in large groups, with no prior development of pair bonds. Two or more males maintain positions beside a female and release **milt** (sperm-bearing fluid) as the female releases her adhesive eggs. Ruby-throated hummingbirds also have a promiscuous mating system.

Monogamous mating systems involve a pair bond between one male and one female. Both the gray partridge and the field sparrow are monogamous species. Monogamy is rare in mammals, but some mammals, such as the red fox, have a monogamous mating system. Monogamy is even rarer in fishes. Mating systems that are mostly monogamous occur in some species, such as some of the African cichlids, in which both parents provide care for the young. However, even among these fishes, the monogamous system is not strictly followed, as a male often accepts a second female.

In a polygamous system, a single individual of one sex mates with numerous individuals of the opposite sex. A **polygynous** species is one in which a single male mates with multiple females. Common examples include ring-necked pheasants, elk, and mule deer. Less common are **polyan-drous** species, in which a single female mates with multiple males. Female Wilson's phalaropes mate with multiple males. In this species, the female is larger than the male and has brighter plumage. The male builds the nest, incubates the eggs, and rears the young alone.

Most animals have developed reproductive strategies that can be categorized along a continuum from production of large numbers of offspring that receive little parental care to production of few offspring that receive a great deal of parental care. This range of reproductive strategies is referred to as the **r-K continuum**. **r-selected species** use a strategy of low energy expenditure for production of gametes or care of offspring, and large numbers of offspring. **K-selected species** use a strategy of high energy expenditure for embryo or egg development and care of offspring, and small numbers of offspring. Organisms can be placed at a variety of points along the r-K continuum. That is, some K-selected species or r-selected species may be farther toward the ends or the middle of the continuum than others. In birds and mammals, K-selected species tend to have longer life spans than r-selected species; this is generally not true for fishes.

The r-selected female common carp simply broadcasts her adhesive eggs over flooded vegetation. The eggs, which are produced in large numbers (as many as two million in one year), are left to hatch, and the young must survive with no parental care. The common carp invests her energy in developing a large number of eggs; thus, only a fraction of a percentage of the eggs needs to survive to perpetuate the species. Toward the other end of the continuum in fishes, smallmouth bass are nest builders, and their nests commonly contain several thousand eggs. The male guards the eggs after spawning, and also guards the school of young after hatching.

The California condor, a K-selected bird, does not become sexually mature until 6–8 years of age, and then incubates only one egg every other year. Development is unusually slow, with the young learning to fly at 5 months of age. Considerable energy is invested in incubation and care of the young by both parents over a period of about 10 months. The mourning dove is a more *r*-selected bird species. Although only two eggs are laid per clutch, mourning doves can hatch up to five clutches over the course of the summer.

The *r*-selected house mouse reaches sexual maturity 35 days after being born and has a **gestation period** (developmental period in the uterus) of 19 days. Litters can contain up to twelve young, and a single female can have as many as six litters in a year. In contrast, *K*-selected African elephants have a maturation period of 8–16 years, a gestation period of over 20 months, and have a single calf in one cycle.

Fishes have evolved a variety of reproductive strategies, ranging from egg laying (longnose suckers, golden trout, and wedgespot shiners) to bearing live young (guppies and western mosquitofish). There is tremendous diversity within the egg-laying fishes. Some build and guard nests (black crappies, black bullheads, and brook sticklebacks), some build nests but do not protect the eggs (brook trout and pink salmon), and others broadcast eggs onto vegetation (common carp and muskellunge) or gravel (white bass and walleyes) and provide no care thereafter. Striped bass have semi-buoyant eggs that will hatch while being carried by a river current. Freshwater drums have eggs that are nearly neutral in buoyancy and hatch while suspended in water. There are also numerous fish species that carry developing eggs in their mouths, or attached to or in some other portion of their bodies.

Birds are egg layers. However, birds may nest in isolation (green-winged teal and upland sandpipers) or in colonies (double-crested cormorants, Adélie penguins, and bank swallows). They may have single (gray partridge and red-tailed hawks) or multiple (mourning doves) broods within a given year. Many species that have single broods, such as most ducks and gallinaceous birds, have the capability to renest if the first nest is destroyed.

Mammals generally bear live young. The nutritious milk provided by most female mammals aids in the development of the young, but at a substantial energetic cost to the female. Once again, there is a great deal of variation in the length of time that mammals may nurse. For example, young Dall sheep begin to feed on grasses and are at least partially weaned within one month of birth. Conversely, brown bear cubs nurse from the time of their birth in January to March, and are not fully weaned until October to December.

Fishes generally reach reproductive maturity based on their size, rather than their age. For example, male walleyes typically are mature by the time they reach 38 centimeters, while females mature by 45 centimeters. In Kansas, a 38-centimeter male is likely to be 2 years old, while a 45-centimeter female is likely to be 3 years old. In northern Canada, where walleye growth rates are much slower, these fish may not mature until they are 6–8 years old, or even older. Conversely, mammals and birds typically reach reproductive maturity based on age. However, age at reproduction can still be influenced by factors such as size and nutritional status, but not nearly to the same extent as in fishes. Most common terns do not begin nesting until they reach 3 years of age; American robins, many species of warblers, and many species of ducks nest at 1 year of age; California condors first nest at 6–8 years of age. Female montane voles can breed at an age of only 21 days, while most female mountain goats do not breed until they are 2.5 years of age or older.

6.11 Feeding Behavior

There are several aspects to consider when describing the feeding behavior of organisms. What an organism eats is referred to as its **food habits**. Food habits vary from species to species and represent information that is important if an organism or its habitat is to be correctly managed. Food habits are covered in more detail in section 9.6.

Feeding behavior involves when, where, and how an organism obtains food, and is also referred to as its **feeding habits.** The "when" can refer either to time of year, because feeding habits can vary by season, or to time of day. Some animals are **diurnal** feeders; they consume food during daylight hours. The term diurnal refers to any activity that occurs during daylight hours. A central stoneroller slowly moving along the bottom of a stream during midday and scraping food organisms from rocks, a scissor-tailed flycatcher capturing flying insects on a warm summer afternoon, and a Franklin's ground squirrel foraging for seeds and insects during the day are all practicing diurnal feeding. Some organisms feed at night and are termed nocturnal feeders. Channel catfish, common barn-owls, and raccoons are all primarily nocturnal feeders. Other animals are termed **crepuscular** feeders; examples include walleyes and common nighthawks. The crepuscular period is the twilight hours of dawn and dusk.

The "where" of feeding strategies is also an aspect of feeding behavior. Some fishes feed in shallow water while others feed in deep water. Some fishes are surface feeders, some feed in midwater areas, and yet others are **benthic** (bottom) feeders. Much information about the feeding behavior of a fish can be ascertained by looking at the size and location of its mouth (fig. 6.13).

Birds also vary greatly with regard to where they feed. Some feed on the ground, others dive underwater to obtain food, some capture their food in midair, and some creep up or down the trunks of trees while feeding. The variety of feeding locations is almost endless. The same can be said for mammals. Moose feed on aquatic plants in marshes, lakes, and rivers; mink feed on aquatic or terrestrial organisms along the periphery of water bodies; fox squirrels feed on acorns and other plant foods in oak forests; and some bat species feed in midair on flying insects.

The "how" aspect of feeding behavior is also variable. Some animals actively pursue their prey, while others lie in wait for prey to come to them. Striped bass pursue their prey, often feeding in schools. Flathead catfish are also predators, but

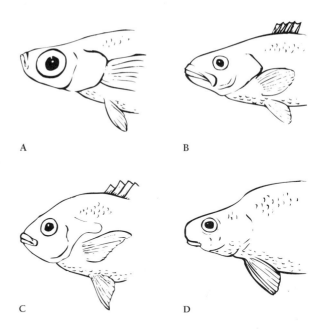

FIGURE 6.13 The location of the mouth in a fish species is related to its feeding behavior. The starhead top-minnow (A) has a supraterminal (or superior) mouth for capturing organisms located above the fish. The smallmouth bass (B) has a terminal mouth that is large enough to capture large prey such as fishes and crayfish. The redbreast sunfish (C) also has a terminal mouth, but it is small and well suited to feeding on insects. Both of these species with terminal mouths primarily feed on items found in front of them. The river carpsucker (D) has a subterminal (or inferior) mouth that is suited to its benthic-feeding behavior.

typically lie motionless waiting for prey to come close enough to be captured. Some predators swallow their food whole, while others bite or tear off chunks of food. Most predatory fishes and many owls ingest prey in one piece (usually head-first); many hawks and most mammalian predators do not, depending on the size of the prey organisms.

Most organisms need to be as efficient as possible in their food intake. The **optimal foraging theory** is applicable to the feeding behavior of both aquatic and terrestrial animals. This theory is based on the premise that natural selection shapes the behavior of an animal in such a way that it does not expend more energy than necessary to obtain food. The most efficient foraging strategy may involve prey type, location, size, or nutrient content. For example, as a northern pike grows larger, it will tend to select increasingly larger prey items. Prey availability and vulnerability may also be important. A large predator may feed on abundant small prey if they are easier to catch than a single large prey item.

The feeding strategies of birds and mammals can also be discussed in terms of the optimal foraging theory. American robins may select earthworms over smaller invertebrates, and downy woodpeckers may concentrate on insect larvae of a particular size to maximize energy intake and minimize energy expenditure. Moose need large amounts of sodium in their diets, and aquatic plants provide good sources of sodium. However, aquatic plants are available only during the summer, and they supply less energy than do terrestrial plants. Thus, moose have to "determine," probably through natural selection processes, how much of their summer diet will be composed of sodium-rich aquatic plants, and how much will be composed of energy-rich terrestrial plants (Stephens and Krebs 1986).

6.12 Sociobiology

Sociobiology is the study of the biological basis of social behavior and the organization of societies. Sociobiology came to the forefront of behavioral science with the publication of a book entitled *Sociobiology: The New Synthesis* (Wilson 1975). Sociobiologists specifically add the genetic component to the assessment of behavior; most behaviors have a genetic basis and are therefore subject to natural selection. The genes underlying a behavior that enhances the likelihood of survival will be passed on to future generations, but the genes underlying an unsuccessful behavior will be lost from

a population because an unsuccessful individual is less likely to reproduce.

Grier (1984) suggested that the major taxonomic groups of organisms can be assessed for the incidence and extent of sociality using seven criteria. The use of these criteria allows a biologist to more thoroughly understand the group behavior of a particular species. The first criterion is related to the number of organisms of the same species in a group: the larger the number, the more social the organism. The second criterion is the length of time or part of the life cycle during which the group remains together. Sibling smallmouth bass remain in a school for no more than the first few weeks of their lives. Conversely, snow geese often remain in family groups from hatching in the spring through the fall migration, wintering, and spring migration. The third criterion relates to how much time or energy is spent on social behavior. Some organisms may actually interact relatively little in a group, while others have substantial and complex interactions. Consider the highly advanced social behavior of the chimpanzee, in which much time is spent on social interactions such as grooming. The fourth criterion concerns whether reciprocal communication is necessary to keep the group together. Honeybees convey messages, but typically without reciprocation. The communication is one-way; honeybee scouts use a dance to tell other bees the direction and distance to a food source, but there is no return message from the honeybees at the hive. More highly evolved social species, such as bighorn sheep, have much more reciprocal communication. When a ram displays an agonistic behavior to another ram during the breeding season, the second ram will either communicate its willingness to fight or display a submissive behavior to avoid a fight. The fifth criterion relates to the existence of social structures in which different individuals have different roles. For example, various species of ants have clearly developed such roles. Some individuals may be workers, other defenders, and yet others may be responsible for reproduction. The sixth criterion relates to the existence of an overlap of generations in which families or parts of families remain together. Gray wolves have such a social structure; even "relatives" such as "uncles" may remain with the pack.

Finally, the criterion least commonly met is the highest level of social behavior: **altruism**. Altruistic behavior is aid-giving behavior that has a cost to the aid-giving individual. This final aspect of social behavior has been a point of contention among sociobiologists. The most extreme example of altruism occurs when an individual sacrifices its life for another. From a genetic or evolutionary perspective, such an extreme level of altruism should result in the loss of the genetic material that allowed such a sacrifice, because that individual would not live to reproduce and replace itself. In some instances, however, **kin selection** has been proposed as a justification for altruistic behavior. The more closely related two individuals are, the more genetic material they share. If individual A lives in a family group and sacrifices its life for the remaining family members, this altruistic act will increase the chances that some of the genetic material individual A shares with its relatives will survive in those relatives and be passed to future generations. A common example of kin selection might involve an "uncle" in a gray wolf pack who is not involved in breeding, but who helps to feed and protect the pups in his family group. Because he is related to at least one of the parents of the pups, his altruistic behavior increases the chances that some of his genes will be passed to future generations. Again, this theory is contentious, and certainly is hard to substantiate through scientific research.

The concept of sociobiology has led to differences of opinion among behavioral scientists. At the simplest level, some biologists believe that it is just a new term applied to an established science. Some of the controversy concerning sociobiology has arisen as sociobiologists have applied their science to human behavior. Often, such application conflicts with religious beliefs; similar conflicts arise with respect to other scientific concepts, such as evolution. In addition, some individuals have argued that the ideas proposed by Wilson concerning genetic determinism in humans could be used to justify racism, sexism, imperialism, and war (Grier 1984), although they were not intended to imply such justifications.

SUMMARY

The behavior of an individual organism is largely dependent upon its senses. The senses of sight, smell, taste, hearing, balance, and touch are commonly understood by humans. Less well understood are the senses that relate to magnetic fields, electrical fields, and solar and lunar influences.

An organism may display innate or learned behaviors. Simple behaviors in simple organisms are almost certainly innate. Many behaviors are actually a blend of innate and learned behaviors.

Precocial offspring are capable of moving and feeding soon after birth or hatching. Altricial offspring are born or hatched in a much more helpless state, and require more parental care at this stage of life than do precocial individuals.

Individual organisms commonly display rhythmic behaviors. Behaviors that vary throughout the day are termed diel or circadian. Circannual behaviors are those that change seasonally or throughout the year. Photoperiod, the proportion of light and dark in a 24-hour period, cues many of the rhythmic behaviors.

Social interactions can be intraspecific or interspecific. Intraspecific interactions are often regulated by dominance hierarchies. If individuals in the social group know their place in a hierarchy, conflicts can be minimized.

Most animals communicate by sight, smell, sound, or taste. Vocalizations can be effective means of communication even in habitats with poor visibility. Visual displays can be effective over long distances in open habitat. Chemical communication has an advantage in that its message can be "sent" for an extended time period.

Migrations are two-way movements with characteristic regularity. Migrations of snow geese in North America, for example, follow a circannual pattern in which the birds summer at northern latitudes and winter at more southerly latitudes.

Schools, flocks, and herds are social groups, and these groups show coordinated movements. Social assemblages such as these should not be confused with aggregations, which are generally gatherings of a less formal and shorter-term nature.

Mating behaviors are obviously social behaviors. Common mating systems include promiscuous systems, in which no pair bonds are formed; monogamous systems, in which one male and one female form a pair bond; and polygamous systems, in which one individual of one sex mates with multiple members of the opposite sex.

Reproductive strategies are highly variable, and fall along an *r-K* continuum. *K*-selected species tend to produce few offspring and provide substantial parental care. At the other end of the continuum, *r*-selected species tend to produce many offspring, but provide little or no parental care.

Feeding habits involve when, where, and how organisms obtain food. They should not be confused with food habits, which are defined as what an organism eats. Diurnal feeders feed during the day, nocturnal feeders feed at night, and crepuscular feeders are those that feed during the twilight periods. The optimal foraging theory indicates that an organism will obtain food in the most efficient manner possible.

Sociobiology is the study of the biological basis of social behavior and the organization of societies. Sociobiologists specifically add the genetic component to the assessment of behavior. Biologists can use several criteria to assess the degree of sociality in a species.

PRACTICE QUESTIONS

1. Make three lists of organisms with which you are familiar that commonly use (1) vocal communication, (2) visual communication, and (3) chemical communication. Be sure to include a variety of fishes, birds, and mammals on each list.

2. Is competition for limited breeding sites among male yellow-headed blackbirds an example of intraspecific or interspecific competition? Is competition between red foxes and coyotes preying upon meadow voles an example of intraspecific or interspecific competition?

3. Are learned behaviors more likely to occur in an insect such as a mayfly or in a chimpanzee? Why?

4. What is the relationship between parental care, as suggested by the *r-K* continuum, and the state of development in early life (altricial or precocial) for a species?

5. Choose an animal species with which you are familiar. How do both the diel and circannual behavior of an individual of that species vary?

6. What is the difference between a school of fish and an aggregation of fish? Give an example of each.

7. How can a dominance hierarchy result in less agonistic behavior? What are the values of reducing agonistic behavior?

8. Why do northern pintails have an annual migration that places them in northern latitudes on the North American continent during the summer and in southern latitudes during the winter?

9. Why is it to the advantage of the various species of Pacific salmon to migrate from the ocean to fresh water for reproduction?

10. How is sociobiology the study of individual behavior? How is it the study of group behavior?

SELECTED READINGS

Alcock, J. 1993. *Animal behavior: An evolutionary approach*. 5th ed. Sinauer Associates, Sunderland, Mass.

Groot, C., and L. Margolis, eds. 1991. *Pacific salmon life histories*. University of British Columbia Press, Vancouver.

MacArthur, R. H., and E. O. Wilson. 1967. *The theory of island biogeography*. Princeton University Press, Princeton, N.J.

Morse, D. H. 1970. Ecological aspects of some mixed-species foraging flocks of birds. *Ecological Monographs* 40:119–168.

Pianka, E. R. 1970. On *r*- and *K*-selection. *American Naturalist* 104:592–597.

Pitcher, T. J., ed. 1993. *Behaviour of teleost fishes*. 2d ed. Chapman and Hall, London.

Smith, J. M. 1964. Group selection and kin selection. *Nature* 201:1145–1147.

Wilson, E. O. 1975. *Sociobiology: The new synthesis*. Harvard University Press, Cambridge, Mass.

Wilson, E. O. 1992. *The diversity of life*. Harvard University Press, Cambridge, Mass.

LITERATURE CITED

Bleckmann, H. 1993. Role of the lateral line in fish behaviour. Pages 201–246 *in* T. J. Pitcher, ed. *Behaviour of teleost fishes.* 2d ed. Chapman and Hall, London.

Bond, C. E. 1979. *Biology of fishes.* W. B. Saunders, Philadelphia.

Davis, W. H., and H. B. Hitchcock. 1965. Biology and migration of the bat, *Myotis lucifugus,* in New England. *Journal of Mammalogy* 46:296–313.

Erkert, H. G. 1982. Ecological aspects of bat activity rhythms. Pages 201–242 *in* T. H. Kunz, ed. *Ecology of bats.* Plenum Press, New York.

Grier, J. W. 1984. *Biology of animal behavior.* Times Mirror/Mosby, St. Louis.

Guy, C. S., D. W. Willis, and J. J. Jackson. 1994. Biotelemetry of white crappies in a South Dakota glacial lake. *Transactions of the American Fisheries Society* 123:63–70.

Horne, A. J., and C. R. Goldman. 1994. *Limnology.* 2d ed. McGraw-Hill, New York.

Howe, F. P., and L. D. Flake. 1989. Mourning dove use of man-made ponds in a cold-desert ecosystem in Idaho. *Great Basin Naturalist* 49:627–631.

Lagler, K. F., J. E. Bardach, R. R. Miller, and D. R. M. Passino. 1977. *Ichthyology.* 2d ed. John Wiley & Sons, New York.

Neumann, R. M. 1994. Growth, distribution, and movement of northern pike in a South Dakota natural lake, with sampling considerations. Doctoral dissertation, South Dakota State University, Brookings.

Stephens, D. W., and J. R. Krebs. 1986. *Foraging theory.* Princeton University Press, Princeton, N.J.

Todd, J. H. 1971. The chemical languages of fishes. *Scientific American* 224 (5): 98–108.

Welty, J. C., and L. Baptista. 1988. *The life of birds.* 4th ed. Saunders College Publishing, Fort Worth, Tex.

Wilson, E. O. 1975. *Sociobiology: The new synthesis.* Harvard University Press, Cambridge, Mass.

7

SAMPLING
THE BIOTA

To study wildlife and fishery resources, biologists often need to sample organisms. A variety of devices and methods are employed for this purpose. Sampling devices vary in overall shape and size, in the mechanics used to attract or retain the animal, in mesh size if netting or screening is involved, in construction material, and in many other characteristics. The permutations are endless and depend only on the needs of the biologist and the imagination and ingenuity that he or she uses. Fishery biologists refer to capture devices as **gear**; wildlife biologists have no such all-encompassing term for the capture devices they use.

This chapter contains more information concerning the capture of aquatic organisms than terrestrial organisms because the capture devices used in fisheries and aquatic work are more numerous and are generally used more often than those employed in the wildlife field. There are a number of possible reasons for this disparity. Terrestrial species can be seen or heard by humans more readily than aquatic animals, and therefore can frequently be monitored without being captured. Because aquatic animals are seen or heard less often, more of the information on those species is based on capture. Also, because of indeterminate growth, a greater variety of sizes of a fish species must usually be captured in order to sample the species. In addition, aquatic animals live in a more three-dimensional world than terrestrial animals, and this difference affects capture methodology.

This chapter describes the most common methods employed to sample fishes, birds, and mammals, as well as some methods used to sample smaller aquatic and terrestrial organisms, such as plankton and insects. There are many less commonly employed methods and devices that are not included here.

7.1 Sampling: Purposes and Problems

Seldom can an entire population be counted or captured for some particular purpose; samples usually must be taken. A **sample** represents a subset or portion of the total number of organisms in a population. In a broader sense, a sample is a subset or portion of the whole. Samples can be obtained by capturing organisms. For example, a sample of a fish population can be obtained by using a specific kind of net or trap. Samples can also be obtained by utilizing noncapture methods. For example, a sample of the birds present in an area can be obtained by listening to and identifying bird songs at sampling sites.

There are a multitude of reasons why we may need to sample organisms. We may need to sample organisms to estimate population number or to index relative abundance. We may need to sample organisms to determine information such as food eaten, food availability, survival rates, parasites present, health or condition, habitat use, movements, home range, or for a variety of other purposes. We may need to capture an animal in order to transport it and release it in a new area. We may need to capture it so that a radio or sonar device can be attached or implanted, or so that the animal can be tagged or marked in some other fashion.

Some sampling methods are lethal. Other sampling methods are nonlethal, making it possible to sample animals with little or no damage to them. The fate of animals sampled by nonlethal means is variable. Some may be captured and immediately returned to the wild after data such as length or weight are obtained or after being marked or tagged. Others may be live-captured, then transported to a laboratory or other holding facility for further study. Some of these animals may be returned to the wild at the conclusion of the study. Some animals may be captured, transported to a laboratory or other holding facility, and killed for the purpose of the study. Study objectives determine how the animal is sampled and its subsequent fate.

Even those sampling methods that are nonlethal often result in **stress** to an organism. Stress results from biotic and abiotic forces that place pressure on the internal stabilizing mechanisms of an organism beyond their ability to maintain homeostasis. In other words, stress is an internal response by an organism to confront external stressing forces. Because stress can affect such important processes as reproduction, growth, and resistance to disease, it is important that it be minimized. Physical injury can also result from nonlethal capture methods. For example, a capture device that damages the fins of a fish or causes the loss of most tail feathers of a bird affects that organism. Stress and physical injury should be kept to a minimum in sampling.

There are ways to reduce physical damage and stress resulting from sampling. For example, head coverings or blindfolds can be placed on large mammals or birds. Many species can be anesthetized upon capture. Damage can also be reduced by avoiding the use of a sampling device in an area containing large numbers of nontarget species that are susceptible to that particular device. Stress and physical damage problems can be encountered in the use of many of the sampling methods and devices discussed in this chapter.

Many of the devices and methods biologists use to sample animals are illegal, except with special sampling permits. State, provincial, territorial, and federal governments all may be involved in regulating who can sample, how, when, and where sampling can be done, and what can be sampled. Biologists must be aware of these regulations.

The device or method used greatly influences what organisms are sampled. All sampling devices and methods are in some way selective; that is, no matter what the device or method, some organisms are going to be sampled more or less effectively than other organisms. In some situations this selectivity is positive; we can choose methods or devices that are either effective or ineffective for particular organisms. For example, a particular type of net with a specific mesh size may be chosen because it will be effective in sampling a particular species and will also be effective in capturing only a particular size of that species. Conversely, a particular kind of device may be chosen because it will be ineffective at sampling an organism that we wish to avoid capturing. Selectivity can also be negative in that **bias** may

result. Bias is defined as the distance of an estimated value or parameter from the actual value or true parameter (see figure 9.1). The farther the estimate is from the actual value, the more biased the information. Bias, especially bias of which we are unaware, can greatly affect the utility of information. For example, while a certain type of net may be quite effective for sampling dollar sunfish, it may be much less effective for sampling largemouth bass. Using a sample obtained with that net to estimate the relative species composition of both fishes would result in biased information. The dollar sunfish would appear more abundant, which would not necessarily be true.

Susceptibility to capture, and therefore bias, may depend on more than just species; it can also be affected by size, sex, behavior, anatomical characteristics, season, or a variety of other variables associated with organisms. For example, adult orangebelly darters may be readily sampled with one type of gear, while young orangebelly darters may require a totally different type of gear for effective sampling. There are a wide array of devices used specifically to sample fish eggs and young; these gears are usually ineffective for capturing adult fishes. A device designed to capture waterfowl during flightless molting periods would be totally different with regard to effectiveness if it were employed when the birds can fly.

There are multiple reasons why sampling devices or methods can be selective. A net with openings, or meshes, of a size that allows some members of the target population to pass through will result in a selective sample. Placing a device where it will capture one sex more often than the other can also give a selective sample. Placing a device in an area only during the day when most of the target organisms are there only at night is another example. Using a capture device that remains stationary when the target organism is not actively moving will be inefficient. In any wildlife or fishery work that involves sampling, the selectivity or effectiveness of the specific method or device used, in relation to the specific animal to be sampled, must be considered.

7.2 Passive Entanglement Devices

Passive capture devices are put into place, essentially remain stationary, and the movement or behavior of the target organism results in its cap-

ture. **Passive entanglement devices** are stationary devices that snare and tangle the animals encountering the device. Examples of such devices are **gill nets** and **trammel nets** used to capture fishes, **mist nets** used to capture birds, and **drive nets** used to capture some mammal and bird species (fig. 7.1). These nets are usually nonlethal to the target organisms. Passive entanglement nets for fishes are suspended in water. To make these nets most effective, a **float line** (a buoyant line attached to the top of the net) and a **lead line** (a weighted line attached to the bottom of the net) are used to keep the net from collapsing while in the water. Maintaining these nets in an open position ensures that the greatest possible amount of net area is available for fish capture. Mist nets and drive nets are set on land and are held open by tying them to trees, bushes, or moveable poles.

Moving organisms encounter these nets and become entangled in them. Gill, mist, and drive

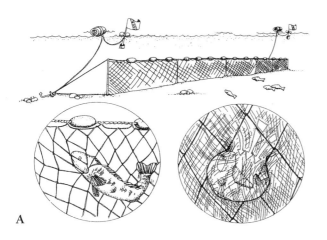

A

B

C

FIGURE 7.1 Passive entanglement devices remain stationary and moving organisms become entangled in them. *(A)* gill net (left), trammel net (right); *(B)* drive net; *(C)* mist net with entangled western tanager. *(Part A drawings courtesy of the American Fisheries Society; part B photograph courtesy of R. Labisky; part C photograph courtesy of K. Jensen.)*

nets have a single panel of netting. When an animal encounters the net, a part of its body may penetrate the mesh, but the entire animal cannot pass through the net. Thus gill coverings, fins, spines, feathers, wings, scales, legs, feet, or other body extensions of the animal may become entangled in the net.

In fishery work, gill nets are sometimes made with numerous sections of different-sized meshes. These nets are called **experimental gill nets** and are used to capture a wider length range of target organisms than is possible with nets of only one mesh size. In addition to net size and mesh size, net material (Collins 1979), color (Jester 1977), and filament diameter (Hansen 1974) can affect what kinds, what sizes, and how many fishes are captured in gill nets.

Trammel nets function on a slightly different principle. They usually consist of three panels of netting, with the inner panel having a smaller mesh size than the outer two; the middle panel also has greater depth and hangs loosely between the two outer panels. Fishes encountering the trammel net pass through the first outer panel, are entangled by the inner panel, and usually pull some of this inner panel of mesh through the second outer panel. The organism is thus entangled in a "pouch" or "bag" of mesh.

While passive entanglement devices are usually intended to be nonlethal, caution must be exercised in their use if mortality is to be minimized. Organisms may be damaged or killed if they remain in the net too long or if they are mishandled after capture. Damage and mortality can be reduced or eliminated by frequent emptying of the nets and careful handling and disentanglement procedures. At times, however, studies require the mortality of the animal.

Passive entanglement devices can be constructed of any net and mesh size, depending on the target organism and purpose of the study. Some nets may be hundreds of meters long, while others may be only 10 meters or less. Similar variation is found in net height. Nets can be constructed of cotton or linen (seldom used anymore), multifilament nylon, monofilament nylon,

or any other appropriate material. Netting made of synthetic material is much more resistant to rotting and tearing than the older cotton or linen nets. Mesh size is the primary factor that determines the size of the animals such nets capture. Smaller meshes are generally more effective at capturing smaller organisms, and larger meshes generally capture larger organisms. In most cases, especially for fishes, the closer the mesh size is to the diameter of the target organism, the more effective the net will be. Additionally, if the correct mesh size is selected, fewer nontarget animals will be captured.

Knowledge of target organism behavior also influences the effectiveness of passive entanglement devices. If low-flying birds are the target organisms, mist nets must be set near the ground. If deep water fishes are the targets, gill or trammel nets need to be set in deep rather than shallow water. Target organism behavior will also dictate whether these nets are set during daylight or at night, in what season they are most effective, and how long they need to remain in place.

At times, the effectiveness of these passive entanglement devices can be enhanced by scaring or driving target organisms in the direction of the nets. Large mammals are often directed into drive nets by humans on foot or horseback or in trucks, snowmobiles, aircraft, or other motorized vehicles. Birds such as sage grouse can be captured by driving hens with broods into mist nets (Browers and Connelly 1986). It is also possible to capture fishes in this manner. An excellent method of capturing adult paddlefish with a gill net is to place the net near the sand bars where these fish congregate, then scare the paddlefish toward the net using a motorized boat.

Another type of passive entanglement device is used primarily by wildlife biologists to capture individual animals; fishery workers have no comparable passive capture devices that are employed to take individual fish. A variety of **leghold traps**, **snap traps** (like the familiar mousetrap), and **snares** occur in this passive capture category (fig. 7.2). These three capture devices are categorized as entanglement devices because they fit better in

A

B

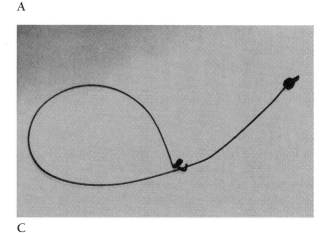

C

FIGURE 7.2 Some passive entanglement devices are used for capturing individual animals. (*A*) leghold trap; (*B*) snap trap; (*C*) snare.

this category than in the category of passive entrapment devices, which capture animals by enclosing them in a specific space (see section 7.3). Leghold traps and snares can be used to capture animals without injury if properly used; snap traps kill the animal captured. Leghold and snap traps come in a variety of sizes, shapes, and construction materials. Padded-jaw leghold traps have been developed to minimize damage to captured animals. The target animal triggers its own capture by stepping upon or in some other way releasing the mechanism that traps it. Snares are usually constructed of steel wires or ropes and are set to capture animals by the leg; they can also be set to capture animals around the neck. When the leg or head of the animal enters the looped end of the snare, the movement of the animal decreases the size of the loop. When the loop is pulled tight,

the animal is caught. A checking device can be placed on a snare to keep the loop from becoming too tight.

Oftentimes with passive capture devices it is necessary to determine how many devices need to be used, in what arrangement they must be placed, and how often the devices need to be checked. Such information is necessary if sampling is to be effective.

It is also imperative that the behavior and habits of the target organism be understood if these devices are to be successful. Setting them in particular places, and in particular ways, will often result in capture of the desired species at the exclusion of others. Individuals who are experienced with traps and snares not only know correct placement, but also know how to use items such as sticks, stones, or logs to direct an animal to the capture device. Some people are sufficiently skilled to capture the target animal by a particular leg.

Often, the effectiveness of traps and snares can be enhanced by providing baits, lures, decoys, or scents to attract the target species. Some of these items, such as fermented eggs or striped skunk musk, can be very odoriferous to humans, while others, such as peanut butter, are relatively innocuous.

A quite different type of snare is composed of a series of monofilament loops set atop a post to snare birds by the feet. This type of entanglement device, when set above a cage containing a prey organism such as a mouse, is called a **bal-chatri trap**. It can be particularly effective for capturing raptors.

7.3 Passive Entrapment Devices

Passive entrapment devices capture organisms that move into an enclosed area by retaining them in that area. Like passive entanglement devices, they remain in a stationary position, thus the term passive. Examples of these devices are **hoop**, **fyke**, **modified-fyke**, and **trap nets**, which are used by fishery biologists (fig. 7.3A, B, C, D); **funnel traps**, which are used to capture organisms such as molting waterfowl, wild turkeys, mourning doves, or gray partridge (fig. 7.3E); **box traps** (sometimes called **live traps**), which are used to

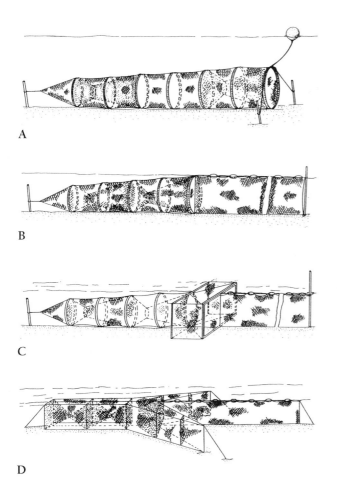

A

B

C

D

FIGURE 7.3 Passive entrapment devices are used to capture a wide variety of aquatic and terrestrial animals. (A) hoop net; (B) fyke net; (C) modified-fyke net; (D) trap net; (E) funnel trap (with wild turkey); (F) box trap; (G) Clover trap (with white-tailed deer); (H) pit trap; (I) corral trap. (*Part A, B, C, and D drawings courtesy of the American Fisheries Society; part G photograph courtesy of J. Jenks; part H photograph courtesy of P. Johnson; part I photograph courtesy of D. Leslie.*)

E

capture a variety of birds and mammals (fig. 7.3F); **Clover traps**, which are generally used to capture deer (fig. 7.3G); **pit traps**, which, depending upon their size, can be used to capture everything from snakes to tigers (fig. 7.3H); and **corral traps**, which are used on a variety of organisms (fig. 7.3I). All of these passive entrapment devices are intended to be nonlethal to the captured animals.

The effectiveness of such passive entrapment devices often depends on the behavior of the organism. For example, fyke netting northern pike in the spring is more successful if the net is placed near the shoreline of a water body, with the **lead net** (directs the animal into the capture part of the device) portion of the fyke net placed so as to direct the fish (fig. 7.4). In spring, northern pike move along the shoreline seeking inflowing streams or flooded terrestrial vegetation for spawning. With correct fyke net placement, the fish are more readily captured. Similarly, a corral trap set for elk works best if placed in an area where natural structures such as rocks or water bodies narrow the escape route; lead nets can also be placed to direct these animals into the trap.

In some situations baits, lures, scents, or decoys can be used to attract target organisms into a passive entrapment device. Clover traps baited with apples can be an effective means of capturing

F

H

G

I

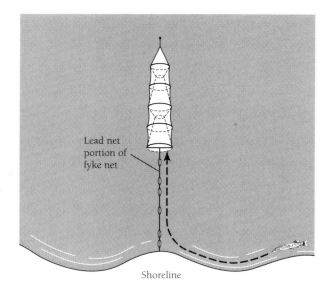

Lead net
portion of
fyke net

Shoreline

FIGURE 7.4   This arrangement of a fyke net and its lead net placed near the shoreline is especially effective for capturing northern pike during their spring reproductive movements.

white-tailed deer. Hanging a bag of stink bait or cheese by-products in a hoop net set for channel catfish can greatly increase the net's effectiveness. A live female duck that serves as a decoy can entice male ducks into a funnel trap during the spring breeding season. A mirror in a box trap can enhance the capture of drumming ruffed grouse. Other attractants that can be used are calls, including chick and prey distress calls, stuffed decoys, and live prey.

It may be necessary for humans to drive or direct target animals to the trap in some situations. This strategy is more effective for some animals than for others. **Big game** (large animals hunted for sport, such as elk, black bears, and bighorn sheep), molting geese, and feral horses are often directed to traps in this manner.

Once in the trap, animals must be retained by the device. There are numerous methods of accomplishing their retention. A net or other blocking device can be put in place, either manually or remotely, to block the exit. In some traps, the opening to the device is designed so that animals can enter easily, but find it difficult to exit. This type of arrangement is demonstrated by the funnel or baffle arrangement of netting in the hoop, fyke, and trap nets used by fishery biologists and

in the various funnel traps used by wildlife biologists. An organism easily follows the large funnel to the small opening into the trap, but rarely finds the opening again from the inside of the trap. Another technique to keep the animal in the trap once it has entered is to have the captured animal trip a release that closes a door or panel on the trap, as in the various box trap designs. Pit traps retain captured animals because they are unable to climb out of a pit into which they have fallen. Pit traps are often constructed with smooth metal sides to increase capture retention.

All passive entrapment devices function by getting the organisms into an enclosed area from which they cannot escape. A fish trap designed to capture a marine fish species may have many lead nets and a containment area that covers many hectares, but it works on the same principle as a small minnow trap. Additionally, a pit trap dug for an animal the size of a tiger works on the same principle as a small pit trap designed for capturing long-nosed leopard lizards.

7.4 Active Capture Devices

Active capture devices require human or mechanical action to move the device; this human-aided movement results in the capture of the target organism. The active capture devices described in this section are generally nonlethal to the target animals.

A common type of active capture gear used in fishery work is the **seine** (fig. 7.5A). Seines have a float line and a lead line, and the ends, especially on smaller seines, are attached to **brails** (poles or

FIGURE 7.5 (*A*) The seine is an effective device for capturing fishes in shallow water areas. (*B*) Adding a bag to a seine can increase its sampling effectiveness. (*Drawings courtesy of the American Fisheries Society.*)

rods), which allow the worker to hold onto and pull the net. Larger seines are often pulled by tractors, other types of motorized vehicles, or winches. Seines come in a variety of shapes and sizes. Sometimes a bag, constructed of the same netting as the seine, is added to enhance its capture effectiveness; such seines are called **bag seines** (fig. 7.5B). Seines are primarily shallow water sampling devices and thus are often called **beach seines**. For these devices to be most successful, the lead line must remain on the bottom of the water body and the float line must not go below the water surface while the seine is pulled to the shore.

Purse seines are used when the net cannot be moved to shore for capture purposes (fig. 7.6). They are primarily used in deep open water, such as large lakes or the ocean, and are put in place with the use of boats. The purse seine is placed in the water in a circular arrangement. Along its

weighted bottom is a purse line, which when drawn together reduces the size of the bottom opening; the top of the purse seine is attached to a float line. When the purse line is completely drawn together, any fishes within the seine are prevented from escaping out the bottom of the net. The enclosed area is then reduced in size by pulling in the seine, and the fishes are ready for removal.

The variety of trawls used in fishery work constitute another type of active capture gear. Trawls are bag- or funnel-shaped nets that are pulled through the water, usually by means of one or more motorized boats. After the trawl has been pulled for some distance, it is brought to the boat and the captured animals are removed through the cod end of the net (the small end opposite the mouth). The mouth or front end of the trawl is held open by various means. **Beam trawls** have a rigid beam across the mouth end of the net (fig. 7.7A). **Otter trawls** have weighted boards (otter boards) that are angled so as to keep the trawl mouth open as it is pulled through the water

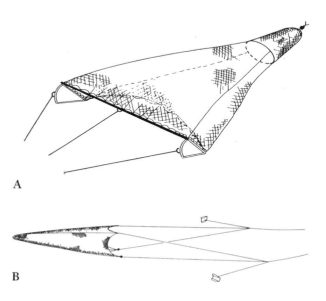

FIGURE 7.7 Trawls can be used to capture a variety of aquatic animals at various water depths. (*A*) beam trawl; (*B*) otter trawl. (*Drawings courtesy of the American Fisheries Society.*)

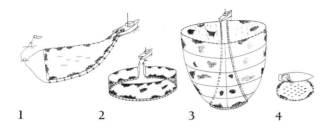

FIGURE 7.6 Purse seines are used in deeper water areas for capturing aquatic animals. Stages 1 through 4 indicate the order in which the purse seine is set and retrieved. (*Drawings courtesy of the American Fisheries Society.*)

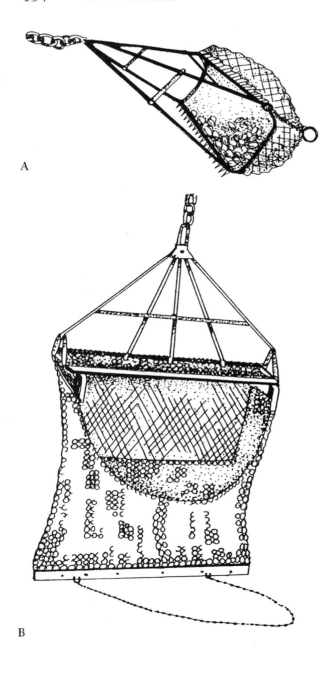

A

B

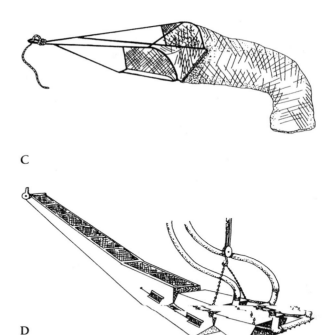

C

D

FIGURE 7.8 Dredges can be used to capture a variety of bottom-dwelling animals. (*A*) oyster dredge; (*B*) scallop dredge; (*C*) crab scrape; (*D*) hydraulic or jet dredge. (*Drawings courtesy of the American Fisheries Society.*)

(fig. 7.7B). Trawls vary in shape and size. Some are used to capture surface and midwater fishes, others are used to capture benthic fishes, and still others are used to capture aquatic animals other than fishes, such as shrimp. Knowledge of target animal behavior, especially location, movement patterns, and gear avoidance behavior, is very important for successful use of these gears.

Dredges are active capture devices used in fishery work to capture a variety of benthic organisms (fig. 7.8). Some dredges are designed specifically to capture mollusks, others are used for crustacean capture, and others are used for a variety of benthic animals. These devices are designed to be dragged over the bottom and usually have blades, rakelike teeth, or other structures that enhance the capture of the target animals.

Some active capture devices are also used in wildlife work. Examples are **cannon nets, rocket**

A

B

C

FIGURE 7.9 Some active capture devices can be used to sample wildlife. (*A*) cannon net; (*B*) drop net (with wild turkeys); (*C*) net gun. (*Part C photograph courtesy of J. Ratti.*)

nets, **drop nets**, and **net guns** (fig. 7.9). Cannon, rocket, and drop nets are used to capture a variety of bird and mammal species; their success is often enhanced by baits or decoys that attract target organisms to the proximity of the net. Cannon nets are lightweight nets that are carried over a group of animals by mortar-type projectiles. Rocket nets are propelled by small rocket-type projectiles. A drop net is suspended above the ground, and when target animals are under the net, it is dropped. Net guns fire projectiles that are attached to a net. Such guns are usually fired from the shoulder or are mounted on helicopters. Net guns have been used effectively on a variety of bird and mammal species.

In addition to these active capture devices, wildlife and fishery personnel use a variety of hand-held nets. These nets, which have variable handle lengths, net diameters, and mesh sizes, can be used from vehicles, boats, or on foot. Such nets are employed to capture a variety of animals, from dip nets used on fishes to lightweight, large-diameter hand-held nets used to capture ground-roosting birds. Vehicle- or backpack-powered lights can be used if the captures take place at night. Loud music has even been used to distract some upland game birds during nighttime lighting and netting operations so that they may be more effectively captured.

Terms such as passive, active, entanglement, and entrapment have always been used in reference to fishery gears. The terms have not generally been applied to capture devices used in the wildlife field. However, these terms do apply to the ways that wildlife sampling devices and methods work. Therefore, it seems appropriate to use these terms in relation to wildlife sampling as well.

7.5 Anesthetics for Sampling Wildlife

Anesthetics are most commonly used by wildlife biologists for live capture of large organisms. Anesthetics are used in fishery work also, but usually after capture. Wildlife anesthetics can be delivered in several ways. The most common method of delivery for larger mammals is a **dart gun** (fig. 7.10A). The darts are syringelike projectiles designed so that a predetermined amount of anesthetic is injected when the animal is struck (fig. 7.10B). The projectile gun is usually an air gun or a powder-charged gun, but a bow and arrow, blowgun, or even the end of a pole or stick can be used to deliver the dart or injection.

When using anesthetics to capture animals, care must be taken to ensure that target animals are not overdosed or in some other way harmed. Television and other media describing wildlife capture activities often depict the darting anesthetization of large animals as a rather simple and foolproof activity. It is seldom mentioned that injury or mortality is not uncommon with such a capture method. Anesthetization is not as simple as it sounds because animal body size must usually be estimated to determine the proper drug dosage, and because there is considerable variation in the response to drugs by individual animals, even within the same species. Even if proper drugs and dosage levels are used, injury and mortality can still occur.

Anesthetics can also be used to capture animals by placing the drug in food. Once the target animal eats the treated food and is anesthetized, the drugged animal is available for study. For example, alpha-chloralose has been used to capture wild turkeys in this manner (Williams 1966).

A variety of drugs are used to immobilize wildlife species, with each having a particular effectiveness depending on the target organism in question. Most of the drugs used are legally controlled substances and can be utilized only by licensed users. Drugs commonly used to immobilize large mammals include carfentanil citrate, ketamine hydrochloride, tiletamine hydrochloride, zolazepam hydrochloride, xylazine hydrochloride,

A

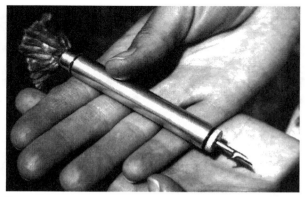

B

FIGURE 7.10 A dart gun (*A*) and dart (*B*) used to anesthetize animals. Care must be taken with this device to prevent injury to the target animal.

acepromazine maleate, diazepam, yohimbine hydrochloride, doxapram hydrochloride, and atropine sulfate. With these and other drugs, it is important to be aware of how they can be administered, the concentration levels needed, their side effects, drugs (if available) that can be used to reverse their action, and other aspects of their usage. It is also a good policy to have a qualified veterinarian as a member of any capture team using anesthetics.

7.6 Toxicants for Sampling Fishes

Toxicants, or poisons, called **piscicides (ichthyocides)** because they affect fishes, are used in fishery work; wildlife biologists have no comparable means of capture. Piscicides are toxicants that, when added to water, result in the immobilization of fishes. This is primarily a lethal capture method.

Toxicant use is one of only two methods available for sampling all of the fishes present in a body of water. The other method available for the complete sampling of fishes is draining a water body.

A variety of piscicides have been used over the years. Most, because of their harmful or potentially harmful effects on the environment and on humans, are no longer legally usable. Currently, the most common legally approved piscicides are **rotenone** and **antimycin**. These are general-purpose toxicants used on a variety of fish species. Rotenone is a naturally occurring material found in some plants and is sold under a variety of brand names. It can be purchased in liquid or powder form and comes in various concentration levels. Rotenone blocks oxygen uptake in the gills of fishes. The effectiveness of rotenone is dependent upon fish size and species and upon water temperature, pH, oxygen concentration, the amount of organic matter, and turbidity. It is used

at parts-per-million concentration levels. Rotenone affects all gill-breathing animals in aquatic environments.

Antimycin, an antibiotic, also inhibits the oxygen uptake of fishes, but at the tissue level. It is sold under the brand name Fintrol. The effects of antimycin are modified by fish size and species, water temperature and pH, sunlight, and fish metabolic activity. It is used at parts-per-billion concentration levels because it is more toxic to fishes than rotenone.

Both rotenone and antimycin can be used somewhat selectively for some fish species in the presence of other fishes because not all species have the same susceptibility to the same concentration levels. Both piscicides can be inactivated by potassium permanganate; however, potassium permanganate is also toxic if concentration levels are too high.

While rotenone and antimycin are general-purpose piscicides, there are two other piscicides that are used for more specific purposes. These toxicants, **TFM (3-trifluoromethyl-4-nitrophenol)** and **Bayer 73**, are used for sea lamprey control. **Lampricides** (kills lampreys) are used primarily on larval sea lampreys in Great Lakes tributary streams. Their primary purpose is to reduce sea lamprey populations in order to reduce the negative effects of this introduced species on other fishes such as native lake trout (see section 10.9).

Toxicants can be applied in a variety of ways, ranging from boat-mounted devices to hand-held sprayers (fig. 7.11). Boat-mounted devices are used in large water bodies; hand-held sprayers can be used in very small, shallow water bodies and in difficult-to-reach portions of large water bodies. At times, even helicopters have been used to apply the toxicant. After the fishes are immobilized by the toxicant, they are quickly collected with hand-held nets (fig. 7.12). Some of those that sink before collection will float to the surface in a day or two.

The use of toxicants has declined in recent years. Problems with toxicant effects on nontarget species, potential effects on human users, aversion to large quantities of dead fishes for aesthetic and

A   B

FIGURE 7.11 Piscicides can be applied in various ways. Two of the more common methods are (A) application from boat-mounted devices (the white material in the boat's wake is rotenone) and (B) application by hand-held sprayers.

ethical reasons, and a general increasing aware-ness of the negative effects of chemical applica-tions have all played roles in reducing piscicide usage. However, under the proper circumstances and when correctly used, toxicants are an effective method for sampling fishes or eliminating fish communities.

Chemical use in fishery (and wildlife) work is a rapidly changing area. Depending upon the chemical and its use, its control in the United States may come under the jurisdiction of the Environmental Protection Agency or the Food and Drug Administration. There are state and federal programs to register and re-register these chemi-cals for use. Regulations regarding which chem-icals can be legally employed, who can apply them, and the animals on which they can be used change from year to year. Biologists must be aware of all current regulations with regard to these chemicals. The chemicals listed as being applica-

ble in fishery work in this section of the book were legal at the time of publication; they may not be legal at some point in the future. The same is true of chemicals utilized in wildlife work, such as wildlife anesthetics (see section 7.5).

FIGURE 7.12 After a piscicide has been applied, the fishes must be collected. This is usually done with hand-held dip nets.

7.7 Electrofishing

The use of electricity to capture animals is a technique primarily employed in fishery work. Electrofishing devices take many forms. Electrofishing units employed in large water bodies are usually mounted on boats (fig. 7.13A). Backpack electrofishing units are usually employed in small water bodies, such as streams (fig. 7.13B). Electric seines are also used in some situations for fish capture (fig. 7.13C). There have been substantial improvements in electrofishing technology in recent years. Most improvement efforts are directed at making these devices more effective, while reducing potential physical damage to fishes and human operators.

Electrofishing devices function by disrupting the nervous systems of fishes. When correctly used, they can temporarily immobilize a fish without causing damage. After the fish is immobilized by the electric current, it can be captured with a hand-held net. Electrofishing is a nonlethal capture method; however, if too much current is applied or if a fish is too close to an electrode (the terminal of the electrical source), ruptured blood vessels, broken vertebrae, or other damage can occur. Such damage may or may not result in fish mortality. The currents most commonly employed are alternating current (AC), direct current (DC), and pulsed DC. The effect of AC is to stun fishes anywhere in the field; DC attracts fishes to

FIGURE 7.13 Electrofishing devices are employed in a nonlethal method of capture. *(A)* boat-mounted unit; *(B)* backpack unit; *(C)* electric seine. *(Part B photograph courtesy of T. Modde; part C photograph courtesy of C. Berry.)*

A

B

C

the positive electrode; pulsed DC extends the duration of attraction to the positive electrode. With most newer units, the operator has control over voltage, amperage, and other electrical field characteristics. This control allows for the selection of electrical characteristics that are particularly effective on various fish species and sizes. The electricity is usually produced by either a gasoline-powered generator or batteries, but hand-operated generators have also been used.

Among the variables that influence the effectiveness of electrofishing devices are differences in water conductivity (its capacity to conduct electricity), fish size, fish behavior, water temperature, and the method of electrical delivery. In addition, fish species, either because of their physiological characteristics or because of the habitat in which they dwell, differ in their susceptibility to electrical currents. Electrofishing devices are primarily used for shallow water sampling.

The use of electricity in proximity to water requires many safety precautions to protect the operator. Special equipment, such as rubber gloves and boots, life jackets, and dip nets with insulated handles, must be used. Circuit-breaker switches and foot-operated pedals are also usually designed into electrofishing systems. Special training concerning essential safety precautions and cardiopulmonary resuscitation also must be conducted. Electrofishing units are dangerous pieces of equipment if improperly handled.

▼

7.8 Sampling Small Terrestrial and Aquatic Organisms

In addition to large terrestrial and aquatic biota, small organisms such as insects or plankton may also need to be sampled. Many different methods must be employed because of the various habitats

in which these organisms are found and their differing behavior patterns. Some burrow into terrestrial soil or aquatic system bottoms. Others fly or swim. Some crawl or walk on the land surface or in aquatic systems. Some inhabit terrestrial or aquatic vegetation. Others live in water for most of their lives before entering a terrestrial life stage.

Terrestrial invertebrates are important to terrestrial wildlife species, but can also influence aquatic animals. For example, terrestrial invertebrates constitute a major portion of the diets of some stream fishes. Conversely, the potential effects of aquatic invertebrates on fishery resources are numerous, but these aquatic animals also influence water-associated wildlife species such as waterfowl and raccoons. In addition, many aquatic animals have terrestrial life phases; these animals can be important to a variety of animals including totally terrestrial ones.

Many terrestrial invertebrates can be sampled by taking a specific measured amount of material such as soil, grass, or tree leaves and counting the animals present in that material. The animals can be hand-separated from the material, but this is a tedious procedure. More commonly, separation is accomplished with a **Burlese-Tullgren funnel** (fig. 7.14). The soil or other medium containing the invertebrates is placed on a screen located in the funnel portion of the device. A heat source at the top of the funnel forces the invertebrates out of the sample into a collecting jar. Numerous modifications can be made to this device; for example, other methods of forcing or attracting animals into the sampling jar can be used.

Pit traps (see section 7.3) can be used for capturing a variety of small terrestrial animals that move along the soil surface. Pit traps can be provided with attractants or even killing agents and preservatives to increase trapping effectiveness.

The kinds and numbers of terrestrial invertebrates found on vegetation can be determined by using tubes, boxes, or nets that totally enclose an item such as a tree branch or a patch of grass. After total enclosure, the animals must be separated from the vegetative material. They can be re-

FIGURE 7.14 Burlese-Tullgren funnels are used to separate small animals from material such as soil, grass, or leaves.

tain sticky substances and traps that use lights, baits, or other attractants.

Many invertebrates spend all of their lives as terrestrial forms; others spend the early stages of their lives in aquatic systems and become terrestrial in a later life stage. These organisms can be collected with some of the devices already mentioned in this section. They can also be captured as they emerge from the ground or the water to begin their next life stage. There are a number of different types of such **emergence traps** that are used to capture both terrestrial invertebrates and those that are emerging from aquatic life stages (fig. 7.16). Some work best in flowing water, while others are designed to capture animals emerging from standing water. Some of the devices float on the water surface; others are sub-

moved by hand picking, or with the funnel device already mentioned. They can also be taken directly from the environment with a device such as a **D-vac sampler** (fig. 7.15). This sampler is essentially a large field vacuum cleaner that collects the animals in a cloth bag located in the collection head.

Terrestrial invertebrates can also be collected with **sweep nets**, which are similar in structure and function to the hand-held nets used to capture fishes, birds, and mammals. Many people remember the hours of childhood enjoyment associated with using these "butterfly" nets.

At times, measurements of the relative abundance of invertebrates found in trees are needed. One method of accomplishing this sampling is by striking tree branches with a pole or stick and collecting the dislodged animals on canvas sheets or in funnel devices located under the trees. Pesticides can also be used to dislodge such organisms.

Aerial insects can also be sampled with **suction traps**. These traps draw in a specific volume of air so that estimates of the total number of collected organisms can be quantified. There are numerous other devices that can be used to capture aerial invertebrates, including traps that con-

FIGURE 7.15 D-vac samplers, which are essentially field vacuum cleaners, can be used to collect terrestrial invertebrates.

FIGURE 7.16 Emergence traps are designed to capture invertebrates as they emerge from an aquatic life stage to begin a terrestrial existence, or as they emerge from the ground. The organisms are collected in the jar.

merged. In addition, some use vacuum devices or sticky substances to increase their effectiveness.

There are many other invertebrate capture devices that could be mentioned. For example, two additional traps designed specifically for capturing flying invertebrates are the **flight-intercept trap** (fig. 7.17A), which captures and holds organisms below the trap in special collection devices, and the **ultraviolet survey trap** (fig. 7.17B), which

A

B

FIGURE 7.17 Specifically designed traps are used to capture both terrestrial and emergent aquatic flying organisms. *(A)* flight-intercept trap; *(B)* ultraviolet survey trap. *(Part A photograph courtesy of P. Johnson.)*

because of the sizes of the organisms captured, are smaller and have finer mesh sizes than the sampling nets previously mentioned in this chapter. While mesh sizes for fish and wildlife capture devices are often measured in centimeters, one centimeter of plankton net could have hundreds of openings or meshes. Even at such fine mesh sizes, mesh size still influences what is captured. Most phytoplankton require a smaller mesh size than do generally larger zooplankton. Some plankters are too small for even the smallest net meshes and must be extracted from water samples using filter paper or by centrifuging. Plankton nets can be pulled by hand or by boat. Some have metering devices to measure how much water has passed through the net. Some devices are used primarily to sample fish larvae and eggs. For example, **meter nets** (fig. 7.18B) are commonly used for fish eggs and larvae.

If the aquatic biota being sampled are not suspended in the water column, other devices need to be employed. A variety of devices can be used to sample benthic organisms or organisms that are attached to submerged objects such as logs, rocks, or vegetation. For sampling benthic organisms in soft bottoms such as mud, sand, or small rocks, a variety of grab samplers can be used; two common ones are **Ekman** and **Petersen grabs** (fig. 7.19). These devices, when lowered to the bottom, grab a portion of the bottom material, which can then be sifted or strained to separate the benthic organisms from unwanted organic and inorganic material. Other devices, such as the **Surber stream sampler**, are used to collect larger organisms attached to bottom objects in flowing water environments (fig. 7.20). The organisms are manually dislodged from rocks, vegetation, or other objects and are then carried by the water current into the sampler, which has been placed downstream. Organisms that are attached to larger objects and cannot be captured by grabs or other samplers can be manually removed from the submerged objects. In addition, biologists can place artificial substrates in the water body, wait for organisms to colonize them, and then remove the artificial substrates for assessment.

A

B

FIGURE 7.18 Fine mesh nets are used to capture smaller organisms. (*A*) Wisconsin plankton net; (*B*) meter net.

attracts night-flying insects with its light. Both are used for terrestrial animals and aquatic emergents. Some aquatic organisms are quite small and require specially designed sampling devices. The **Wisconsin plankton net** is a device commonly employed for capturing plankton (fig. 7.18A), but others, such as the **Juday net** and the **Clarke-Bumpus sampler**, are also used. These devices,

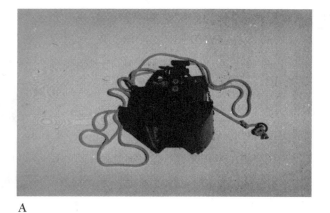

A

B

FIGURE 7.19 Several types of grabs can be used to sample bottom material and the organisms associated with that material. (*A*) Ekman grab; (*B*) Petersen grab.

FIGURE 7.20 Surber stream samplers are used to collect larger benthic invertebrates in flowing water environments.

7.9 Combining Sampling Devices

There are times when sampling devices are best used in combination. In sampling with electrofishing gear, there are situations in which portions of the water body being sampled are blocked with nets to reduce fish escapement. Nets are also commonly used with piscicides to define the sampling area and prevent fish entry or escape. Fishes captured in purse seines are often removed from the seine using hand-held or mechanically operated dip nets. Brown bears may initially be captured using a snare, then anesthetized. House Finches may be initially captured in a funnel trap, then removed from the trap with a hand-held net.

7.10 Noncapture Sampling

Not all sampling requires the actual capture of organisms. **Drag lines** are ropes or cables that can be pulled by two people or two vehicles across fields to flush ground-nesting birds, which can then be counted or their nests examined (fig. 7.21). These lines are specifically designed to eliminate damage to any nests they may encounter. In water, photoelectric cells can be used to count fishes as they pass through a confined area. **Hydroacoustics** are also used in fishery work and

FIGURE 7.21 Drag lines are used to flush ground-nesting birds so that nests can be located or birds counted. (*Photograph courtesy of the U.S. Fish and Wildlife Service.*)

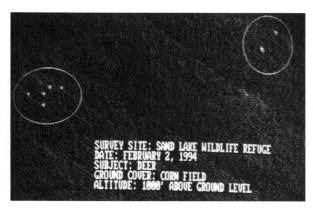

SURVEY SITE: SAND LAKE WILDLIFE REFUGE
DATE: FEBRUARY 2, 1994
SUBJECT: DEER
GROUND COVER: CORN FIELD
ALTITUDE: 1000' ABOVE GROUND LEVEL

FIGURE 7.22 Infrared photography, which detects heat, can be used to count animals. The large light splotches in the encircled areas are white-tailed deer. This photograph was taken under snow cover conditions. (*Photograph courtesy of J. Jenks.*)

involve the use of sonar. Some hydroacoustic devices are very technologically advanced, and the use of hydroacoustics for activities such as fish enumeration and behavioral studies is likely to increase as these devices become even better developed. Video or film photography, either normal or infrared, can also be used to sample organisms (fig. 7.22). Cameras can be manual, time-lapse, or activated by the target animal. Birds, mammals, and even fishes can be viewed with optical equipment, such as binoculars and spotting scopes, or the unaided eye. Some optical equipment can be quite sophisticated, such as **starlight** or **night-vision scopes** (fig. 7.23). Fishes can be observed with a variety of underwater viewing devices, or observed by scuba or snorkel-equipped divers.

Counts of some animals can be made from vehicles, aircraft, or with satellite imagery. A common practice, especially in the wildlife profession, is to sample with the use of lights. Using lights at night to find and count animals such as white-tailed deer, American alligators, swift foxes, and coyotes is a common sampling method. Sampling may involve seeing the whole animal or just the light reflecting from its eyes.

Birds and mammals can be counted in a variety of ways that do not even require the actual viewing of the animal. For example, counts or indices of abundance can be obtained from bird songs or calls, ruffed grouse drumming, coyotes calling, elk bugling, or white-tailed deer droppings (**scats**, **fecal pellets**, or **pellets**). In addition, information on some birds and mammals can be obtained from their tracks. The counting of tracks is often enhanced by smoothing dirt and adding scent attractants to the area, as in work with coyotes or red foxes. Track counting under

FIGURE 7.23 Night-vision or starlight scopes can be used to see animals under low-light conditions.

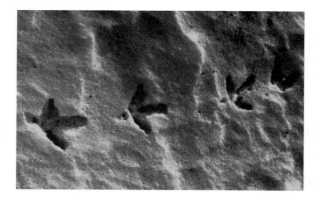

FIGURE 7.24 Tracks can be used as an indicator of abundance for various animals. These are ring-necked pheasant tracks in snow. (*Photograph courtesy of South Dakota Tourism.*)

snow cover conditions can also be an effective sampling method (fig. 7.24). For smaller organisms such as mice, carbon-blackened paper attached to trees or located in other areas the mice inhabit can be used to record tracks. When a mouse moves across the paper, it leaves footprints in the carbon black.

7.11 Additional Sampling Devices

Not all sampling devices are designed and made specifically for scientific use. Under certain circumstances, regular fishing gear, used by the biologist with either natural baits or artificial lures, can be effective for fish capture. Even spearguns used in conjunction with scuba or snorkel diving can be effective. In addition, sport and commercially captured fishes can yield a variety of valuable data. Useful data can also be obtained from wildlife species taken by hunters, trappers, or biologists with shotguns, rifles, bows, traps, or other legal devices. The use of sport-harvested fishery and wildlife biota for data collection has the added benefit of reducing the cost of sampling. Even road-killed organisms can provide useful information. For example, in many states it is common to check road-killed deer to determine the number of young that pregnant females are carrying. Such information is helpful in determining the health of the deer herd and potential population trends.

Numerous other devices and methods not mentioned in this chapter have been employed to capture animals involved in wildlife and fishery work. These vary from the use of explosives in fishery work to the use of hand-held snare poles to capture wildlife. In addition, sampling of terrestrial and aquatic invertebrates and fish larvae and eggs is carried out by means of devices and methods too varied and too numerous to be discussed here. We reiterate, however, that all capture methods are selective and can introduce bias into the information obtained. We must also reiterate the importance of our responsibilities to these animals for humane care and handling. Various professional societies have developed protocols and procedures for correct animal handling. The topic of animal welfare is discussed in section 16.6. In addition, there are regulations, at various government levels, concerning the welfare of captured and held animals.

SUMMARY

A sample is a subset of the organisms in a population. Who can sample; how, where, and when sampling can be done; and what animals can be sampled is regulated. Many of the specialized devices used in wildlife and fisheries for sampling or capturing animals require special permits.

All sampling methods are selective and can introduce bias into the information obtained. Sampling devices can be selective for such variables as species, size, sex, and age. Regardless of whether a sampling device is lethal or nonlethal, animal welfare must be considered during any sampling activity.

Sampling devices tend to be either passive or active. Passive devices are put into place and the target animals move or are driven into the devices. Active devices are themselves moved, and this movement results in animal capture. Passive entanglement devices capture animals by entangling some body part such as a leg, wing, or fin. Passive entrapment devices capture animals by enclosing the entire animal in a confined area from which it cannot escape.

Anesthetics are primarily used by wildlife biologists for the live capture of animals. Anesthetizing drugs are usually delivered with syringelike projectiles delivered by a dart gun. The use of toxicants for sampling is primarily a fishery activity. General-purpose toxicants such as rotenone and antimycin are used, as are toxicants specific to sea lampreys. The use of chemicals in wildlife and fishery work is controlled by a variety of regulations, which change frequently. Biologists must be aware of current regulations. Electrofishing is primarily used in fishery work. This method is generally used in shallow water and results in the live capture of fishes.

Specific devices and methods are used to sample small aquatic and terrestrial organisms, such as insects and plankton. These range from nets that strain small organisms from water to vacuum devices used to sample insects from trees.

Not all sampling requires that animals actually be captured. At times all that is needed for sampling is to see or hear an animal or observe some evidence that it is or was present.

PRACTICE QUESTIONS

1. Identify a dozen reasons why a fishery biologist might need to capture a plains minnow.

2. Why might the capture of a young wild turkey require a different sampling method or strategy than the capture of an adult wild turkey?

3. List four types of passive entanglement devices and describe how they capture an animal.

4. List four types of passive entrapment devices and describe how each may be selective for a particular organism.

5. List four types of active capture devices and describe how they capture an animal.

6. Why is the use of anesthetics delivered by dart guns unpredictable in regard to the effects on the target animal?

7. How do AC and DC differ in their effects on fishes?

8. Why has the use of piscicides as a sampling technique decreased in recent years?

9. Describe a dozen different ways that animals can be sampled without capturing them.

10. Describe four devices used to capture or sample aquatic invertebrates.

11. Describe four devices that can be used to capture or sample terrestrial invertebrates.

SELECTED READINGS

Backiel, T., and R. L. Welcomme, eds. 1980. *Guidelines for sampling fish in inland waters.* EIFAC/T33. Food and Agriculture Organization of the United Nations, Rome, Italy.

Davies, W. D., and W. L. Shelton. 1983. Sampling with toxicants. Pages 199–213 *in* L. A. Nielsen and D. L. Johnson, eds. *Fisheries techniques.* American Fisheries Society, Bethesda, Md.

Hayes, M. L. 1983. Active fish capture methods. Pages 123–145 *in* L. A. Nielsen and D. L. Johnson, eds. *Fisheries techniques.* American Fisheries Society, Bethesda, Md.

Helfman, G. S. 1983. Underwater methods. Pages 349–369 *in* L. A. Nielsen and D. L. Johnson, eds. *Fisheries techniques.* American Fisheries Society, Bethesda, Md.

Hubert, W. A. 1983. Passive capture techniques. Pages 95–122 *in* L. A. Nielsen and D. L. Johnson, eds. *Fisheries techniques.* American Fisheries Society, Bethesda, Md.

Murkin, H. R., D. A. Wrubleski, and F. A. Reid. 1994. Sampling invertebrates in aquatic and terrestrial habitats. Pages 349–369 *in* T. A. Bookhout, ed. *Research and management techniques for wildlife and habitats.* The Wildlife Society, Bethesda, Md.

Pond, D. B., and B. W. O'Gara. 1994. Chemical immobilization of large mammals. Pages 125–139 *in*

T. A. Bookhout, ed. *Research and management techniques for wildlife and habitats.* The Wildlife Society, Bethesda, Md.

Reynolds, J. B. 1983. Electrofishing. Pages 147–163 *in* L. A. Nielsen and D. L. Johnson, eds. *Fisheries techniques.* American Fisheries Society, Bethesda, Md.

Schemnitz, S. D. 1994. Capturing and handling wild animals. Pages 106–124 *in* T. A. Bookhout, ed. *Research and management techniques for wildlife and habitats.* The Wildlife Society, Bethesda, Md.

Snyder, D. E. 1983. Fish eggs and larvae. Pages 165–197 *in* L. A. Nielsen and D. L. Johnson, eds. *Fisheries techniques.* American Fisheries Society, Bethesda, Md.

Thorne, R. E. 1983. Hydroacoustics. Pages 239–259 *in* L. A. Nielsen and D. L. Johnson, eds. *Fisheries techniques.* American Fisheries Society, Bethesda, Md.

Wetzel, R. G., and G. E. Likens. 1991. *Limnological analyses.* 2d ed. Springer-Verlag, New York.

LITERATURE CITED

Browers, H. W., and J. W. Connelly. 1986. Capturing sage grouse with mist nets. *The Prairie Naturalist* 18:185–188.

Collins, J. J. 1979. Relative efficiency of multifilament and monofilament nylon gill net towards lake whitefish (*Coregonus clupeaformis*) in Lake Huron. *Journal of the Fisheries Research Board of Canada* 36:1180–1185.

Hansen, R. G. 1974. Effect of different filament diameters on the selective action of monofilament gill nets. *Transactions of the American Fisheries Society* 103:386–387.

Jester, D. B. 1977. Effects of color, mesh size, fishing in seasonal concentrations, and baiting on catch rates of fishes in gill nets. *Transactions of the American Fisheries Society* 106:43–56.

Williams, L. E. Jr. 1966. Capturing wild turkeys with alpha-chloralose. *The Journal of Wildlife Management* 30:50–56.

8

DETERMINATION AND USE OF AGE, GROWTH, AND SEX INFORMATION

Information on age, growth, and sex, provides insight into the status and health of populations of wild animals. For example, age and growth information can be used to evaluate various harvest strategies. If a fish population is being over-harvested, it will be characterized by young, rapidly growing fish and few large individuals. If age and growth data collected for a fishery reveal this pattern, biologists can then implement appropriate measures, such as harvest regulations or habitat enhancement, to improve the quality of that fishery. Perhaps, while fishing, you helped to provide such data for evaluating a fish population when biologists checked your fish for length and weight and collected a few scales.

Wildlife biologists can determine the relative productivity of various populations, such as American wigeon, fox squirrels, or elk, by determining the proportion of the population that occurs in various age categories. You may have assisted in collecting such data by mailing waterfowl wings to the U.S. Fish and Wildlife Service or deer or elk incisor teeth to a state, provincial, or territorial wildlife agency. The coloration and shape of various feathers on duck wings in the fall can be used to determine whether a duck was hatched the preceding summer or more than a year earlier. Likewise, growth rings on the roots of deer or elk incisors can be used to estimate the age of the elk or deer. The pre- and post-hunting season sex composition of harvested animal populations, such as ring-necked pheasants in which males are selectively harvested, can be used to estimate the percentage of the population killed by hunters. In projects involving banding, marking, tagging, or radiotelemetry, it is often necessary to determine sex or age of an animal at the time of capture.

Age, growth, and sex information can be used in a variety of ways by biologists to understand and conserve wildlife and fishery resources. This chapter is a review of some of the more common methods used in obtaining age, growth, and sex data and includes discussions of the application of such information.

8.1 Age Category Terminology

The terminology used in grouping animals by age can be confusing, and there are differences among the terms used by wildlife and by fishery biologists. When fish population structure is evaluated, biologists generally place fishes in either age groups or year classes. **Age groups** represent the number of years a fish has lived. A fish in its first year of life is a member of age group 0, a fish in its second year of life is a member of age group 1, and so on. Thus, the age group of an individual fish changes each year. Commonly the "group" portion of the term is deleted; thus fishes in age group 1 are usually called age-1 fishes. All fishes are assumed to have a 1 January "birthday" (more properly "hatch-day" for most fishes). In other words, regardless of whether a fish was hatched in March or August of a year, it will become an age-1 fish on 1 January of the next year. Alternatively, fishery biologists can place fishes in **year classes** instead of age groups. If a fish was hatched in 1993, then that fish would be a member of the 1993 year class. In two years, four years, or nine years, it would still be a member of the 1993 year class.

The term **young-of-the-year**, which refers to animals in the early portions of their lives, is commonly used in both fishery and wildlife work. In fisheries, there is an effort to replace the term young-of-the-year with the term age 0. The term **subyearling** is also used in fisheries to denote a fish of this age. A variety of other terms, such as **fry** and **fingerling**, are commonly used to refer to young fishes. However, these terms are variously defined, often have different meanings to different people, and should not be used unless they are specifically defined. Piper et al. (1982) defined fry as fishes between the time of hatching and the time at which they reach 25 millimeters in length, and fingerlings as fishes between 25 millimeters

FIGURE 8.1 Parr marks (arrows) are characteristic of young salmonids such as this rainbow trout. The parr marks on this fish are fading as it reaches adulthood.

and the length at age 1. Newly hatched fishes are also often referred to as **larval fishes.** Various terminologies are used for phases of larval fish development (Snyder 1983), but no one system has been universally accepted.

In addition, specific terminology is used for young members of the salmon family. **Parr** refers to the salmonid life stage from the time of first feeding until fishes become sufficiently pigmented to obliterate their parr marks. **Parr marks** are vertical pigmentation bands found on the sides of young salmonids (fig. 8.1). Some fish species other than salmonids also have parr marks. Salmonids that migrate from fresh water to the ocean become **smolts** at the time of physiological adaptation to life in the marine environment. The parr marks are typically lost at this time.

The age group system used on fishes, with the assumed 1 January birthday, is not normally applied to wildlife, although that system would work for species whose age can be determined in years. The year class system is sometimes used for wildlife in a manner identical to that in fisheries. For example, one may find reference to the 1989 year class of elk, meaning the elk that were born in that year. Most often in wildlife, the term **age class** is used to refer to chronological age in years and to various age or life stage groupings; the term has a very broad usage. For example, individuals less than 1 year of age may be grouped as age class 0–1, age class 0, age class 0.5, or age class 1. To further confuse the issue, the term year class in wildlife is sometimes used interchangeably with the term age class. In this text, year class will be used as defined in fishery work and will not be used interchangeably with other terms.

The term cohort is sometimes used synonymously with age class, year class, and age group. However, while a **cohort** can be a group of individuals in a population born or hatched during a one-year period, a cohort can also be defined on the basis of shorter or longer time periods. For example, larval fish cohorts may be defined as narrowly as those hatched on one day.

The terms **immature** or **juvenile, subadult,** and **adult** are also used in both wildlife and fisheries, further illustrating the broad usage of a variety of terms having to do with age classification. In wildlife, the terms immature and juvenile are used interchangeably for animals that are too young to breed and can be distinguished from adults based on external characteristics. For example, a three-year-old bald eagle lacks a white head and tail and is easily distinguishable from an adult; thus, it is classed as a juvenile or immature bird. In the fall, a greater white-fronted goose that hatched in late June lacks the dark markings on the breast and belly and the white on the front of the head found on subadults (one and two years old) and adults; thus, it is classed as a juvenile (fig. 8.2). In wildlife, the term subadult is used specifically for individuals that are too young to breed but are externally indistinguishable from adults. For example, two-year-old Canada geese are classed as subadults because they look like adults, but usually do not breed until three or four years of age, at which time they are adults. Two-year-old black bears are subadults because they look like adults but are not reproductively active. In fisheries, the terms immature, juvenile, and subadult tend to be used interchangeably; the wildlife usage of these terms is usually more

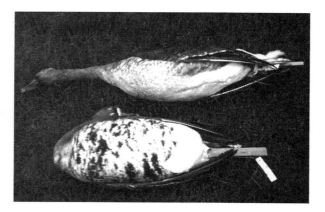

FIGURE 8.2 During fall migration, juvenile greater white-fronted geese (top) lack the dark markings on the breast and belly and white on the front of the head found in subadults (one- and two-year-olds) and adults (bottom).

birds younger than one year old (hatched that year), while birds one year old and older are grouped as AHY. For species such as northern bobwhites, which reproduce or attempt to reproduce at one year of age, the terms HY and juvenile or immature can be used interchangeably to describe those that have not reached breeding age; the terms AHY and adult can be used after they reach sexual maturity, at approximately one year. The term HY is more specific than juvenile because some birds are still juveniles after 1 year.

<hr/>

8.2 Determination of Age

The most common methods of age determination are those that use hard body parts such as scales, feathers, and bones or soft body parts such as reproductive organs. Age determination for fishes and many mammals is usually possible to the nearest year. In most birds, the methods of age determination are primarily effective in separating individuals that have not reached reproductive maturity from those that are adults (reproductively mature). An overview of age determination techniques is provided in this chapter, but biologists should consult more advanced books and recent journal publications to determine the most appropriate methods of age determination for particular species. It is quite common for particular aging methods to vary in effectiveness from one species to the next. For example, the use of scales is a good aging technique for a fish species such as the bluegill, while the use of bones is sometimes more effective for a species such as the northern pike.

specific. In both fisheries and wildlife the adult age classification includes animals that have reached normal breeding age or size.

In ungulates, animals in their second summer and fall after birth are sometimes called **yearlings,** a term that is also an age classification. For example, a male elk in its second summer or fall may be referred to as a yearling. The term yearling is also used in fisheries to donate an age-1 fish. Even **pre-breeding** and **post-breeding** categories are sometimes used as age classifications.

Natal down is the short, fluffy feathering on newly hatched birds. It is especially abundant on newly hatched waterfowl, gallinaceous birds, and other precocial birds. Some altricial birds, particularly hawks and owls, also have abundant natal down. The term **juvenal plumage,** with the -al suffix, is used to describe the plumage following natal down. A rose-breasted grosbeak hatched in early June would have a drab brownish plumage in July that is properly called juvenal plumage; this bird would be classed as a juvenile or immature bird because it has not reached sexual maturity and can be distinguished from the adult.

In birds, the terms **hatching year** (**HY**) and **after-hatching year** (**AHY**) are commonly used as age classifications. The category HY includes all

In both wildlife and fisheries, it is often difficult to correctly determine the age of an organism based on an aging structure. Regardless of which structure is chosen for aging, biologists must validate the accuracy of the technique on a case-by-case basis (Beamish and McFarlane (1983). One way in which biologists do so is by marking known-aged animals so that they may be identified later and compared to animals whose ages have been determined through the technique in question. Animals reared in captivity are also used to validate aging techniques. There are varoius other ways to validate age information.

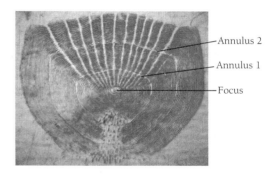

FIGURE 8.3 Impression of a white crappie scale on an acetate slide. This fish was age 2, and was collected in the fall, after completing most of its third-year growth. Thus, the scale has two visible annuli. (*Photograph courtesy of C. Soupir.*)

8.3 Aging Methods

Scales and bones in fishes, teeth and bones in mammals, and feathers and bones in birds are all hard body parts that can be used to estimate age. Fishes are commonly aged using the annual growth rings on scales (**scale annuli**) (fig. 8.3). Scales have concentric markings (**circuli**) around the center (**focus**). Most fish species living in temperate regions grow more rapidly during the warmer periods of the year when the fish feed more heavily and their body temperatures and metabolic rates are higher. These periods of greater growth are characterized by widely spaced scale circuli. When growth is slow, usually during the colder periods of the year, the scale circuli become crowded and form an annual growth ring (**annulus**). Therefore, a fish captured in the summer of 1990 that had three annuli on its scales would be placed in age group 3 (an age-3 fish) and in the 1987 year class. Some fishes can be aged accurately by this method, but sometimes aging is complicated by difficulty in deciphering the annual rings, especially in older fishes; **false annuli** or **spawning checks** (scale growth variations caused by periods of environmental stress or at spawning time) can make aging difficult. Such false annuli can also be found on some other hard parts used to age animals. One of the primary advantages of using scales is that they can be removed with little or no damage to the fish; scales are regenerated in the area from which they are removed.

Scales are more easily aged for fish species living in areas where seasonal temperature changes are substantial than for those near the equator where seasonal temperature differences are minimal. Scales are also lacking in some fish species and can be very difficult to use in other species.

A variety of other hard body parts can also be used for fish aging. The bones of the inner ear (otoliths) are used for some species, while other bones such as spines, **cleithra** (part of the pectoral fin support), and **branchiostegal rays** (part of the hyoid, or throat, support system) are also used. Once again, latitude can affect which structure is used to age fish. At southern latitudes, for example, ages assigned to black and white crappie otoliths were more precise than ages for scales (Schramm and Doerzbacher 1982; Boxrucker 1986; Hammers and Miranda 1991). For populations at a more northerly latitude, however, ages assigned to black crappie scales and otoliths were similar (Kruse et al. 1993). Thus, biologists at northern latitudes may be able to use the nonlethal scale method for crappies, while those at southern latitudes may have to extract otoliths.

In paddlefish and sturgeons, the ossified and cartilaginous parts of the jaw and skull have growth rings that can be used for age estimation. Bone or cartilage density differences between periods of fast growth and slow growth are indicative of different growth periods. By viewing sections of these hard body parts with reflected or passed-through light, bone or cartilage density differences can be seen and ages determined. In most cases the use of bones requires killing the fish. In catfish species, however, the annual rings on a cross section of the pectoral spine can be used to determine age without killing the fish. Another disadvantage of the use of bones for aging is that they often take longer to use than scales, as they must be sectioned and in other ways prepared for viewing. While most of these methods yield information on fish age in years, some hard parts of young fish, such as otoliths, can also be used to determine days of age. Again, validation is necessary.

In a number of mammal species, annual deposits (**cementum annuli**) on the teeth provide an aging method somewhat similar to scale reading in fishes (fig. 8.4). In deer, for instance, the roots of the incisors (front teeth) can be sectioned, stained, placed on slides, and the age estimated; molars and premolars can be used for species such as grizzly bears and coyotes. Differences in cementum deposition between the periods of slowest growth (winter) and the remainder of the year cause the development of growth rings or annuli that can be counted like tree rings or fish scale an-

exterior of root

dentine

Illustrates 5 annuli developed after loss of milk tooth; 6th annulus has not yet developed.

FIGURE 8.4 Longitudinal section of the root of an incisor tooth of a white-tailed deer, illustrating the cementum annuli of a 6.5-year-old deer. The oldest annulus (band of cementum) is nearest the dentine, and the most recently formed annulus is nearest the exterior of the root. This age calculation takes into account one year prior to milk tooth loss. *(Photograph courtesy of L. Rice.)*

nuli. Cementum annuli closest to the dentine of the tooth root are the oldest, and those nearest the exterior of the root are the most recently formed. This method seems simple but in reality is quite complex due to various aberrations, such as false annuli, that require experience to decipher. Similar problems occur in aging other hard parts of animals.

Another common mammalian aging technique is the use of annual rings on the horns of mountain goats and bighorn sheep. These growth rings, which are visible on the surface of the horn, also reflect slow versus rapid growth periods during the year (fig. 8.5). The rings closest to the skull are the most recently formed. This method works well for several years, but in older animals, nine years of age, for example, the rings become difficult to decipher near the skull. Also, the end portions of the horns may be broken or worn, making age determination difficult.

In some species, such as mink, raccoons, and long-tailed weasels, there is a bone in the penis (**baculum**) that grows with age and can be used to estimate age in years. In these species, collections of bacula from pen-reared or marked known-aged animals can provide a reference to use in estimating the ages of individuals from wild populations.

Antler growth is primarily determined by nutrient levels in the diet and soils; therefore, antler size is generally not a very dependable source of age information. For example, in some areas where the quality of the forage is high, 2.5-year-old white-tailed deer males may have four or more points (**tines**) on each antler, and 3.5-year-old males can have impressive **beams** (the main trunk of the antlers) with lengthy tines that hunters would define as trophy quality. In other areas with poorer nutrient levels in the soil and forage, 2.5-year-old males seldom have more than two tines on each antler and have very little antler mass; 3.5-year-olds may have three or four tines per side and still have little antler mass or size. In addition, injury to the growing antlers may give rise to extra tines. Studies of captive deer have clearly demonstrated variation in antler development due to nutrition as well as genetics. A biolo-

FIGURE 8.5 The annual rings (arrow shows a distinct annual ring) used in age determination can be seen on this horn from an eight- or nine-year-old bighorn sheep. The rings near the skull are the newest, but become crowded and more difficult to distinguish in older males.

gist who is familiar with age and growth characteristics of a particular deer population may be able to infer something about age from antler size. However, in general, antler growth is not an acceptable method for estimating deer age.

Considerable variation can also occur in antler growth in species such as elk and moose. General age trends of elk through the first few years of life, as indicated by number of tines and beam length and diameter, may be evident to persons familiar with antler growth in a particular population. In most elk herds, very few 1.5-year-old males have developed forks in their antlers. However, in some cases 1.5-year-old male elk have four or five tines per antler. A **raghorn**, or a 2.5-year-old male elk, can usually be recognized because its antler structure is lighter and smaller than that of an older elk.

Milk teeth (the first set of teeth in mammals) are replaced and permanent teeth erupt at predictable and characteristic times in mammals. Tooth eruption and replacement charts can be used to

determine the ages of many ungulates to within an accuracy of several months. For example, the presence of three-cusped fourth premolars instead of adult, two-cusped fourth premolars reveals that a deer is less than eighteen months of age (fig. 8.6). Likewise, the two middle incisors (anteriormost teeth of the lower jaw) are replaced by larger adult incisors by six months of age (Severinghaus 1949; Dimmick and Pelton 1994). Wear on the crests of the molars and premolars and the development of **dentine lakes** (areas where surface enamel is worn away and dentine is exposed) can provide useful information for age determination in deer, elk, and other ungulates (fig. 8.7). In using the tooth wear method, the biologist must recognize that local differences may occur due to differences in diet abrasiveness that can affect tooth wear. Tooth wear and eruption patterns are particularly useful if jaws from known-aged animals are available for comparison. Some of the advantages of using tooth eruption and wear data for aging include their availability at road checks of harvested animals and in other field situations, their usefulness on live-trapped animals, and the value of immediate age estimates as a public relations tool.

In young mammals, the long bones, such as the femur and humerus, are characterized by a cartilaginous zone near the ends of the bone where growth occurs. The caps or ends of these long bones are ossified, as is most of the remainder of the bone. When a long bone is x-rayed, this **epiphyseal cartilage** separating the **epiphysis** (the end of a long bone) from the **diaphysis** (the shaft of a long bone) is visible as a distinct line in the young of mammals such as raccoons, gray squirrels, and eastern cottontails (fig. 8.8). Adults of these species exhibit no visible cartilage line in the long bones.

Feathers found on the wings are often used to classify birds as HY or AHY. Commonly used wing feathers are the **primary feathers (primaries), secondary feathers (secondaries),** and **coverts;** coverts include the **primary coverts,** the **greater, middle,** and **lesser secondary coverts,** and the **marginal coverts.** Figure 8.9 illustrates the locations of these feathers on the wing of a duck.

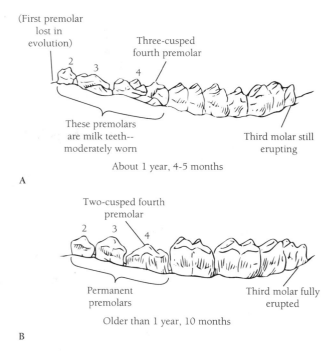

(First premolar lost in evolution)

Three-cusped fourth premolar

2 3 4

These premolars are milk teeth-- moderately worn

Third molar still erupting

About 1 year, 4-5 months

A

Two-cusped fourth premolar

2 3 4

Permanent premolars

Third molar fully erupted

Older than 1 year, 10 months

B

FIGURE 8.6 A three-cusped fourth premolar and an erupting third molar on the lower jaw distinguish a white-tailed deer approximately 1 year, 4–5 months of age (A) from a deer older than 1 year, 8–10 months (B), which has a two-cusped permanent fourth premolar and a fully erupted third molar. (*Adapted from Larson and Taber 1980.*)

In HY gallinaceous birds, with the exception of pheasant species, the first eight primary feathers (P) of the wing are replaced during **molt** (loss of feathers in process of renewal) from the one closest to the body (P1) out to P8, but the last two juvenal primaries (P9 and P10) are retained until the following summer's molt. The characteristics of these two primaries, which are retained through the bird's second summer or even into early fall, can be used to distinguish HY from AHY birds. The ninth and tenth primaries of gallinaceous birds in juvenal plumage tend to be more pointed than the recently replaced P9 and P10 of adults; P9 and P10 feathers in juveniles also appear more worn and faded than recently replaced adult primaries in the fall. In addition, the ratio of the diameter of the **calamus** (the bare shaft of the feather below the feathering) of primary feathers P9 and P8 (the diameter of P9 divided by the diameter of P8) is lower (not as close to 1) in the juveniles of some gallinaceous birds, such as sharp-tailed grouse, than in adults (Caldwell 1980).

At bag check stations for sharp-tailed grouse and other grouse species, biologists often look for blood in the base of the calamus of P9 and P10, which indicates a newly grown primary. HY sharp-tailed grouse have no blood in the calamus of P9 and P10; AHY grouse have blooded primaries (assuming that the adult had replaced P9 and P10 by the time the hunting season started). Grouse, quail, partridge, and other gallinaceous

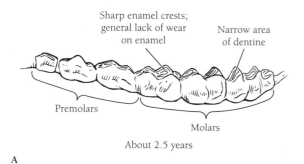

Sharp enamel crests; general lack of wear on enamel

Narrow area of dentine

Premolars

Molars

About 2.5 years

A

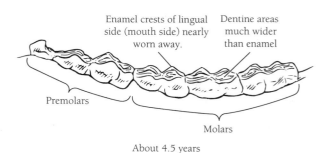

Enamel crests of lingual side (mouth side) nearly worn away.

Dentine areas much wider than enamel

Premolars

Molars

About 4.5 years

B

FIGURE 8.7 The enamel is less worn down and the dentine lakes (dark areas on dorsal surface) are smaller on the premolars and molars (lateral view) of a 2.5-year-old white-tailed deer (A) compared with those of a 4.5-year-old deer (B). The enamel crests and height of the tooth also wear away rapidly with increased age. This method of aging is subject to many inaccuracies due to regional differences in tooth wear caused by diet and soil type. (*Adapted from Larson and Taber 1980.*)

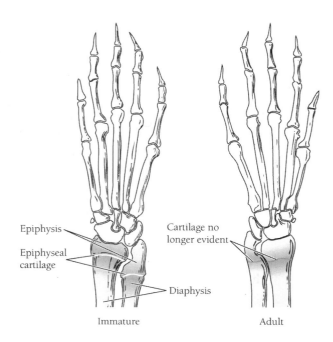

Epiphysis

Epiphyseal cartilage

Cartilage no longer evident

Diaphysis

Immature   Adult

FIGURE 8.8 The epiphyseal cartilage can be seen in the forelimb of an immature raccoon, but is absent in an adult raccoon. (*Adapted from Sanderson 1950.*)

an example in learning what the outer two primaries should look like in an adult.

In addition to the other methods used for gallinaceous birds, male wild turkeys are often aged from the **rectrices** (large tail feathers). The rectrices of AHY wild turkey males are all of similar length, while the central rectrices are noticeably longer than the other rectrices in HY turkeys (fig. 8.10); these differences can be observed from a distance. Adult male wild turkeys also have longer spurs, generally a longer **beard** (a specialized group of hairlike feathers on the breast), and a broader and more colorful patch of greater secondary coverts than do HY males.

Feather characteristics can be used to age other bird species; a few more examples follow. In HY northern bobwhites, scaled quail, and Gambel's quail, the primary coverts are buffy or white tipped, more ragged at the tip, and less rounded than those of adults. Immature gray-cheeked

birds, excluding pheasant species, in the process of replacing P9 or P10 are adults. Thus, if a ruffed grouse wing has a P10 in the process of replacement, this bird can be classified as AHY. The biologist can then look at the characteristics of the already replaced P9 in this AHY bird and use it as

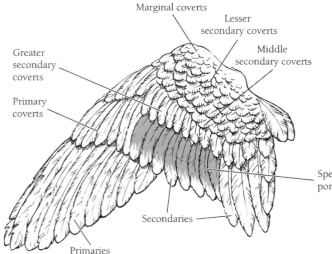

Marginal coverts

Lesser secondary coverts

Greater secondary coverts

Middle secondary coverts

Primary coverts

Speculum (colored portion of secondaries)

Secondaries

Primaries

FIGURE 8.9 The feather groups identified on this duck wing are commonly used in age and sex determination. (*Adapted from Carney 1992.*)

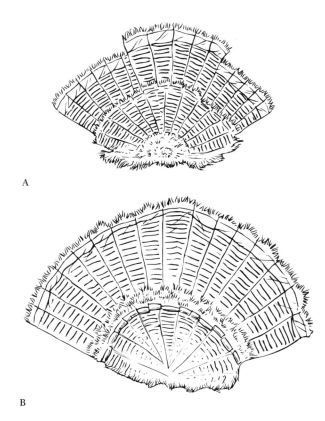

A

B

FIGURE 8.10 The central rectrices of HY wild turkeys (A) are noticeably longer than the other rectrices, while those of AHY birds (B) are equal in length.

thrushes have terminal buff spots on the greater secondary coverts, while adults lack these spots. The secondary coverts in an immature black-and-white warbler are brown, in comparison to black in adults. Mourning doves with cream or white-tipped coverts (including all of the types of wing coverts lying over the primaries and secondaries) are HY birds, but individuals lacking cream or white-tipped coverts could be HY or AHY birds. If mourning doves have already molted the coverts, biologists can look at the last few primaries, which are molted later than the last coverts molted. If the HY primaries remain, they will usually have lighter tips than the uniformly gray primaries of adult mourning doves. If all the primaries have been molted (by about five months of age) and re-

placed, the age of the mourning dove cannot usually be determined from the wings. Detailed information on age determination in mourning doves was presented by Mirarchi (1993).

Waterfowl biologists can use a combination of wing feather characteristics (shape and color) to categorize ducks as HY or AHY individuals. Ducks do not replace most of their HY flight feathers (primaries and secondaries) until the second summer of life. However, the most **proximal** (near the body) secondaries of HY birds, which are usually longer and differently shaped and colored than the other secondaries, are usually molted with the body feathers during the first fall (Carney 1992). These proximal secondaries, because of their distinctive shape, length, and often color, can be used to separate HY from AHY ducks until they are molted in mid-fall. The juvenal wing coverts (referring to the primary and various secondary coverts) are replaced in the first fall, winter, or spring and, because of age-related differences in color and shape, may also be useful in determining age. For example, the middle and secondary coverts on the forewing of male American wigeons are white in adults; these feathers have considerable dusky coloration mixed with white in HY individuals.

In addition, the rectrices of HY ducks and geese are notched until they are replaced sometime in the fall or winter (fig. 8.11). The notching in the rectrices of HY waterfowl is caused by abrasion or wearing of the downy tips of these feathers. Wear on the rectrices of adult waterfowl can sometimes cause the feathers to look notched like those of HY birds.

Various other hard features, including additional tooth eruption patterns, fusing of bones in the skull, molt patterns on the pelts of mammals such as muskrats, and additional feather and spur characteristics in some birds may also provide age information on wildlife species. The fissures between many skull bones in juvenile carnivores, for example, are much more visible than in adults. In addition, in HY ring-necked pheasants, the lower bill will bend or break more easily than in AHY birds.

Notched tip and protruding shaft

Pointed tip, no notch

A   B

FIGURE 8.11 *(A)* The rectrices in the tail of an HY Canada goose illustrate the typical notched tip and protruding shaft. *(B)* The rectrices of an AHY Canada goose have a pointed tip (no notch).

Some age determination methods depend on soft body parts. The weight of the eye lens, for example, increases with age in mammals. If lenses are collected from recently killed specimens, their dry weight can be used as an indication of age, particularly in medium-sized mammals such as the raccoon. The protein tyrosine increases in eye lenses throughout life and has been used to age both large and small mammals (Dapson and Irland 1972; Ludwig and Dapson 1977).

In birds, a dorsal pocket in the **cloaca** (the common chamber where the intestinal, kidney,

and reproductive canals empty) termed the **bursa of Fabricius** is present in the young, but disappears or becomes shallow in adults (fig. 8.12). The bursa is thought to be important to the immune system of young birds. The bursa can be used to estimate HY or AHY status for several months after fledging. For example, the bursa in a juvenile ring-necked pheasant male through the first November of life is normally greater than 10 millimeters in depth, while the bursa in an adult male is less than 8 millimeters. Bursal depth is also an effective means of determining HY or AHY status in ducks and geese until at least late fall or winter.

The shape and size of the penis, a structure located on the wall of the cloaca in ducks and geese and a few other birds, changes with age and can be used to distinguish between HY and AHY male birds. In ducks, the age difference is very marked, and the method works well on live or recently killed birds. Another method of aging that uses soft body parts involves the oviduct opening in female waterfowl, which is easier to detect in adults that have produced eggs than in immature waterfowl.

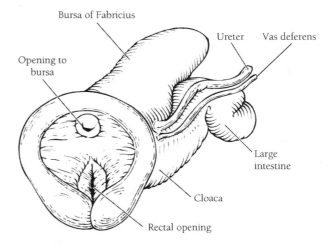

Bursa of Fabricius

Ureter

Vas deferens

Opening to bursa

Large intestine

Cloaca

Rectal opening

FIGURE 8.12 The bursa of Fabricius is a dorsal pocket in the cloaca of birds that is present in HY birds into late fall or winter, but is absent or much shallower in adults. (*Adapted from Godin 1960.*)

In some animals, the developmental stage of the gonads, either ovaries or testes, can indicate age. For example, an immature brook trout will have undeveloped gonads, while a mature brook trout, even if similar in body length to the immature fish, will have developed gonads. This difference in gonadal development is especially apparent in or near the breeding season. Similar differences in gonad development between immature and adult mammals and birds also occur.

Other aging methods are also used. Because fishes exhibit indeterminate growth, a procedure called **length-frequency analysis** can be used in fisheries. Fish of similar age in a particular water body tend to be of approximately the same size. If the lengths of a large number of fish from a particular population are plotted on a graph (fig. 8.13), there will be a tendency for like-aged fish to occur within similar length ranges. This method is more effective for some species than for others, and is usually most effective in distinguishing younger fish (perhaps up to age 1 or 2).

Many additional characteristics can be mentioned that separate the young of a species from more mature individuals. The spots on a deer fawn or young American robin or the off-white col-

oration of a HY white-phase snow goose are examples. Other examples of such characteristics are parr marks on young salmonids, eye (iris) and mouth lining color changes in some birds, development of the scrotum or teats in small mammals, the **breeding tubercles** (bumplike, epidermally derived skin structures usually used during reproductive behavior) found on fishes such as the male hornyhead chub, and a variety of other structural or color differences exhibited by either mature or young organisms.

8.4 Use of Age Information

Age information can be used in a variety of ways. The most common uses involve measurements of population age structure and growth. A graph of the age structures of two white-tailed deer populations was presented in figure 3.7 along with an account in section 3.7 of the way that information was used to interpret the effects of harvest strategies on populations of white-tailed deer. The age structure information used to evaluate harvest strategies in studies such as this one is gained through collection of incisors on an annual basis from hunters followed by age determination from cementum annuli. The hunter records the area of the kill and the sex of the deer, and encloses an incisor in an envelope that is mailed to a biologist.

Similarly, figure 3.8 depicts the age structures of two different fish populations, one with consistent and the other with inconsistent recruitment. Fishery biologists also assess mortality from age structure. Decreased proportions of older individuals can indicate excessive mortality. Figure 8.14A depicts the age structure of a sample collected from a lightly exploited northern pike population that might occur in southern Minnesota. The same hypothetical population, but with excessive angling mortality, is depicted in figure 8.14B. Age determination alone is seldom used in fishery work other than for assessment of age

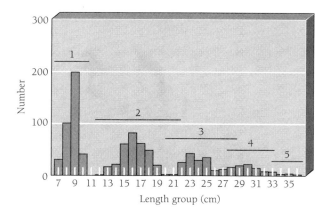

FIGURE 8.13 Length frequency for a black crappie sample obtained during the spring using a modified-fyke net. The horizontal lines indicate the length range for each age group (1–5). As the fish age, there is more overlap in length between age groups (cm = centimeters).

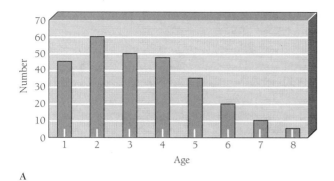

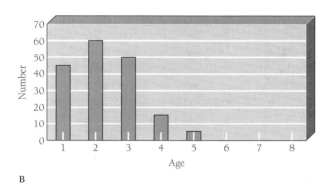

FIGURE 8.14 Age structures of (A) a lightly exploited population and (B) a heavily exploited population of northern pike.

structure; in most cases growth information is also determined.

Age structure data are also used in the management of populations of large animals such as mountain lions, black bears, and American alligators. If such populations are being hunted excessively, their age structures are likely to exhibit a lack of older and larger animals in the harvest. Thus, hunting quality may be deteriorating. Of course, biologists managing such a population would also be closely observing recruitment rates.

Hunting regulations are sometimes based on visible age indicators. For example, in many bighorn sheep populations, only rams with a ¾ curl or greater (horn curl equal to or greater than ¾ of a full circle) are allowed to be harvested. Thus, biologists may base the number of permits on the number of ¾ curl rams in the population to allow some harvest and still maintain an adequate num-

ber of older rams for aesthetic (watchable wildlife), breeding, and future harvest purposes.

Each year the U.S. Fish and Wildlife Service contacts a number of waterfowl hunters to request their assistance in saving and returning duck wings and goose rectrices for later identification, aging, and sexing (ducks only) by waterfowl biologists. These data provide information on the composition of the harvest by species as well as by age and sex. The ratio of young to adults, when corrected for age-related differences in vulnerability to hunting, provides an indication of annual recruitment. For most waterfowl species, recruitment information is also obtained through the annual aerial surveys undertaken by the U.S. Fish and Wildlife Service and the Canadian Wildlife Service.

For many species, such as fox squirrels, ringnecked pheasants, eastern cottontails, and sage grouse, the long-term data obtained from harvested individuals have provided a general model of expected mortality or turnover rate in the population. The base of an **age pyramid**, representing animals less than 1 year of age, is generally broad, with a decline in subsequent layers or age classes (fig. 8.15). In age pyramids for birds, the age categories used are generally HY and AHY (just two tiers) because of the difficulty of estimating the ages of AHY birds to the nearest year. In most gallinaceous birds and in many small birds such as red-winged blackbirds and American robins, 50%–80% of the total population in the fall may be in the HY age category. Similar age pyramid characteristics are typical for a wide variety of short-lived species, including mourning doves, muskrats, and eastern cottontails.

In terms of management value, information on age ratio (ratio of age class 0 individuals to all those in age class 1 or higher) may prove worthwhile in long-term studies relating environmental conditions to recruitment. There are, however, pitfalls in collecting age ratio data in harvested species because the young are generally more vulnerable to hunting or trapping than the adults. Similar biases may also exist in age ratio data obtained from live capture and release of game and

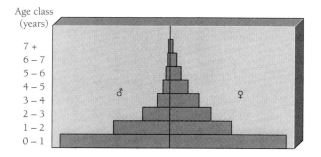

Age class
(years)

7 +
6 – 7
5 – 6
4 – 5
3 – 4
2 – 3
1 – 2
0 – 1

♂   ♀

FIGURE 8.15 Age classes of white-tailed deer can be organized into a pyramid to illustrate the mortality rates in both the male and female segments of the population. Males have higher mortality than females after the first year due to hunter selection for antlered deer.

nongame species. The data are still valuable, but the bias should be considered. In addition, if the adults have unusual and undetected spring or summer mortality so that few survive to reproduce in comparison to previous years, the ratio of young to adults might appear excellent and yet total recruitment could be low.

In a ring-necked pheasant population, the ratio of young to adults can be used to estimate the production per hen. Further, the ages of HY birds, based on primary molt progression, can be used to determine a **hatching curve** (plot of the number of clutches hatched in relation to the progression of the reproductive season). All of this information is biologically interesting. However, the benefits and costs of gathering annual age ratio data from some harvested species, such as the ring-necked pheasant, should be considered. Only male ring-necked pheasants are usually hunted. As the ring-necked pheasant hunting season progresses, the males become increasingly more difficult for hunters to flush within shooting range, and hunter effort and success usually decline. Adequate numbers of males will remain for breeding purposes in this polygynous species, regardless of season length or population numbers at the beginning of the season. The length of the season and bag limit are therefore of little consequence to year-to-year ring-necked pheasant populations

in this naturally short-lived species. A biologist would be likely to recommend a long season and the same year-to-year bag limits (such as three per day) for a male-only season regardless of what population estimates in midsummer indicated. In most cases, sufficient past data on age ratios in the ring-necked pheasant harvest would be available; unless for some specific purpose or study, collection of further age ratio data would be unnecessary.

Age ratios in sharp-tailed grouse and other upland game populations may have little practical value unless biologists have a specific purpose in mind. Such a purpose might be to inform hunters of recruitment rates as a means of explaining their success or lack of success during the season. However, if one is evaluating the influences of long-term weather conditions, possible habitat changes, or other factors that may influence recruitment, the expense of collecting age structure data may well be justified.

▼

8.5 Measurement of Growth

Growth determination usually involves age data because growth is generally assessed per unit of time (per year, for example). Thus, an increase in weight or length per individual per unit of time is a measure of growth. Increase in length per unit of time is the primary measure of growth in fishes, while weight is more commonly used in wildlife species. Growth is most often determined for a particular year, but may also be determined on a daily basis, or in some cases, by growing season.

Fishery biologists can back-calculate body length at specific ages from hard body part lengths because the growth of fish body length and the growth of hard body part length are generally proportional (fig. 8.16). Assume that a redear sunfish is 150 millimeters in total length and has two annuli. When a scale from this fish is measured under magnification, it is found that the distance

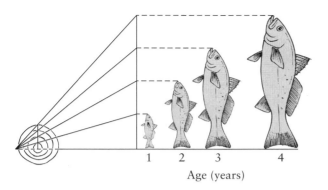

FIGURE 8.16 Scale growth and body growth in fishes are often directly proportional; therefore, body growth in length can be determined for past years using back-calculation from scale growth.

from the focus to the edge of the scale is 100 millimeters, and that the distance from the focus to annulus 1 is 40 millimeters (or 40% of the total scale length). It can then be calculated that this redear sunfish was approximately 60 millimeters in length (0.40 × 150 millimeters) at the time the first annulus was formed. A variety of scale magnification techniques are used by biologists; scales are magnified so that the annuli can be more easily located and measurements more accurately made. The total scale length and length to annulus 1 are really much smaller than the 100 and 40 millimeter figures provided; these lengths occur because the scale is magnified. It does not make any difference what magnification is used because the proportional relationship between the two measurements is what matters. However, the same magnification must be used throughout the procedure.

The methods used by fishery biologists to back-calculate length at a specific age are actually more complex than this simple redear sunfish example would indicate. Body length and hard body part size are essentially directly proportional, but hard body parts do not necessarily start to grow or appear before the fish has developed some length. For example, a newly hatched fish has body length, but its scales may not yet be formed.

This difference requires adjustments when back-calculating length at a particular age. A variety of length back-calculation methods take these differences into account, but they are beyond the scope of this text. Busacker et al. (1990) described some of these methods. It is important to remember that the use of hard body parts in back-calculating fish lengths provides growth information with regard to prior fish length, not prior fish weight.

In wildlife, growth is usually determined by evaluating the weight of the animal in relation to age and sex. Captured small mammals, such as northern grasshopper mice, can be placed in a cloth bag, weighed, and the weight determined by subtracting the bag weight. Likewise, captured ducks, songbirds, and many other small to intermediate-sized animals can be weighed in this manner. In large ungulates, weight along with age and sex is sometimes obtained for field dressed animals at bag check stations. Weights can be taken for any fish species, either by direct weighing or weighing in a water-filled container, with the weight of the container and water subtracted.

8.6 Use of Growth Information

Growth rate information is particularly important in fisheries and of lesser importance in wildlife because fishes, unlike wildlife, exhibit an indeterminate growth pattern (see section 1.3). In some fishery applications, age is determined only in order to obtain growth data. Following are examples of the potential uses of growth information.

Fish growth is often a reflection of food availability in relation to population size. A decrease in food base (prey) or an increase in predator fish abundance can lead to predator populations with slower growth; the converse is also true. Biologists can collect annual growth rate information and, from these data, determine the adequacy of the prey base for the number of predator fish present. Sampling of prey fishes to determine their growth rates and age structure would be needed to more fully understand the status of the predator-prey relationship. If the predator fish population is exhibiting slow growth rates and prey species are found to be in low supply relative to the predator population, the biologist can modify harvest regulations or, if the predator is a stocked fish, modify future stocking rates. In addition, habitat modifications might be implemented to address such a problem.

Fishery biologists commonly use comparative data to assess fish growth. Many conservation agencies have determined average growth rates for fish species in their region. In addition, regional summaries of growth data are available for many fish species (Carlander 1969, 1977). For example, smallmouth bass average 107, 202, 282, and 346 millimeters at annuli 1 through 4 in the southeastern United States. If a smallmouth bass population in a Tennessee lake were found to reach 390 millimeters at age 4, then by regional standards, those fish would be considered fast growing.

Growth rates for wildlife species can also yield useful management information, particularly if comparisons can be made with populations from other areas or years. In wild ungulates, overpopulation can lead to poor nutrition and animals that lack the size and weight of similar-aged individuals under less crowded circumstances. For instance, weights of male elk in Yellowstone National Park might be compared with those of male elk of the same age in the Snowy Range near Laramie, Wyoming, to evaluate the growth rates of individuals within the two populations. Such comparisons can provide information on relative growth rates, which may reflect the availability of nutrients in particular habitats or during specific periods of time. Growth rate reductions in lesser snow geese, as reflected in adult size and weight, have been related to the depletion of forage plants in the Arctic nesting grounds of some populations of geese (Cooch et al. 1991). Thus, the biologist, by observing growth rates (or simply weights) and maintaining long-term records, can identify situations in which the habitat of a species is declining in quality. High bag limits for lesser snow geese were initiated in the early 1990s in the central region of the United States in an attempt to control populations and reduce habitat destruction on the breeding grounds caused by too many lesser snow geese; declines in HY and AHY weights of geese originally alerted biologists to this problem.

Growth information on birds and mammals can be obtained from harvested animals, but this practice is not widespread. In general, body size is relatively constant within a species in a region, and thus weight is not usually an important criterion for biologists or hunters and is usually not recorded. Weights in many birds and mammals, as in fishes, can vary markedly during the day depending on feeding patterns; weights can also vary during the year in relation to physiological changes and energy needs associated with events such as reproduction and migration.

8.7 Determination of Sex

The sex of an organism can be determined through dissection to observe the presence of testes or ovaries. Internal examination to determine sex is necessary when an organism lacks external **sexual dimorphism** (differences in form or shape due to sex) or **sexual dichromatism** (differences in color due to sex). Even organisms that are not sexually mature can often be sexed based upon internal structures. The testes of fishes are usually elongate in shape, paired, and located in the upper portion of the body cavity. The ovaries of fishes are also dorsal, usually longitudinal (lengthwise direction), and paired. As a fish matures, the ovaries can expand to constitute as much as 70% (more typically 5%–15%) of the total weight in some species (Lagler et al. 1977). Mature testes seldom constitute more than 4%– 5% of the total body weight. When the ovaries are mature and expanded, gross examination of the gonads will

FIGURE 8.17 This ovary (arrow) in a female black crappie is filled with developing eggs. *(Photograph courtesy of K. Pope.)*

readily identify the sex of the fish (fig. 8.17). In sexually immature fishes or during nonbreeding periods when the gonads regress, microscopic examination may facilitate differentiation between ovaries and testes.

Unlike those of most mammals, bird testes are internal and may be cooled by their position in relation to the air sacs. The paired testes in birds are located dorsally near the anterior portion of the kidneys. They are whitish and generally shaped liked a kidney bean. They are readily identifiable during the breeding season, when they are enlarged. During the nonbreeding season the testes are small, but can usually be located and identified.

The ovary in birds is composed of many **follicles,** each of which may develop into an ovum for release into the oviduct. Almost all birds have only a single ovary and oviduct (on the left side), an adaptation that saves weight and may reduce cracking of eggs in the lower oviduct. During the breeding season, the ovary in birds resembles a cluster of grapes (fig. 8.18); follicles from which ova have been released resemble a grape in which the skin has been split and the pulp removed. The oviduct is enlarged and easily identifiable during the reproductive season. During the nonbreeding season the ovary and oviduct are greatly reduced in size and are more difficult to identify; the quiescent ovary resembles a cluster of sand granules. By first locating the ovary in reproductively active

birds, a biologist can obtain the experience needed to identify ovaries in reproductively inactive birds. In general, ovaries are more difficult to identify during the nonbreeding season than are testes.

In mammals, internal identification of sex is seldom necessary due to the dimorphism of external features. However, if needed for sex identification, the uterus in females is normally identifiable and is located just below the bladder in the abdominal region.

External sex organs can usually be employed to determine sex in mammals, although identification can be difficult during the nonbreeding season in some mammals, such as deer mice. During the breeding season, the scrotum in most rodents swells greatly as the testes descend through the **inguinal canal** to enter the scrotum and expand to a large and easily identifiable size. In many mammals, such as raccoons and coyotes, there should be no difficulty in identifying the scrotum or the penis with its bony inclusion (baculum). In studies of a number of small mammal species, the condition of the scrotum has been used to identify the period of reproductive activity. Likewise, the enlarged teats of female mammals

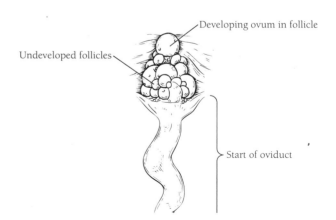

FIGURE 8.18 The ovary and oviduct of a female bird are greatly enlarged during the breeding season. The same structures can be difficult to locate during nonbreeding periods. This drawing depicts the ovary and oviduct of a female mallard.

are evident during the reproductive season, especially during lactation.

There are a variety of other external characteristics that allow for sex determination in various mammal species. During much of the year, the presence or absence of antlers differentiates adult males from females and juveniles in most deer species. The size of the horns in pronghorns, much larger in males than in females, can be used to differentiate the sexes. The size and shape of horns also varies between male and female bison.

Although not as common as in birds and fishes, sexual dichromatism is exhibited by some mammals. The dark neck patch and snout (mask) of male pronghorns, for example, is distinct from the lighter color and pattern of the female. The mandrill, a species of baboon found in equatorial West Africa, exhibits reds and blues in the area of the nose, face, and buttocks; these colors are more intense in adult males than in females.

Sexing by external appearance is possible in birds that display sexual dichromatism. However, winter plumage (basic plumage) and juvenal plumage can at times make it difficult or impossible to determine sex in many species. In some HY birds, differences in feather color or shape can provide clues as to sex. Duck wings, even if in drab juvenal plumage, can generally be identified by sex. Adult duck wings can be particularly useful in determining the bird's sex, based on differences in color, color pattern, or feather shape. For example, color of the **speculum** (a colored, sometimes iridescent patch on the secondaries; see fig. 8.9) can be used to sex some duck species such as northern pintails. The color and shape of the elongated proximal secondaries can be used to determine sex in many duck species (Carney 1992).

Feather size and color can be used to determine sex in many species that are superficially similar in their plumage. In greater prairie-chickens (fig. 8.19) and sharp-tailed grouse, the barring patterns on the tail and crown (top of head) can normally be used to determine sex. The **pinnae feathers** (elongated feathers on the neck) (fig. 8.19) of greater and lesser prairie-chickens are longer in males than in females. The presence of

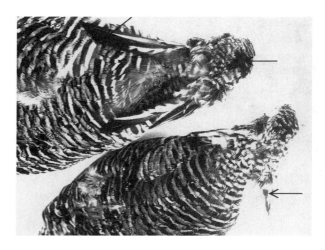

FIGURE 8.19 The dark crown (arrow) and the length of the pinnae feathers (arrows) can be used to determine sex in greater prairie-chickens. (Top) male; (bottom) female.

two dots on the rump feathers of a ruffed grouse indicates a male; females have a single dot. Male ruffed grouse also have longer rectrices than similar-aged females. In northern bobwhites, the white on the head of the male versus the buff on the female distinguishes the sexes. In mourning doves, the pinkish rosy hue on the facial and throat area and pinkish iridescence on the neck are usually more distinct during the breeding season in males than in females. There are numerous other methods for determination of sex in birds that have sexually similar plumages as adults or must be sexed while in the juvenal plumage (Canadian Wildlife Service and U.S. Fish and Wildlife Service 1977; Dimmick and Pelton 1994).

Some birds also have other characteristics that can be used in sexing, such as the beard, spur, and **wattle** (fleshy lobe on throat) of male wild turkeys, bill color in many ducks, or the external air sacs in greater prairie-chickens. The presence or absence of a penis in ducks and geese can be used to determine sex.

Many fishes likewise can be sexed by external appearance, especially during the breeding season. Male orangespotted sunfish have bright orange spots, while the spots of females are rusty orange in color. Mature fathead minnow males have dark coloration compared to females. In many salmonids, males develop prominent hooked jaws (**kype**) and a humped back during the spawning season. In some fishes, such as the channel catfish, bluegill, and muskellunge, differences in the shape, size, and direction of the urogenital opening can be used to determine sex. The swollen body shape of a female fish carrying eggs can also be used to determine sex in some situations. The presence of breeding tubercles on various parts of the body, especially during the breeding season, is characteristic of males in many fish species. In some fishes the size and shape of the **genital papillae** (fleshy protuberances through which eggs and sperm are discharged) also differ between males and females. Fin shape and size can also vary. The male fantail darter has knoblike structures on the spine tops of the dorsal fin, while the mature Arctic grayling male has a larger dorsal fin than the female. There are a variety of color, fin shape, body shape, and other sexual differences exhibited by numerous fish species.

8.8 Use of Sex Information

The ratio of antlered to antlerless individuals in the harvest of some large ungulates can provide important information under certain harvest regulation schemes. For instance, if biologists were concerned about overcrowding of white-tailed deer in an area, information on the antlerless kill under an any-deer permit system would be very important. Do hunters, given the opportunity, take enough does to control or reduce recruitment the following year, or do they concentrate heavily on antlered deer, having little effect on population numbers? Biologists have usually found that any-deer permits do not result in the doe harvest level needed to control prolific deer populations. Antlered deer harvest in a deer season in which both antlered and antlerless deer are legal may exceed

70%, thus defeating the primary purpose of any-deer permits if population reduction is the goal.

Comparisons of pre- and postseason data on sex ratios have long been of value for making a change-in-ratio estimate of kill during seasons in which harvest is directed at a particular sex (Hanson 1963). As an example, if the preseason sex ratio in a ring-necked pheasant population were approximately 1 male:1 female and the postseason ratio were 1 male:5 females, the approximate percentage of kill on males could be estimated as follows: $([5 - 1] \div 5) \times 100 = 80\%$. The change in ratio is proportional to the kill of males, assuming that hunter kill of females is negligible. Appropriate corrections can be made to adjust for illegal hen kill.

Data on sex ratios in the harvest are of questionable value in many cases. Sex in many game species, such as the northern bobwhite, cannot be readily determined by hunters until the animals are harvested. In northern bobwhites, the sex ratios in the population and in hunter-killed birds should be near 1:1, and annual collection of these data would seem to be of little value. The same could be said for sex ratios in many other fish and wildlife species.

Sex ratio data have limited utility in fishery work, except in culture situations. However, because fishes can vary by sex in regard to food habits, movements, size, growth, longevity, or other characteristics, there are situations in which information on sex can be useful. In many fishes, including northern pike, the females tend to be larger than similar-aged males.

Just being able to sex an organism can have much utility. Fishery or wildlife biologists may specifically desire information on one sex. In biotelemetry studies, for example, it is usually important to identify the sex of an individual at the time of initial capture. Banding data from birds can be used to estimate sex-specific mortality if sex is identified at the time of banding. Regulations that require the protection of one sex, such as male-only seasons, have been immensely successful; however, such regulations are possible only if the sexes are readily identifiable.

8.9 Management Implications of Age, Growth, and Sex Information

We have illustrated that information on growth rates, which depends on a knowledge of age, is important in fisheries and less important in wildlife management. Determination of age and growth are both highly important in the evaluation and management of fish populations (see section 9.4). In fishes, growth, movements, behavior, and susceptibility to anglers may be quite different by sex; therefore, biologists may also be interested in the sex composition of a fish population or the sex of individual fish being used in research.

Age and sex ratios as well as weights (growth information) can be important to wildlife biologists, but, as in fisheries, a specific objective should be identified before allotting time, equipment, and other resources toward the collection of such data. As previously indicated, the sex and age ratio in the harvest can be very important in the management of large ungulates such as white-tailed or mule deer. However, sex ratio data for most smaller game species are usually not needed for their management. As in fisheries, knowing the sex and age of animals being studied in behavioral research, for example, in a radiotelemetry study, is highly important.

In many cases there are public relations benefits of sufficient importance to warrant collection of age, growth, and sex data on any animal population. A variety of age, growth, and sex information on wildlife and fish populations can be used in newspaper articles and other public forums to gain support for a management plan, such as adoption of a length limit for a fish species or the harvest of additional cow elk, that may not be initially accepted by the public. Face-to-face contact between biologists and users during the collection of such data can also serve as part of a public relations and law enforcement effort. Knowledge of how to sex and age fish and wildlife species can be used in other ways by professionals when working with the public. People are often interested in the age or sex of a fish, bird, mammal, or other wild organism, or how rapidly it is growing. The ability to explain and provide such information is a public relations asset. In addition, anglers, hunters, or members of the general public that receive assistance in determining the sex and age of an animal will gain respect for the professionalism of biologists. This respect can provide credibility to biologists when they address more important and controversial issues. Thus, it is important for wildlife and fishery personnel to become competent in sexing and aging animals for more than just conservation and research purposes.

It is also important that more biologists, particularly those in administrative positions, define specific and measurable goals before committing resources to the collection of sex, age, and growth data on wildlife and fish populations (see section 20.6). This should not be construed to mean that there is little biological utility for such data; however, it does mean that it is not information that should be routinely gathered without thought to its purpose.

▼ SUMMARY

Wildlife and fishery biologists collect age, growth, and sex information from populations to provide information needed in management, research, and public relations. Fishes are generally grouped in age groups or year classes. Fishes in their first year of life, for example, would be in age group 0, while fishes in their second year of life would be in age group 1. The year class of a fish is simply the year in which it was hatched; year class is sometimes used in the same manner for wildlife. The term age class is often used in wildlife. This term can be used in a manner similar to the use of age groups in fisheries, but is also used much more broadly in wildlife. In many wildlife species, particularly birds, the exact age in years after the first year cannot be accurately estimated. Therefore, age class designations such as juvenile and adult or, in birds, hatching year (HY) and after-hatching year (AHY) are used.

Age is generally determined from hard body parts such as scales, bones, spines, teeth, and feathers, or from soft body parts such as the bursa of Fabricius. Annuli on scales and annual rings on spines of fishes or the roots of mammal teeth are commonly used in age determination. In several mammal species the baculum changes in size and shape with age and can be used to determine age. Antlers are difficult to use in age determination because their development is strongly influenced by nutrition and genetics. Tooth replacement patterns and the presence or absence of the epiphyseal cartilage in long bones can be used in age assessment in mammals. In birds, the characteristics and replacement patterns of feathers are often used to determine age.

The bursa of Fabricius, a soft body part, is reduced in depth or disappears as birds mature. Some birds, such as waterfowl, have a penis on the wall of the cloaca that increases in size and changes in shape as the bird matures; it can also be used in sex determination. Colors in some mammals and many fishes and birds change with age.

Other characteristics, such as breeding tubercles and kypes in some fishes and enlarged external testes in small mammals, provide information on maturity and thus can be used for age assessment. In some water bodies, length-frequency analysis of a fish population can also be used to determine age for younger age groups.

Age data provide information on recruitment rates in various populations. Such information may be valuable in evaluating the effects of various harvest or management plans on a population. Long-term data on age in relatively stable populations can be used to develop age pyramids reflecting mortality rates in various age groups or classes. Age ratio data may be influenced by greater vulnerability of juveniles to various capture and harvest methods.

Growth is generally measured in terms of length or weight increase per unit of time. Growth in fishes is generally measured in terms of length, although weight is also sometimes used. Because fishes exhibit indeterminate growth, growth information is used much more often in fishery work than in wildlife work. Growth data in terms of relative weights per age class can provide measures of the relative health of some wildlife populations, such as elk or lesser snow geese.

Sex determination can be made from external differences in some species, especially during breeding seasons, or may need to be determined by internal examination. Most small mammals have a readily visible scrotum, testes, or teats during the breeding season unless they are juveniles. Cloacal characteristics are commonly used in juvenile waterfowl to determine sex. Many birds are sexually dichromatic, and even juvenile birds, or birds in which both sexes are the same color as adults, may have certain feather color or body shape differences that can be used in sexing. Many fishes have color or shape differences, especially during the breeding season, that distinguish males from females.

Vertebrate species that have been harvested or otherwise killed can be dissected to determine or confirm sex. The testes, ovaries, and accessory structures (such as the oviduct in birds) are well developed during the reproductive season, but are more difficult to locate during the nonbreeding season.

Biologists may need to determine the sex of target organisms because various species exhibit differences between sexes with regard to aspects such as movements, food habits, and habitat use. Biologists may also need to identify the sexes during marking or observation. Sex determination can also be important in the evaluation of certain harvest strategies designed to selectively harvest one sex.

For a biologist working with the public, it is important to be proficient in age, growth, and sex determination. Information on both game and nongame species regarding age, growth, and sex is often requested when a biologist interacts with human users.

PRACTICE QUESTIONS

1.   Incorporate the terms age group and year class into one or two sentences about a fish to illustrate how these terms are used.

2.   Define the terms juvenile, subadult, and adult and give examples of their use as applied to wildlife species. How is the term juvenal used?

3.   What do the abbreviations HY and AHY mean?

4.   Why are the scales of many fish species useful in age and growth determination?

5. Describe how cementum annuli are used in age determination in mammals.

6. How can the long bones in mammals be used in age determination?

7. Discuss how P9 and P10 feathers are useful in age determination in most gallinaceous birds.

8. What is the bursa of Fabricius, and how is it used in age determination?

9. How can length-frequency analysis be used to obtain age information for a population?

10. List three ways in which information on the age structure of a population can be used by biologists.

11. Why can fish scales be used to back-calculate fish length at specific ages?

12. How can growth rate information be important in managing a fishery? How can it be important in managing a population of elk or other large ungulates?

13. Provide two examples each (other than those used in the text) of sexual dichromatism in fishes, birds, and mammals.

14. How useful are antlers for age determination in white-tailed deer, mule deer, and elk? Why?

15. How can information on the sex of captured or harvested animals be useful in wildlife and fisheries?

16. How might age, growth, and sex information sometimes be of value in public relations?

17. Why is it important to validate age estimates?

SELECTED READINGS

Bagenal, T. B., ed. 1974. *The ageing of fish.* Gresham Press, Old Woking, Surrey, England.

Dimmick, R. W., and M. R. Pelton. 1994. Criteria of sex and age. Pages 169–214 *in* T. A. Bookhout, ed. *Research and management techniques for wildlife and habitats.* The Wildlife Society, Bethesda, Md.

Jearld, A. Jr. 1983. Age determination. Pages 301–324 *in* L. A. Nielsen and D. L. Johnson, eds. *Fisheries techniques.* American Fisheries Society, Bethesda, Md.

LITERATURE CITED

Beamish, R. J., and G. A. McFarlane. 1983. The forgotten requirement for age validation in fisheries biology. *Transactions of the American Fisheries Society* 112:735–743.

Boxrucker, J. 1986. A comparison of the otolith and scale methods for aging white crappies in Oklahoma. *North American Journal of Fisheries Management* 6:122–125.

Busacker, G. P., I. R. Adelman, and E. M. Goolish. 1990. Growth. Pages 363–387 *in* C. B. Schreck and P. B. Moyle, eds. *Methods of fish biology.* American Fisheries Society, Bethesda, Md.

Caldwell, P. J. 1980. Primary shaft measurements in relation to age of sharp-tailed grouse. *The Journal of Wildlife Management* 44:202–204.

Canadian Wildlife Service and U.S. Fish and Wildlife Service. 1977. *North American bird banding techniques.* Vol. 2. Populations and Surveys Division, Canadian Wildlife Service, Ottawa.

Carlander, K. D. 1969. *Handbook of freshwater fishery biology.* Vol. 1. Iowa State University Press, Ames.

Carlander, K. D. 1977. *Handbook of freshwater fishery biology.* Vol. 2. Iowa State University Press, Ames.

Carney, S. M. 1992. *Species, age, and sex identification of ducks using wing plumage.* U.S. Fish and Wildlife Service, Washington, D.C.

Cooch, E. G., D. B. Lank, R. F. Rockwell, and F. Cooke. 1991. Long-term decline in body size in a snow goose population: Evidence of environmental degradation. *Journal of Animal Ecology* 60:483–496.

Dapson, R. W., and J. M. Irland. 1972. An accurate method of determining age in small mammals. *Journal of Mammalogy* 53:100–106.

Dimmick, R. W., and M. R. Pelton. 1994. Criteria of sex and age. Pages 169–214 *in* T. A. Bookhout, ed. *Research and management techniques for wildlife and habitats.* The Wildlife Society, Bethesda, Md.

Godin, A. J. 1960. A compilation of diagnostic characteristics used in aging and sexing game birds and mammals. M. S. thesis, University of Massachusetts, Amherst.

Hammers, B. E., and L. E. Miranda. 1991. Comparison of methods for estimating age, growth, and related population characteristics of white crappies. *North American Journal of Fisheries Management* 11:492–498.

Hanson, W. R. 1963. *Calculation of productivity, survival, and abundance of selected vertebrates from sex and age ratios.* Wildlife Monographs 9.

Kruse, C. G., C. S. Guy, and D. W. Willis. 1993. Comparison of otolith and scale age characteristics for black crappies collected from South Dakota waters. *North American Journal of Fisheries Management* 13:856–858.

Lagler, K. F., J. E. Bardach, R. R. Miller, and D. R. M. Passino. 1977. *Ichthyology.* 2d ed. John Wiley & Sons, New York.

Larson, J. S. and R. D. Taber. 1980. Criteria of sex and age. Pages 143–202 *in* S. D. Schemnitz, ed. *Wildlife management techniques manual.* 4th ed. The Wildlife Society, Bethesda, Md.

Ludwig, J. R., and R. W. Dapson. 1977. Use of insoluble lens proteins to estimate age in white-tailed deer. *The Journal of Wildlife Management* 41:327–329.

Mirarchi, R. E. 1993. Aging, sexing, and miscellaneous research techniques. Pages 399–408 *in* T. S. Baskett, M. W. Sayre, R. E. Tomlinson, and R. E. Mirarchi, eds. *Ecology and management of the mourning dove.* Stackpole Books, Harrisburg, Pa.

Piper, R. G., I. B. McElwain, L. E. Orme, J. P. McCraren, L. G. Fowler, and J. R. Leonard. 1982. *Fish hatchery management.* U.S. Department of Interior, Fish and Wildlife Service, Washington, D.C.

Sanderson, G. C. 1950. Techniques for determining age of raccoons. *Illinois Natural History Survey Biology Notes,* no. 45. Urbana, Ill.

Schramm, H. L., Jr., and J. F. Doerzbacher. 1982. Use of otoliths to age black crappie from Florida. *Proceedings of the Annual Conference of the Southeastern Association of Fish and Wildlife Agencies* 36:95–105.

Severinghaus, C. W. 1949. Tooth development and wear as criteria of age in white-tailed deer. *The Journal of Wildlife Management* 13:195–216.

Snyder, D. E. 1983. Fish eggs and larvae. Pages 165–197 *in* L. A. Nielsen and D. L. Johnson, eds. *Fisheries techniques.* American Fisheries Society, Bethesda, Md.

9

POPULATION ASSESSMENT

A biologist must understand certain aspects of a population or community before implementing any management strategy. Chapter 8 described the methods used to assess age, growth, and sex information for wildlife and fishery populations. This chapter builds on that material by describing other population assessment techniques.

Biologists often are responsible for conserving a particular resource. This may involve a specific component of the biota, such as largemouth bass, black phoebes, or mountain lions, or it may involve entire animal communities or ecosystems. Regardless of the biota involved, some type of assessment of populations or communities is usually necessary. The type of assessment needed depends on the organisms involved and the information required. Even after initial assessment leads to the development and implementation of a management strategy, reassessment is needed to determine the success or failure of the implemented strategy (see section 20.6).

As an example, assume that you are working for a conservation agency as a biologist in charge of a private pond program. A pond owner calls and is concerned because his or her pond contains only small largemouth bass. A corrective management strategy cannot be suggested until the population is understood. Does the pond contain a high density of largemouth bass that are slow growing (a stunted population)? If so, some small fish may need to be removed so that the growth rates of the remaining fish will increase. Alternatively, the pond may contain a low density of small fish with fast growth rates because largemouth bass harvest by anglers has been excessive. In this case, the last thing the pond needs is further largemouth bass removal. Fish will need to be released after capture rather than harvested, at least initially. Finally, after a period of time, the success of the chosen management strategy will have to be evaluated, and if the strategy was not successful, modifications will have to be made.

This chapter describes some of the more common techniques that are used to assess the biota and reviews some of the methods by which organisms are marked, tagged, or banded so that they can be identified during assessment. Most of the examples given involve population assessment. While biologists are also interested in communities and ecosystems, many assessment tools are most useful for assessing populations. Biologists must understand populations before they can understand communities, and understand communities before they can understand ecosystems.

9.1 Accuracy and Precision

Two characteristics of data describe their value: **accuracy** and **precision** (fig. 9.1). Accuracy refers to the closeness of a measure to the true value. The collection of accurate (unbiased) data requires knowledge and planning. For example, if knowledge concerning the size structure of a white crappie population in a Kansas reservoir is needed, fish can be sampled with a modified-fyke net. If the sampling is done during August, when most of the larger white crappies are offshore and are not likely to be caught with this near-shore gear, then the size structure and average size of white crappies in that reservoir will be underestimated. The sample will be biased rather than accurate.

Before proceeding to a discussion of precision, it will be useful to describe some basic statistical parameters that are often calculated for wildlife and fishery data. Table 9.1 includes the weights measured for five adult female sage grouse. The sigma (Σ) sign refers to the sum of the numbers.

The **range** for this data set is the difference between the largest and smallest observations in a sample.

$$\text{range} = 1.42 - 1.16 = 0.26 \text{ kilograms}$$

The range is a crude measure of variation because it is solely dependent on extreme or outlying values.

The **arithmetic mean** ($\overline{Y}$) is obtained by summing all the observations, then dividing by the

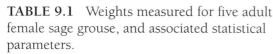

TABLE 9.1 Weights measured for five adult female sage grouse, and associated statistical parameters.

Sample (*n*)	Weight (kilograms) = Y	Y^2
1	1.16	1.35
2	1.27	1.61
3	1.31	1.72
4	1.39	1.93
5	1.42	2.02
n = 5	$\Sigma Y = 6.55$	$\Sigma(Y^2) = 8.63$

number of observations to obtain an average, and is calculated as follows:

$$\bar{Y} = \frac{\Sigma Y}{n} = \frac{6.55}{5} = 1.31 \text{ kilograms}$$

The **standard deviation (SD)** is the average deviation between multiple observations and the mean. It is a measure of the dispersion or variation of the individual items within the sample. The SD is obtained as follows:

$$SD = \sqrt{\frac{\Sigma(Y^2) - \frac{(\Sigma Y^2)}{n}}{n-1}}$$

$$= \sqrt{\frac{8.63 - \frac{(6.55)^2}{5}}{4}}$$

$$= 0.11$$

Another measure of variation that is commonly used is the **standard error (SE)** of the mean. It is needed to obtain a confidence interval, as discussed below, and is obtained as follows:

$$SE = \frac{SD}{\sqrt{n}} = \frac{0.11}{\sqrt{5}} = 0.05$$

A **confidence interval (CI)** can be calculated for many statistical parameters to assess their reliability. A CI is the interval within which a parameter has a certain probability of occurring. A 95% CI indicates that if a biologist takes 100 random samples from a population, the measured value will fall within the range indicated by the CI 95 times. The 95% CI for the sage grouse mean weight above is calculated as follows:

$$95\% \text{ CI} = \bar{Y} \pm 2.776 \text{ (SE)}$$

$$= 1.31 \pm 2.776 \text{ (0.05)}$$

$$= 1.31 \pm 0.14$$

The 2.776 value for a sample size of 5 was obtained from a Student's *t*-distribution table, which can be found in any common statistics book (for example, Sokal and Rohlf 1981; Moore 1995). When sample size is infinite, the value multiplied by the SE to obtain a 95% CI is 1.960. To obtain an 80% CI when sample size is infinite, the value multiplied times the SE is 1.282.

The term precision refers to the repeatability of a measure; that is, the closeness of repeated measures to one another (fig. 9.1). Statistics such as the standard deviation about a mean are measures of precision. In general, the lower the variation about the mean for a sample, the greater the precision, and the more likely that a similar result will be obtained if another sample is collected in the same way. Compared with the data collected by scientists working in the controlled conditions of a laboratory, wildlife and fishery field data tend to be quite variable. Thus, substantial sampling effort and a careful study design are often required if

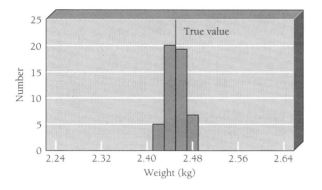

A

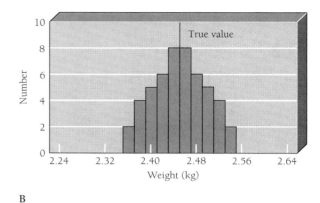

B

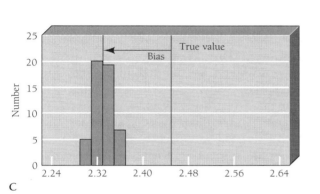

C

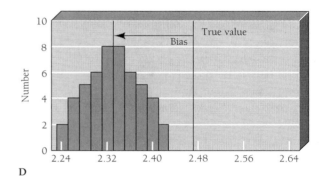

D

FIGURE 9.1 The difference between accuracy and precision. For each of the four figures, assume that a different biologist made fifty repeated measures for a single variable, in this case the weight of an adult male sage grouse that had an actual weight of 2.45 kilograms (kg). Each of the four biologists took the measurements at a different location, under different wind conditions, and used a different scale. Sample A is accurate and has high precision; B is accurate, but has low precision; C is biased (inaccurate), but has high precision; D is biased (inaccurate) and also has low precision.

the level of precision is to be acceptable (see section 9.7).

One method by which the precision of a sample mean can be assessed is the use of confidence intervals. In the sage grouse mean weight example (table 9.1), the 95% confidence interval was 1.31 ± 0.14, or $1.17-1.45$ kilograms. That is, if this same sage grouse population were sampled 100 times, we would expect to obtain a mean

weight between 1.17 and 1.45 kilograms 95 times out of 100. Only 5 times out of 100 would we expect to obtain a mean weight that was smaller or larger than the confidence interval. Biologists often use 95% confidence interval values for research purposes, while 80% confidence intervals may be used for management purposes. However, the actual confidence interval used may vary depending on study type and objectives.

9.2 Assessment of Population Number

Both wildlife and fishery biologists commonly assess the number of individuals in a population. Wildlife biologists may estimate the number of mountain goats or lodgepole chipmunks in an area and summarize population density in terms of the number of individuals per square kilometer or some other unit of area. Fishery biologists commonly discuss the number of fish per surface hectare of water or the number of fish in a particular portion of a stream. Assessments can also be accomplished for harvested organisms such as commercially captured fishes, shellfishes, or furbearing mammals. Some of the procedures for analyzing such data are quite complex. The type of procedure used depends on the methods of counting and the reasons for making the counts.

To assess a population or portion of a population, either a **census** or a **survey** can be conducted. The terms census and survey are often used interchangeably, which leads to needless confusion; these two terms have quite different meanings. A census is a complete count of the organisms present or harvested, while a survey is a partial count (sample). The results of a survey can be expanded into an estimate or may simply be used as an **index**. An index is a value, often a ratio, used in lieu of the actual number. For example, the number of longnose dace collected per hour of electrofishing in a stream can be used as an index to population density. An index can be obtained with less effort and expense than a population census. Both censuses and surveys can be conducted using either capture or noncapture techniques.

A census can be obtained for an entire population or for a portion of a population. Often, the harvested portion of a population is censused. If harvested organisms are involved, fishery biologists undertake **creel censuses**, while wildlife biologists undertake **bag censuses**. Creel or bag censuses do not provide actual numbers of the population present, just numbers of organisms that are harvested. To do a creel or bag census, biologists obtain information from all anglers or hunters affecting the population in question over a defined time period. Common examples would include all fishes harvested from a reservoir in a 12-month period, or the number of northern bobwhites harvested on a state-owned game production area during the fall hunting season. In addition to the total number of organisms harvested, a variety of information types, such as public opinion data, can be obtained through these censuses.

Censuses can also involve the counting of all organisms in a population, not just those harvested by hunters or anglers. A fishery biologist may census a population of longnose gar by draining a body of water and counting all the individuals of that species. A wildlife biologist can census the population of whooping cranes in North America because the population size is small and the birds are highly visible and accessible on their wintering grounds; therefore they can all be counted. At times, censuses of a portion of a population can be conducted. For example, biologists may census a breeding population of trumpeter swans.

Total censuses of animal populations are rarely undertaken because of the expense and logistical problems involved. However, censuses can provide valuable information on the actual number of organisms present in a population. Such data are often important for a thorough understanding of the biology of a species. For example, a census of brook trout populations in a series of Wyoming streams would allow a biologist to evaluate the environmental factors that are conducive to desirable population characteristics with much more confidence than would a partial count conducted by means of surveys.

Creel surveys or **bag surveys** are done more often than creel or bag censuses because they are less expensive and easier to conduct. Such surveys are typically expanded to an estimate of total harvest. Because surveys are only partial counts, biologists must use a variety of statistical methods to extrapolate an estimate of the total harvest. For example, assume that a technician is assigned to randomly check the number of ruffed grouse harvested by hunters on a public hunting area in northern Wisconsin. If the technician is at the checking station one out of every four days throughout the entire sixty-day hunting season, and finds that a total of 800 ruffed grouse were harvested on those fifteen days (60 days ÷ 15 days = 4), then the total harvest for the entire season in that area was probably about 3,200 birds (4 × 800 = 3,200).

This ruffed grouse example is overly simple compared with most field situations. Hunting effort is generally not randomly distributed; for example, more hunting usually occurs on weekends than on weekdays. Thus, biologists often make separate estimates for weekend and weekday harvest. In this type of survey, **random-stratified samples**, are collected from separate predetermined categories (stratifications). In this case, the stratification is the division of the survey into weekend and weekday periods; the random portion of the description refers to the randomly chosen sampling days within each of the two stratifications. Fishery biologists also use random-stratified creel surveys. There are a variety of ways that samples can be stratified, including weekday–weekend, day–night, or shore anglers–boat anglers.

Regardless of the method used for such extrapolations, projected estimates should always include an associated range of values within which the probable true values occur. Most commonly, this takes the form of a confidence interval.

Surveys are also undertaken to index the number of organisms in a population. Wildlife and fishery biologists differ somewhat in their use of the term index. In the wildlife profession, an index is usually a count or estimate of some aspect related to animal number. For example, to index muskrat density, a biologist might count muskrat houses. The more muskrat houses, the higher the likely population density. Other examples of noncapture wildlife indices include mourning dove "coo" counts, deer track counts, coyote scent post surveys, and ring-necked pheasant brood counts. Fishery biologists often use the catch per unit of effort (CPUE) of fish captured with a particular sampling gear as an index to population density. Examples of such indices include the number of adult white catfish caught per net night in a hoop net, the number of age-0 walleyes caught per hour of night electrofishing during the fall, and the number of Utah chubs caught per hour a gill net is set.

Indices are not usually used to estimate total populations. Rather, they are used as comparative data over time or space. For example, a count of ring-necked pheasant broods along a particular road route during a particular time of day and year can be compared with brood counts done on that same route in previous years, or can be compared with brood counts at other locations along a similar length of route. Even when indices are used, statistical methods should be employed to place a range of values about the average catch or observation so that sample precision can be assessed. Again, confidence intervals about a mean provide a common method of documenting the precision of such samples.

Some survey methods are designed to estimate the total population of organisms. **Population estimates** are made by obtaining a sample of the population and expanding those data into an estimate of total population number. Both capture and noncapture techniques are available. Wildlife and fishery biologists each have somewhat different problems in observing the organisms for which they wish to make an estimate. Most wildlife species are mobile and can thus leave or enter a particular area, making it difficult to assess the population. Fish species are confined within the boundaries of water bodies, but are usually not as visible to the human eye, making them difficult to count.

Population estimates can be obtained by capturing organisms. The simplest population estimates are single mark-and-recapture estimates. Consider a small field of grassland that is surrounded by well-cultivated row crops on all four sides. The objective is to obtain a population estimate for meadow voles found on that piece of grassland. Four hundred box traps are placed in a square grid at 15-meter intervals over the entire field. The next day, 77 meadow voles are removed from the traps, marked with a dye or paint, and released. After three days, meadow voles in the field are again trapped. Sixty-eight meadow voles are captured, and 21 of them bear marks that indicate they were among the 77 originally captured. The population estimate is obtained using the following formula:

$$N = \frac{(M)(C)}{R}$$

where

N = population estimate
M = number of meadow voles marked in the initial sample
C = total number of voles in the second sample
R = number of voles in the second sample that have marks (recaptures)

The estimated meadow vole population is thus $N = (77)(68) \div 21 = 249$. The 95% confidence interval for the population estimate is obtained using the following formula:

$$95\% \ CI = N \pm 1.96 \sqrt{\frac{(M^2)(C)(C-R)}{R^3}}$$

$$= 249 \pm 1.96 \sqrt{\frac{(772)(68)(68-21)}{21^3}}$$

$$= 249 \pm (1.96)(45.2)$$

$$= 249 \pm 89$$

That is, if we sampled this population 100 times, 95 times we would expect the population estimate to be between 160 and 338 (249 ± 89). There are a number of important assumptions that must be met for such a population estimate to be valid (Van Den Avyle 1993).

The single mark-and-recapture population estimate described above is called the **Lincoln method** by wildlife biologists and the **Petersen method** by fishery biologists. A more reliable population estimate can usually be obtained by using different methods to sample animals for the first and second captures. Consider the meadow vole example. If meadow voles are initially captured in box traps, those individuals may be less likely or more likely to enter the traps a second time, causing a bias in the population estimate. Thus, a wildlife biologist might use box traps initially to capture the meadow voles and pit traps for the second (recapture) sample.

A variety of more complex methods can be used to estimate population number (Ricker 1975; White et al. 1982; Van Den Avyle 1993; Lancia et al. 1994). At times, these more complex methods may provide more precise and accurate estimates. The statistical techniques used for population estimation are chosen based on the type of method by which a particular population is sampled. For example, unique and complex estimation methods are typically used for commercially harvested fish species.

One example of a noncapture population estimation technique is **spot-mapping**, which is commonly used for songbirds that sing or call within established territories. This technique involves plotting the locations of individual birds on a gridded map during repeated visits by a biologist to a study area. Clusters of location sites then become apparent on the map. The total number of clusters in the study area equals the number of clusters completely inside the area plus the sum of the fractional cluster parts on the boundaries. For example, a cluster on the boundary that has 40% of observations inside the study area boundary and 60% outside would add 0.4 to the sum of fractional clusters. The estimated number of birds in the study population is then obtained by multiplying the number of clusters by the mean number of birds per cluster, which is commonly two if a breeding pair is present. As with population estimation techniques that involve captured organisms, a series of important assumptions must be met for an estimate obtained from the spot-mapping method to be valid. A variety of other noncapture population estimation techniques, such as drive counts, sighting probability models, and line transects, were described by Lancia et al. (1994).

Not all methods of population assessment are easily categorized as either a survey or a census. For example, a technique called cove rotenone sampling is commonly used to sample reservoir fish populations in the southern United States. A small cove, often 1–2 hectares in surface area, is blocked with a net, and a piscicide such as rotenone is applied. Fishes are collected for three days after the addition of the toxicant. If all the fishes in the cove were collected, the sample would be a census of the fishes in that cove. However, recovery of fishes after piscicide treatment is generally incomplete. Some fishes do not float, others may be small and are therefore overlooked, and others may be eaten by animals such as mink, turtles, or American crows before recovery can occur. If a correction factor for nonrecovered fishes is applied to the sample, it becomes an estimate, and therefore would be categorized as a survey. If the fish numbers in a cove or coves were used to index or estimate populations for the entire reservoir, this count would also be a survey. Wildlife biologists also sometimes undertake complete censuses on multiple small areas, and then extrapolate from these counts to obtain an estimate for a larger area.

Animal populations can also be estimated or counted by a variety of types of remote counting methods. In fishery work, hydroacoustic techniques, which use sonar signals and the returning echoes from targets (fishes) struck by these sound waves, are used to count fishes. Wildlife population estimates or counts can be obtained by a variety of **remote sensing** methods, ranging from visual surveys from aircraft to remote photography or thermal imagery (fig. 7.22). In its broadest sense, remote sensing is the measurement or acquisition of information about some property of an object or phenomenon by a recording device that is not in physical or intimate contact with the object or phenomenon under study. The use of such methods for counting organisms is increasing. Other methods of sampling

animals were discussed in chapter 7; most of those methods can be used for population assessment purposes.

9.3 Assessment of Natality/Recruitment, Growth, and Mortality

Neither fishery nor wildlife biologists are concerned only with the number of organisms in a population. They also measure the dynamic rate functions: natality/recruitment, growth, and mortality. Growth assessment was previously discussed in sections 8.5 and 8.6.

Wildlife biologists consider natality (birth or hatching rate) to be a measure of population health in many species. In general, more young are produced in populations that are in better health. Actual counts of young per female may be difficult to obtain because most wild animals bear (or hatch) and rear their young in relatively secluded locations. Thus, natality is often indexed as the number of young observed per female. Brood counts are undertaken by means of aerial surveys for many North American duck populations; roadside counts are also done for many upland game birds. Natality indices for large mammals such as white-tailed deer and elk can be obtained using doe:fawn or cow:calf ratios. Often such data are collected on standard survey routes during late summer, fall, or winter when the animals are less secretive and more visible. Other methods of assessment include young:adult ratios in the harvest, examination of road-killed mammals for pregnancy rates, studies of nesting success, and biotelemetry studies to monitor reproduction by individual animals.

Fishery biologists are usually concerned with the measurement of recruitment (individuals that survive to reproductive size) rather than natality.

At times the number of eggs per ovary are counted to obtain an estimate of **fecundity**. Fecundity is the potential number of reproductive products or units that can be produced by an organism. The term **year-class strength** is commonly used by fishery biologists to refer to the abundance of a particular year class of fish. Year-class strength can serve as an index of recruitment. Because very young fish are likely to experience extremely high mortality rates, year-class strength is generally not determined until this period of extreme mortality has passed. A fishery biologist generally tries to determine the year-class strength of a fish species as early as possible in the life of that particular species, but year-class strength is set earlier in life for some fish species than for others. For example, walleye year-class strength is generally thought to be set by the late summer or fall of the first year of life. Thus, gill netting, bottom trawling, or night electrofishing in the late summer or fall are commonly used to determine an index to walleye recruitment (e.g., the number of age-0 walleyes caught per gill net night in a specific mesh size, the number of age-0 walleyes caught per distance or time a trawl is towed, or the number of age-0 walleyes caught per distance or time an electrofishing boat is operated). For other species, such as largemouth bass, year-class strength may not be established until spring, when fish have survived to reach age 1. Thus, the number of age-0 largemouth bass caught per seine haul in July may not serve as a reliable index to recruitment because these abundant young bass may still suffer very high or even complete mortality before the next spring. The number of age-1 largemouth bass caught per hour (or shoreline distance) of electrofishing in the spring is more likely to serve as a reliable index.

Mortality rates are determined by fishery and wildlife biologists using similar techniques. However, wildlife biologists are more likely to calculate and report survival rates than mortality rates. In either case, remember that the mortality and survival rates always add to 100%. That is, if total mortality is 40%, then 60% of the organisms must obviously have survived.

The total mortality of an animal population can be determined by analyzing its age structure. Figure 9.2 depicts the age structures for samples from two populations of cutthroat trout. Population A has a higher mortality rate than population B, as evidenced by the more rapid decline in fish numbers for each age group. Mortality rates can be calculated from these data by the **catch-curve method** (Ricker 1975; Van Den Avyle 1993), based on the declining abundance of successive age groups. In figure 9.2, the total annual mortality rate for population A is 44% (survival = 56%),

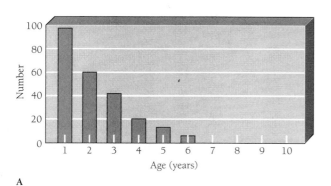

A

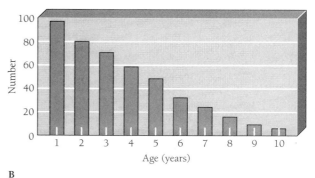

B

FIGURE 9.2 Age structures for samples from two populations of cutthroat trout. Population A has a higher mortality rate than population B, as evidenced by the more rapid decline in fish numbers for each age group.

TABLE 9.2 Life table for a population of fox squirrels.

Age (years)	Number[a]	Number of deaths	Mortality rate	Survival rate
0–1	88	43	43/88 = 0.49	45/88 = 0.51
1–2	45	24	24/45 = 0.53	21/45 = 0.47
2–3	21	14	14/21 = 0.67	7/21 = 0.33
3–4	7	5	5/7 = 0.71	2/7 = 0.29
4–5	2	2	2/2 = 1.00	0/2 = 0.00

[a]Number of squirrels alive at the beginning of each age interval.

meaning that on the average, 44% of the population died in a one-year period. The mortality rate for population B is only 27%.

Wildlife biologists use similar data to develop **life tables** (Caughley 1966; Johnson 1994). A life table is simply a summary of the survivorship or mortality of a population from one age class to the next (table 9.2).

Biologists also monitor natural and harvest mortality for harvested organisms. Natural mortality and harvest mortality together constitute total mortality for such organisms (see section 3.3). Information from marked animals can provide a basis for harvest mortality estimates. Wildlife biologists use banding analyses to determine harvest mortality for waterfowl and other migratory game birds. The relative return rates for bands also reflect survival to the years following banding and can be used to estimate survival rates for banded game and nongame birds (see section 9.9). Estimation of mortality or survival rates from banding data is challenging, and the statistical analysis involved is often complex. Similarly, if a fishery biologist tags a certain number of the catchable-sized Arctic char in a population, the percentage of those tags returned after one year provides a minimum estimate of harvest mortality.

9.4 Assessment of Population Structure

Population quality can be interpreted either as the sizes and ages of animals in a population or total population number. Fishery biologists commonly discuss quality in terms of the size structure of a population. Wildlife biologists most often use number as a measure of quality. Other aspects such as antler size or body weight can also be used as a measure of quality for some animals.

Fish populations that consist primarily of small individuals are usually considered low-quality populations even if many fish are present. Fishery biologists commonly have a management objective specifying that a certain percentage of the individuals in a game or food fish population be over a certain length. Thus, fishery biologists managing under the concept of optimum sustained yield are generally concerned with the sizes of fish in a population as well as fish number.

Fishery biologists often begin an assessment of population quality by determining population structure and plotting age structure (see fig. 9.2)

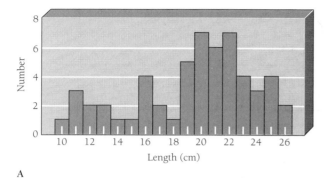

A

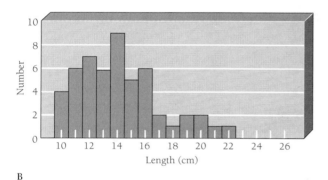

B

FIGURE 9.3 Length frequencies for samples from two rock bass populations. The proportional stock density (PSD) for population A is 71, while the PSD for population B is 14. Minimum stock length for rock bass is 10 centimeters (cm), and minimum quality length is 18 centimeters.

$$PSD = \frac{\text{Number of fish} \geq \text{quality length}}{\text{Number of fish} \geq \text{stock length}} \times 100$$

Stock length is typically the minimum length of a fish that can provide any recreational value to anglers. Fish smaller than this length are rarely caught by anglers. Quality length is typically the minimum length of a fish species that anglers will harvest. Fish shorter than quality length are likely to be released after being caught. Stock and quality lengths are set by species and are based on standard percentages of world-record length for each fish species (Gabelhouse 1984). For example, bluegill stock length is 8 centimeters and quality length is 15 centimeters. For northern pike, stock length is 35 centimeters and quality length is 53 centimeters.

Minimum stock length for rock bass is 10 centimeters, while minimum quality length is 18 centimeters. Thus, the PSD for population A in figure 9.3 is 71; that is, there are 55 fish that are of stock length, and 39 of those are also of quality length ($39 \div 55 \times 100 = 71$). The PSD for population B is 13 ($7 \div 52 \times 100$). Populations with a high PSD are dominated by large individuals, while populations with a low PSD are dominated by small individuals.

Wildlife biologists less often discuss quality in terms of size structure, again reflecting a difference in philosophy between the fishery and wildlife fields. Wildlife biologists more often equate quality with the number of organisms present. While there may be a desire by ring-necked pheasant hunters to bag a bird with long tail feathers, wildlife biologists do not manage populations with a strategy aimed at producing a certain percentage of birds with 55-centimeter tail feathers. Similarly, wildlife biologists do not manage Canada goose populations to ensure that 20% of the harvested geese exceed a particular weight.

Body size management may not be a common wildlife management objective, but it can be an indirect objective for some other management criterion, such as antler size. For example, in many

and length frequency (fig. 9.3). In figure 9.2, the cutthroat trout in population A have a higher mortality rate than those in population B. Therefore, population B has an older age structure than population A. The rock bass in population A of figure 9.3 have a broader length range and reach larger sizes than those in population B, and most anglers would prefer to catch more of the larger fish found in population A.

Proportional stock density (PSD) (Anderson 1976) is an index to the size structure of a fish population that is often used to assess the quality of fish populations. This index is the percentage of stock length fish that are also of quality length.

geographic locations, yearling white-tailed deer bucks often have small antlers, such as single spikes or small forks, while older bucks often have three or more tines per antler. The antlers typically become heavier and larger during the first several years of a buck's life. While a conservation agency might manage white-tailed deer under a strategy that relates to age, the goal of that strategy would probably be to increase the number of deer with large antlers. The management objective for bucks might be a harvest composed of 60% yearlings (age class 1.5) and 40% older bucks (age class 2.5 and older). Of the 40% older bucks, the objective might be to harvest 25% that are age class 4.5 or older. Biologists would monitor the harvest by determining buck age from incisor teeth submitted by hunters or by using some other aging technique. If age class 2.5 and older bucks composed less than 40% of the harvest for a hunting season, the agency would be likely to decrease the number of buck licenses that are issued the next year.

9.5 Assessment of Physiological Status

Various physiological indices or indicators can be used to provide information for population assessment. For example, placental scars in some mammals, such as bobcats, appear as dark bands on the wall of the uterus. Because each fetus leaves a distinct scar, these marks can be used as a physiological index to reproduction. Likewise, a hormone-secreting body (**corpus luteum**) develops in the ovary and remains throughout pregnancy in mammals. Even after pregnancy, corpora lutea can be identified by slicing the preserved ovary into 1-millimeter-thick sections; they provide an index to pregnancy rates and numbers of fetuses per female in many ungulate populations, such as elk and mule deer. Measures of fat in the body cavity,

in bone marrow, or around an organ such as the kidney can provide a physiological index to nutritional status in animals as diverse as white-tailed deer and rainbow trout. Analysis of urine samples collected from snow can be used to obtain information about the physiological condition of ungulates in the winter. Physiological indices can also be quite simple. For example, the mean weight of harvested yearling white-tailed deer bucks can provide a meaningful index to the nutritional status of the herd.

Blood characteristics can also reveal information about the nutritional status of an animal. A physiological index to stress can be obtained from blood chemistry or adrenal gland weight. Fish culturists can monitor blood chemistry to determine the health of fish being reared at a hatchery.

Fishery biologists often use fish weight-length data to assess physiological status. Fish tend to become progressively heavier as length increases, which results in a curvilinear relationship between length and weight (fig. 9.4). Fish in a stunted population tend to weigh less at a given length than fish in a fast-growing population. Thus, fishery

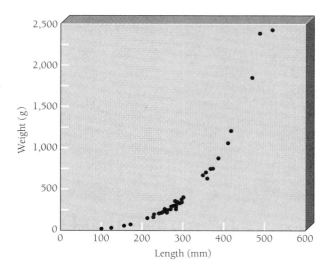

FIGURE 9.4 The relationship between weight in grams (g) and length in millimeters (mm) for a largemouth bass population.

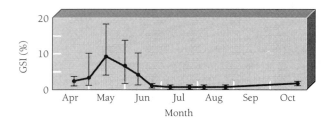

FIGURE 9.5 Gonadosomatic index (GSI) values for female gizzard shad collected from Melvern Reservoir, Kansas. The filled circle represents the sample mean; the vertical bar represents the range. (*From Willis 1987.*)

biologists use fish weight-length data to calculate **condition** (relative plumpness). Several different types of condition indices are used (Murphy et al. 1991). One commonly used condition index is **relative weight**, in which the actual weight of a fish is divided by a standard weight selected to represent an "optimal" weight for a fish of that species at that length, and multiplied by 100. Fish with condition values of less than 100 are "thin," and those with values over 100 are "plump." Interpretation of condition indices is complicated by the fact that fish condition typically varies by season, by sex, and even by time since last feeding.

Seasonal changes in reproductive status in fishes can be monitored using the **gonadosomatic index** (**GSI**). This index is obtained by dividing the weight of the ovaries or testes by the body weight of the fish and multiplying by 100. GSI in most fishes increases rapidly just prior to spawning and declines after spawning (fig. 9.5). Seasonal changes in reproductive status for birds and mammals can also be monitored through changes in gonad size, weight, or appearance; the closer to the time of reproduction, the heavier the gonad weight.

9.6 Assessment of Food Habits

Very commonly, animal populations are assessed with regard to their diets. The term food habits refers to the kinds and amounts of foods eaten by an organism. Feeding habits involve how, when, and where organisms obtain their food, and were discussed in section 6.11. To understand and manage wildlife or fishery resources, knowledge of food habits is essential.

To determine food habits, samples of digestive tract contents usually are obtained; however, observation of animals that are feeding may provide information on the foods selected if an observer can view the animal at close quarters without disturbing it. In birds, food samples are normally obtained from the esophagus or crop, which are both located in the initial portion of the digestive tract (see fig. 5.8), to avoid biases caused by differential digestion rates for various food items. Another alternative is available in the case of raptors that regurgitate "pellets" containing remains of prey organisms. Raptor pellets commonly contain bones and other hard parts from the small animals the birds eat; these prey items can be identified by characteristics such as skulls, teeth, and feathers. Similarly, diet analyses for herbivores such as bighorn sheep can be conducted by analyzing their fecal pellets. However, fecal pellet analysis in ungulates is strongly biased toward the most indigestible items in the diet. The esophagus or crop of a bird can sometimes be flushed to obtain diet samples without killing the bird. The upper digestive tract of some shorebirds, such as pectoral sandpipers or long-billed dowitchers, can be pumped without injury to the bird. In some cases, as with small mammals such as Pacific jumping mice, black-eared mice, or yellow-nosed cotton rats, the animal is usually killed to obtain its digestive tract contents.

Fish digestive tract contents can also be obtained by lethal or nonlethal techniques. Nonlethal techniques include inducing regurgitation, flushing the stomach with a stomach pump, or inserting glass or plastic tubes into the stomach and withdrawing most of the contents. These nonlethal methods do not always ensure that the total stomach contents are removed. Lethal techniques for obtaining fish digestive tract contents involve killing the organism so that all of the contents can be removed.

Foods eaten can vary by season, by habitat, by body size, and sometimes by time of day. Thus, one of the biggest challenges in completing a thorough food habits analysis for an animal population is to adequately sample individuals of all sizes, in all habitat types, at all times of the day and night, and in all seasons. A food habits analysis done in only one habitat type during one time of year is inadequate because it will provide only a partial documentation of food habits, or information on only a single "window" in time. For example, to study the food habits of flathead catfish in a reservoir, a biologist needs to sample all sizes of the fish, in deep and shallow water, in open water and along logs or boulders, throughout the entire year. Likewise, mule deer that occupy ponderosa pine forest areas in the summer have different diets than those summering in piñon-juniper forest areas; mule deer wintering in piñon-juniper forest areas are likely to have different diets than those living in the same area during summer. Finally, food habits vary depending on food availability. For example, the most abundant prey of sand shiners in one stream may be quite different than that in another stream if the two streams have different prey availabilities.

Both wildlife and fishery biologists tend to analyze food habits using similar methods. Food habits are generally described in terms of numerical analysis, frequency of occurrence analysis, gravimetric (weight) analysis, or volumetric analysis; each of these methods is described below.

In numerical analysis, the food items found in sampled animals are counted, and the percentage by number of each food type found in each animal or the entire collection of animals is then calculated. For example, 17% of all food items, by number, in a sample of digestive tract contents collected from brown trout might be mayflies. The weakness of this method is that the food most commonly ingested may not be the most important nutritionally. A rock bass may consume 99% zooplankton, but one ingested crayfish may provide much more energy to the fish than all the zooplankton combined. Similarly, a gray wolf would obtain more energy from eating one snowshoe hare than one Arctic ground squirrel.

Frequency of occurrence is the percentage of the sampled organisms that consumed at least one of a particular food type. For example, a 25% frequency of occurrence for a particular kind of plant eaten by muskrats means that 25% of the muskrats sampled contained that plant. This technique overemphasizes the use of persistent or highly abundant food items. For example, all golden trout in a California stream at a certain time of year may contain at least one caddisfly because those insects are extremely abundant at that time. However, the greatest proportion of the diet by weight or volume may actually be composed of large stoneflies that are hatching at the same time. The frequency of occurrence analysis might indicate that both food types are used equally, or even that the less used food is found in more animals, but in reality they may have quite different levels of use.

In gravimetric analysis, food habits are expressed as the percentage of the total weight of food found in the animals that one food type composes. This method generally overemphasizes the use of large food items.

Finally, volumetric analysis expresses food habits as the percentage of total food volume that one food type composes. This method overemphasizes the use of large food items that are slowly digested.

Food habits studies generally need to be supplemented with information on the kinds and amounts of foods available, frequency of feeding, the nutritional value of the food eaten, and food preferences. In addition, food habits may change depending on the type and amount of competition that occurs or the vulnerability of prey to particular predators. More in-depth studies of food habits may examine daily food requirements, seasonal food requirements, efficiency of food conversion, digestion rates, and indices that compare food eaten with food available.

9.7 Sampling Design

Because field data tend to be variable, sample size and subsequent statistical validity are constant concerns for field biologists. Fishery biologists using catch per unit of effort as an index to population density must use sufficient sampling effort to obtain reliable estimates. For example, compare the two gill net samples for American shad in table 9.3. Both samples have a mean catch of 15 fish per net. However, the sample from population A has much more variation among individual net catches. The standard deviation for population A is 12.7, while it is only 3.7 for population B. The 95% confidence interval for the mean catch per net for population A is 15 ± 13, while it is only 15 ± 4 for population B. As one more example, suppose that the same mean catch per net (15) and variation (i.e., the same standard error) were obtained for population A, but that twenty nets were used instead of six. The 95% confidence interval for such a sample would be 15 ± 11. As these examples demonstrate, the precision of samples is highest when the variation among individual net catches is low, and when sampling effort is adequate.

Biologists are also concerned with **sample replication**. Replication can refer to repeated measures of samples or experimental units. However, Hurlbert (1984) warned that treating replicate samples as if they represented replicate experimental units is actually **pseudoreplication**. For example, assume that a study is designed to determine whether controlled burns affect the population density of chestnut-collared longspurs in grassland habitat. First, it will be necessary to obtain multiple samples on both burned and unburned areas so that the abundance of the birds, as well as the variation about that measure (such

TABLE 9.3 Catches of American shad in experimental gill nets from two populations. The mean catch per net is 15 for both samples, but the sample for population A has much higher variation about the mean.

Net number	Population A	Population B
1	29	10
2	1	20
3	24	15
4	2	18
5	8	12
6	26	15
Mean ($\bar{Y}$)	15	15
Standard deviation (SD)	12.7	3.7
Standard error (SE)	5.2	1.5
95% confidence interval	± 13	± 4

as the standard deviation or confidence interval), can be determined. However, the results of a study involving only a single burn would not be reliable. A biologist cannot simply obtain multiple samples on a single site prior to and after the burn. It is important that the burned areas and unburned areas be replicated at more than one site. If consistent trends in chestnut-collared longspur population density occur after burns are conducted on multiple study sites, and no increases in longspur density are observed at multiple unburned study sites, then biologists will be more confident that the effect was indeed caused by the burns.

Because of the variation common in field sampling data, most biologists use **standardized sampling** procedures. Standardized sampling means that sampling devices are chosen that will effectively sample the target species, and that the same sampling devices are used (as much as possible) at the same sites, at the same times of year, and from year to year. The proper method for selection of sampling sites is much debated. Some biologists subjectively choose sampling sites that they believe will allow effective sampling of the target species, while other biologists believe that sampling sites should be randomly selected. Even among those who prefer random site selection there is disagreement. Some biologists randomly choose the initial sampling sites, then return to those sites for samples in subsequent years. Still other biologists prefer to select sampling sites randomly each year.

The purpose of standardized sampling is to minimize variation due to sampling device efficiency and seasonal changes in sampling data. Long-term trends in population structure and dynamics can then be more reliably monitored. For example, wildlife biologists can monitor the relative abundance of coyotes using a scent post survey. Small, flat containers of scent are placed along dirt or gravel roads. All marks and tracks are smoothed from the soil surrounding the scent container so that coyotes that visit the site will leave recognizable tracks. In general, the more coyotes present in an area, the greater the number of scent sites that will be visited and the more visits per

site. A wildlife biologist performing such a scent post survey would want to use the same scent, at the same locations, at the same time of year from year to year. In this way, trends in population abundance could be more reliably determined.

Fishery and wildlife biologists realize that measurable objectives must be set for management strategies so that they can be evaluated (see section 20.6). Each management strategy should be developed with a desired result in mind. For example, if sampling data indicate that the proportional stock density of a largemouth bass population in a lake is only 15, a fishery biologist might recommend that a length limit be placed on largemouth bass in that lake. The intent of this management strategy is to improve the size structure of the largemouth bass population. Therefore, the biologist might set a measurable objective of improving PSD to 40–70. Measurable objectives can take many forms. Biologists may want to manage a white-tailed deer herd for a hunter success of 65%, a walleye population for an angling catch rate averaging 0.2 fish per hour, or an eastern bluebird population for densities reaching a certain number of birds per unit of habitat area. Unless objectives are measurable, the success or failure of a management strategy cannot be determined.

9.8 Modeling

A **model** is a conceptual or mathematical description used to portray a complex system. Modeling is an increasingly important tool in the assessment of animal populations. The development of microcomputer software and hardware capabilities has made modeling a common tool for biologists.

Wildlife biologists often use a model such as RAMAS (Ferson et al. 1990) to model fluctuations in population number by age class. In this particular model, biologists must input initial abundance, emigration and immigration rates, fecundity, survival schedules, and density dependence information. The model then predicts abundance as a function of time, age-class distributions for each generation, and so forth. As another example, the bioenergetics of fish food habits were related to fish growth in a model developed by Hewett and Johnson (1987).

Models are especially useful for predicting the likely effects of management strategies prior to their implementation. For example, Clark et al. (1980) developed a model to predict the effects of various types of length limit regulations on brook and brown trout populations in Michigan. By using a computer model to eliminate management strategies that are unlikely to be successful, a biologist can save both time and money. However, it is essential that computer models be validated by field testing. This component of computer modeling is sometimes overlooked. Often, models become more accurate over time as more accurate data are entered. The more accurate the data entered into the model, the better its predictive capabilities.

9.9 Purposes of Marking, Tagging, and Banding

Animals are marked, tagged, or banded by biologists for a variety of reasons. Any of these techniques can be used in mark-and-recapture studies or any other study that requires the identification of individual organisms or groups of organisms. For example, mortality/survival rate information is commonly calculated from marked, tagged, or banded animals. Mark, tag, or band returns can be used to obtain estimates of **exploitation** (harvest by humans) for organisms that are utilized for sport or commercial purposes. If a certain number of fish are tagged in a lake, the percentage of those fish harvested by anglers in one year can provide

an estimate of annual exploitation. This assumes that the tags have not influenced the natural mortality of the fish or the vulnerability of the fish to angling, and that anglers return tags from all tagged fish. Many exploitation studies provide rewards for tag, band, or mark returns as an incentive for people to participate in the program.

The rates at which waterfowl and other migratory game bird bands are recovered over the years following banding can also be used to estimate both exploitation and overall survival rates. Band recovery rates (proportion of bands recovered and reported to the banding agency) and band reporting rates (proportion of recovered bands that hunters actually returned to the banding agency) can be used to estimate percentages of a population harvested by hunters. An estimation of harvest rate is obtained by dividing band recovery rates by band reporting rates in a particular year. Harvest rate is corrected using an estimate of nonretrieved birds killed by hunters to obtain the exploitation rate. Annual survival rates from all causes can also be estimated. The proportion of bands recovered from year to year from a particular banded population reflects the relative proportions of the population surviving and dying. For example, if 10% of all bands recovered from gadwalls banded as flightless ducklings in central Canada during a particular year were recovered during or after the fourth year post-banding, this would indicate that about 10% of the population was surviving to at least the fourth fall after hatching.

Banding, tagging, and marking programs can also provide information on animal movement. For example, an American avocet banded as a young bird by wildlife biologists in Minnesota and recaptured in Louisiana provides information on migration. Similarly, a channel catfish tagged in the South Dakota portion of the Missouri River might be recaptured at the confluence of the Missouri and Mississippi rivers near St. Louis, Missouri. The problem with movement information provided by banding, marking, and tagging data is that only the minimum distance moved can be documented. Perhaps the channel catfish in the above example actually had been down the Mississippi River to New Orleans and back to St. Louis before it was recaptured.

Other information can also be obtained from tagging, banding, or marking. For example, preferred habitat can be determined, maximum life span can be estimated, and growth or age estimation methods can be validated.

9.10 Marking Methods

For the purposes of this text, the term marking will include both temporary and permanent marks placed on organisms. Examples of marking include toe clips on Mexican voles, fin clips on brindled madtoms, ear notches on mule deer, tattoos on bobcats, heat brands on white-tailed deer, freeze brands on walleyes, fluorescent pigments sprayed on fathead minnows, and subcutaneous latex injections on young sockeye salmon.

Marking often involves some form of mutilation, and biologists that use such techniques may be questioned by those who support animal rights (see section 16.6). In addition to ensuring that marking does not result in unnecessary damage or discomfort, a biologist needs to be certain that it does not influence the behavior of the marked organism. Such changes in behavior could bias the results of a study.

Marks can be used either to identify a group of organisms or to identify a relatively small number of individual organisms. Thus, marks could be used by a biologist to assess the performance of two strains of rainbow trout that are stocked into a stream. One strain could be marked with a fin clip or freeze brand on the left side of the body, and the other strain marked on the right side of the body. By clipping different fins or branding at different body locations, the biologist could even identify individual fish. However, the number of such mark combinations

available for individual fish identification would be relatively limited.

Fin clipping is the simplest method of externally marking a fish. When a fin is clipped near the body of a fish, the fin often does not regenerate. If the fin does regenerate, the mark can generally be recognized by the irregularities in the regenerated fin rays (fig. 9.6). Toes clipped from small mammals, such as Mexican voles, do not regenerate, and these marks are easily recognizable throughout the life of the animal.

Often, only short-term marks may be needed. Fishery biologists conducting a Petersen mark-and-recapture population estimate often use a handheld paper punch to mark fishes. This method creates a clean mark in a fish fin that is quick and easy to make, causes little damage to the fish, and is easy to recognize for a short time period.

Paints, dyes, and other substances are also used for marking organisms. Hair or feather dyes or patches of paint can be used for temporary marking of terrestrial animals, and dyes and stains can be used for temporary marking of fishes. Paints, dyes, and chemical substances are most often used for mass marking when it is not important to recognize individual organisms. Some of these marks are relatively short-lived. Fluorescent pigments, embedded in the skin of fishes using compressed air and observed under ultraviolet light, can last up to a year. If fish are immersed in a chemical called oxytetracycline, the

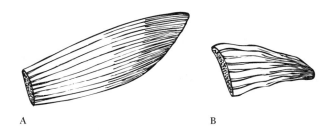

A   B

FIGURE 9.6 (*A*) A Normal fish fin. (*B*) A fin that has regenerated after clipping.

resulting mark can be detected for some time on scales; however, the oxytetracycline is eventually lost through exposure to sunlight. Some marks have longer durations. If fish are immersed for a sufficient time or fed oxytetracycline in a prepared diet, a permanent mark will be visible when bones such as vertebrae are sectioned and viewed under ultraviolet light.

FIGURE 9.8   Numbered ear tags can be used to individually recognize white-tailed deer or other large ungulates. *(Photograph courtesy of D. Naugle.)*

9.11 Tagging and Banding Methods

Tags and bands are differentiated from marks in that they involve the physical attachment of an artificial structure to the organism. Examples include neck collars on giant Canada geese (fig. 9.7), aluminum leg bands on American black ducks or house finches, ear tags on white-tailed deer (fig. 9.8), neck bands or collars on wolverines, and external dart or anchor tags on muskel-

lunge. These examples are all external tags, but internal tags can also be used. Body cavity tags are simply small metal tags containing numbers, addresses, or other information that are inserted into the body cavity of an animal. Small coded wire tags are commonly inserted into the snouts of fishes; such tags can contain a substantial amount of information.

Tags can be used for either group or individual recognition. Neck collars placed on a group of Canada geese may be all of one color to allow recognition of a group of captured and released birds, and the collars may also be individually numbered so that individual birds can later be recognized. Coded wire tags can also be used to differentiate groups of fish. For example, if two groups of striped bass were reared to different sizes in a hatchery prior to stocking in a reservoir, two types of coded wire tags could be implanted in the snouts of the different groups. When striped bass were collected in subsequent years, a biologist could then determine which size of fish survived at a higher rate. However, a large number of individual codes can also be inscribed on this tiny tag; thus, individual fish can also be recognized if desired.

Captured birds are often tagged by placing an aluminum band around a leg (fig. 9.9A). Bands generally are numbered and carry a return address.

FIGURE 9.7   A giant Canada goose with a neck collar. *(Photograph courtesy of S. Vaa.)*

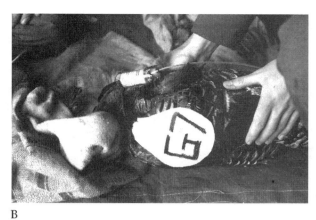

A

B

FIGURE 9.9 Birds can be identified with a band on a leg (*A*) or a patagial tag on the wing (*B*). The wild turkey in (*B*) is also wearing a radio transmitter with an antenna.

A band is properly fitted when it can move freely up and down the leg and turn smoothly, but cannot be pulled down over the toes. Because many individuals and organizations are involved in bird banding, the U.S. Fish and Wildlife Service and the Canadian Wildlife Service issue banding permits and keep banding records for all migratory birds. Migratory birds include a variety of birds ranging from waterfowl species to songbirds that are **Neotropical migrants** (birds that migrate be-tween the Nearctic and Neotropical regions). Birds can also be identified with back streamers, wing tags (fig. 9.9B), and nasal tags.

A variety of tags can be used for identifying fishes (fig. 9.10). These tags can be brightly colored to attract the attention of anglers, and can carry a return address. If the tags are individually numbered, individual fish can also be monitored.

A similar array of external devices can be used on mammals. For example, neck collars are often used on large mammals, including types that are self-adjusting for young ungulates. Ear tags are used on bats, and plastic ear tags are used on wild ungulates (fig. 9.8) just as they are on domesticated livestock. Reflective tape can be placed on tags or collars to allow for better nocturnal observation.

U.S. Fish & Wild Service 003

Dart tag

Stainless steel wire

U.S. FWS 000428

Petersen disk tag

Carlin tag

U.S. FWS 000428

Strap (opercle) tag

U.S. Fish & Wild Service 00425

Anchor tag

FIGURE 9.10 Some common types of external tags used on fishes.

9.12 Choosing Appropriate Marks, Tags, or Bands

Biologists must consider numerous factors when choosing marks, tags, or bands. A mark, tag, or band should not affect the behavior of an organism. While applying a temporary dye to a fish probably would not affect its swimming capability, clipping the entire tail fin could! As mentioned earlier, many marking, tagging, and banding programs are undertaken to determine mortality rates; thus, the chosen mark, tag, or band should not cause mortality or result in an increased mortality rate for the organism in question. For example, a tag or band that is too large for an organism could cause mortality. When fluorescent pigment is sprayed on rainbow trout, the air pressure must not be high enough to kill fish. A bright orange ear tag placed on a marsh rabbit could be easily spotted by a biologist. However, it might also be easily spotted by an avian predator, and the tagged group of marsh rabbits might have higher predation rates than unmarked individuals. The chosen mark, tag, or band also should not affect the likelihood of recapture. For example, metal jaw tags on fish might make them more susceptible to capture in gill nets than untagged fish.

9.13 Biotelemetry

Biotelemetry involves electronic devices that are placed on or in an animal and relay information about that animal to biologists. Biotelemetry usually requires two devices, one placed in or on the animal to transmit information and the other retained by the biologist to receive the information. If the transmitters provide information on location, biologists can determine daily movement patterns, seasonal movement patterns, whether an animal establishes a home range, and a variety of other types of information. The biologist can follow all of the animal's movements, not just determine minimum distances moved as with tagged, marked, or banded organisms. Transmitters with unique capabilities that provide much information about animals are available. For example, some transmitters can send a motion signal that

differs from the signal transmitted when the animal is at rest. By this means, biologists can determine not only animal locations but also activity patterns. Mortality-sensing collars, activated by a drop in body temperature, are also of value, particularly for predation studies. Some transmitters used for fishes can relay information on fish depth. More sophisticated transmitters can even relay biological information such as body temperature, heart rate, or feeding activity.

Biotelemetry allows for the study of organisms from sufficient distances to prevent disturbance of their normal activities. In fishes, biotelemetry also allows remote tracking of an organism that is not normally visible to humans. The same can be said for nocturnal terrestrial animals; however, even terrestrial animals active during daylight are often monitored using biotelemetry. Advances in biotelemetry equipment in recent years have been extraordinary. Increased miniaturization, more effective placement, longer transmitter life, and increased dependability have all added to its usefulness.

Wildlife biotelemetry studies most commonly use externally attached radio transmitters. These transmitters are usually battery powered, with life spans for most ranging from a few days to perhaps two years. Some transmitters are solar powered. Radio transmitters must have antennas; however, some antennas are quite small and internal and thus might not be easily seen. Transmitters used on large animals, such as black-tailed deer or collared peccaries, are generally attached with neck collars (fig. 9.11A). Radio transmitter systems that can remotely inject a drug into an animal are available, making it possible to anesthetize and recapture the animal at will. Transmitters for most birds are placed on the back or breast area and partially covered by feathers (fig. 9.11B). They can be attached by a variety of methods, including harnesses around the body or wings, glue, suturing to the skin, and ponchos or necklaces fitted around the neck. Transmitters glued to tail feathers have proven valuable in bald and golden eagle studies.

A

B

FIGURE 9.11 Externally attached radio transmitters are used in determining animal locations and activity patterns in wildlife biotelemetry studies. *(A)* Collar on white-tailed deer; *(B)* antenna wire leading from transmitter on eastern meadowlark *(Part A photograph courtesy of D. Naugle; Part B photograph courtesy of D. Granfors.)*

An antenna and receiver are used to receive the radio signal from the transmitter; the antenna can be handheld (fig. 9.12A), mounted on a vehicle (fig. 9.12B), carried on or in an aircraft or mounted on a stationary tower. Small transmitters used for songbirds typically have ranges from 200 meters up to 2 kilometers. Transmitters used on larger animals can be much more powerful. For example, transmitters used on white-tailed deer that are being tracked from the ground commonly have a range of 3–4 kilometers. Some animals, such as polar bears, have been tracked by satellite. Satellite collars are much larger and more expensive than ordinary radio collars. Satellites can record animal locations at all times of day and under all weather conditions. However, the location data currently lack sufficient accuracy for some study purposes, and thus satellite collars have been used in only a small number of studies to date.

Radio transmitters can provide signals of different wavelength bands that can be used to identify different individuals within the same study. Radio signals are directional, so a biologist can determine the direction in which an animal is located by rotating the antenna until the signal is strongest. This characteristic allows determination of animal locations by triangulation. If a raccoon with a transmitter is located within a particular area, biologists can obtain directional bearings from two or three different known locations, either simultaneously or in a very short period of time before the animal moves. On a map of the area, lines are drawn from the known locations along the directional bearings obtained for the animal. The intersection of those lines provides the projected location of the animal.

Fish biotelemetry studies can be done with radio transmitters, but ultrasonic transmitters that can send sound waves through water are also used. Ultrasonic transmitters (fig. 9.13) are more effective than radio transmitters in waters that have high amounts of total dissolved solids, including salt water, and in deep water. Radio signals do not pass through such media very well.

Ultrasonic transmitters send sound waves, and underwater hydrophones (fig. 9.14) are used to receive their signals. The life expectancies and ranges of these transmitters vary with battery size and functions performed. Obviously, a larger fish

A

B

FIGURE 9.12 Receiver antennas receive radio signals from transmitters and enable biologists to track and record the activities and locations of monitored animals. (A) Handheld unit; (B) vehicle-mounted unit. (*Part B photograph courtesy of D. Naugle.*)

FIGURE 9.13 Insertion of an ultrasonic transmitter into a striped bass × white bass hybrid.

can carry a larger transmitter with a larger power source than can a smaller fish. For general purposes, most ultrasonic transmitter signals cannot be received at distances of more than 1–2 kilometers.

Ultrasonic transmitters can be "coded" so that individual fish can be recognized. For example,

the sound pulse pattern for one fish could be a series of four repeated signals such as beep, beep, pause, beep. A second fish could be distinguished if its transmitter had a repeated pattern of beep, pause, beep, pause.

Ultrasonic signals are blocked or reduced by physical obstacles such as aquatic plants, lake features such as islands or points, areas of large temperature differences, turbulent waters, and outboard motors. Radio signals are better suited for waters with low levels of dissolved solids, shallow water, and turbulent water. Radio transmitters are well suited for use in large, shallow rivers and for fish species such as channel catfish or saugers that may make long movements. In such cases, fish can be located with an antenna that is handheld, mounted on a boat, or mounted on an aircraft that flies along the river. Radio signals can even be received through ice. Different individual fish can be distinguished by different radio frequencies.

Both ultrasonic and radio transmitters are often implanted within the body cavity of a fish, but external attachment is also possible. Ultrasonic transmitters do not need an antenna, while radio transmitters must have some type of antenna. If the antenna can be extended outside the body of a fish, the range at which the signal can be received is greatly extended.

FIGURE 9.14 A hydrophone and receiver for an ultrasonic biotelemetry system.

Fish locations can be obtained through triangulation, as in wildlife work. However, recent advances in technology have increased the techniques available for determining animal location. The **long-range navigation-C (LORAN-C) system** of radio stations developed for human air navigation can be used for this purpose. For example, if a boat is positioned over a fish with a transmitter, a LORAN-C receiver can be used to obtain map coordinates for that fish. The use of **global positioning systems (GPS)** is also becoming popular for location determination on the ground and in the air. Global positioning systems utilize earth-orbiting satellites for position location.

SUMMARY

To understand population assessment, some knowledge of basic statistics is essential. Accuracy refers to the closeness of a measure to its true value, while precision refers to the closeness of repeated measures to one another. Measures that are inaccurate are also referred to as biased. The average for a series of measures is termed the arithmetic mean, and statistical parameters such as standard deviation and standard error describe the amount of variation about the mean.

Biologists can determine population number through a census, which is a complete count of all organisms in a population. However, censuses are rarely undertaken because of the time and expense involved. Biologists more often use surveys, which are partial counts of a population. The data from a survey can be expanded to an estimate of the total population number, or simply used as an index to relative population abundance. Estimates based on survey data should be accompanied by a measure such as a confidence interval that will allow the reliability of the estimate to be assessed.

Wildlife biologists often index natality as number of young per female or number of eggs hatched per nest. Fishery biologists who wish to assess recruitment must know the age at which mortality of young fish is similar to that of adults. Mortality rate, equal to 100% minus survival, is often determined from population age structure. Life tables can be used to calculate mortality by age class for a population.

Population structure refers to the age structure or size structure of a population. Fishery biologists are often concerned with the quality of a fish population, which is related to its size structure. Wildlife biologists rarely manage for body size, but do at times manage for size of structures such as antlers.

Monitoring the physiological status of individuals and populations can also provide useful vital statistics. For example, blood chemistry can reveal much about nutritional status. Body condition is also often monitored.

Food habits are the kinds and amounts of food eaten. Digestive tracts of organisms are commonly sampled to obtain food items. Numerical analysis, frequency of occurrence analysis, gravimetric analysis, and volumetric analysis are typically used to summarize food habits data. However, each of these methods has its own set of biases, and careful interpretation is essential.

Field biologists typically find high variation among individual samples. Thus, it is important to measure this variation, to design sampling programs carefully, and to use sufficient units of sampling effort to obtain reliable data.

Models are simplified conceptual or mathematical descriptions of complex systems. The use of models allows prediction of the effects of management strategies before they are implemented. However, the reliability of models should be assessed with actual field trials. Reliability increases as more accurate data are entered into the model.

Marking, tagging, and banding can be used for procedures such as population estimates and mortality/survival estimates. Marks can be physical or chemical, and are most often used for group recognition. Tags and bands are often individually numbered, allowing recognition of individual organisms. In addition, tags and bands can be attached externally or internally.

Biotelemetry allows the activity of individual organisms to be monitored. Sophisticated transmitters can even relay such information as body temperature, heart rate, and feeding activity. Most wildlife biotelemetry studies involve the use of radio transmitters. Fishery workers also use radio systems, but ultrasonic transmitters can be more effective in some aquatic environments.

PRACTICE QUESTIONS

1. Why is it important to obtain some measure of variation about a sample mean?

2. What is the difference between a creel survey and a creel census? On the same body of water, which would be most expensive to obtain? Why?

3. How does a mortality rate calculated from age structure using the catch-curve method differ from mortality information generated from a life table?

4. Cutthroat trout are quite vulnerable to sport angling, and are easily overexploited. Will an overexploited cutthroat trout population have a low or high proportional stock density? Now assume that you have found a harvest regulation tool that will improve the size structure of that overexploited population. Will the proportional stock density now increase or decrease?

5. What is the difference between numerical analysis of food habits data and frequency of occurrence analysis?

6. What is the difference between food habits and feeding habits?

7. What parameters would need to be available to develop a realistic computer model that would predict the age structure of a fish population? Hint: Think of the three dynamic rate functions.

8. In what kinds of studies would group marking techniques be suitable, and under what circumstances would individually recognizable tags be necessary?

9. Biotelemetry studies can provide much information on the behavior of individual organisms. Choose a fish or wildlife species with which you are familiar. Make a list of the different types of behaviors that could be monitored using biotelemetry.

SELECTED READINGS

Anderson, R. O., and S. J. Gutreuter. 1983. Length, weight, and associated structural indices. Pages 283–300 *in* L. A. Nielsen and D. L. Johnson, eds. *Fisheries techniques.* American Fisheries Society, Bethesda, Md.

Bowen, S. H. 1983. Quantitative description of the diet. Pages 325–336 *in* L. A. Nielsen and D. L. Johnson, eds. *Fisheries techniques.* American Fisheries Society, Bethesda, Md.

Harder, J. D., and R. L. Kirkpatrick. 1994. Physiological methods in wildlife research. Pages 275–306 *in* T. A. Bookhout, ed. *Research and management techniques for wildlife and habitats.* The Wildlife Society, Bethesda, Md.

Haufler, J. B., and F. A. Servello. 1994. Techniques for wildlife nutritional analyses. Pages 307–323 *in*

T. A. Bookhout, ed. *Research and management techniques for wildlife and habitats.* The Wildlife Society, Bethesda, Md.

Johnson, D. H. 1994. Population analysis. Pages 419–444 *in* T. A. Bookhout, ed. *Research and management techniques for wildlife and habitats.* The Wildlife Society, Bethesda, Md.

Lancia, R. A., J. D. Nichols, and K. H. Pollock. 1994. Estimating the number of animals in wildlife populations. Pages 215–253 *in* T. A. Bookhout, ed. *Research and management techniques for wildlife and habitats.* The Wildlife Society, Bethesda, Md.

Ney, J. J. 1993. Practical use of biological statistics. Pages 137–158 *in* C. C. Kohler and W. A. Hubert, eds. *Inland fisheries management in North America.* American Fisheries Society, Bethesda, Md.

Nielsen, L. A. 1992. *Methods of marking fish and shellfish.* Special Publication no. 23. American Fisheries Society, Bethesda, Md.

Samuel, M. D., and M. R. Fuller. 1994. Wildlife radiotelemetry. Pages 370–418 *in* T. A. Bookhout, ed. *Research and management techniques for wildlife and habitats.* The Wildlife Society, Bethesda, Md.

Van Den Avyle, M. J. 1993. Dynamics of exploited fish populations. Pages 105–135 *in* C. C. Kohler and W. A. Hubert, eds. *Inland fisheries management in North America.* American Fisheries Society, Bethesda, Md.

White, G. C., and W. R. Clark. 1994. Microcomputer applications in wildlife management and research. Pages 75–95 *in* T. A. Bookhout, ed. *Research and management techniques for wildlife and habitats.* The Wildlife Society, Bethesda, Md.

Willis, D. W., B. R. Murphy, and C. S. Guy. 1993. Stock density indices: Development, use, and limitations. *Reviews in Fisheries Science* 1:203–222.

Winter, J. D. 1983. Underwater biotelemetry. Pages 371–395 *in* L. A. Nielsen and D. L. Johnson, eds. *Fisheries techniques.* American Fisheries Society, Bethesda, Md.

LITERATURE CITED

Anderson, R. O. 1976. Management of small warm water impoundments. *Fisheries* (Bethesda) 1 (6): 5–7, 26–28.

Caughley, G. 1966. Mortality patterns in mammals. *Ecology* 47:906–918.

Clark, R. D. Jr., G. R. Alexander, and H. Gowing. 1980. Mathematical description of trout-stream fisheries. *Transactions of the American Fisheries Society* 109: 587–602.

Ferson, S., F. J. Rohlf, L. Ginzburg, G. Jacquez, and H. R. Akakaya. 1990. *Modeling fluctuations in age-structured populations: RAMAS/age 2.0.* Applied Biomathematics, Setauket, N.Y.

Gabelhouse, D. W. Jr. 1984. A length-categorization system to assess fish stocks. *North American Journal of Fisheries Management* 4:273–285.

Hewett, S. W., and B. L. Johnson. 1987. *A generalized bioenergetics model of fish growth for microcomputers.* WIS-SG-87-245. University of Wisconsin Sea Grant Institute, Madison.

Hurlbert, S. H. 1984. Pseudoreplication and the design of ecological field experiments. *Ecological Monographs* 54:187–211.

Johnson, D. H. 1994. Population analysis. Pages 419–444 *in* T. A. Bookhout, ed. *Research and management techniques for wildlife and habitats.* The Wildlife Society, Bethesda, Md.

Lancia, R. A., J. D. Nichols, and K. H. Pollock. 1994. Estimating the number of animals in wildlife populations. Pages 215–253 *in* T. A. Bookhout, ed. *Research and management techniques for wildlife and habitats.* The Wildlife Society, Bethesda, Md.

Moore, D. S. 1995. *The basic practice of statistics.* W. H. Freeman, New York.

Murphy, B. R., D. W. Willis, and T. A. Springer. 1991. The relative weight index in fisheries management: Status and needs. *Fisheries* (Bethesda) 16 (2): 30–38.

Ricker, W. E. 1975. *Computation and interpretation of biological statistics of fish populations.* Bulletin 191. Fisheries Research Board of Canada, Ottawa.

Sokal, R. R., and F. J. Rohlf. 1981. *Biometry.* 2d ed. W. H. Freeman, San Francisco.

Van Den Avyle, M. J. 1993. Dynamics of exploited fish populations. Pages 105–135 *in* C. C. Kohler and W. A. Hubert, eds. *Inland fisheries management in North America.* American Fisheries Society, Bethesda, Md.

White, G. C., D. R. Anderson, K. P. Burnham, and D. L. Otis. 1982. *Capture-recapture and removal methods for sampling closed populations.* LA-8787-NERP. Los Alamos National Laboratory, Los Alamos, New Mexico.

Willis, D. W. 1987. Reproduction and recruitment of gizzard shad in Kansas reservoirs. *North American Journal of Fisheries Management* 7:71–80.

In this chapter we describe some of the common practices used by biologists to manage the biota. Although it might seem initially that a discussion of such management practices would be extensive and complicated, much of the management of biota involves management of their habitat and human users, which we will discuss in chapters 15 and 17, respectively.

Biota management techniques are often directed at the population level. For example, if Cooper's hawks have been eliminated from an area and suitable habitat is available, restoration of the population by reintroduction is an option.

Biota management techniques can also be directed at the community level. For example, a newly impounded reservoir has an existing fish community that is adapted to the environment of a river, not the environment of a reservoir. Biota management in this situation might involve the addition of a number of new fish species and other organisms to exploit the variety of newly available niches in the reservoir.

Biota management techniques can also be directed at the ecosystem level. However, at that level, biota management, habitat management, and human user management are usually combined.

The word stocking has a negative connotation to some people. Negative aspects of stockings are discussed in sections 4.5, 10.8, and 14.9. However, in many situations stocking is the only alternative available to the biologist. Almost all biological systems in North America have been perturbed by humans, some severely. In many cases, stocking is an effort to ameliorate those perturbations. Restoration biology (see section 2.10), for example, is based in part on stocking, as is conservation biology (see section 1.6). In North America, current population distributions of species such as white-tailed deer, wild turkeys, and cutthroat trout have been restored through extensive stocking programs. Many species, both game and nongame, could be added to this list of successful restorations.

This chapter will focus on partial removal and complete removal of populations, the introduction of organisms, and the culture and propagation of fish and wildlife species, including their subsequent performance upon release. In addition, a brief discussion of diseases will be presented.

10.1 Partial Population Removal

Partial population removal may involve the removal of some individuals from a population that the biologist is trying to improve, or it may involve the removal of other species that are adversely affecting the target population or populations. Removal of a portion of a population can be accomplished in numerous ways. For example, problem populations can be partially removed by selective piscicide kills, by mechanical removal, or by commercial harvest. In fishery work, removal of a portion of a population or community is called **partial renovation** (**partial rehabilitation**). Wildlife biologists have no comparable terminology, but partial wildlife renovations are possible.

Fishery biologists often use piscicides for the selective removal of problem species; this method is essentially unavailable to wildlife biologists. Gizzard shad, for example, often become overly abundant in reservoirs in the southeastern United States. At high population levels, gizzard shad compete with the young of other fish species for food but produce few offspring that can serve as prey for predatory fishes. Gizzard shad are sensitive to toxicants such as rotenone and antimycin at levels that will not affect most other fishes. Although the entire gizzard shad population is not eliminated with low-level toxification, a 75%–90% reduction of the adults will result in reduced competition with other fishes and increased reproduction by the surviving adult shad.

Selective removal using toxicants can also be accomplished by taking advantage of fish behavior.

For example, stunted bluegill populations can be selectively controlled in this manner. A biologist can wait until the height of the bluegill spawning season, then apply piscicides to the shallow water areas used by bluegills for spawning. Most of the fishes in this habitat area at this time of year will be bluegills. Some other fishes will be killed, but the target bluegills will incur the majority of the resulting mortality.

A strategy of selectively controlling problem fish species with species-specific toxicants would be highly useful if environmentally safe, target-specific toxicants were available. Unfortunately, there are few such chemicals. The primary selective toxicants currently available are used for sea lamprey control in tributaries of the Great Lakes (Meyer and Schnick 1983; see section 7.6) and they are not completely species specific. An ecologically safe, selective toxicant for a species such as the common carp would be invaluable for both fishery and wildlife purposes because of the environmental problems caused by this fish (see section 1.5); however, such a selective toxicant has yet to be developed (Marking 1992).

Mechanical removal can also be used to affect populations of organisms. This technique usually involves traps, seines, or other capture devices. Wildlife biologists, for example, may wish to reduce predator populations in relatively small areas in order to increase populations of ground-nesting birds. Some waterfowl species, such as mallards, gadwalls, and Canada geese, will nest in high densities on islands or in other areas where mammalian predator populations have been substantially reduced; while total elimination of predators may be a management goal, it is unlikely to be successful (fig. 10.1). Biologists may selectively trap in these areas to remove mink, red foxes, raccoons, striped skunks, and other predators that destroy clutches or kill nesting females. Waterfowl generally have high nesting success in these predator-reduced habitats.

High-density populations of species such as the white-tailed deer, elk, and Canada goose can cause problems in a variety of ways. For example,

FIGURE 10.1 This electric barrier was used to prevent terrestrial predators from reinvading an area after they had been removed by trapping. In this case, waterfowl nest density and nesting success increased on the peninsula protected by the barrier.

such populations can damage agricultural crops, natural vegetation (fig. 10.2), or urban shrubbery. Their depredations can result in substantial economic and environmental damage. Mechanical removal with a lethal method, such as shooting, can be an effective control method. However, public opinion or a dense human population in an area can preclude the use of shooting as a control measure. Under these circumstances, trapping or the use of anesthetics, followed by removal and relocation, may be used for population reduction. Such procedures can be costly, and there is also the problem of finding suitable relocation sites. Often these relocated animals die after release because of the stress involved in capture and transport, or because they are unable to compete with existing organisms in the area to which they were moved.

Coyotes, especially when abundant, may prey upon domesticated livestock. Some conservation agencies hire trappers and aerial shooters that respond to such damage complaints. Beavers can cause problems when they damage trees or agricultural crops, build dams in unwanted locations, or plug culverts. Again, control methods involve substantial worker hours to trap the beavers or remove dams. In most such damage situations,

FIGURE 10.2 A browse line caused by elk. *(Photograph courtesy of L. Parker, Wyoming Game and Fish.)*

removal of the specific offending individuals is more cost-effective and less environmentally damaging than indiscriminate removal of the offending species.

Most mechanical removals in fisheries are attempts to reduce the abundance of fishes such as common carp, bigmouth and smallmouth buffaloes, white suckers, various bullhead species, or stunted panfish species. Most of these species can develop high standing stocks, and will convert a substantial proportion of the food energy available in the ecosystem to their own biomass. In such situations, recruitment of other fish species often declines. Although improvement in some fish populations has been documented after mechanical reduction of competing populations (Johnson 1977), many such attempts have been unsuccessful. Mechanical reduction of fish populations is quite labor intensive, and rarely is it economically feasible to continue the reduction until a sufficient biomass of the problem species has been removed. In addition, high recruitment is a common response to reduced population abundance in these fish populations. Thus, the population can quickly return to problem status unless an additional management strategy, such as predator fish protection with a length limit (see section 17.5), is used in conjunction with the removal.

Commercial harvest can be an alternative means of partial removal of organisms. It seems an attractive alternative, where and when feasible, to allow private individuals to harvest and sell the problem organisms, rather than having a conservation agency pay for mechanical removal. In wildlife, the selective trapping of furbearers is a commercial harvest method used for partial removal. There have been successful cases in which commercial fishing removed a sufficient biomass of problem species to allow increased abundance and growth of other fishes (Stephen 1986). However, it is difficult to remove enough of the problem species population so that improvements in target populations can be achieved. Often, just when the commercial harvest is reaching a sufficient proportion of the standing stock of the problem species, catches begin to decline and the commercial harvesters want to move on to other areas where a greater profit can be made. At that time, the only alternative left to a conservation agency might be to subsidize continued commercial exploitation, and the cost-benefit ratio of such operations is often questionable.

10.2 Complete Population Removal

Fishes can be totally eliminated from a water body by means of certain toxicants or by draining. Because the use of toxicants is such a drastic and expensive measure, they are used much less frequently now than in the past. Wildlife managers essentially have no comparable methods of total removal of a population or community. Attempts can be made to use capture devices to completely remove a wildlife population, as in the example of the mammalian predators in section 10.1, but the costs are so high that only under special circumstances is this even attempted.

Total renovation (**total rehabilitation**) in fishery work refers to the elimination of an entire fish community. Total renovation, like partial renovation, is a fishery term with no equivalent wildlife term. This biota management method is most often applied in a situation in which a particular fish species is overly abundant. For example, common carp may have been introduced into a water body through an accidental inclusion with a purposeful fish stocking or as an accidental introduction with released bait from a minnow bucket; they may also have naturally dispersed into an area. When common carp reach high standing stocks, they route much of the productivity of an aquatic ecosystem into their population biomass, and densities and biomasses of other fishes generally decline. At times, the most appropriate solution to this problem may be the elimination of the entire fish community, followed by restocking of desired fish species. Similarly, panfishes can overpopulate and stunt. In some cases, total population renovation is the most realistic management alternative. Obviously, the larger the water body, the less likely it is to be toxified or drained.

Total renovations should be considered a management alternative of last resort. There are numerous negative effects associated with total renovations. If the totally renovated water body is restocked with small fishes, several years generally must pass before a fish community develops that will resemble a natural community and provide recreational fishing opportunities. If larger fishes are stocked, the restocking effort is much more expensive; such a restocking may not even be possible due to a lack of adequately sized fishes available for restocking. In addition, a biologist needs to be certain that the original problem will not simply recur in a few years. If common carp were the original problem, total renovation may be useless if there is no way to prevent the carp from reentering the water body and reestablishing a population in a few years.

Another negative aspect of total renovations is that the problem species is not the only organism removed. Consideration should be given to restoration of other native organisms eliminated by the renovation, not just the target game fishes and their primary prey species. Restoration of all the native organisms eliminated would ensure the species diversity needed for the aquatic ecosystem to function correctly. Complete restoration is seldom, if ever, feasible, especially if invertebrates, key to many food webs, are included in the restoration.

Total renovations have sometimes been used by fishery biologists to allow for the restoration of native species. For example, introduced brook, rainbow, and brown trout have often reduced or eliminated native cutthroat trout populations from Rocky Mountain streams. Some conservation agencies use a toxicant such as rotenone to eliminate the fish community from a reach of stream, and then restore the native cutthroat trout. Good sites for such renovations are stream stretches above a natural barrier to fish movement, such as a waterfall. Some of the initial restorations of greenback cutthroat trout were completed using this technique. This subspecies of cutthroat trout, native to the eastern slope of the Rocky Mountains in Colorado, was at one time believed to be extinct. Remnant populations were found and brought into a fish hatchery. Progeny produced at the hatchery were stocked above barriers in streams that had been chemically renovated. Greenback cutthroat trout have now expanded to the point that catch-and-release fishing is allowed in some places in Colorado. Biologists in the Appalachian Mountains, where the brook trout is native, have undertaken similar renovations to remove introduced rainbow trout and restore brook trout.

Depending on the size of a stream and the complexity of the native fish community in that stream, restocking of nongame fishes after total renovation may be necessary. This is especially true if renovations are done above natural barriers. When renovations are undertaken on a reach of stream with no barriers to fish movement, native fishes often recolonize the reach from above and below the treatment area.

Common carp are also viewed as a detriment by most managers of waterfowl refuges (Crivelli 1983). Common carp are well suited to the shallow, vegetated waters that refuge managers maintain for waterfowl production. When common carp populations attain high standing stocks, the quality of waterfowl habitat is reduced. Waters typically become more turbid, and stands of aquatic plants such as sago pondweed and wild celery, as well as associated aquatic invertebrates that are used as food sources by many waterfowl and other species, are reduced. Total renovation of the fish communities in waterfowl refuge pools is a tool that can be used by waterfowl biologists, especially in situations in which the water can be drained from part or all of the system.

Total renovation as a biota management tool is primarily limited to fishery work because fish communities are generally limited by the boundaries of a water body. One of the rare attempts at total elimination of a species in the wildlife field occurred when biologists attempted to eliminate introduced Arctic foxes from islands in the Aleutian chain (Springer et al. 1977). These islands are isolated and have distinct boundaries, which allowed this unique wildlife renovation attempt. Arctic fox predation on Aleutian Canada geese was a direct cause of the decline of the geese and their listing as an endangered species. After Arctic fox populations had been greatly reduced or eliminated, Aleutian Canada geese from captive flocks were restored to selected islands. An interesting sidelight to this story is that biologists found it necessary to include a few adult wild birds with each release of captive-hatched goslings. The captive-hatched birds had unusually high mortality because of a tendency to migrate straight south over the Pacific Ocean. The wild birds traditionally had migrated to the continental coastline before turning south, and the adult wild birds led the captive-hatched birds along this route.

10.3 Other Types of Population Control

Other attempts at population control have been made that do not involve killing or moving organisms. Most of these biota management techniques involve interference with reproduction or attempts at scaring or repelling organisms from an area. For example, insects, striped skunks, and coyotes have been sterilized and released into the wild. These individuals' unsuccessful attempts at reproduction can result in lower overall numbers of offspring in a population. Similarly, contraceptives can be delivered to organisms in the wild through direct injection or in baits (Turner et al. 1992). Attempts to interfere with reproduction in order to control population numbers are more likely to be successful in small populations located in confined areas. It is difficult and expensive to apply these techniques over large geographic areas. The use of contraceptives is also an area of contention with animal rights and even some animal welfare proponents (see section 16.6).

Fishery biologists sometimes rely upon the release of polyploid or monosex fishes to prevent future reproduction by a released species (see section 4.8). Many states in the United States limit introductions of grass carp to triploid fish. Grass carp are herbivores and can be used for vegetation control. However, they are an exotic species in the United States, and their natural reproduction is not desired. Stocking of only triploid fish can ensure that these introductions do not result in the subsequent development of wild populations. Unfortunately, not all states had this requirement, and reproducing populations of grass carp now occur in the United States (Brown and Coon 1991; Raibley et al. 1995).

Another method of population control is the introduction of forms for which appropriate reproductive habitats are not available. For example, stocking a Pacific salmon species into a water body that has no salmon spawning habitat will ensure that the population can eventually be eliminated by cessation of stocking. Similarly, a wildlife species such as the chukar can be released into certain areas for sport hunting with the knowledge that the correct habitat, including reproductive habitat, is absent and the species will not develop a self-sustaining population.

Selective control of some problem organisms may involve attempts to repel or scare them from

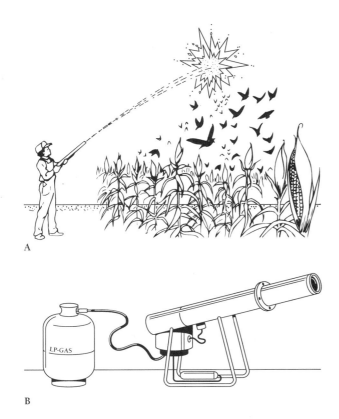

FIGURE 10.3   Devices used for scaring away problem wildlife. (*A*) Shell crackers, fired from a 12-gauge shotgun, produce an aerial explosion. (*B*) Propane cannons, also known as liquid propane gas exploders, also make loud sounds at periodic intervals.

an area. Blackbirds migrate in large, mixed-species flocks that can cause substantial economic damage to agricultural crops. Attempts to control these birds have involved repellents, fright calls, and scare devices such as shell crackers and propane exploders (fig. 10.3), but the effectiveness of these techniques is often limited. Similarly, white-tailed deer can cause economic damage to orchards, residential plantings, and other plant life. Repellents, screens, fences, scare devices, and even the spreading of human hair from a barber shop are all techniques that have been used in attempts to reduce damage. Again, the effectiveness of these techniques typically is limited.

While part of the work of a wildlife biologist involves protecting wildlife from people, it is also

necessary at times to protect people from wildlife. Mountain lion harvest, and trapping and relocating, have been increased in some areas to reduce the likelihood of attacks on humans. An increased number of human attack and threat situations have occurred in areas such as Montana, Colorado, and California as mountain lion populations have increased and dispersed into areas of dense human populations, and as human populations have spread into traditional lion habitat.

Management techniques used to reduce human conflicts with grizzly bears include proper disposal of domesticated livestock carcasses, redistribution of **carrion** (dead animal flesh), protection of beehives with specialized electric fence systems, education of landowners, and proper garbage storage and disposal programs. Relocation of nuisance grizzly bears to areas outside of their known home ranges has proven to be generally ineffective and no more than a temporary solution.

In fishery work, electrical barriers (Verrill and Berry 1995) and mechanical barriers have been used to keep fishes from moving into specific areas (fig. 10.4). The long-term success of such

FIGURE 10.4 An electrical barrier used to prevent common carp movement at Heron Lake, Minnesota. The cement pad visible on the far bank runs along the stream bottom and contains electrodes. (*Photograph courtesy of C. Berry Jr.*)

barriers is sometimes questioned because flooding or failure of the mechanical devices, as well as the human habit of either intentionally or unintentionally introducing fishes, can render them ineffective. Electric barriers have also been used in wildlife work to protect ground-nesting birds from predators (see figure 10.1). In these areas, predators such as striped skunks and red foxes are removed, and the electrical barriers slow reinvasion.

10.4 Stocking Types

Fish stockings are generally categorized as **introductory**, **maintenance**, or **supplemental stockings**. Introductory stockings involve the release of fishes into a new or renovated body of water or the introduction of a new fish species (either native or introduced) into waters with an existing fish community. Maintenance stockings are those made to sustain a fish population that for some reason has no, or perhaps very limited, natural reproduction. Supplemental stockings are those made to augment a naturally reproducing population. Maintenance and supplemental stockings can be difficult to differentiate; in reality, there probably is a continuum across these two definitions.

Stockings of wildlife organisms can fit into the categories described for fishes; however, these terms are generally not used to describe wildlife stockings. The introduction of gray partridge into North America could be termed an introductory stocking. The restoration of wild turkeys to an area they had previously inhabited is also an example of introductory stocking. Maintenance stockings are quite rare in wildlife. However, some conservation agencies or private hunting preserves may release adults of a species that cannot reproduce in a certain area. Ring-necked pheasants, for example, typically do not recruit in the wild in the southeastern United States, but the species can be pen-reared and released in a private hunting preserve. Supplemental stockings of wildlife species are also undertaken at times. Examples might include attempts to supplement existing northern bobwhite or ring-necked pheasant populations with pen-reared birds.

10.5 Introductory Stockings

One task of the biologist is to introduce native species into new or renovated water bodies or into appropriate terrestrial habitats. While quite common in fishery work, such introductions are currently less common in wildlife. There are numerous reasons for this difference. Fishery biologists have two tools available for the total renovation of fish communities (toxicants and draining); thus, restockings are commonly made after total renovations. Wildlife biologists have no comparable management tools. Also, fishes tend to be less mobile than most wildlife species. When a water body is newly constructed or renovated, often the only means for the reasonably rapid colonization of fishes is stocking. Conversely, when terrestrial habitat is developed, wildlife organisms are more likely to disperse naturally into the newly created habitat areas.

These differences do not mean that introduction of native species does not occur in wildlife, just that it occurs less often. For example, river otters and martens have been restored by state conservation agencies in locations from which they were **extirpated** (eliminated from a portion or all of their native range). The reestablishment of breeding populations of giant Canada geese in north-central states is another example of a successful restoration. This restoration program has been so successful that those states now allow regulated sport hunting of giant Canada geese. Extensive restoration programs in various portions of their native ranges have been completed for animals such as elk, pronghorns, white-tailed deer, mountain goats, and bighorn sheep. Much of the current abundance of large mammals in the United States is the result of restoration programs that were completed earlier in this century. In a comparable management program, fishery biologists have used cultured paddlefish to restore populations in Missouri (Graham 1986).

Peregrine falcons have been restored to many areas of the United States by a technique known as **hacking**. Offspring produced by captive birds are reared and fed with as little direct contact as possible with humans. They are then moved to the wild, fed in an elevated caged nest, again with as little human contact as possible, then allowed to fly free. Typically, food is supplied at the hacking site until the birds have learned to fend for themselves.

Because introductory stocking of native species is a biota management tool most often used in fisheries, the remainder of this section will pertain to fishes. Introductory stockings are more common in artificial (human-created) water bodies than in natural water bodies. Natural water bodies typically contain a complement of fish species that evolved in that ecosystem. Thus, biologists generally need to do little in the way of biota management to improve the native fish communities in these waters. In fact, this natural species diversity tends to make it more difficult to affect fish communities by stocking in natural water bodies. Introductory stockings have been made in natural lakes, especially when these habitats have been modified by human activities. The introduction of some Pacific salmon species into the Great Lakes is an example of such stockings.

Artificial water bodies often have no complement of fish species that evolved in the system. Artificial water bodies exist along a continuum that ranges from highly stable waters quite similar to a natural lake to impoundments with high rates of water exchange that may actually be little more than a wide spot in a river. The fish community that lived in a river before it was dammed is likely to contain relatively few species that will thrive in a reservoir.

When making an introductory stocking into a new or renovated water body, the biologist must consider the available habitat and stock species best suited to the available niches. A small water body has fewer niches, and it is generally best to keep the fish community simple. Over much of the United States, largemouth bass are introduced into small water bodies as a predator and bluegills as a panfish and prey species. However, the available habitat must be considered. Largemouth bass and bluegills are appropriate species for introduction into small water bodies that have relatively clear (e.g., typically transparency of 45 centimeters or more), warm water for most of the year and contain appropriate coverage of submergent aquatic plants. If aquatic plants are overly abundant, the bluegills will have too much cover and will avoid predation by the largemouth bass, leading to a stunted bluegill population. If aquatic plants are too sparse, largemouth bass recruitment may be low in many parts of the United States. In these cases too few largemouth bass will be present to prey upon the bluegills, again leading to a stunted bluegill population. If a small, warm water body is constantly turbid, perhaps because of suspended colloidal clay particles, channel catfish may be a better choice for introduction. In cold water bodies, or those that have colder water in deep areas, coldwater species such as rainbow trout can be stocked.

In larger artificial water bodies, the diversity of available niches is greater and the biologist can typically introduce a wider array of fish species; some may even be transplants or exotics. One can use a large warmwater reservoir as an example for this type of stocking strategy. First, consider the top-level predators available for such a system. For shallow warmwater habitat, the largemouth bass immediately comes to mind. However, both the smallmouth and spotted bass can also be used in this capacity. Largemouth bass tend to congregate in areas with aquatic plants, flooded timber, or some other type of structure, while spotted and smallmouth bass tend to use rocky habitat. However, this niche segregation is commonly not clear-cut. Biologists usually would also want to introduce open water predators into a large reservoir. Some fish species commonly chosen to fill this niche include white bass, striped bass, striped bass × white bass hybrids, and walleyes. Deep water predators also need to be considered. The flathead catfish in southern states (a warmwater species) could fill this niche.

The next category of fish species to consider for this reservoir is a group that is difficult to categorize: these fishes can be termed mid-level piscivores. Fishes such as the yellow perch, black crappie, and white crappie occur in this category. Although they do eat some prey fishes as adults, during much of their lives they feed on aquatic or terrestrial invertebrates. These panfishes are extremely important fishes in most systems. While the top-level predators are "glamour fishes," panfishes might be called the "bread-and-butter" or "family" sport fishes. They support a tremendous portion of the sport fish harvest across the United States and Canada because they are easily caught, can be caught in large numbers, and are low on the food chain and thus can have relatively high standing stocks. In addition, they serve as prey for the higher-level predators.

The biologist should now consider a lower trophic level, the insectivores. Some fishes obtain most of their food energy from aquatic or terrestrial insects; common examples include bluegills and rainbow trout. Actually, many fishes feed primarily on insects at certain life stages and at certain times of year. For example, during a summer hatch of a large mayfly such as *Hexagenia,* channel catfish, largemouth bass, black and white crappies, yellow perch, walleyes, and white bass will all feed on this abundant food source.

Finally, prey organisms (commonly fishes or invertebrates) need to be considered. In the past, there was a tendency for fishery biologists to stock top-level predators without being certain that the system could provide a sufficient prey base. This rarely occurs today. The particular prey fishes used depend on the geographic location and habitat. Gizzard and threadfin shad are important prey fishes in the southeastern and midwestern United States. Many minnow species are important prey

fishes. Emerald and spottail shiners serve as prey for predators in South Dakota reservoirs, while red shiners are more likely to be common in Kansas reservoirs, and golden shiners are common in New England. The term **forage fish** is sometimes used synonymously with the term prey fish. However, animals such as cows and horses eat forage, while predators eat prey; thus the term forage is inappropriate

The species stocked will vary according to water type. Obviously, warmwater species will be stocked in warm water and coldwater species in cold water. Stocking may include introduced forms if no native species can fill the available niches. **Two-story fisheries** can also exist. In large, deep reservoirs that have both warm and cold, well-oxygenated water strata, a biologist can develop a warmwater or coolwater fishery in the surface waters and a coldwater fishery in the colder bottom waters. Coldwater species such as rainbow trout can be used in the coldwater habitat, while warmwater species such as largemouth bass or coolwater species such as walleyes can be used in the warmer water habitat. At northern latitudes, the construction of a reservoir may result in new warmwater habitat. The coldwater natives will be present to populate the bottom story of the fishery, while introduced species may be stocked into the upper warmer water habitat.

Introductory stockings of either native or introduced fishes for food web improvement are appropriate in some circumstances. Prey fish stockings are often undertaken to improve food webs, especially if predators have already been stocked without prior consideration of the prey base. It is important to note, however, that some prey fish stockings have provided mixed results at best, and that caution should be exercised. Biologists sometimes find that fish stocked as prey for adult predators actually compete with young predators. For example, threadfin shad introduced as a prey species often compete directly with young largemouth bass and black crappies.

Stockings of either native or introduced fishes can be used to develop trophy fisheries by introducing such species as muskellunge or striped bass. However, many such populations may not be sustained by natural reproduction, and maintenance stockings may be necessary.

10.6 Introductory Stockings of Exotic and Transplanted Species

You will recall that exotic species were defined as organisms introduced from another zoogeographic region. Transplants were defined as organisms moved outside their native range but within the same zoogeographic region (see section 2.10). Both exotic and transplant introductions have resulted in successes; such introductions have also resulted in ecological disasters (see section 1.5). Most of the exotic and transplant introductions that have proven to be both successful and beneficial have involved organisms introduced into areas that had been so changed by habitat modification that native species could no longer exist. Such introductions are a common form of biota management.

The Missouri River in North and South Dakota was once a relatively shallow flowing water environment characterized by highly turbid water. The construction of a series of dams on that river resulted in a primarily nonflowing environment with deep, clear water. Many fishes native to the original river could not adjust to these habitat changes, and some new habitats were produced that no native species could inhabit. Essentially, new habitats were formed in a region that had no native fish species able to exploit those new habitats. A variety of coldwater salmonid species, including both trout and Pacific salmon, were introduced into some reservoirs on this river system, where they have been successful and beneficial. Prey species such as rainbow smelt and the invertebrate *Mysis relicta* were also introduced because native prey

species could not exploit all of the new ecological niches.

A wildlife example of biota management using exotics or transplants that has been both beneficial and successful is the introduction of exotic ring-necked pheasants. In some areas of the United States, the native gallinaceous birds could not exist well in landscapes changed by human activities, primarily agriculture. New habitats and landscapes had been produced, and native forms could not fill these new ecological niches. The ring-necked pheasant could, and its introduction has been a success. The introduced gray partridge has also been successful and is even more adaptable to intensive agriculture than the ring-necked pheasant.

Introducing exotics and transplants where new habitats have not been formed can be a risky business. An example of an introduction into a habitat that had not been extensively modified and in which native species still existed took place in the Black Hills of South Dakota and Wyoming. The Black Hills had habitats that would support trout species, but trouts had apparently not dispersed naturally into that region. Brook, brown, and rainbow trout were introduced into the Black Hills, and a successful and beneficial trout fishery was produced. The success of these introduced forms, however, has resulted in population decreases of native forms, primarily minnows and suckers. Sport fishery advocates generally accept such decreases in the population abundance of nongame species as long as nongame species do not become threatened or endangered. However, not all agree with such stocking policies, and much greater scrutiny is now common before such introductions are made.

Other introductions of trouts in the United States have also had mixed and sometimes primarily negative effects (fig. 10.5). The brook trout's native range extends from the Appalachian Mountains to the northeastern United States and southeastern Canada. Naturalized brook trout populations in the Rocky Mountains have, through competition, reduced or eliminated native subspecies of cutthroat trout, especially in small high-elevation streams. The rainbow trout is native to

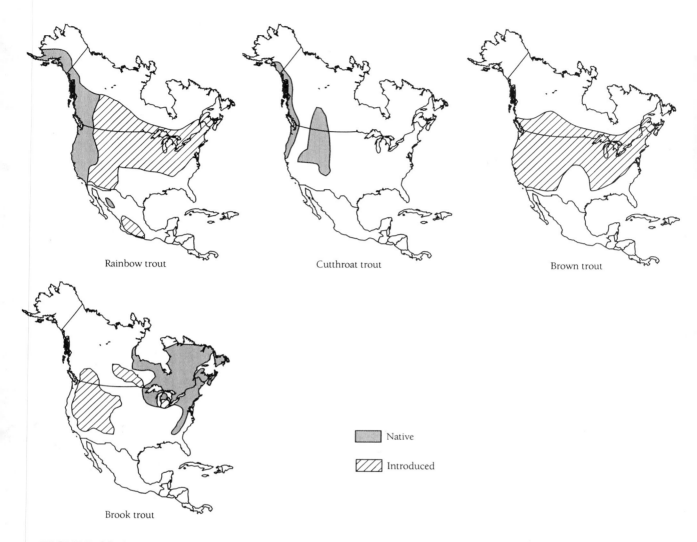

Rainbow trout

Cutthroat trout

Brown trout

Brook trout

Native

Introduced

FIGURE 10.5 Distribution maps for rainbow, cutthroat, brown, and brook trout in North and Central America. Cutthroat trout have been transplanted outside of their native range, but such stockings are rare and generally localized. The brown trout is native to Europe and Asia, and is an exotic species in North America.

the extreme western drainages along the Pacific coast of North America. These fish have become naturalized in some Rocky Mountain and Appalachian streams, where they compete with native species. The brown trout is native to Europe, and thus is an exotic species in North America. Because it can tolerate warmer water temperatures than brook, cutthroat, or rainbow trout, it has been considered a desirable species for introduction in streams where water temperatures are marginal for other trouts. However, brown trout are piscivorous and can negatively affect native trouts and other species through competition and predation.

European starlings and house sparrows are exotic species from the Palearctic region that have adapted to environments in the United States and Canada. Both exotic species often outcompete native species, resulting in declining populations of what are generally considered to be more desirable species.

Many exotic ungulates have been introduced into North America and other areas of the world, and most have created problems through competition with native species, hybridization with natives, and the introduction of diseases. For example, escaped red deer have come into contact with and interbred with native elk in some areas. However, these hybrids currently have limited distribution. Barbary sheep introduced into Texas are carriers of diseases that can affect native bighorn sheep populations. Several western states have forbidden the importation of exotic ungulates.

New Zealand has incurred serious vegetation losses and erosion problems due to introductions of exotic ungulates. Likewise, native species in New Zealand face serious competition from exotic fishes and wildlife. Many islands, including Hawaii and Jamaica, have lost or are losing substantial numbers of native bird species due to competition with or predation by exotic birds and mammals.

10.7 Supplemental and Maintenance Stockings

While introductory stockings are often justifiable and successful, supplemental and maintenance stockings must be more carefully assessed. For example, a study of walleye stockings in 125 water bodies over a 100-year period indicated that 48% of introductory stockings, 32% of maintenance stockings, and 5% of supplemental stockings were successful (Laarman 1978). Stocking success depended more on environmental and biological conditions at the time of stocking than on the number and size of walleyes stocked. Two conclusions can be drawn from this study of walleye stockings. First, it appears that maintenance stocking of walleyes into waters that contain no suitable habitat for reproduction can create sport fisheries. Second, supplemental stocking of walleyes to increase the density of a population that has some amount of natural reproduction is not likely to succeed.

Maintenance and supplemental stockings are also attempted in wildlife, although these specific terms are rarely used to describe such introductions. Private individuals commonly release 8–10-week-old ring-necked pheasants into areas with established wild pheasant populations. In a few cases, natural resource agencies, due to public and political pressure, have also released young pen-reared ring-necked pheasants. However, where wild ring-necked pheasants are already established, such releases have repeatedly failed to boost long-term population levels. In most release areas, the habitat is already suboptimal, or wild ring-necked pheasants would naturally be more abundant. Stocked birds are highly vulnerable to predation; however, predator control of sufficient intensity to benefit the released birds is time-consuming and expensive, and it is seldom attempted or accomplished. In addition, where intensive predator control is attempted, private individuals may illegally kill raptors and other beneficial organisms that can reduce stocking success. In recent years, public opposition to native predator control has increased. When habitat is improved, as illustrated by the substantial increases of ring-necked pheasants after the implementation of the Conservation Reserve Program (see section 1.3), wild pheasants quickly expand their numbers in response to the enhanced habitat conditions, and no stocking is required. Stocking in such situations may even be harmful to existing populations because of the poor genetic fitness of the pen-reared birds and potential disease problems.

Spring release of pen-reared adult hens has also generally failed to provide appreciable increases in subsequent fall ring-necked pheasant numbers. Spring release of such birds is also extremely expensive (Hessler et al. 1970; Hill and Robertson 1988). Pheasant preserve operators generally release mature males shortly before or at regular intervals during the hunting season to minimize natural mortality of stocked birds and to increase the percentage of stocked birds that are harvested.

Supplemental and maintenance stockings can take two different forms, **put**, **grow**, **and take stockings** or **put and take stockings**. When subcatchable-sized rainbow trout are maintenance stocked into a remote mountain lake with no inlet or outlet stream, the fish use the natural productivity of the lake to grow, and then are harvested by anglers when they reach catchable size. This is a put, grow, and take fishery and is quite different from fisheries in which catchable-sized rainbow trout are stocked into high-use streams and lakes. Catchable-sized stockings are put and take stockings; these fish are immediately available to anglers. Frequent stockings, often several per year, may be necessary to maintain populations in high-use situations in which fish harvest is important.

Many waters stocked with catchable-sized trout are quite popular with tourists. Anglers drive large, expensive recreational vehicles from many parts of the country to camp and fish in these areas. Supplemental stockings of catchable-sized fish to sustain such fisheries are often criticized. However, if the income from license sales to non-resident anglers that harvest the catchable-sized fish is greater than the cost of the culture and stocking program, the catchable-sized stockings probably will be economically justified. This is especially true if the overall economic impact of the tourism industry in these areas is considered. Other anglers may prefer to fish for wild trout, even if most or all of the fish are protected by harvest regulations and have to be released; such anglers can seek areas set aside for this type of angling experience.

Urban fishery programs are highly reliant upon stocking. Approximately 75% of the people in the United States currently live in cities with populations of 100,000 or more. Providing fishing opportunities to these people has become an important mission of many conservation agencies. These fisheries typically sustain high levels of fishing effort; as much as 30,000 hours of angling effort per hectare of water per year has been documented (Alcorn 1981). At such high levels of fishing effort, natural reproduction and growth cannot sustain fishable populations when the fish caught are removed (harvested), so stocking of catchable-sized fishes is a common management strategy. The fish species stocked is often less important to the urban angler than the opportunity to catch fish; thus, many urban programs are based on fishes such as common carp, buffaloes, or bullheads.

The terms put and take and put, grow, and take are typically not used by wildlife biologists, but these terms are applicable to wildlife stockings. Hunting preserves that release pen-reared adult birds or mammals prior to a hunt are practicing put and take stocking. Some conservation agencies sponsor such practices in areas with substantial levels of hunting effort. Put, grow, and take stockings are less often used, primarily because the release of young birds or mammals that will grow and be harvested at a later date is typically not a very successful strategy.

10.8 Philosophical Considerations for Stocking Programs

Stocking is often viewed by the public as the most important and effective management tool available to biologists. It seems to represent some form of magical solution to all fishery and wildlife problems. If fishing or hunting success declines, the public will often request more stocking. It does not seem to matter if fishing has declined because a fish species has overpopulated and stunted, or wildlife populations have declined because of a lack of habitat; more stockings are still requested. It may take a major public relations effort by biologists to convince the public that stocking in certain situations is unnecessary, useless, or even harmful. In some cases, stocking is a valid management tool. However, it is certainly not the answer in many situations.

In many instances, stocking represents the treatment of a symptom rather than the cause of a problem. Stocking cannot address habitat degradation, overexploitation, poor population structure, or other problems that affect populations and communities. Aside from the fact that stocking is often unsuccessful, it can also cause additional problems in some situations. There is increasing evidence that inappropriate stockings can harm rather than help a resource. A good case in point involves the various species of Pacific salmon. There is concern, and some evidence, that the release of hatchery-produced Pacific salmon in the native range of these fishes may actually harm native stocks, perhaps through the introduction of inferior genetic material (Hilborn 1992). Fishery biologists in British Columbia were sufficiently concerned about the conservation of genetic material to change their method of stocking **steelheads**. The steelhead is a rainbow trout form that normally lives part of its life in marine waters. The new stocking programs in British Columbia involve the capture of small numbers of returning steelheads from each of a series of rivers, spawning and rearing those fish in individual lots, and then stocking offspring only in the stream from which their parents were collected. Pacific salmon and steelhead population declines will be more effectively addressed when their causes, such as dams that hinder migrations of adults up the river or juveniles down the river, lost or degraded spawning habitat, and other human-influenced limiting factors, are addressed. The current arguments primarily used against addressing causes rather than symptoms are economic. However, in the long term, addressing causes typically makes more economic sense than treating symptoms.

A situation similar to that of the Pacific salmon and steelhead may have been averted in the wildlife field. In the late 1980s, a plan was developed to stock massive numbers of mallards into North America. The North American population of mallards was declining, and the proposed solution was to raise large numbers of mallards in captivity for release. Fortunately, the plan has not been implemented.

A variety of problems can be caused by stocking in the wrong situation. As mentioned above, the introduction of inferior genetic material into a native population may decrease the survival characteristics of that population. For example, after a strain of rainbow trout has been raised in fish hatcheries for generations, intentional and unintentional artificial selection is likely to have favored those fish adapted to hatchery existence rather than to life in the wild. If any of that strain of rainbow trout survive when stocked in the wild and reproduce with native fish, the native gene pool may be diluted. Another problem is that many introductions are made from a small gene pool. For example, if five sibling wild turkeys are trapped in one area and relocated to another to establish a new population, inbreeding will occur. The turkeys thus may not have sufficient genetic diversity to flourish in their new environment. Even if the same number of nonsibling wild turkeys were relocated to the new area, the small number of individuals might still provide minimal genetic diversity.

Another important concern about stocking programs is their influence on species diversity. For example, consider what can happen when a predatory species such as the walleye is stocked into a water body in which it was previously not present. The stocked walleyes may have an effect on other fishes and a variety of other aquatic organisms in the water body. In addition, walleyes may escape from that water body into other water bodies. Now other potential problems arise. A substantial number of large, predatory fish may have moved into an area such as a small stream in which no such top-level predators may have previously existed. What effect will these new predators have on native stream species? To add to the problem, the fish community in that small stream may have already been reduced because of habitat degradation. Thus, the influence of a stocking program on species diversity in the stream becomes a thorny issue, probably a much more important one than the effects of introduced walleyes in a human-created habitat such as a reservoir.

Wildlife biologists do not receive nearly the demand from the public for stockings that fishery biologists do. This is not to say that the wildlife field is immune to the problem. For example, there is still considerable demand for northern bobwhite and ring-necked pheasant stockings.

Biologists must carefully define the reason for a stocking program. Specific management objectives need to be developed, and those objectives need to be measurable (see section 20.6). In addition, all of the biological ramifications of the stocking must be taken into consideration. What effects will it have on other organisms? Economic issues must also be considered. What is the final cost of the organism caught or harvested? Are the benefits greater than the costs of the stocking?

Some stockings are done for political rather than biological reasons. For example, a biologist may be successful in stocking catchable-sized rainbow trout in an urban stream in the southern United States during the winter. A popular fishery may develop, but the cost of the program may outweigh the return from license sales and other economic benefits. The biologist may wish to cease the program, but may find that influential political figures and the general public in the city expect the stockings to continue. The outraged biologist may approach his or her supervisors, and be even further upset when an administrative decision is made to continue the program. Who is right and who is wrong in this situation? This is a difficult question to answer. Perhaps the administrators decided that bearing the cost of the stocking program was a lesser evil than bearing the costs of a political battle and negative public opinion that could carry over to larger issues.

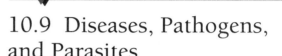

10.9 Diseases, Pathogens, and Parasites

A **disease** is any departure from health; it is a particular destructive process in an organ or organism with a specific cause and symptoms. **Disease agents** can be physical, chemical, or biological factors that cause a disease. Disease-causing biological agents can also be called **pathogens**. This section of the text will address a few of the biological factors that result in disease. Particular pathogens cause particular diseases. Pathogens and parasites encompass a wide variety of organisms, including viruses, bacteria, protozoans, fungi, and helminths. Some are external, while others are internal. Only a few will be highlighted in this text.

Diseases can be a concern in all aspects of biota management. Obviously a biologist would be concerned about disease outbreaks in wild populations. In addition, diseases can be even more of a concern when stocking or accidental introduction is involved. Any type of fish or wildlife stocking, be it introductory, supplemental, or maintenance, creates the opportunity for pathogens or parasites to spread to a new environment. This problem can be especially serious when exotic organisms are involved because they may transfer a disease against which native organisms have no defense.

The potential for the spread of pathogens or parasites is intensified in the case of stockings because most stockings use hatchery-reared fish or pen-reared wildlife. Most parasites in the wild affect only a few of the organisms in a population. In an artificially reared or held population, the crowded organisms are much more susceptible to disease outbreaks and an outbreak is likely to affect a high percentage of the population. The stress that results from crowding creates greater opportunities for large-scale eruptive disease outbreaks (**epizootics**) as well as low-level, chronic disease problems (**enzootics**).

Mammals are subject to a variety of diseases that can cause problems in wild and captive populations. Bison are known to carry the bacterial disease brucellosis. For example, bison in Yellowstone National Park have apparently carried the disease since 1917, with little effect on the herd (Meagher 1978). The disease is also present in the Jackson Hole elk herd that ranges in the same area. Brucellosis can have serious economic effects on ranchers that raise domesticated cattle. Lawsuits have been filed over the transmission of the disease from bison to domesticated cattle in the Yellowstone Park area; the resulting decisions have basically said that transmission was probable but not proven. Most bison herds in the United States are maintained in brucellosis-free condition.

White-tailed deer are susceptible to outbreaks of epizootic hemorrhagic disease (EHD). The causative agent for this disease is a virus, and outbreaks commonly occur in the southeastern United States and sporadically in other states and Canada. There is no effective control for this disease, but outbreaks are usually associated with high white-tailed deer densities, evoking similarities with cultured fish and pen-reared wildlife.

The possible introduction of tuberculosis into wild populations of ungulates from infected private game farm animals is a source of concern to wildlife biologists in both the United States and Canada. It could also create a public health problem for humans. Laws regulating the transport of domesticated livestock do not necessarily apply to game farm ungulates, and disease tests developed for domesticated livestock do not necessarily work well for other species. Another disease concern in the United States and Canada is the possible introduction of the meningeal worm into western North America. Introduction of this parasite through game farm animals could have catastrophic effects on western elk and moose populations.

Birds are also subject to a wide variety of diseases that are important in wild and captive populations. Botulism, caused by the anaerobic bacterium *Clostridium botulinum,* causes severe disease outbreaks in waterfowl. Sick birds cannot hold their necks erect, so the disease is also called "lim-ber neck" (Bellrose 1976). The disease cycle commonly includes fly maggots, so it is important that carcasses of waterfowl killed by the disease be gathered and burned. Avian cholera, a bacterial disease, annually kills thousands of waterfowl during spring migrations.

Another common disease in birds is aspergillosis. This disease, which is prevalent in many birds, including gallinaceous birds and waterfowl, attacks the respiratory system. The causative agent is a fungus that often occurs on moldy grain. A viral disease that primarily attacks waterfowl is duck virus enteritis. The pathogen responsible for this disease is a herpesvirus.

Birds can even be parasitized by other birds. The brown-headed cowbird parasitizes the nests of many other bird species. This bird does not build its own nests; instead, the females lay their eggs in the nests of various other birds and allow these other species to raise the brown-headed cowbird young. The young brown-headed cowbirds are aggressive and often are the most successful young in the host bird's nest.

Fishes are subject to a wide complement of diseases. Some of these occur primarily in aquaculture situations, while others can at times be found at epizootic levels even in wild populations. Among fishes, disease outbreaks are again most common under stressful or crowded conditions. In addition, poor nutrition can lead to stress, which again increases the likelihood of disease outbreaks.

Examples of bacterial diseases of fishes include bacterial kidney disease (BKD) and furunculosis, which affect salmonids, and columnaris, which commonly affects warmwater fishes such as those of the sunfish family. Viral diseases include infectious pancreatic necrosis (IPN), which affects salmonids throughout the world, and channel catfish virus disease (CCVD), which obviously is of concern to the channel catfish aquaculture industry in the United States.

The technology for diagnosing bacterial and viral fish diseases has improved substantially over the last decade, especially for early diagnosis of these diseases. However, treatment capabilities have actually declined over the same time period because of the loss of chemicals that had been approved for such uses. Two U.S. federal agencies regulate which chemicals can be used for fish disease treatment. The U.S. Food and Drug Administration is concerned about the effects of the chemicals used on human health, while the U.S. Environmental Protection Agency is concerned with the environmental effects of those chemicals. Thus, chemicals that have proven carcinogenic to humans or those that have proven harmful to the environment can no longer be used for fish disease treatment.

External parasites can also create fish health problems. Again, large outbreaks of parasites are more likely under crowded or stressful conditions. A common fish health problem in the channel catfish aquaculture industry and at times in wild populations is outbreaks of "ich" (the causative agent is the protozoan *Ichthyophtherius multifilis*). The same disease is commonly found in aquarium fishes; "ich" can actually affect a wide variety of warmwater fishes. *Trichodina* is another external parasite, and commonly affects both salmonids and warmwater fishes. Treatment capabilities for external parasites are much better than for bacterial or viral diseases.

Some states require inspection of fishes transported across state lines in an attempt to reduce the transfer of fish diseases. However, state regulations vary widely, and no central federal authority has emerged to regulate such transport.

Parasites do not necessarily have to be small organisms. The parasitic sea lamprey is an accidental introduction that gained access to the Great Lakes west of Lake Ontario when the construction of the Welland Canal connected the lakes to the Atlantic Ocean. This canal was intended to allow passage of commercial shipping into the lakes. However, the sea lamprey became an uninvited guest that survived and reproduced. The sea lamprey was one of several problems that led to the demise of lake trout and other native fishes in the Great Lakes. Today, sea lampreys are partially held in check by continual application of piscicides (see section 7.6). Otherwise, the popular chinook and coho salmon fisheries that have developed in

many of these lakes could not exist at their current levels, and rehabilitation of native lake trout stocks would be unlikely.

The effects of diseases on biota represent an important concern for wildlife and fishery biologists. The above examples are just a few of the many that could have been mentioned. Actually, it is often difficult or impossible to identify a single causative agent for an epizootic. For example, bighorn sheep are susceptible to lungworms. Under certain conditions, the lungworm infestation can lead to an epizootic, with bighorn sheep actually dying of pneumonia, which may be caused by bacterial or viral infection. These lungworms, bacteria, and viruses have been found in apparently healthy as well as diseased bighorn sheep. The "certain conditions" that lead to a disease outbreak are complex and interrelated (Figure 10.6). Stressed

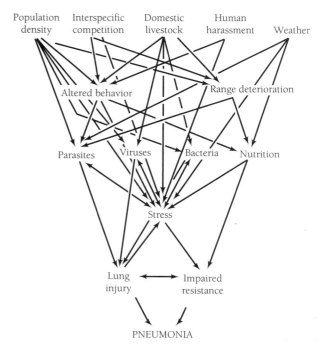

FIGURE 10.6 The causes of an epizootic in a population of wild organisms can be complex. This diagram depicts the variety of interrelated factors that can cause the lungworm-pneumonia complex in bighorn sheep. *(From Wobeser 1994.)*

animals are more likely to become sick, and sources of stress can include competition with other bighorn sheep or domesticated livestock, harassment by humans, weather, and range condition and forage quality.

▼ SUMMARY

At times, biologists find it necessary to remove part or all of an animal community. Partial removals are often undertaken by fishery biologists when a particular population has become overabundant, has slow growth rates, and its size structure is poor. A somewhat comparable management technique used by wildlife biologists might be extensive predator removal on a waterfowl nesting island.

Complete population renovations are attempted primarily by fishery biologists. At times, an entire fish community may be eliminated, perhaps due to circumstances such as the overabundance of a species such as the common carp. Complete renovations have also been used to expand the population size of threatened or endangered species.

Population control has also been attempted through interference with the reproductive process by releasing sterile organisms or by use of contraceptives. Attempts to interfere with reproduction are more likely to succeed in small populations in confined areas. When fishery biologists prefer that a species not become naturalized, they can produce and stock polyploid or monosex fish, which cannot reproduce.

Stocking of organisms can be categorized as introductory, maintenance, or supplemental. These terms are commonly used by fishery biologists; while not used by wildlife biologists, they are still applicable to that field. Introductory stockings involve the release of organisms into a new or renovated water body or newly created terrestrial habitats, or the introduction of a new species into an existing community. Maintenance stockings are made to maintain a population that for some reason has no or very limited natural reproduction. Supplemental stockings are made to augment naturally reproducing populations; they are generally less likely to be successful than introductory or maintenance stockings.

Examples of introductory stockings include the restoration of river otters to Missouri rivers and streams from which they had been extirpated and the introduction of brook trout to the Rocky Mountains. Maintenance stockings of walleyes are commonly made in many lakes and reservoirs. Supplemental stockings of organisms such as ring-necked pheasants generally are unsuccessful, and can lead to problems of reduced genetic fitness and disease transfer. Biologists should avoid unnecessary and unwise stocking of organisms. However, stocking can be an effective tool for the management and restoration of game and nongame species.

Diseases can account for a substantial portion of the natural mortality in wild populations, and thus are of concern to biologists. Transfer of pathogens and parasites during fish stocking or wildlife introduction may be an even greater concern, especially when exotic organisms are involved. Exotic organisms may transfer a pathogen or parasite for which native organisms have no defense.

PRACTICE QUESTIONS

1. Renovations of impoundments, such as those carried out using a fish toxicant, are less frequently undertaken by fishery biologists today than in the past. Why might the use of this technique be declining?

2. What are some of the advantages and disadvantages for both mechanical removal and chemical removal undertaken for a partial rehabilitation of a fish population?

3. What are some of the advantages and disadvantages of having a mechanical removal of fishes completed by commercial harvest rather than by a state, provincial, or territorial conservation agency?

4. Why are supplemental stockings of ring-necked pheasants and northern bobwhites typically not used to bolster wild populations?

5. List several situations in which biologists might undertake population control to protect humans from wildlife.

6. What are the differences between introductory, maintenance, and supplemental stockings? Give an example of each for both a fish and a wildlife species.

7. Why does a fishery biologist need to consider available habitat and fish niches when deciding on the complement of fish species to stock into a new reservoir?

8.   What is the difference between a put and take stocking and a put, grow, and take stocking?

9.   What are some of the potential disease problems that might arise if game ranching of large ungulates is not undertaken in a careful manner?

10.   If an exotic fish species were proposed for introduction, would the use of polyploid individuals provide any unique opportunities to safely study the introduction? Why?

SELECTED READINGS

Allen, L. J., ed. 1984. *Urban fishing symposium proceedings.* Fisheries Management Section and Fisheries Administrators Section, American Fisheries Society, Bethesda, Md.

Behnke, R. J. 1992. *Native trout of western North America.* Monograph 6. American Fisheries Society, Bethesda, Md.

Bolen, E. G., and W. L. Robinson. 1995. *Wildlife ecology and management.* 3d ed. Macmillan, New York.

Carlson, C. A., and R. T. Muth. 1993. Endangered species management. Pages 355–381 *in* C. C. Kohler and W. A. Hubert, eds. *Inland fisheries management in North America.* American Fisheries Society, Bethesda, Md.

Heidinger, R. C. 1993. Stocking for sport fisheries enhancement. Pages 309–333 *in* C. C. Kohler and W. A. Hubert, eds. *Inland fisheries management in North America.* American Fisheries Society, Bethesda, Md.

Hygnstrom, S. E., R. M. Timm, and G. E. Larson, eds. 1994. *Prevention and control of animal damage.* University of Nebraska, Cooperative Extension Service, Lincoln.

Kohler, C. C., and J. G. Stanley. 1984. A suggested protocol for evaluating proposed exotic fish introductions in the United States. Pages 387–406 *in* W. R. Courtenay Jr. and J. R. Stauffer Jr., eds. *Distribution, biology, and management of exotic fishes.* Johns Hopkins University Press, Baltimore, Md.

Li, H. W., and P. B. Moyle. 1993. Management of introduced fishes. Pages 287–307 *in* C. C. Kohler

and W. A. Hubert, eds. *Inland fisheries management in North America.* American Fisheries Society, Bethesda, Md.

Noble, R. L. 1981. Management of forage fishes in impoundments of the southern United States. *Transactions of the American Fisheries Society* 110: 738–750.

Piper, R. G., I. B. McElwain, L. E. Orme, J. P. McCraren, L. G. Fowler, and J. R. Leonard. 1982. *Fish hatchery management.* U.S. Fish and Wildlife Service, Washington, D.C.

Teer, J. G., L. A. Renecker, and R. J. Hudson. 1993. Overview of wildlife farming and ranching in North America. *Transactions of the Fifty-eighth North American Wildlife and Natural Resources Conference,* 448–459.

Wiley, R. W., and R. S. Wydoski. 1993. Management of undesirable fish species. Pages 335–354 *in* C. C. Kohler and W. A. Hubert, eds. *Inland fisheries management in North America.* American Fisheries Society, Bethesda, Md.

LITERATURE CITED

Alcorn, S. R. 1981. Fishing quality in two urban fishing lakes, St. Louis, Missouri. *North American Journal of Fisheries Management* 1:80–84.

Bellrose, F. C. 1976. *Ducks, geese, and swans of North America.* 2d ed. Stackpole Books, Harrisburg, Pa.

Brown, D. J., and T. G. Coon. 1991. Grass carp larvae in the lower Missouri River and its tributaries. *North American Journal of Fisheries Management* 11:62–66.

Crivelli, A. J. 1983. The destruction of aquatic vegetation by carp. *Hydrobiologia* 106:37–41.

Graham, L. K. 1986. Establishing and maintaining paddlefish populations by stocking. Pages 96–104 *in* J. G. Dillard, L. K. Graham, and T. R. Russell, eds. *The paddlefish: Status, management, and propagation.* Special Publication no. 7. North Central Division, American Fisheries Society, Bethesda, Md.

Hessler, E., J. R. Tester, D. B. Siniff, and M. M. Nelson. 1970. A biotelemetry study of survival of pen-reared pheasants released in selected habitats. *The Journal of Wildlife Management* 34:267–274.

Hilborn, R. 1992. Hatcheries and the future of salmon in the northwest. *Fisheries* (Bethesda) 17 (1): 5–8.

Hill, D., and P. Robertson. 1988. Breeding success of wild and hand-reared ring-necked pheasants. *The Journal of Wildlife Management* 52:446–450.

Johnson, F. H. 1977. Responses of walleye (*Stizostedion vitreum vitreum*) and yellow perch (*Perca flavescens*) populations to removal of white sucker (*Catostomus commersoni*) from a Minnesota lake, 1966. *Journal of the Fisheries Research Board of Canada* 34:1633–1642.

Laarman, P. W. 1978. Case histories of stocking walleyes in inland lakes, impoundments, and the Great Lakes: 100 years with walleyes. *American Fisheries Society Special Publication* 11:254–260.

Marking, L. L. 1992. Evaluation of toxicants for the control of carp and other nuisance fishes. *Fisheries* (Bethesda) 17 (6): 6–13.

Meagher, M. M. 1978. Bison. Pages 122–133 *in* J. L. Schmidt and D. L. Gilbert, eds. *Big game of North America: Ecology and management.* Stackpole Books, Harrisburg, Pa.

Meyer, F. P., and R. A. Schnick. 1983. Sea lamprey control techniques: Past, present, and future. *Journal of Great Lakes Research* 9:354–358.

Raibley, P. T., D. Blodgett, and R. E. Sparks. 1995. Evidence of grass carp (*Ctenopharyngodon idella*) reproduction in the Illinois and upper Mississippi Rivers. *Journal of Freshwater Ecology* 10:65–74.

Springer, P. F., G. V. Byrd, and D. W. Woolington. 1977. Reestablishing Aleutian Canada geese. Pages 331–338 *in* S. A. Temple, ed. *Endangered birds: Management techniques for preserving threatened species.* University of Wisconsin Press, Madison.

Stephen, J. L. 1986. Effects of commercial harvest on the fish community of Lovewell Reservoir, Kansas. Pages 211–217 *in* G. E. Hall and M. J. Van Den Avyle, eds. *Reservoir fisheries management: Strategies for the 80s.* Reservoir Committee, Southern Division, American Fisheries Society, Bethesda, Md.

Turner, J. W. Jr., K. M. Liu Irwin, and J. F. Kirkpatrick. 1992. Remotely delivered immunocontraception in captive white-tailed deer. *The Journal of Wildlife Management* 56:154–157.

Verrill, D. D., and C. R. Berry Jr. 1995. Effectiveness of an electrical barrier and lake drawdown for reducing common carp and bigmouth buffalo abundances. *North American Journal of Fisheries Management* 15:137–141.

Wobeser, G. A. 1994. *Investigation and management of disease in wild animals.* Plenum Press, New York.

11

ENDANGERED AND THREATENED SPECIES

The disappearance of species and the evolution of new species is a natural process that has occurred throughout the history of life on earth. Rates of species extinction have varied over different geologic periods, but generally, until the last few hundred years, extinctions have occurred primarily as a result of natural factors. Extinction rates today, however, are being greatly accelerated by human activities. In the last two centuries, the rate of extinctions has risen far above the rate at which new species are evolving.

It is probable that hunting by humans in prehistoric times, and more recently in regions for which historic information is not available, caused the extinction of several large mammals (Ehrlich and Ehrlich 1981). Mastodons, for example, are now extinct. The specific role that humans played in their extinction is unknown, but spear points are often found in mastodon bones. The bones of many extinct species have been found in the debris left by early humans. The moas, such as the slender moa and lesser megalapteryx, were large **cursorial** (adapted for running) birds of New Zealand. Moas may have survived into the early 1800s and were probably brought to extinction by the hunting activities of indigenous peoples. Other recent extinctions due to human influences include the passenger pigeon of eastern North America, the heath hen, which lived along the east coast of North America from Massachusetts to Maryland, the blue pike of the Great Lakes region, the Steller's sea cow of the Bering Sea, and the great auk of the North Atlantic.

Unfortunately, extinctions due to human influence are increasing rapidly. Thousands of species are being lost annually from tropical rain forests alone. Lugo (1988) projected that 20%–50% of the plant and animal species on the earth in the 1970s could be lost by the year 2000. Many species exist that are unknown to us; they could be lost without our even being aware of their disappearance.

There are many causes of extinction, including natural causes, unregulated harvest by humans, and introduced organisms, but by far the most common cause of extinction is habitat loss or modification. Most habitat loss or modification is ultimately due to human population growth, ever-increasing human demands for space and natural resources, and, in some cases, inappropriate uses of technology. Extinctions due to natural causes are inconsequential at this time in the history of the earth compared with human-related species losses. We seem to be rapidly increasing our ability to destroy without accompanying increases in our wisdom to preserve. Extinction is not retractable; it is permanent and final.

Human demand for natural resources and the by-products of human life combine to challenge our abilities to conserve biodiversity on the planet (see section 2.1). The use of the term biodiversity is usually restricted to wild organisms; in a broader sense, the gene pools of all organisms, even domesticated or cultivated ones, also contribute to biodiversity.

The purposes of this chapter are to discuss endangered and threatened species, the causes of their declines, and plans for increasing their populations.

11.1 Why Be Concerned with Extinction?

To many people, the loss of little-known species, even in the great numbers currently being lost in tropical rain forests, is of little consequence. A large, familiar species, such as the bald eagle, is likely to gain greater public support and interest when its population drops to low levels than, for instance, a small, unknown insect, reptile, or plant.

There are two categories of values that can be associated with any organism. The first category includes the values of that organism to humans. Animals and plants can have economic, scientific, ecological, recreational, aesthetic, and moral values to humans. The second category includes values that do not relate to humans. Our anthropocentrism often causes us to believe that everything revolves around human needs. A species has value in itself; it does not need to be of direct use to humans to have value.

If the value of a species to humans were the only issue of importance to people, then the scientific and medical values that many species provide to humans should convince us of the benefits of preserving species. A variety of organisms have contributed greatly to humans; some of these have been quite inconspicuous. The little-known rosy periwinkle, for example, a tropical plant, has provided the constituents for much-improved treatment of Hodgkin's disease (Myers 1983). What if humans had caused the extinction of the rosy periwinkle before its medical value was discovered? Penicillin was derived from a mold, hardly an organism that elicits great love from humans. Physiological studies of other animals have contributed greatly to human medical science. The loss of species may destroy entire gene pools that hold the potential to unlock secrets important to our own health and survival. We know little about most species, and yet each may have scientific and medical benefits to humans.

In the development of domesticated animals and plants we have manipulated the gene pools of the original wild organisms and selected for characteristics useful to humans. Domesticated organisms have lost portions of their original gene pools that could have been important to the development of valuable characteristics in future domesticated varieties. For instance, a domesticated animal or plant may lose certain immunities to disease, adaptations to extreme cold, or the ability to adapt to greater salinity levels. Once a portion of the gene pool is lost from a domesticated organism and its wild ancestor is extinct, those genetic characteristics are gone forever. Further, many wild organisms could one day provide the basis for additional valuable domesticated varieties.

Such issues demonstrate the economic importance of all species to humans. Additionally, all species function interconnectedly in the web of life. Life on earth is one large food web, of which humans are a part; think of it as a pyramid of cans or a house of cards. Who knows what levels above may crash if a supporting can or card is pulled? What if many cans or cards are removed?

It seems appropriate here to describe the economic values of some species that have faced extinction, are in decline, or are of unknown status. Bison are providing a source of low-fat meat of increasing economic importance and also hold promise for improving grazing systems, yet bison populations once reached levels near extinction. Aesthetics also provide economic value. The whooping crane, for example, is viewed by thousands of people at the Aransas National Wildlife Refuge in south Texas; many of those people spend money for tourism services. Some developers are finding that the value of property increases as they coordinate human space needs with those of species such as nesting bald eagles or dusky seaside sparrows. Undisturbed areas left for these other species may ultimately add considerable value to properties that are developed for housing and other human purposes.

Wild fishes, birds, mammals, and other organisms are important to human well-being from an aesthetic viewpoint (fig. 11.1). Millions of people are involved in observing "watchable" wildlife and fishery resources. Underwater observation of the many colorful marine organisms found on coral reefs and in other coastal areas is becoming increasingly popular. National, state, provincial, territorial, and local parks continue to be extensively used; a primary reason for their appeal is their wild animal populations. People obtain much everyday enjoyment from the natural elements of the ecosystems in which they live. We spend great sums of money on mental health care, and yet the mental health benefits of the plant and animal communities that surround us are often taken for granted. How important are songbirds in a backyard or fox squirrels at a feeder to senior citizens

FIGURE 11.1 People derive much enjoyment from observing birds and other wildlife.

with limited mobility beyond their own homes? How much would the average citizen miss the chirping of crickets, the songs of birds, the choruses of frogs, and the rustling of cottonwood leaves if they were lost? These are the sounds of life. When birds, mammals, and other wildlife disappear from backyards, parks, or favorite picnic areas, people quickly express their displeasure. Natural resource biologists continually receive questions and calls from the public requesting information about wild organisms. Recently, one of the most frequent questions has been, "Where are the birds?" When people do not hear or see birds in places where they expect them, they express their concern. Too often we forget how much most people really care about their natural environment.

One does not even have to see or hear wild organisms; just the knowledge that they are present has value. For example, relatively few people will ever observe firsthand the wildlife that abounds in places like the Arctic National Wildlife Refuge in Alaska or the Serengeti region of Africa, but the knowledge that those places exist has value

to many people (see section 16.3). Extinctions degrade and destroy such natural environments.

An appreciation of life also engenders a respect and moral responsibility for all life. To many people, it is morally wrong to negatively influence other species to such an extent that their populations decline to dangerously low levels or reach extinction. As with all moral issues, however, what may appear to be simple is really quite complex. While the vast majority of people would agree that humans should not allow their actions to cause the extinction of species such as the peregrine falcon or the cheetah, can the same be said of a bacterium that causes death or disease in humans? The attitude of humans towards extinction of a disease organism is understandable. Our respect for the right of most species to exist, however, should not be in question. Other species do not exist just for the well-being of humans. To quote Aldo Leopold, "In short, a land ethic changes the role of *Homo sapiens* from conqueror of the land-community to plain member and citizen of it. It implies respect for his fellow-members, and also respect for the community as such" (Leopold 1949).

11.2 Endangered Species Legislation

In 1964, a Committee on Rare and Endangered Wildlife Species was formed in the U.S. Department of the Interior. This committee was responsible for producing the first list of species threatened with extinction; this publication is known as the *Red Book*. In 1966 the Endangered Species Preservation Act was passed in the United States, but it included few restrictions to protect species in danger of extinction. In 1969 the Endangered Species Conservation Act was passed. This law showed promise by broadening the scope of conservation to include species throughout the world. However, the 1969 law still lacked specific measures to protect species; this legal weakness was largely due to the lobbying efforts of special interest groups. In 1973, the current Endangered Species Act (ESA) became federal law in the United States, and with the addition of amendments, it remains largely intact today (U.S. Fish and Wildlife Service 1992). The ESA includes substantial criminal penalties and fines for violations of its provisions; without penalties and fines, laws such as the ESA would be unenforceable.

The ESA provides for the listing of species as endangered or threatened and specifies measures to be taken to protect and restore them. As defined in the ESA, an **endangered species** is a species in danger of becoming extinct throughout all or a portion of its range. A **threatened species** is a species that is likely to become endangered throughout all or a portion of its range. Thus, endangered and threatened are legal terms defined by the ESA and should be used only to describe species listed under that law as endangered and threatened. The Secretary of the Interior and the Secretary of Commerce act through the U.S. Fish and Wildlife Service (USFWS) and the National Marine Fisheries Service (NMFS) to authorize the listing of species. The Secretary of Commerce is specifically authorized to list marine mammals, while all other organisms are under the jurisdiction of the Secretary of the Interior. Additions to the lists are made through these agencies after consideration of the status of a particular species. The ESA also allows for species nominations by the public.

The term "portion of its range" in the ESA allows for the listing of populations that face local extinction, even though the species may be abundant in other parts of its range. The ESA also includes a "similarity of appearance" clause that can be used to protect an unlisted species if that species cannot be readily distinguished from a listed species in a problem area. For instance, if greater prairie-chickens were listed as endangered for a geographic area where sharp-tailed grouse also occur, the act could require the cessation of hunting of sharp-tailed grouse in that area because hunters cannot readily distinguish the two species in flight.

Along with the listing of species, current rules require designation of **critical habitat** at the time a species is listed as endangered or threatened, or soon after. Critical habitat is the geographic area and ecosystem essential for the survival of the listed species. Where critical habitat designation causes substantial economic problems, it can be delayed while still listing the species. The ESA also provides a process, in the form of species **recovery plans**, for bringing endangered or threatened species back to healthy population levels. This process usually involves stabilizing critical habitat and, where needed, obtaining additional habitat. The ESA allows for acquisition of critical habitat using federal funds, such as those available through the Land and Water Conservation Fund Act of 1965 (Rohlf 1989).

The ESA requires other federal agencies or others involved in federally funded projects to ensure that their actions are not detrimental to endangered or threatened species. Thus, if timber sales in a national forest or the development of a federally funded dam threaten the critical habitat of a threatened or endangered species, the federal agency responsible must respond to the problem or be in violation of the ESA. The line between violation of the ESA by a federal agency and its normal operating procedures in the course of such activities can at times be unclear. The USFWS and/or the NMFS are responsible for advising other federal agencies as to when they must conduct a biological assessment to evaluate the effects of their proposed actions on listed species. If the proposed actions of a federal agency are found to jeopardize a listed species or its critical habitat, the agency must develop contingency plans to reduce the damage. The USFWS and/or NMFS must issue a written opinion with regard to agencies and projects that may affect listed species.

The ESA prohibits international trade in endangered or threatened species or parts thereof through its implementation of the Convention on International Trade in Endangered Species of Wildlife Fauna and Flora (CITES). The CITES agreement also provides for the regulation of imports and exports of certain species not in the endangered or threatened categories. A less well known agreement, the Convention on Nature Protection and Wildlife Preservation in the Western Hemisphere, was also activated in part through the ESA. This convention, an agreement among various Western nations, was signed in 1940, but was not implemented at that time. However, in 1982, Congress directed the Secretary of the Interior and the Secretary of State to put this convention into effect, especially as it relates to migratory birds (Rohlf 1989).

Species can be placed on the ESA list of endangered and threatened species even if they do not occur in the United States. Over half of the species currently listed are foreign. The development of recovery plans and the designation of critical habitats for foreign species are of lower priority than are those for organisms of special concern within the United States. However, as noted previously, just listing a foreign animal or plant provides it with legal protection in terms of importation and international trade restrictions that fall under jurisdiction. Migratory animals that spend part of the year outside the United States, such as the whooping crane, do receive high priority.

Under the ESA, the U.S. government can cooperate with states that develop appropriate endangered species programs and can provide federal cost-sharing of up to 90%. Thus, most states have developed their own lists of organisms that may be threatened, endangered, or of special concern. Species on state lists may or may not be federally listed as threatened or endangered. California, for example, lists the gila woodpecker as endangered and the pink salmon as a species of special concern; neither of those species was federally listed as of 1995. The least shrew and lake sturgeon are listed as endangered in Pennsylvania, but currently are not federally listed. State lists of endangered, threatened, or special concern species

can be obtained by contacting the state agency responsible for wildlife and fishery resources. In most cases, biologists are employed by each state to work with the endangered species program in that state.

The status of federally listed species in the United States must be reviewed at five-year intervals. Species are removed from the list if they become extinct, if they recover sufficiently to no longer fit the endangered or threatened categories, or if data are presented indicating that the population is healthier than was indicated by the earlier assessment. The ultimate goal of the ESA is to remove species from the list due to population recovery.

The listing of a species as endangered or threatened is fairly difficult, even for a high-profile organism. Real problems arise when the organism is small, unknown, and seemingly valueless, such as an insect, plant, or small mammal. Public support for such organisms often diminishes when a highway project or some other human activity is negatively affected by the ESA. The federally funded Tellico Dam in Tennessee, for example, was started and nearly completed prior to the passage of the ESA in 1973. A new species of fish, the snail darter, was identified in 1973 in a portion of the Little Tennessee River that was to be inundated by the dam; no other populations of this fish were known to exist. Federal court rulings stopped dam construction for a period of years because the ESA prohibited the destruction of the habitat of an endangered species. Eventually, under considerable pressure, an exemption was granted and the dam was completed. Relocated snail darters from the Little Tennessee River have become established in other waters, and additional populations have now been located.

Mandatory reductions in the harvest of old-growth forests on federal lands to protect the habitat of the threatened northern spotted owl have stirred similar levels of controversy concerning the ESA. The snail darter and northern spotted owl examples are indicative of the controversy that can develop over the ESA when its provisions have immediate economic repercussions. A delicate balance must be struck between maintaining and enforcing a strong and viable ESA and maintaining sufficient public support for its continued authorization and funding.

The ESA has other problem areas that attract many complaints. Subspecies listing is a contentious issue. Some subspecies exist in very limited areas or at small population levels, even though the species as a whole has a large range and healthy populations. Another contentious issue involves small populations of a species, often at the edge of the species' range. While the species may be common in other localities, it may be found in only limited numbers at such a location.

In 1995, the ESA was targeted for elimination or significant change by many people. While some modification of the legislation would not eliminate its usefulness, loss or major modification of the ESA would reduce its utility in recognizing the importance of endangered and threatened species. Unfortunately, too many humans think in terms of short-term economic gain at the expense of long-term sustainability. Some groups will continue to attack legislation such as the ESA; other groups will continue to work at broadening and strengthening such legislation.

Canada does not have a single act dealing with endangered species. There are twelve acts, including the Canada Wildlife Act, the Migratory Birds Convention, and the Canadian Environmental Protection Act, that can apply to the conservation of threatened or endangered species. Canadian federal government legislation does not require the development of a listing of endangered or threatened species. The provinces of New Brunswick, Quebec, Ontario, and Manitoba have endangered species legislation; this legislation authorizes, but does not compel, action to conserve endangered species. Canada is considering federal legislation directly addressing endangered species conservation. In doing so, it is reviewing federal endangered species legislation from countries such as the United States, Australia, and Japan. Canada is also looking closely at controversial situations, such as the snail darter and north-

ern spotted owl cases, that have developed in the United States under an endangered species program that legally compels federal action. While having no single endangered species act, Canada is active in international conventions that relate to the welfare of endangered and threatened species, such as CITES.

11.3 Causes of Extinction or Endangerment of Populations

Extinctions and serious population declines among wild organisms occur due to a variety of human influences, including destruction or alteration of habitat; addition of toxic substances to the environment; direct killing or capture of excessive numbers of organisms in an uncontrolled manner; and competition with, predation by, or hybridization with exotic, transplanted, or even native organisms.

The causes of extinction or endangerment are highly variable, depending on the particular organism, and usually more than one factor is responsible for an organism's decline. As previously indicated, habitat destruction is the single most important cause of extinction and endangerment. For example, in coastal Texas, most of the suitable habitat within the original range of the endangered Attwater's greater prairie-chicken (a subspecies of greater prairie-chicken) has been destroyed due to agricultural activities, suburban development, and brush encroachment. Black-footed ferrets are endangered due to extensive removals of their primary prey species, the black-tailed prairie dog, tilling of their original, grassland habitat, and possibly disease. Colorado squawfish are endangered due in large part to water projects on the Colorado River and its tributaries (fig. 11.2). However, transplanting of northern pike into the Colorado River drainage must also be viewed as a negative influence on the Colorado squawfish due to com-

FIGURE 11.2 Colorado squawfish have suffered from degradation of the Colorado River habitat and the introduction of northern pike. *(Photograph courtesy of the U.S. Fish and Wildlife Service.)*

petition and, for smaller squawfish, predation. The large and beautiful Lahontan cutthroat trout (a subspecies of cutthroat trout) of northwestern Nevada and northeastern California is threatened due to dam construction and reduced water levels related to agricultural and urban water needs, competition from introduced trouts, and hybridization with transplanted rainbow trout (Gerstung 1988). Other subspecies of cutthroat trout, including the greenback cutthroat trout of the South Platte River and Arkansas River drainages of eastern Colorado (Stuber et al. 1988), are threatened for similar reasons.

The effects of introduced organisms on native species have been especially evident in island habitats. Feral goats and sheep on the Hawaiian Islands have caused difficulties for native species by destroying vegetation that the native species depend upon; the list of endangered species for the Hawaiian Islands is long. The introduction of predators such as the Indian mongoose on Jamaica and other islands to control introduced Norway and black rats has led to the destruction of many species of native reptiles, amphibians, and birds. The introduction of Arctic foxes to the Aleutian Islands for commercial fur farming caused the near extinction of the Aleutian Canada goose (see section 10.2).

Another serious concern involves the introduction of closely related forms that can crossbreed with species native to an area. In North America, game ranching is becoming increasingly popular, and the red deer of Europe, a subspecies of elk, is a popular large ungulate on these ranches. Game ranches can sell the meat, antlers,

velvet, (highly vascularized tissues that cover developing antlers) and hunting opportunities provided by these animals, but they are required by state laws to keep them within fenced areas. These fenced areas can include extensive ranches or smaller areas, but regardless of the situation, experience has shown that red deer do occasionally escape; native animals can also enter these fenced areas. Biologists are well aware that fences function at less than 100% efficiency in holding animals within or outside particular boundaries. When red deer escape, they may move long distances, join native elk herds, and eventually hybridize with elk. Red deer characteristics can thus enter native elk herds, and could lead to the extinction of native subspecies of elk in North America (see section 4.7). Stopping this threat has proved extremely difficult because of economic and political influences. It would seem appropriate that the red deer be prohibited as a game-ranch animal in areas of North America near any population of wild elk.

Exotics and transplants are not the only threats to native organisms. Many native species have been widely stocked into areas without proper consideration of the genetic characteristics and genetic integrity of the original populations. Wild turkeys and largemouth bass, for example, are commonly stocked into areas that already contain those species. Stockings such as these can decrease the genetic integrity of natural populations.

Pollution can also lead to endangerment or extinction. Brown pelicans in the southeastern coastal areas of North America were brought to near extinction due to environmental contamination with chlorinated hydrocarbon insecticides such as DDT (fig. 11.3). These pesticides are concentrated through the successive trophic levels of a food web, a process known as **food chain biomagnification** (**biological amplification**). In the brown pelican, a higher-level predator, pesticides reached levels that caused serious eggshell thinning and nesting failure. Bald eagles and peregrine falcons also suffered reproductive failure due to food chain biomagnification of chlorinated hydrocarbons, primarily DDT.

FIGURE 11.3 Brown pelicans suffered nesting failures and were reduced to dangerously low numbers prior to the banning of chlorinated hydrocarbon insecticides such as DDT. *(Photograph courtesy of the National Park Service.)*

FIGURE 11.4 The dodo of the island of Mauritius was driven to extinction in the 1600s by Europeans and the predators and scavengers they introduced.

Uncontrolled commercial harvest and illegal harvest have led to extinction for some species and to serious population reductions for others. The dodo of the island of Mauritius in the Indian Ocean was extirpated by European travelers and settlers and the animals they introduced in the 1600s (fig. 11.4). Market hunting was a major component in the incredible decimation and extinction of the passenger pigeon, possibly the most abundant bird in North America at one time. Hunting for feathers and meat was important in the reduction of the whooping crane. Killing of birds for their feathers also caused substantial reductions of species such as the snowy and great egrets. The paddlefish, already a fish of special concern in many states, has been further reduced by illegal capture with gill nets for commercial sale of their eggs. The eggs are used to make caviar. The northern sea lion is currently threatened, at least partially because of the value of its hide. The once endless herds of bison in North America were brought close to extinction due to extensive uncontrolled slaughter. This wasteful reduction in bison populations was purposefully carried out as government policy to bring hardships to indigenous peoples dependent on the bison for food, clothing, and shelter. Bison now appear safe from

extinction, but the population still is based on the small gene pool of those few animals that remained before restoration began. Small populations can lead to inbreeding and a lack of genetic heterozygosity, which in turn can lead to reduced adaptability (see section 4.6).

11.4 Recovery Plans and Results

Recovery plans have been written and fully or partially implemented for many ESA listed species. These plans specify reasonable actions that can be taken to allow for the recovery or protection of listed species. Recovery plans are published by the USFWS or NMFS, but may be written by consultants, state agencies, recovery teams (composed of a variety of professionals and agencies interested in the species), or others. For example, the USFWS published a recovery plan for the Colorado squawfish in 1978, and revised recovery work plans for that species are developed annually (U.S. Fish and Wildlife Service 1990). The recovery plan includes a status report, in which the status of the Colorado squawfish and the reasons for its decline are reviewed and its habitat requirements, range of distribution, limiting factors, and other aspects of its life history are addressed. The recovery plan lists objectives for specific streams or recovery areas that contain appropriate or degraded Colorado squawfish habitat. The plan calls for monitoring of populations and additional research, revision and enforcement of the existing laws protecting Colorado squawfish and their habitat, restoration of habitat and assessment of the damage caused by existing water development projects, restoration into the historic range of the species, and the development of criteria upon which upgrading (moving from endangered to threatened) or delisting will be based. However, such a recovery plan does not ensure that adequate funding to implement the plan will be available.

The primary goal of any species recovery plan is to upgrade the status of the species or remove it from the endangered or threatened list. The bald eagle has recovered dramatically in most regions of North America (U.S. Fish and Wildlife Service 1990). As of June 1994, the USFWS had proposed upgrading the status of the bald eagle from endangered to threatened in all of the lower 48 states except in the Southwestern Recovery Region, primarily Arizona and New Mexico. This upgrading reflects the success of the recovery program. The banning of chlorinated hydrocarbon insecticides in North America was critical to the recovery of populations of bald eagles, brown pelicans, and other long-lived predatory birds. There has also been a concerted effort to educate the public and to protect bald and golden eagles from shooting and poisoning by vandals. Law enforcement efforts to reduce the illegal possession and sale of eagle feathers remain important to the restoration and maintenance of eagle populations. Another raptor, the peregrine falcon, is still listed as endangered, but its status is also improving due to reductions in the use of chlorinated hydrocarbon insecticides, increased protection from illegal taking,

and the restoration of populations using captive breeding and hacking techniques (see section 10.5). The Arctic subspecies of peregrine falcon is doing particularly well and may soon be removed from threatened status.

The whooping crane (fig. 11.5) is sometimes viewed as a success story, and yet the entire wild population is still about two hundred birds. This is a great improvement, however, over a population that had fallen below forty individuals. The recovery plan for whooping cranes involves protection of the wild population from disturbance and habitat destruction in Wood Buffalo National Park in Canada and the maintenance of two captive breeding flocks. Just learning enough about whooping crane behavior to breed the birds successfully in captivity was in itself a long and difficult project. At one point biologists attempted to establish a second flock of whooping cranes in an area near Yellowstone National Park using sandhill cranes as foster parents. Whooping cranes lay two eggs, but seldom rear more than one chick. A number of "extra" eggs from Wood Buffalo National Park were flown to southeastern Idaho for placement with sandhill crane foster parents. The young whooping cranes hatched, fledged, and several migrated successfully, but the attempt to develop a reproducing population was not successful.

FIGURE 11.5 Whooping cranes winter on the Texas coast and nest at Wood Buffalo National Park in Canada; about two hundred currently exist in the wild. (*Photograph courtesy of T. Stehn, U.S. Fish and Wildlife Service.*)

FIGURE 11.6 Black-footed ferrets are being bred in captivity for restoration into the wild. (*Photograph courtesy of the U.S. Fish and Wildlife Service.*)

The greenback cutthroat trout is still threatened in Colorado, but its status is improving through restoration into streams and lakes, particularly in Rocky Mountain National Park. Recovery plans call for maintenance of captive breeding stocks in several separate locations and monitoring of the remaining populations and their habitats. While removal of introduced trout species from all of the greenback cutthroat trout's original habitat is not feasible, the recovery plan calls for total or partial renovation of some streams that this subspecies once occupied (see section 10.2).

The recovery plan for the threatened Ozark cavefish calls for locating existing populations and protecting cave habitats and their water recharge areas. The status of this fish is improving largely due to the discovery of new populations. The endangered Pahrump poolfish of the Pahrump Valley of Nevada is also improving in status, largely due to transplants of the species into additional desert springs.

The black-footed ferret, which is native to areas containing black-tailed prairie dog colonies in the northern Great Plains, was thought to be extinct by the 1970s (fig. 11.6). Loss of black-tailed prairie dog colonies due to control measures and black-footed ferret disease problems both contributed to ferret declines. In the early 1980s a rancher near Meeteetse, Wyoming, identi-

fied a black-footed ferret killed by a ranch dog. This find led to the discovery of a substantial black-footed ferret population scattered over a number of black-tailed prairie dog towns in the area. Due primarily to canine distemper, the Meeteetse population of black-footed ferrets declined a few years after it was discovered. The remaining black-footed ferrets were captured and held in captivity to develop a captive breeding program. The captive black-footed ferrets were kept in three separate locations in Wyoming, Nebraska, and Virginia to reduce the chances of loss of the entire species, and even the ferrets at individual sites were kept relatively separated, except for breeding purposes, to reduce the chances of the spread of disease. The captive population is growing, and the recovery plan calls for restoration into widely dispersed black-tailed prairie dog towns within the original range of the species (U.S. Fish and Wildlife Service 1990; Clark and Harvey 1991). The search for additional wild populations of black-footed ferrets continues. The continued control of black-tailed prairie dog populations on private and many public lands illustrates how difficult it is to implement long-term solutions even with a recovery plan in place.

The recovery plan for the California condor (fig. 11.7) also called for the capture of the last few individuals remaining in the wild, and for the development of a captive breeding program. The captive breeding populations were divided between the San Diego and the Los Angeles zoos to reduce the chances of loss of the entire population. Captive breeding has been successful, and the species is increasing in captivity. A strategy for releasing California condors into the wild was rehearsed using captive-bred Andean condors (U.S. Fish and Wildlife Service 1990). Captive-bred California condors released into the wild will face the problem of encroaching human populations throughout the Sierra Nevada and coastal mountain ranges of California. Concern also remains over the availability of sufficient and nontoxic carrion sources for wild populations once they become established. Secondary poisoning from animals killed by toxicants and even the in-

FIGURE 11.7 The California condor is being bred in captivity and released into suitable areas to restore populations in the wild. (Photograph courtesy of the U.S. Fish and Wildlife Service.)

gestion of lead from unrecovered deer following hunting seasons pose threats to restoration success.

Unfortunately, the status reports on federally listed species are seldom encouraging (U.S. Fish and Wildlife Service 1990); most listed organisms are declining, stable, or of unknown status. Funding for efforts to evaluate populations, determine the needs of listed species, and develop workable recovery plans is in short supply. While great effort and expense is being directed toward some recovery programs, much more needs to be done.

Table 11.1 provides a few other examples of vertebrates listed as endangered or threatened. The table also gives the location where the organism is found, the causes of its decline, and plans for recovery.

11.5 Toward a Broader Approach: Biodiversity

With large numbers of endangered and threatened species listed in the United States, a staggering list on a worldwide basis, and many species not even identified, the limited success of our efforts to save individual species often seems discouraging. Given our past record of recovery success, the massive destruction of tropical rain forests and

other ecosystems, and the continued increase in the human population, one can understand that discouragement.

It is difficult to argue against efforts expended on the recovery of individual species listed as endangered or threatened. However, an ultimate slowing of the rate of extinctions can probably best be accomplished by broader approaches by which entire ecosystems, landscapes, and their biodiversity are conserved (Noss 1983). The Attwater's greater prairie-chicken in southeastern Texas is a case in point. Losses of native grassland to agriculture and development are the ultimate causes of the decline of this bird. The importance of native grassland regions in Texas to a variety of other grassland organisms needs emphasis. The Attwater's greater prairie-chicken is really a symbol of the loss of native grasslands in Texas, just as the northern spotted owl may reflect the loss of old-growth forest ecosystems in the Pacific Northwest. Species such as these are often referred to as **indicator species**. Indicator species are key organisms that serve as indicators of the status of ecological conditions in an area.

The problem of extinction ultimately can be solved only by maintaining adequate amounts of each ecosystem and landscape to support the native plant and animal communities of that ecosystem or landscape. Maintenance of an ecosystem can often best be accomplished through human use of that system on a truly sustainable basis. Thus, encouragement of well-managed grazing by domesticated livestock on lands containing Attwater's greater prairie-chickens could be considered as one strategy for maintaining that grassland ecosystem in a form usable by those birds. In California, water flowing into Mono Lake has been diverted for use in Los Angeles, changing the entire nature of this unique lake ecosystem; too much water is being removed. The lake is particularly rich in invertebrates that are sought by resident and migratory shorebirds. Mono Lake is one of the largest fall stopover and molting sites for birds such as eared grebes and Wilson's phalaropes and is important to the welfare of those populations. The strongest support

TABLE 11.1 Status, location, causes for decline, and recovery efforts for selected organisms listed as endangered (E) or threatened (T) under the U.S. Endangered Species Act of 1973.

Species or subspecies	Location	Causes of decline	Recovery efforts
Mammals			
West Indian (Florida) manatee (E)	Caribbean area, southeastern United States, and South America	Human disturbances, including injury or death from power boats	Establishment of sanctuaries and protective zones
Florida panther (E)	Southern Florida	Loss of habitat due to large human populations; killing of panthers in earlier years	Acquisition of Florida Panther National Wildlife Refuge
Columbian white-tailed deer (E)	Restricted sites in western Oregon and lower Columbia River on Washington border	Habitat loss to agricultural and residential development	Land acquisition; conservation easements
San Joaquin kit fox (E)	San Joaquin Valley and foothills in California	Loss of habitat to irrigated farms; habitat fragmentation; earlier losses due to poisoning	Land acquisition and land exchanges with the Bureau of Land Management
Birds			
Hawaiian goose (E)	Hawaiian Islands	Introduction of exotic predators; encroachment on habitat by humans	Captive propagation and reintroduction into the wild; control of introduced predators
Florida scrub jay (T)	Florida	Encroachment of housing and agriculture; fragmentation of habitat	Consultation with developers to reduce unnecessary habitat loss; recovery plans call for several preserve areas
Piping plover (E) (interior population)	Great Lakes area and Northern Plains region	Human disturbance of nesting beaches, sandbars, and islands; flooding due to water releases below some dams; predation on nestlings	Posting and recreational closure of nesting areas; releases from some dams are being regulated to avoid flooding nests
Kirtland's warbler (E)	Reproduces in northern Michigan	Jack pine habitat reduced due to fire exclusion policies in national forest; increased nest parasitism due to spread of brown-headed cowbirds	Controlled burning to create early successional jack pine; control of brown-headed cowbirds

Species or subspecies	Location	Causes of decline	Recovery efforts
Reptiles and amphibians			
Kemp's ridley (sea turtle) (E)	Gulf of Mexico and tropical Atlantic	Capture on nesting beaches for meat and parts; incidental capture in shrimp trawls	Protection of nesting beaches near Santa Cruz, Mexico; required use of turtle excluder devices on shrimp trawls
Desert tortoise (T)	Desert Southwest including areas in California, Nevada, Arizona, and Utah; also in Mexico	Disease; predation by birds; vandalism and collection by people; livestock grazing; disturbance of habitat by vehicles	Designation of Desert Wildlife Management Areas where domestic livestock grazing will be removed or restricted; restrictions on off-road vehicles; closure of roads and trails in some areas
Santa Cruz long-toed salamander (E)	Santa Cruz, California	Disking and filling of breeding ponds; livestock grazing; salt intrusion into wetlands	No information
Fishes			
Apache trout (T)	Arizona	Hybridization and competition with introduced trout; habitat destruction	Removal of introduced trout and restocking with Apache trout stocks; barrier construction on streams to isolate the species from other trout species
Woundfin (E)	Restricted drainages in Utah, Arizona, and Nevada	Competition with the introduced red shiner	Eradication of the red shiner and reintroduction of woundfin
Pallid sturgeon (E)	Mississippi and Missouri rivers	Changes in river ecosystem related to dam construction; interruption of migratory patterns; changes in spawning and nursery habitat	Artificial propagation; efforts to make river flows more natural

Source: U.S. Fish and Wildlife Service 1990.

for Mono Lake and its wildlife fauna has come indirectly through the efforts of trout anglers to maintain stream flows through legal mandates. User groups can be powerful supporters of ecosystem preservation in many such cases (Mono Basin Ecosystem Study Committee 1987).

The rapid loss of species caused by the clearing of tropical rain forests can hardly be managed on a species-by-species basis. This extensive destruction of an ecosystem and the resulting broad reduction in biodiversity have repercussions yet unforeseen. Many of the species lost in tropical ecosystems are little known or even unknown to humans.

Likewise, the tilling of grasslands, especially tall-grass prairies (see section 12.1), in North America to such an extent that only remnant pieces of these habitats remain has extinguished a variety of grassland life forms from much of their original range. This too is a biodiversity and ecosystem problem. It is true that the planting of trees on farmsteads throughout the original tall-grass prairie has resulted in new animal species being present and has contributed to overall species diversity. However, our responsibilities in terms of conservation entail maintaining the diversity of life native to these grasslands. To maintain this biodiversity, portions of the original ecosystem need to be maintained, or in some cases restored, throughout the areas where native tall-grass prairie once occurred.

Native fishes and other aquatic life likewise can be viewed as part of the original ecosystem in which they resided. They are as integral a part of a grassland, tropical rain forest, or other ecosystem as terrestrial organisms. What happens to the upland ecosystem most certainly relates to the fate of the organisms in, or associated with, aquatic systems.

Private organizations such as The Nature Conservancy, the Sierra Club, the National Audubon Society, the National Wildlife Federation, and the Wilderness Society have made great efforts to protect ecosystems and biodiversity. These organizations often have specific policies in regard to environmental issues. Their published policies on

issues such as human population numbers, environmental degradation, the marine environment, wetlands, and endangered species specifically relate to ecosystems and biodiversity. The Nature Conservancy has been involved in purchasing habitat critical to a variety of species since even before the ESA was passed. Its program of using private funds to purchase important habitat has been exemplary in protecting blocks of ecosystems and landscapes and the species associated with them. The lands it purchases may be given to public agencies for management if their management practices fulfill the original goals of the purchase.

Even when blocks of an ecosystem can be maintained or restored, there are many additional questions to answer. What size blocks of such habitat are needed to be truly functional? Perhaps populations of organisms known to have existed there can be restored, but what of organisms that we do not know were there? Do they even exist now? Total restoration may be impossible, but partial restoration is better than nothing, and may be the best that can be hoped for in some situations.

The establishment of large blocks of land to protect ecosystems and species is an attractive conservation measure, but political and economic realities make it a difficult and complicated one. Private landowners often consider government purchases of land to protect species and habitat, even on a willing-seller basis, to be a threat to private property rights. Regional or local government entities may oppose purchases of land by federal government agencies because of actual or perceived losses of tax revenue. For example, both state and county governments have in several instances restricted or attempted to restrict purchases of Waterfowl Production Areas by the USFWS in states such as North Dakota. Interestingly, large federally funded projects such as dams and canals for power generation or irrigation, which also involve federal land purchases, are often strongly supported by state and local governments. In many cases biologists working for natural resource agencies are prohibited from openly opposing such projects.

Even on federal lands in the United States, such as those managed by the U.S. Forest Service and the Bureau of Land Management, protection of blocks of habitat remains difficult. For example, designation of wilderness or scenic and wild river areas, protection of remnant mature forest stands, and reduction of grazing by domesticated livestock are all measures intended to protect ecosystems. All of these actions have economic, political, social, and other implications and involve a variety of human user groups; some of those user groups have strongly opposing viewpoints. Thus, protection of portions of ecosystems on public lands can be controversial. Conflict resolution is often required if such areas are to be protected (see section 20.5).

Conservation of tropical rain forests seems especially appropriate because of the potential for losses of large numbers of species in these habitats. However, it is much easier to promote this important cause than to accomplish such a task. Biologists involved in protecting tropical rain forests encounter many impediments that people in North America should readily understand. Those people currently replacing tropical rain forests with agricultural crops or harvesting rain forest timber at nonsustainable rates can easily point to the large-scale land modifications made to North American ecosystems, such as grasslands and forests, in the past. In terms of habitat and species losses, the past destruction and modification of some of our North American ecosystems has interesting similarities to the current destruction of tropical rain forests. The reasons for our large-scale ecosystem modifications were economic; it seems the removal of tropical rain forests is also economically motivated. While North Americans are now more aware of and appreciative of the consequences of habitat loss than we were during earlier periods of our history, any effort at restoring even fairly small areas of habitat to more natural conditions is still opposed by many people.

The development of land use practices compatible with the long-term maintenance of ecosystems is critical to species protection. For example, game ranching of wild ungulates in Africa

and ranching of domesticated livestock or bison in North America both are long-term means of harvesting grass and protecting ecosystems if done on a sustainable basis. Well-managed ranchlands are remarkable examples of the potential for protecting ecosystems while reaping economic returns. Efforts to develop economic activities that are compatible with the maintenance of ecosystems offer hope for protecting those ecosystems. While it is important, for example, to protect remnant old-growth forest stands, it is also important to continue to utilize other portions of the forest for logging on a sustainable basis. The economic benefits of such logging can be an important ingredient in conserving forest ecosystems for multiple purposes. Recreational expenditures for hiking, camping, hunting, rock climbing, fishing, mountain biking, and other activities on public lands have become increasingly important to many local economies. When these benefits are recognized, they encourage public support for conservation of the ecosystems that provide them. Conservation of tropical rain forests is much enhanced when forest products, including lumber, are valuable enough to be harvested on a sustained yield basis; however, management for optimum sustained yield must include long-term consideration for unborn generations of humans and for all organisms dependent on the tropical rain forest. The rapid growth of human populations will continue to challenge our ingenuity as we attempt to protect ecosystems from degradation and subsequent catastrophes in terms of species losses.

▼ SUMMARY

Extinctions of species have always occurred naturally, but the rate of extinctions has increased greatly due to human influences over the last two hundred years. A variety of human influences cause extinctions of species, but the greatest single cause is the alteration and destruction of habitat.

Species are valuable to humans for a variety of economic, scientific, ecological, recreational, aesthetic, and moral reasons. Wild organisms are also an integral part of the ecosystems in which they exist. Regardless of their value to humans, they are part of the working mechanisms of an ecosystem and have value in and of themselves.

Federal laws in the United States designed to protect endangered species were passed in 1966 and 1969, but were ineffective in species protection. The Endangered Species Act (ESA) of 1973 has been more effective in protecting endangered and threatened species in the United States. The listing of a species as endangered or threatened under the ESA allows for the development of a recovery plan and the designation of critical habitat. The ESA requires that the actions of federal agencies not threaten or endanger a species. International trade in endangered or threatened species or their parts is prohibited by the ESA primarily through its implementation of the Convention on International Trade in Endangered Species of Wildlife Flora and Fauna.

Recovery plans may involve the purchase of critical habitat, protection of habitats and organisms from human disturbance, release of captive-reared organisms into former habitat, establishment of new populations, and other means of reducing the chances of extinction. Recovery plans are designed to upgrade the status of a species from endangered to threatened and eventually to allow the delisting of the species.

Most recovery plans are species specific. In the long term, a more fruitful means of protecting the biodiversity of the earth and reducing the rate of extinctions lies in maintaining, and where necessary restoring, ecosystems and landscapes. The political and economic realities of maintaining and restoring ecosystems, even in the United States, greatly challenge our ingenuity. However, many ecosystems can be maintained by land use practices that offer economic returns to humans while conserving the ecosystem and its unique species.

PRACTICE QUESTIONS

1. What are some of the values of wild organisms to humans?

2. What is biodiversity? How does it differ from species diversity?

3. What is an endangered species? What is a threatened species? Identify two of each. Try to use organisms not discussed in this text.

4. Who authorizes the listing of marine mammals under the ESA? Who authorizes the listing of all other organisms?

5. What is the similarity of appearance clause of the ESA?

6. What is critical habitat? Provide an example for an endangered or threatened organism of your choice, preferably one other than those discussed in this text.

7. What is a recovery plan? Give an example.

8. What are the benefits of the Convention on International Trade in Endangered Species of Wildlife Fauna and Flora?

9. What are some human-related causes of extinction of organisms?

10. Briefly discuss the causes of extinction or endangerment for two organisms of your choice, preferably organisms other than those discussed in this text.

11. What are the benefits of an ecosystem approach to saving threatened or endangered organisms?

SELECTED READINGS

Clark, T., and A. Harvey. 1991. Implementing recovery policy: Learning as we go? Pages 147–163 in K. A. Kohm, ed. *Balancing on the brink of extinction: The Endangered Species Act and lessons for the future.* Island Press, Washington, D.C.

DeBlieu, J. 1991. *Meant to be wild: The struggle to save endangered species through captive breeding.* Fulcrum Publishing, Golden, Colo.

Ehrlich, P., and A. Ehrlich. 1981. *Extinction: The causes and consequences of the disappearance of species.* Random House, New York.

Head, S., and R. Heinzman, eds. 1990. *Lessons of the rainforest.* Sierra Club Books, San Francisco, Calif.

Johnson, J. E. 1987. *Protected fishes of the United States and Canada.* American Fisheries Society, Bethesda, Md.

Kohm, K. A., ed. 1991. *Balancing on the brink of extinction: The Endangered Species Act and lessons for the future.* Island Press, Washington, D.C.

Myers, N. 1983. *A wealth of wild species: Storehouse for human welfare.* Westview Press, Boulder, Colo.

Ono, R. D., J. D. Williams, and A. Wagner. 1983. *Vanishing fishes of North America.* Stone Wall Press, Washington, D.C.

Wilson, E. O., ed. 1988. *Biodiversity.* National Academy Press, Washington, D.C.

LITERATURE CITED

Clark, T., and A. Harvey. 1991. Implementing recovery policy: Learning as we go? Pages 147–163 *in* K. A. Kohm, ed. *Balancing on the brink of extinction: The Endangered Species Act and lessons for the future.* Island Press, Washington, D.C.

Ehrlich, P., and A. Ehrlich. 1981. *Extinction: The causes and consequences of the disappearance of species.* Random House, New York.

Gerstung, E. R. 1988. Status, life history, and management of the Lahontan cutthroat trout. Pages 93–106 *in* R. E. Gresswell, ed. *Status and management of interior stocks of cutthroat trout.* American Fisheries Society, Bethesda, Md.

Leopold, A. 1949. *A Sand County almanac and sketches here and there.* Oxford University Press, London.

Lugo, A. E. 1988. Estimating reductions in the diversity of tropical forest species. Pages 58–70 *in* E. O. Wilson, ed. *Biodiversity.* National Academy Press, Washington, D.C.

Mono Basin Ecosystem Study Committee. 1987. *The Mono Basin ecosystem: Effects of changing lake level.* National Academy Press, Washington, D.C.

Myers, N. 1983. *A wealth of wild species: Storehouse for human welfare.* Westview Press, Boulder, Colo.

Noss, R. F. 1983. A regional landscape approach to maintain diversity. *BioScience* 33:700–706.

Rohlf, D. J. 1989. *The Endangered Species Act: A guide to its protections and implementation.* Stanford Environmental Law Society, Stanford Law School, Stanford, Calif.

Stuber, R. J., B. D. Rosenlund, and J. R. Bennett. 1988. Greenback cutthroat trout recovery program: Management overview. Pages 71–74 *in* R. E. Gresswell, ed. *Status and management of interior stocks of cutthroat trout.* American Fisheries Society, Bethesda, Md.

U.S. Fish and Wildlife Service. 1990. *Report to Congress: Endangered and threatened species recovery program.* U.S. Department of the Interior, Fish and Wildlife Service, Washington, D.C.

U.S. Fish and Wildlife Service. 1992. *Endangered and threatened wildlife and plants.* 50 CFR 17.11 & 17.12. U.S. Department of the Interior, U.S. Government Printing Office, Washington, D.C.

THE
HABITAT

12

HABITAT TYPES

Habitat, as defined in section 2.5, is the place where an organism lives, and includes both biotic and abiotic components. It would be difficult to manage populations and communities effectively without a knowledge of general and specific habitat types. Habitat classifications can range from general categories, such as grasslands or large reservoirs, to very specialized systems, such as those describing the specific habitat needed by canvasbacks for over-water nest sites or the specific feeding site needs of rainbow darters. While a general habitat may occur within the overall range of tolerance for a species, the species may also need specific habitats that can fill particular requirements within narrower ranges of tolerance. For example, a fish species may need general habitat with water temperatures within the range of 2°–20°C during the year, but it may also require water that is between 12°C and 13°C during a specific time of the year for successful reproduction.

Terrestrial biologists most often classify habitat by its vegetative characteristics; aquatic biologists rely more on the structural characteristics of a water body for habitat classification. This general difference in the way wildlife and fishery biologists view habitat may be one of the reasons that they differ somewhat in the ways in which they categorize themselves. Wildlife biologists are more apt to classify themselves by species or groups of species, for example, as elk, raptor, Neotropical migrant, or furbearer biologists. Less often will they refer to themselves as grassland or desert wildlife biologists. Conversely, fishery biologists are more likely to identify themselves with a habitat type, for example, as stream, reservoir, or pond biologists; they are less likely to categorize themselves as walleye or rainbow trout biologists. There are obviously many exceptions to this generalization, such as wetland biologists and salmon biologists, but in general this difference in categorization holds true.

As our understanding of biological systems increases, there will probably be fewer differences in the ways in which wildlife and fishery biologists categorize themselves. The advent of ecosystem management, landscape ecology, and other more holistic views of habitat and organismal interaction should result in a broader outlook. It should be recognized, however, that in most cases, habitat and its management account for a substantial portion of the effort expended by biologists, regardless of whether they consider themselves to be fishery, wildlife, landscape, restoration, or conservation biologists, or any other type of natural resource-oriented biologist.

Terrestrial and aquatic habitats are not really separate systems. While they may be discussed separately for the sake of simplicity, they are closely interconnected. For example, the aquatic habitats of a stream and the terrestrial watershed of that stream are ecologically inseparable; each affects the other in myriad ways.

The primary purpose of this chapter is to identify and classify some of the habitats found in aquatic and terrestrial ecosystems. General habitat delineations as well as more specific habitat types will be addressed.

12.1 Terrestrial Habitats

As mentioned in section 2.1, a broad form of terrestrial habitat classification is by biome or major ecosystem. Biomes are large geographic areas that are primarily recognizable by the type of vegetation present. Climate and soil influence vegetation type, and vegetation and physical factors influence the types of animals present. Biomes are repeated on a global basis; a biome found in North America is also often found in Europe, Asia, or other regions of the world. North American biomes will be highlighted in this text. Figure 12.1 depicts the distribution of the major biomes of the world. There are numerous biome classification systems, and biome names, numbers, and descriptions vary

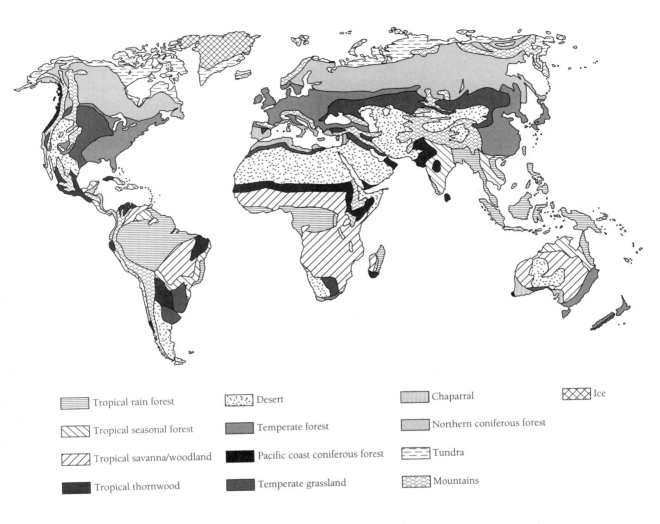

FIGURE 12.1 Most of the major biomes of the world are present in North America. These biomes repeat themselves in other parts of the world.

with these systems. The classification system used in this text is just one model.

In North America, **tundra**, also referred to as **Arctic tundra**, is primarily found in northern Canada and Alaska (fig. 12.2). Tundra is characterized by short, cool summers and long, cold winters. The growing season is short, and the vegetation consists primarily of grasses, sedges, lichens, and low shrubby plants. A distinctive charac-

teristic of tundra is **permafrost**, the constant presence of frozen ground, usually a short distance below the soil surface, during the entire year. Permafrost impedes the growth of trees by limiting root systems.

Many animals of the Arctic tundra are present only during the warmer seasons of the year. Migrating birds, such as snow geese, other Arctic geese, and many species of shorebirds, such as the

FIGURE 12.2 Arctic tundra is the northernmost North American biome. *(Photograph courtesy of J. Jenks.)*

lesser golden-plover, reproduce in this biome. There are also year-round residents, such as caribou, brown lemmings, Arctic foxes, muskoxen, and rock ptarmigan, that are adapted for living through the long, harsh winters.

Arctic tundra is a wet place, even though it has relatively low levels of precipitation. Two reasons why it remains so wet are reduced surface evaporation from the soil and reduced transpiration from plants; both result from the cool climate and short growing season. Permafrost also increases wetness by reducing moisture percolation through the soil.

Tundra can also be found at high elevations at lower latitudes. Biomes found at high elevations are called **altitudinal biomes**. The tops of high mountains in areas such as Colorado and Montana contain what is called **alpine tundra**. Alpine tundra often differs from Arctic tundra in regard to the amount of permafrost and precipitation levels. Because of its high winds, wide temperature fluctuations, and high levels of ultraviolet radiation, alpine tundra may have a climate even more severe than that of Arctic tundra.

The tundra is a fragile environment because of permafrost and the short growing season. Human activity in the tundra can easily cause long-term environmental problems. In addition, low species diversity contributes to the fragility of this biome. Food webs tend to be simple; thus, they are easily

disturbed. The destruction or reduction of any one segment of the biota can have substantial effects on other trophic levels that depend on the lost or reduced plant or animal population.

In North America, **northern coniferous forest** (**boreal forest** or **taiga**) occurs south of the tundra (fig. 12.3). This biome is present in broad expanses, primarily in Canada and Alaska. Northern coniferous forest overstory vegetation consists primarily of spruce, fir, and pine trees. These trees produce dense shade year-round, which often results in poor development of understory layers. This biome is warmer than the tundra and usually has higher levels of precipitation. Most of the precipitation falls as snow. The soils of northern coniferous forests are relatively infertile and have limited use for agricultural purposes, but these forests are among the great timber-producing areas of the world.

The animals of the northern coniferous forest include both migrants and year-round residents. Large numbers of migratory birds use the area seasonally. Many birds and mammals, such as moose, lynx, spruce grouse, snowshoe hares, three-toed woodpeckers, and wolverines, are year-round residents. As in most cold areas, reptiles and amphibians are few. This biome is sometimes referred to as the "spruce-moose" biome.

FIGURE 12.3 The northern coniferous forest biome is circumpolar in the northern hemisphere. *(Photograph courtesy of J. Riis.)*

FIGURE 12.4 The temperate forest biome is found over most of the eastern United States. (*Photograph courtesy of M. Brown.*)

Northern coniferous forests are also present as altitudinal biomes. In North America, these **montane coniferous forests** can be found much farther south than Canada or Alaska; mountainous regions such as those in Washington, New Mexico, and the Adirondack Mountains of New York contain montane coniferous forests.

In North America, the **temperate forest** biome occurs primarily below the northern coniferous forest biome in the eastern portion of the continent (fig. 12.4). Most of this biome is characterized by deciduous trees. The dominant trees are oaks, maples, beeches, and other hardwoods, although some areas also have large numbers of evergreen trees. Precipitation is generally higher than in the northern coniferous forest, and the growing season is longer. These climatic conditions result in dense vegetation. The temperate forest biome is characterized by many year-round resident animal species, including white-tailed deer, wild turkeys, gray squirrels, blue jays, raccoons, and various reptiles and amphibians. Human activities have eliminated or reduced numerous species, such as the black bear, in many areas of this biome. Many bird species inhabit these forests; the majority of them migrate south during the winter. Much of the temperate forest biome in North America has been converted to agricultural cropland.

The varying climatic conditions over the range of the temperate forest biome result in different vegetation patterns and thus the presence of different animal communities. For this reason, the temperate forest biome is often divided into several subareas. For example, the **southeastern mixed forest** ranges from Virginia through the Carolinas and Alabama and contains many conifers. The **outer coastal plain forest**, which ranges from Florida to Louisiana, contains vegetation such as mangroves and live oaks.

Most of the **temperate grassland** biome in North America is found west of the temperate forest and south of the northern coniferous forest (fig. 12.5). Temperate grasslands generally have relatively low precipitation compared with temperate forests and are characterized by wet and dry cycles. Summers are hot and winters cold. Periodic droughts and fires retard tree growth. In contrast to the multiple vegetation layers (trees, shrubs, grasses, etc.) found in most forests, temperate grassland vegetation usually occurs in a single lay-

FIGURE 12.5 Temperate grassland covers most of central North America.

er. This layer, however, is characterized by many plant species. A variety of grasses and other low-growing plants are important in this biome.

Grasslands vary depending primarily on rainfall and are often classified by grass height. **Short-grass prairies** are relatively dry and have short vegetation compared with **tall-grass prairies**. Intermediate between the two are **mixed-grass prairies**. Portions of eastern Colorado and eastern New Mexico represent areas of short-grass prairie. Central Kansas and most of North Dakota are mixed-grass prairie areas. Central Illinois and much of Iowa represent tall-grass prairie regions. Characteristic grassland animals include western meadowlarks, pronghorns, black-tailed prairie dogs, badgers, burrowing owls, and greater and lesser prairie-chickens.

Because rainfall levels are low in temperate grasslands, nutrients are not easily leached out of the soil. Grasslands therefore tend to have very rich, productive soils, and are heavily used for agricultural production. Most of the tall-grass prairie in North America has been converted to grain production; mixed- and short-grass prairies are usually used for domesticated livestock grazing or grain production.

The **desert** biome is arid, usually having less than 25 centimeters of precipitation in a year (fig. 12.6). In North America, the desert biome occurs primarily in the western and southwestern regions. Both high-altitude and low-altitude deserts are present, and each has characteristic vegetation patterns. High desert (sometimes called shrub-steppe), such as that in the Great Basin of the United States, is characterized by sagebrush, low shrubs, and intermixed grasses and forbs. The low deserts of southwestern North America have been subdivided into three subareas: the Chihuahuan, Mojave, and Sonoran deserts. These subareas vary in precipitation levels, vegetation, and animal life. For example, the Sonoran desert in southern Arizona is characterized by mesquite, creosote bush, and a variety of cacti, while the Chihuahuan desert of Mexico is rich in cacti and other plant forms that conserve water.

FIGURE 12.6　The desert biome is a dry area with a harsh climate. This photograph shows Sonoran desert in southern Arizona. (*Photograph courtesy of R. Labisky.*)

The desert biome has a harsh climate, and many of the plants and animals that live there have special adaptations for water conservation. Common desert animals include mule deer, black-tailed jack rabbits, numerous hawks, thrashers, and a variety of reptiles.

Although dry, some desert areas have very fertile soils. In many of those areas, irrigated agriculture has been developed to a high degree. This development, along with grazing of domesticated livestock and the exclusion of fire, has had substantial effects on the desert biome and on some of its native organisms.

These five biomes (tundra, northern coniferous forest, temperate forest, temperate grassland, and desert) cover most of the area of North America. However, other biomes of smaller size are also present.

One of the smaller North American biomes is the **chaparral** biome. It is characterized by a mix of trees and shrubs, often with hard, thick evergreen leaves. The coastal area of southern California is representative of this biome. Many people consider the chaparral biome as having an ideal climate. Dry, warm summers and short, mild winters are the norm, with most precipitation occurring in the cooler months. Chaparral, however, evolved with fire and tends to burn easily. In fact, the seeds of many plants in this biome need fire to break the seed coat so that germination can occur. Despite this tendency and need for chaparral to burn, people continue to build homes and other structures in this biome, want the vegetation to remain the same, and then are surprised when fires occur.

The animals of the chaparral biome include black-tailed deer, California quail, and an abundance of other birds, mammals, and reptiles. The wildlife of this biome has been affected by human activities, especially fire exclusion.

Another small North American biome is the **Pacific coast coniferous forest** (fig. 12.7). This biome, which is often included in the northern coniferous forest biome or even in the temperate forest biome, occurs along the Pacific coast from Alaska to northern California. Cool climates and abundant rainfall make this a highly productive timber area. Sitka spruce, giant redwoods, and Douglas firs are all commercially important trees

FIGURE 12.8 Tropical rain forests are currently being lost at a rapid rate. *(Photograph courtesy of C. Johnson.)*

found in this region. Because of their timber value, old-growth forests in this biome are the focal point of a major ecological controversy. The continuing concern over old-growth forests and northern spotted owls (and other native biota) in the Pacific Northwest represents a contentious issue that is mentioned several times in this text. Portions of this biome are wet enough to be called **temperate rain forests**. The Olympic National Forest in Washington is an example of such a rain forest.

The southern tip of Florida represents another small North American biome, the **tropical seasonal forest**. This forest type is characterized by wet and dry seasons and by hot temperatures year-round. It contains both evergreen and deciduous tree species.

Another major biome that should be mentioned is the **tropical rain forest** (fig. 12.8). Although it is not present in North America, large areas of Central America and southern Mexico consist of this forest type, as do large areas of southeast Asia and equatorial South America and Africa. The global importance of this biome necessitates its mention in this text. Tropical rain forests are characterized by heavy rainfall, usually in excess of 220 centimeters per year. While these forests appear to be exceedingly productive, most nutrients are retained in the lush vegetation. The soils, because of the high rainfall, tend to be ra-

FIGURE 12.7 The Pacific coast coniferous forest biome contains some of the largest trees in the world. *(Photograph courtesy of K. Jenkins.)*

pidly leached of nutrients. Human attempts at agriculture in tropical rain forest areas are generally short-lived because the soils are poor and their nutrients are rapidly lost. A large variety of birds, reptiles, and bats are found in tropical rain forests, along with many species of monkeys and invertebrates.

While North America has no tropical rain forest, it is affected by human activities that are degrading this biome. The effects of this degradation on North America are diverse, ranging from potential climate changes to reduced populations of Neotropical migrants. In addition, tropical rain forests are a major repository of worldwide species diversity.

The biomes that have been described should not be thought of as monolithic blocks of habitat. In all of these biomes there are interspersions of a variety of habitats. Grasslands have corridors of trees along rivers or on other wetland margins, temperate forests have grassland openings, and all biomes can have mountainous areas that contain islands of various habitat types not found in the surrounding area. In addition, human activities have modified many portions of these biomes into varied landscapes that are quite different from their original states.

Biomes do not end abruptly. Ecotone areas, sometimes quite unique ones, characterize their edges. The cross timbers area of Oklahoma, where the transition between temperate grassland and temperate forest occurs, is one example. Another example is the mosaic of scrubby trees and open areas lying between the northern coniferous forest and the tundra. These edge areas can have their own unique animal and plant communities. Species diversity is often quite high in such areas because organisms from both adjacent biomes may be present, as well as organisms specifically adapted to the ecotone area.

Some wildlife species are quite adaptable and are found in numerous biomes, while other species are relegated to particular portions of certain biomes. The pronghorn, for example, is an animal of open spaces. While common in some of the short- and mixed-grass prairies, it is not well-adapted to tall-grass prairies. Pronghorns need to be able to see for long distances, primarily for protection from predators, and tall grass is as much a barrier to their existence as forests. Other animals, such as the American crow and the mountain lion, are very adaptable and can be found in several biomes.

There are numerous other delineation and classification schemes for characterizing broad types of terrestrial habitat. The zoogeographic regions discussed in section 2.10 are an example. Another excellent way of classifying the biotic world is by **ecoregion**. This classification scheme is based on regional similarities associated with soils, land use, geology, land surface form, climate, and potential natural vegetation. Boundaries between ecoregions are quite arbitrary. Ecoregion characteristics also affect water bodies located in a particular region, and thus this scheme also has utility with respect to classifying or identifying aquatic systems.

12.2 Terrestrial and Aquatic Transitional Habitats

Upland areas are terrestrial habitats, while streams, lakes, and most other water bodies are aquatic habitats. Located between the two are transitional habitats. Such transitional habitats include the riparian zones along rivers and streams, the wet margins of lakes, ponds, and other standing water bodies, and a variety of marshes, bogs, swamps, coastal wetlands, and beach-ocean interfaces. **Marshes** are low, treeless wet areas characterized by sedges, rushes, and cattails. **Bogs** are wet areas characterized by floating spongy mats of vegetation often composed of sphagnum, sedges, and heaths. **Swamps** are wet areas that usually contain standing trees. **Coastal wetlands** are transitional areas between land and the oceans. Each of these transitional areas contains characteristic plants and animals that thrive in terrestrial-aquatic interface habitats. Transition zones are usually not abrupt, but can be represented by a gradient or continuum, becoming either more terrestrial or more aquatic depending upon the distance from each of the two primary habitat types. However, some aquatic-terrestrial interfaces can be fairly abrupt, as in the case of many areas of the beach-ocean interface.

The organisms associated with these aquatic-terrestrial transition areas are many. A variety of shorebirds, such as the American avocet, Baird's sandpiper, and greater yellowlegs, are dependent on transitional areas. Some mammals, such as mink and raccoons, frequent the terrestrial-aquatic edge. Some organisms, such as beavers and muskrats, tend toward the aquatic end of the gradient in these habitats, as, of course, do fishes. In addition, these transitional areas often have abundant populations of amphibians and reptiles. Many animals spend the majority of their lives in transitional areas.

Riparian zones and the shorelines of lakes, ponds, and other standing water bodies are transitional areas that are generally considered to be primarily wildlife habitats. Transitional areas such as bogs, marshes, swamps, and coastal wetlands are primarily aquatic habitats and are important to fishes, other primarily aquatic animals, and a variety of terrestrial wildlife species. Cowardin et al. (1979) divided wetland and deep water habitats into five different systems. Their classification scheme divides these habitats into system, subsystem, and class groupings (fig. 12.9). This text will highlight the system portion of the classification. **Marine** (open-ocean areas) systems can contain transitional habitats at the beach-ocean interface. **Estuarine** (transitional areas between marine and freshwater areas) systems will be discussed in section 12.5. The freshwater portion of the Cowardin et al. classification scheme contains three major categories: **riverine**, **lacustrine**, and **palustrine** systems. Riverine systems are associated with rivers and streams, while lacustrine systems are associated with lakes and ponds; these two categories will be discussed in sections 12.3 and 12.4 as aquatic habitats. The palustrine systems (bogs, marshes, and swamps) are discussed here as transitional habitat.

While any watered area, regardless of depth or size, can be described as a wetland, biologists usually use the term **wetland** when referring to bogs, marshes, swamps, or coastal wetlands. Wetlands are generally described as areas where the water table is near ground level or just shallowly covers the land surface (fig. 12.10). Various characteristics can be indicative of a wetland area, including the presence of **hydric plants** (plants associated with water), the presence of **hydric soils** (soils developed in wet areas), or the presence of water saturation or coverage during some portion of the year. Some wetland areas have visible water at all times of the year, while others, over the course of a year or some other time period, sometimes more closely resemble a terrestrial habitat. For many wetlands, a dry period is an important element in maintaining productivity and vegetative characteristics. Just because a wetland is dry does not mean it is not a wetland. The fact that a wetland may contain no water for appreciable periods of time is one of the reasons why there is currently a great deal of public debate as to what constitutes a wetland. This controversy creates difficulties in efforts to protect wetlands, and finding a definition of wetlands that everyone will agree upon may not be possible.

Regardless of whether wetlands contain water at all times or for only brief periods of time, their utility as habitat is unquestioned. Wetlands not only serve as habitat for aquatic-oriented organisms, but also represent a habitat commonly used by many terrestrial organisms. In addition to providing habitat for wild organisms, wetlands are important in floodwater retention, nutrient cycling and trapping, groundwater recharge, microclimate modification, water purification, providing forage and water for domesticated livestock, and other beneficial functions (see section 14.6).

System   Subsystem   Class

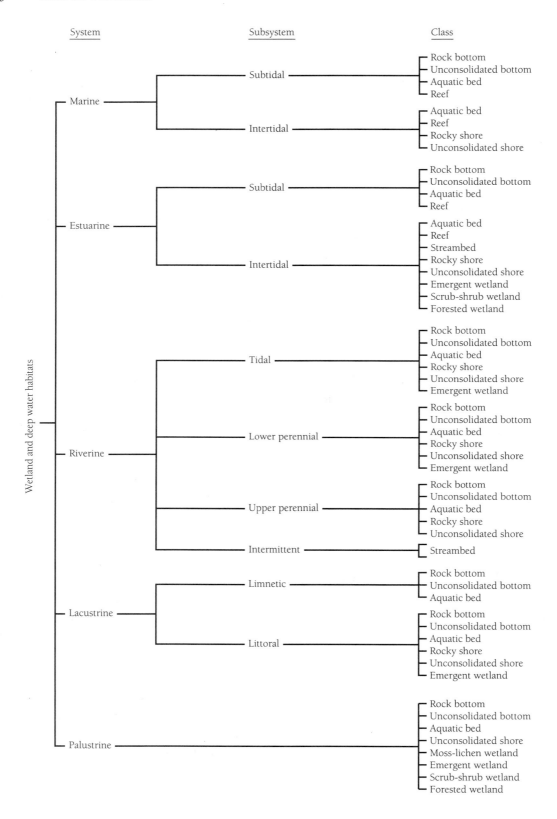

The beach-ocean interface is a major transitional habitat that is quite different from most of the areas discussed in this section. This habitat contains its own unique assemblage of organisms, ranging from marine mammals and seabirds to marine fishes and invertebrates. This habitat zone is of concern to many public agencies as well as the general public. It is an area rich in commercial products, such as fishes and shellfishes, and is heavily used for consumptive and nonconsumptive outdoor pursuits such as fishing, swimming, sunbathing, and viewing wildlife. It is biologically important because of its diverse flora and fauna. The variety of shorebirds, seabirds, and marine mammals that are present make this a particularly rich habitat.

FIGURE 12.10 Wetland areas are characterized by the presence of hydric plants and hydric soils.

12.3 Standing Water Habitats

Natural lakes, small impoundments, ponds, and reservoirs are called standing water, or **lentic**, habitats. The use of these terms does not mean that these water bodies have no current or water movement. Water, either surface or underground, moving into these bodies results in current, as does wind action and the heating and cooling of water. There is, however, usually little unidirectional water flow compared with that found in streams and rivers. **Lotic** systems, such as rivers and streams, have a stronger unidirectional water flow regime; these systems will be discussed in section 12.4. Within the lentic water body category, some water bodies have a considerable degree of unidirectional water flow. For example, natural water bodies can have inlets and outlets

◀ **FIGURE 12.9** *(opposite)* The Cowardin et al. classification divides wetland and deep water habitats into different systems, subsystems, and classes. *(Adapted from Cowardin et al. 1979.)*

that result in constant or temporary water flows, while artificial lentic water bodies created by human activities such as damming rivers or streams often have various levels of unidirectional water flow. Lentic and lotic environments occur on a continuum; some water bodies are almost totally lentic or lotic, while others occur at various intermediate levels. Lentic water bodies are also referred to as lacustrine systems.

There is a great deal of variation in the terminology associated with lentic systems. Terms such as pond and lake are often used interchangeably and can lead to confusion. There are no universally accepted standardized definitions to describe, for example, when a pond is large enough to be called a small impoundment, or what size a water body must be to be classified as a reservoir. Some biologists state that for a water body to be called a lake, variations in water temperature with depth or windswept shorelines must be present. No classification system will satisfy everyone. We will describe these lentic systems with just one of the many classification schemes that can be used.

Natural lakes, as the name implies, are naturally formed lentic water bodies. No sizes are usually defined. Natural lakes can vary in size from very large water bodies, such as Lake Superior or Lake Champlain, to rather small water bodies, such as the many small natural lakes (often called ponds) found in New England. Natural lakes can be formed in numerous ways, the most common of which are glacial activity, movements of the earth's crust, and changes in river courses.

Deep natural lakes tend to be **oligotrophic**, or low in nutrients. Oligotrophic lakes are rather unproductive and usually have very transparent water. Their lack of nutrients may result from their being located in unproductive areas or because they are so recent successionally that nutrient levels have not had time to increase. **Eutrophic** natural lakes tend to be productive and shallow. It is important to remember that the terms oligotrophic and eutrophic do not refer just to natural lakes, and that the terms are often used somewhat subjectively. There is a continuum of productivity levels in water bodies, and there is no set point that everyone agrees upon at which the eutrophic and oligotrophic categories begin and end. Numerous attempts at such divisions have been made, but these categorizations vary.

The remaining lentic systems are artificial (human constructed) water bodies. Ponds, small impoundments, and reservoirs are **impoundments** because they result from human activities that block natural watercourses such as rivers and streams. As mentioned, it is difficult to categorize water bodies by size, but the following system is sometimes used. **Ponds**, the smallest impoundments, are less than 4 hectares in surface area. The previously mentioned use of the term pond to describe a small New England natural lake leads to a certain amount of confusion. **Small impoundments** range from 4 to 40 hectares in surface area. **Reservoir** surface areas usually exceed 40 hectares. Obviously, a 100,000-hectare reservoir may be quite different from a 100-hectare reservoir. What is important to remember about these impoundments is that artificial lentic systems occur in a wide variety of sizes and vary greatly with respect to why and how they are constructed and how much water flows through them.

As already defined, impoundments are constructed by damming watercourses. In the case of ponds, the damming can involve small intermittent streams or small drainage areas. In the case of large reservoirs, a large river such as the Columbia, Colorado, or Savannah might be dammed. At times, earth moving, other than that associated with dam construction, can be used to deepen or modify such water bodies during construction.

There are numerous reasons why impoundments are constructed. However, whether they are constructed for domestic or agricultural water supplies, flood control, electricity generation, navigation, recreation, or a combination of uses, they still represent aquatic habitat that can have varying degrees of utility for both aquatic and terrestrial communities.

Other artificial water bodies result from excavation activities. Gravel pit, strip mine, quarry, and borrow pit (excavations along roadways) water bodies are the result of excavation. These excavations are not purposely constructed for holding water; they are by-products of human activities usually related to mining or road construction (fig. 12.11). Some other excavated water bodies, such as dugouts (fig. 12.12), are specifically constructed to hold water for domesticated livestock. However, dugouts can dewater small wetlands and negatively affect some organisms.

Ponds are the smallest artificial water bodies (fig. 12.13). Because they are small, they tend to have a limited number of ecological niches, and

FIGURE 12.11 Borrow pits are excavations along highways from which fill is taken for road building. They are not purposely constructed for holding water. Note the interstate highway in the background.

FIGURE 12.12 Dugouts are constructed primarily for watering domesticated livestock, but also provide aquatic habitat for some other species.

FIGURE 12.14 Reservoirs vary greatly in their size, depth, and rates of water volume exchange.

therefore have rather low species diversity. In some ways this makes them easy to manage, but in other ways their simplicity makes their management more difficult. Low species diversity may simplify food webs, but this simplicity can also cause problems because a change in one portion of the web can quickly and drastically affect other portions.

Small impoundments and reservoirs are larger than ponds and usually have greater species diversity (fig. 12.14). Some are sufficiently deep to have water temperature gradients from the surface to the bottom. There is a great deal of variation among these water bodies in the amount of water that flows through them. Many small impound-

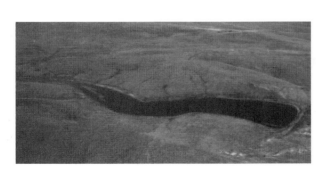

FIGURE 12.13 Ponds are small artificial water bodies that serve numerous functions. *(Photograph courtesy of T. Hill.)*

ments have substantial amounts of water inflow and outflow only during brief portions of the year. Most large reservoirs have constant inflow and outflow year-round. In some reservoirs the total volume of the water body may be exchanged in a few days; these impoundments are more lotic in nature. In other reservoirs the total exchange of water volume may take two or three years; these impoundments more closely resemble lentic systems. The rate of water exchange has important effects on the development of aquatic communities (Ploskey et al. 1984).

All lentic bodies, whether natural or artificial, can contain a variety of different zones or habitats. The number of zones depends on water depth, water body size, and numerous other physical and chemical characteristics. The terminology used to describe the various zones present in a lentic water body is quite variable. The following system is simplified, but it is representative of the terminology commonly used for freshwater lentic environments (fig. 12.15).

Most lentic systems have a **littoral zone**—a peripheral shallow water area. In many cases, this zone extends from the shoreline to the limit of rooted aquatic vegetation (fig. 12.15A). In some water bodies the entire body is shallow enough to support rooted aquatic vegetation; in this case the entire water body consists of littoral zone. In

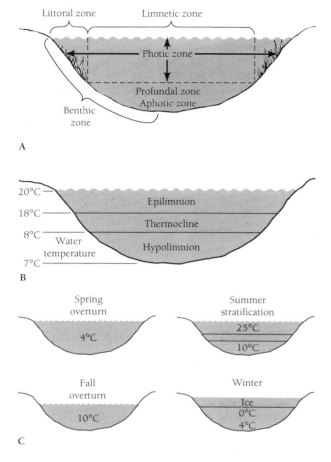

A

B

C

FIGURE 12.15 Various zones may be found in fresh-water lentic systems. (*A*) Zones determined by water depth and sunlight penetration. (*B*) Zones determined by water temperatures. Water temperature in relation to depth in a stratified water body during summer is shown. (*C*) Seasonal patterns of thermal stratification for water bodies in northern regions that stratify.

deeper water bodies, the littoral zone is usually represented by a band of rooted aquatic vegetation. along the shoreline. Using such vegetation to describe the littoral zone, however, can lead to confusion. Consider the shallow area of a water body that is too turbid or windswept to support rooted aquatic vegetation. This shallow area still represents a littoral zone, even though no rooted aquatic vegetation is present.

In lentic water bodies that have deeper water beyond the littoral zone, the open water area is called the **limnetic zone**. The limnetic zone includes all of the water volume outside the littoral zone that is lighted. The total lighted area of a water body is called the **photic** (**euphotic** or **trophogenic**) **zone**. This lighted area serves as an environment where phytoplankton can live and photosynthesize. If a water body is deep enough to have an unlighted portion below the limnetic zone, this area is called the **profundal zone**. The profundal zone is an unlighted **aphotic** (**tropholytic**) **zone**, and does not support photosynthesis.

In any lentic or lotic water body there is also a bottom or **benthic zone**. This zone, depending on water depth and other characteristics, provides habitat for organisms that do not suspend in the water column.

Lentic water bodies can also be characterized by layers with different water temperatures and oxygen levels. Such **thermal stratification** occurs where there is sufficient water depth, seasonal temperature change, and protection from wind (fig. 12.15B). In stratified water bodies, three layers are present during summer. The upper layer is called the **epilimnion**; the middle layer, where the water temperature drops abruptly, is the **thermocline** (**metalimnion**); the deepest and coldest layer is the **hypolimnion**. Stratification results from the relationship of water density to water temperature. Except near freezing, the warmer water is, the lighter it is. Waters of different temperatures, and therefore different densities, do not easily mix. Thus, the thermocline, the area of abrupt temperature change, acts as a barrier to water mixing. The epilimnion, because it is in the photic zone and therefore has photosynthetic activity, and because it is open to wind action and other mechanical agitation, can have its oxygen supply replenished. The hypolimnion is sealed from this mixing zone by the thermocline and, depending on the productivity of the water body and the relative volumes of the hypolimnion and epilimnion, can contain either adequate or inadequate amounts of dissolved oxygen for oxygen-demanding organisms. The hypolimnion is likely to retain dissolved oxygen during the entire year in oligotrophic water bodies; this is less likely to occur in eutrophic water bodies.

As autumn approaches and the water temperature in the epilimnion decreases, the thermocline erodes. Eventually, water temperatures throughout the water body equalize, allowing complete mixing of the water. This phenomenon is called **fall overturn**. The water body at this time has uniform temperatures and dissolved oxygen concentrations from surface to bottom.

In cold climates, especially when ice cover is present, **inverse stratification** generally occurs in winter. Water attains its maximum density at 4°C. Below 4°C, it becomes lighter and rises to the surface. That is why ice forms on the top of a water body, not at the bottom. After ice formation, the coldest water is near the surface, and the warmest water is in the deeper areas of the water body. In spring, when the ice melts and water temperatures equalize, a **spring overturn** occurs. Figure 12.15C depicts the seasonal changes that occur in a stratified water body in northern regions.

The unique relationship of water density to water temperature has important ramifications for aquatic ecosystems. If water at 0°C were more dense than water at 4°C, and if ice therefore formed at the bottom of water bodies rather than at the top, aquatic ecosystems would be very different than they are. Try to envision the effects on organisms if water bodies froze from the bottom up rather than from the top down. In fact, bottom ice can occur, and most aquatic organisms are ill prepared to handle its effects. Ice crystals from above the water surface can fall into streams and form ice masses on stream bottom rocks and other substrates. This ice, when firmly attached, is called **anchor ice**. It can dam up and divert water from the streambed. Disoriented fishes and aquatic invertebrates can wander into these diverted channels and be stranded and die when the ice dams melt and the water returns to the normal stream channel. Needham and Jones (1959) reported that in some high Sierra Nevada streams, such mor-

tality can have substantial effects on fish and insect populations. Anchor ice can also damage the benthic community portion of the food web of a stream.

The various zones and zone characteristics present in water bodies have many ramifications for fisheries and even wildlife systems. Littoral zones serve as habitat for shallow water fish species, such as the sunfishes. Limnetic zones provide habitat for open water fishes, such as the shads. Benthic zones are inhabited by bottom-dwelling species such as bullheads and suckers. Some wildlife, such as belted kingfishers, moose, and raccoons, feed in the littoral zone, while birds capable of diving, such as oldsquaws, may use the limnetic zone. Several species, including lesser snow geese and mallards, use the isolated open water of the limnetic zone to escape disturbances during fall migration.

Thermal stratification can provide warmwater, coolwater, and coldwater habitats for aquatic organisms if sufficient dissolved oxygen is available. In this case, two-story fisheries, as described in section 10.5, can be present, with warmwater and coolwater fishes inhabiting one temperature layer and coldwater fishes utilizing another.

In reservoirs, thermal stratification can also have substantial effects on the organisms present downstream from the dam. The area immediately below the dam is called the **tailrace** (fig. 12.16). It can support different types of organisms depending on whether water is released from the warmer or the cooler layer of the reservoir. For example, trout fisheries have been developed in warm climates in tailrace areas where cold, well-oxygenated hypolimnetic water is released through a dam. Tailrace areas in colder climates may provide ice-free waters below dams. Such areas can be attractive wintering places for animals, such as Canada geese, that would normally be forced to migrate because of ice cover. Tailrace water releases are not always beneficial, however. For example, if a stream or river has a native biota of warmwater fish species and other organisms dependent on warm water, and the water released

FIGURE 12.16 Tailrace areas can be ecologically very different than the lotic systems originally present.

from a reservoir is cold, the native species will be reduced.

The construction of ponds, small impoundments, and reservoirs can provide habitat types not normally present in an area. The large reservoirs on the Missouri River in North Dakota, South Dakota, and Montana have provided a coldwater habitat not natural to that area. Trout and salmon, which could not otherwise live in the area, have been introduced into these new habitats. The construction of ponds and small impoundments in many areas of North America has likewise provided new habitat for some species in areas where such habitat was not previously available. Numerous native and introduced fish species have been placed in these artificial habitats. Ponds and small impoundments have also been particularly beneficial to wildlife such as waterfowl and shorebirds.

There are some drawbacks to these human-induced changes, however. One of these is their negative effects on native communities. For example, the new habitats made available by the construction of reservoirs on the Missouri River were produced at the expense of existing habitats. What was once a free-flowing lotic system is now a series of essentially lentic systems. Many of the fishes adapted to the original lotic habitat, such as paddlefish, sicklefin chubs, blue suckers, and silver chubs, have been diminished. At the locations of the reservoirs, extensive areas of riparian habitat that were important to terrestrial species are also greatly modified or destroyed. When a tree-lined prairie river is impounded, for example, it loses the trees in its riparian zone. This belt of trees may have served as a corridor for movement for many species dependent on trees as well as providing habitat for resident species. In addition, not only the immediate area of a reservoir is affected. The modification of water flow changes downstream habitat in various ways for both terrestrial and aquatic organisms. For example, the rates at which islands and sandbars are formed or eliminated are changed, the scouring effects of floods are changed, the seasonal fluctuations in water flows are modified, the continual sediment enrichment of flood plains is ceased, and downstream habitats are affected by the modified water temperatures of reservoir outflows. Thus, ecological effects go well beyond the impoundment and can affect many aquatic and terrestrial organisms beyond its immediate locale.

Artificial aquatic systems are generally much more difficult to manage on a long-term basis than natural systems due to their rapidly changing nature. Natural aquatic systems tend to be stable systems; artificial systems generally progress rapidly through various successional stages and tend to be unstable.

12.4 Flowing Water Habitats

Lotic habitats range from very small streams, ones that a person could step across, to rivers the size of the Mississippi. The dominant feature of this type of habitat is a fairly consistent, unidirectional water flow. Because of this water flow, lotic systems are dynamic and are constantly changing.

A stream order classification is commonly used to describe lotic waters (fig. 12.17). In this system, small unbranched tributaries are designated as order 1 streams. To be an order 1 stream, a watercourse must be large enough to appear on a 7.5-minute quadrangle map. When two order 1 streams join, an order 2 stream is formed. This increasing order continues through large rivers such as the Mississippi, which may reach a stream order number as high as 7 or 8. The utility of this classification system is based on the assumption that stream order number is directly proportional not only to stream size but also to such characteristics as water discharge rate and watershed area.

Because they exhibit differences in depth, length, width, area, flow volume, shoreline characteristics, stream gradient (rate of descent to sea level), bottom type, and water temperature, lotic systems are quite variable. Lotic systems can experience large seasonal fluctuations in dissolved oxygen content as well as water temperature; however, compared with lentic water bodies, water temperature and dissolved oxygen concentrations in lotic systems are much more uniform through-

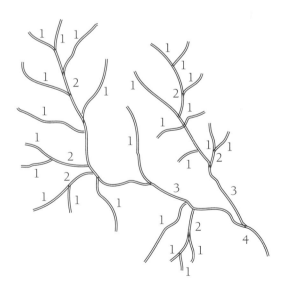

FIGURE 12.17 Stream order designations provide information on much more than stream size alone.

out the system at any specific point in time. Lotic systems are also usually shallower than lentic systems. Because water is constantly moving through them, their bottom structure is less stable than that of lentic systems. In addition, nutrients in lotic systems are constantly moving downstream and must be replaced; nutrients entering a lentic system tend to stay there. The productivity of lotic systems is more closely tied to the productivity levels of the surrounding terrestrial area than is that of lentic systems. This tie between lotic and terrestrial systems is especially evident in arid regions, where not only the visible water in a stream but also the nonvisible underground portion of the stream contributes to the terrestrial system. In such situations this nonvisible water may be the basis for lush vegetation along the stream margin, which provides benefits to many terrestrial and aquatic animals.

There are two broad habitat types commonly found in most headwater lotic systems: **riffles** and **pools** (fig. 12.18). Riffles are generally shallow and narrow, and their bottoms are usually characterized by gravel and large rocks. Because of their relative shallowness, current speed in riffles is greater than that in pools; this swift current sweeps away most material of smaller particle size. The swifter the current, the larger the rocks in a riffle. Pools tend to be deeper, wider, more soft-bottomed, and have less current flow than riffles. In some situations pool areas can closely resemble lentic habitat. Some biologists classify the areas between the fastest portion of a riffle and a pool as **raceways**, **runs**, or **chutes**. These areas are intermediate in depth and current compared with riffles and pools, and often have their own unique assemblages of plants and animals.

Because of the differences mentioned, the fishes and other organisms in a lotic system can vary greatly from the pool to the riffle areas. For example, fishes such as orangethroat darters, mountain madtoms, and desert suckers that need gravel or rocks for nest sites or other habitat needs are more often associated with riffle areas. Fishes that are less well adapted to current, such as green sunfish, flat bullheads, and brassy minnows, are more likely to be found in pool areas. Some fish species can be found in the pool areas during one stage of life and in the riffle areas during another life stage. It is common to see quite different animal and plant communities in the two habitat types.

Lotic systems occur across a continuum ranging from headwater mountain streams that resemble constant riffle habitat to sluggish, nearly stationary downstream rivers that are almost totally pool habitat. In addition, pool-riffle relationships are not always linear. A stream or river can have a pool area and a riffle area side by side. Backwater areas, areas connected to the lotic system but without unidirectional current, may also be found in association with rivers and streams (see fig. 12.18). These areas have essentially no water flow, and the fishes and other biotic assemblages found in them often include species more commonly associated with lentic systems.

Although there is usually a constant unidirectional water flow in lotic systems, this flow is not uniform. As mentioned, the shallower and narrower riffles have faster flow rates than the deeper, wider pools. Also, the closer to the bottom or sides of a lotic system, the slower the current; this pattern is a result of frictional effects. Few organisms are adapted to life in swift current areas. Most fishes and other animals found in lotic systems associate with the bottom or with structures that reduce current flow. Most venture into the swift portion of the current only on occasional forays.

In addition, most lotic systems do not travel in a straight line; they meander in a series of curves. Wherever a stream or river curves or changes direction, one can find different current speeds and therefore different habitats. The outer segment of a curve (the **erosional zone**) tends to be deeper than the inner segment (the **depositional zone**) because the swifter current in the outer segment is able to move larger bottom particles. This is also the reason why such outer segments tend to have firmer bottoms. The inner

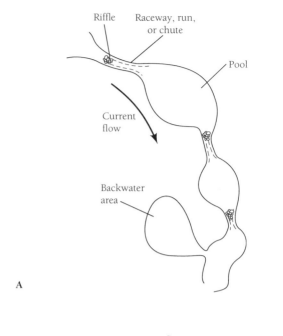

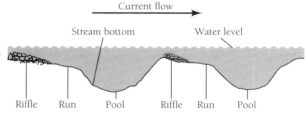

FIGURE 12.18 Lotic systems contain two broad habitat types: riffles and pools. Backwater areas may also be found in association with lotic systems. (*A*) Top view; (*B*) side view.

portion of the curve generally is shallower, softer-bottomed, and has a slower current speed.

One other characteristic of lotic systems that should be mentioned is the effect flowing water can have on the feeding behavior of organisms. Because lotic systems flow, food items can be moved by water current. Some organisms, such as brook trout, have feeding strategies based on allowing the current to move food to them. Such a strategy is beneficial to an organism because it does not have to expend as much energy obtaining food.

12.5 Estuarine Habitats

An **estuary** is the place where a river enters the ocean. Estuarine habitats are often associated with coastal wetlands. Estuaries are areas of transition between freshwater and marine environments. Although marine habitat is not discussed in this text, estuaries, coastal wetlands, and beach-ocean interfaces are; their roles as transitional zones between freshwater and marine systems and between aquatic and terrestrial habitats make their inclusion important.

An estuary can be a rather small area where a single stream enters the ocean, or it can be a sizable area, such as Chesapeake Bay, where a number of freshwater sources enter the ocean. Estuaries can experience extreme fluctuations in salinity. In addition, estuaries can be affected by tides. The density difference between salt and fresh water can result in the formation of a **salt wedge**, which provides a structure similar in some ways to the thermal stratification found in lakes and impoundments. The salt wedge, because of its higher salinity, is heavier than the overlying fresh water and can extend upstream beneath the fresh water. This wedge can advance and retreat with tidal movement. At the interface where the fresh and salt water masses meet, small particles and other materials are formed into larger aggregates. This nutrient-rich material contributes to the high productivity of estuaries. Also contributing to this productivity is the constant movement and flushing of water resulting from river currents and tides.

The high nutrient input into estuaries generally benefits these areas and has an important effect on estuarine production. The word "generally" is used because excessive nutrient input into any water body is not beneficial (see section 14.2). Estuaries are very rich habitats, and many marine and terrestrial vertebrate and invertebrate species use them as adult habitat and as nursery areas for their young. Because of their ecological importance, the current decline in the quality and quantity of estuarine habitats is of special concern to natural resource advocates. Pollution and human building activities continue to reduce the amount of estuarine area and its utility to native organisms.

Coastal wetlands, while not always associated with estuarine areas, are also very important transitional zones. They are high in primary productivity, provide habitat for many year-round resident and migratory terrestrial and aquatic species, and have a natural ability to lessen flood and storm damage, minimize coastal erosion, and improve water quality. They, along with estuaries, serve as spawning grounds and nurseries for over 75% of the commercially important marine fish and shellfish species.

12.6 Reproductive Habitats

The previous portions of this chapter have addressed general aquatic and terrestrial habitat types. Located within each of these general habitats are many more specific habitats that organisms use. While the previous categories, such as deserts and rivers, are general in nature, organisms require more specific or functional habitat units. These specific habitat requirements can relate to organismal needs associated with reproduction, feeding, shelter, concealment, rest, or to other habitat characteristics that a particular species may require. In addition, habitat needs for a particular species can vary depending upon the age or sex of the organism, the time of year or time of day, and the proximity of one type of habitat to another.

The availability of appropriate reproductive habitat is crucial for most species. Optimal reproductive habitat can provide a variety of benefits to a species. It may shield the eggs or young from predators, protect them from adverse environmental elements such as silt, sunlight, or rain, or protect them from damage by the parents.

In aquatic systems, reproductive habitat needs usually revolve around a number of habitat characteristics. For example, a suitable spawning substrate is often vital for fish reproduction. Some species, such as the northern pike and the brook stickleback, need vegetative material as spawning substrate. Central stonerollers and pink salmon require gravel substrates. Some fishes, such as the johnny darter, spawn on the undersides of rocks. Others, such as the freshwater drum, require no substrate at all and produce eggs that are free-floating in the water. Water depth and current also influence the suitability of reproductive habitat. Some fishes, such as the orangespotted sunfish, prefer lentic water, often less than 1 meter in depth. Others, such as the lake trout, usually require lentic water and can spawn at water depths of up to 36 meters. The mountain sucker requires swift-moving shallow water, while the sauger can spawn in sluggish rivers in water over 4 meters in depth.

To understand just how specific habitat requirements for fish reproduction may be, consider the example of the brown trout. A few of the specific habitat requirements for optimal reproduction in this species include water temperatures of 6.5°–13.0°C during embryo development, a minimum of 10 milligrams per liter of dissolved oxygen, 40–70 centimeters per second average current velocity over the spawning area, and no more than 10% of substrate particles less than 3 millimeters in diameter (Raleigh et al. 1986).

Birds also have specific needs for reproductive habitat. The western meadowlark is a grassland ground nester, sage thrashers nest in low-lying shrubs 1.0–1.5 meters tall, northern orioles nest in treetops, and yellow-headed blackbirds nest on emergent vegetation in wetlands. Even within a grouping such as ground-nesting birds, there is great variation: piping plovers use almost barren sandbars, common nighthawks utilize sparse ground vegetation or gravel, killdeers nest in sparse ground vegetation, and grasshopper sparrows place their nests in dense ground vegetation.

Some species, such as redheads and canvasbacks, construct nests of emergent vegetation that float on water. Many birds, such as the gray partridge, need solitary locations for nesting, while others nest only in colonies. Some birds, such as double-crested cormorants and great blue herons form nesting colonies called **rookeries** and need such aggregations to be successful (see fig. 6.12).

An example of specific nesting habitat needs is provided by those of the common snipe in North America. The optimal breeding habitat of this species is restricted to organic soils, primarily the early successional stages of bogs. The shallow fringe portions of these specific bog areas are considered to be the best-quality breeding habitat (Fogarty et al. 1977).

Because most mammals bear live young, they need no specific sites for laying eggs, but they still have a need for sites in which to bear young and keep them safe after birth. Southern flying squirrels need tree cavities, muskrats need dens, Gunnison's prairie dogs need their burrow systems, and mule deer seek specific types of vegetational cover for fawning. Consider the dens where kit foxes bear and raise their young. These dens usually have more entrances than the dens the kit foxes use for shelter during the nonbreeding season. Around the nursery dens, the kit fox family also tramples dried vegetation around the den opening; this forms a disturbed apron for up to 10 meters around the den. In addition, numerous prey remains are scattered about the den opening (O'Farrell 1987).

Organisms may have other reproductive habitat needs in addition to locations in which to nest or place young. Greater and lesser prairie-chickens gather on communal areas called **booming grounds** to perform their courtship displays. Sharp-tailed grouse perform such reproductive behavior on **dancing grounds** (fig. 12.19A), while sage grouse perform on **strutting grounds** (fig. 12.19B). Such places where these species gather to perform mating displays are sometimes collectively referred to as **leks**. Ruffed grouse need drumming logs, and western grebes need sufficient open water to perform their breeding chase.

A

B

FIGURE 12.19 Grouse gather to perform courtship displays. *(A)* Sharp-tailed grouse on dancing grounds; *(B)* sage grouse on strutting grounds. *(Part A photograph courtesy of South Dakota Tourism; Part B photograph courtesy of L. Parker, Wyoming Game and Fish.)*

A fish such as the bluntnose minnow also needs sufficient space of the right kind to perform its reproductive behavior, as does a mammal such as the mountain goat.

▼

12.7 Feeding Habitats

Animals may also require specific types of habitat in which particular types of food are produced. If a mammal needs specific grasses or other plants for food, habitat to support such plant life must be present. If a fish needs aquatic invertebrates that are associated with rocky substrates, rocks must be present in its feeding area. If a bird needs specific seeds or invertebrates as a food source at a particular time, habitat that can provide such food must be present. Feeding habitat requirements may also entail vegetation or other structural elements that allow a predator to get close to or to see prey. Some raptors tend to sit atop posts or dead trees because these habitat features allow them to survey their feeding area more efficiently. A chain pickerel tends to remain motionless near a vegetated area awaiting prey species that venture away from their vegetative hiding places. In some cases feeding habitat may be used year-round, while in others such habitat may be used for only part of the year. For example, migrating birds may use a particular habitat for only a few days or hours of the year.

Requirements for feeding habitat can be quite specific. For adult shortnose sturgeon, optimal feeding habitat during summer must have, among other characteristics, a water temperature between 11°C and 22°C, current velocities between 15 and 45 centimeters per second, and bottom substrates ranging from vegetation to sand (Crance 1986). Habitat that falls outside of those ranges represents suboptimal habitat.

Optimal feeding habitat for polar bears must contain seals along with ice types that allow the bears to hunt the seals successfully. These ice types include areas where ice is continually forming and refreezing (providing openings in the ice), edge habitat that includes pressure ridges (providing places to hide while stalking prey), and stable flat ice habitat suitable for the seals (McLaren 1958).

Feeding habitat requirements for birds can also be very specific. For example, Brown and Fredrickson (1986) reported that white-winged scoters feed primarily on scuds (a type of small aquatic crustacean). In their study in Saskatchewan, Brown and Fredrickson found that one species of scud provided 97% of the diet of female white-winged scoters during nesting and brood rearing. Obviously, these quantities of scuds would be found only in specific types of habitat.

12.8 Cover Habitats

Habitat that is used by organisms for shelter, concealment, or escape purposes is often referred to as **cover**. Organisms use cover in numerous ways. Endotherms, such as birds and mammals, can use cover as shelter to prevent excess gain or loss of body heat (fig. 12.20). Being able to avoid wind, rain, snow, hail, and other adverse elements may be essential for survival. Fishes also use cover as shelter for energy conservation purposes. A mountain whitefish remaining behind a rock or other stream obstruction usually does so to avoid stronger water current.

Cover is used for concealment and escape by many organisms, both aquatic and terrestrial. A small green sunfish hiding in aquatic vegetation is using that habitat to conceal itself. If it is outside of this vegetation and becomes alarmed, it will seek such a place in which to escape. A white-tailed deer will use a thick stand of timber to conceal itself or escape to when disturbed. A ring-necked pheasant will conceal itself in heavy cover or escape to such cover when threatened.

Requirements for escape cover may be quite specific and vary by species. Pulliam and Mills (1977) described escape behavior and its relation to habitat for several bird species in Arizona. In the area they studied, a variety of birds inhabit open grassland containing scattered mesquite trees. Species such as white-crowned sparrows, chipping sparrows, rufous-crowned sparrows, and brown towhees are highly dependent on tree or shrub cover for escape; they tend to stay close to the mesquite trees. Vesper sparrows often use mesquite for escape cover, but may also feed far from mesquite trees and use ground cover (grass) for camouflage or escape. Grasshopper sparrows escape to dense ground cover when necessary and use their cryptic coloration and solitary behavior to escape predators; they seldom use mesquite trees for escape cover. Chestnut-collared longspurs escape to more open ground cover and make strong use of camouflage and aerial group flights to avoid predators.

FIGURE 12.20   Escape cover is especially important to wild animals. The dense cover shown in this photo could be used by a variety of species. (*Photograph courtesy of J. Jenks.*)

Fishes, too, can have particular cover requirements. For example, the black bullhead thrives in lotic waters where the percentage of pool and backwater areas during average summer flows is between 50% and 75% of the habitat present. This species also thrives where cover (vegetation and debris) within pools, backwaters, or littoral areas during summer represents 25% of the habitat, and where average current velocity at 0.6 meters of water depth during average summer flow is approximately 4 centimeters per second. Additionally, black bullheads do best where the littoral area represents 25% of a water body during summer (Stuber 1982).

Cover requirements for mammals can also be quite specific. For example, river otters use partly submerged trees and logjams as shelter. Cavities among tree roots, shrubs, and grass also provide escape cover (Melquist and Dronkert 1987).

The particular attributes needed in concealment and escape cover can vary depending on what is being avoided. If airborne predators are a threat to a bird or mammal, some type of overhead cover is sought. In the case of fishes, all that may be needed are small crevices in rocky areas, or thick vegetation that restricts predator maneuverability.

An additional example of cover habitat specialization is demonstrated by white-tailed and mule deer. White-tailed deer have a running locomotion mode that allows them to travel rapidly along established trails in dense cover; this strategy prevents pursuing predators from taking "shortcuts" in pursuit of the deer. Mule deer have a running locomotion mode that allows them, while moving in trailless open country, to change direction rapidly while being pursued. Thus, each species is specialized for a particular cover situation.

Cover needs can vary by season. Fishes often move to deeper water areas during cold periods of the year and shallow water areas during the summer. In winter the deeper water areas often provide, among other things, warmer water. As mentioned earlier, water is its densest at 4°C; therefore, a water body near freezing will have its warmest water near the bottom. Elk and deer use different habitats in winter than in summer. They often migrate to higher-altitude areas in the summer and to lower elevations during winter. In the western mountainous regions of North America they often select a winter range that is kept snow-free by the high-velocity warm winds characteristic of such areas. Daily movements among different cover habitats also occur. Any person who has angled for fishes readily learns that their habitat use can vary by time of day. For example, the knowledgeable angler trying to catch large brown trout is well aware that such fish will tend to seek cover and deeper areas during daylight and move to open, shallower areas to feed near dusk or at night.

12.9 Other Habitat Requirements

In the wildlife field, organisms are often described as using loafing or resting cover. These areas are used in a variety of activities. For example, sharp-tailed grouse use loafing cover to sun or dust themselves, while a Brandt's cormorant may rest on a rock while drying its feathers. Specific habitat requirements for other activities, such as defeca-

tion, watering, and mineral seeking, are common among wild animals. It should be noted that in many areas, sites that are particularly attractive for homesites and other human activities are the same ones that contain critical habitat components for wild animals. Deer winter ranges and riparian zones are examples of habitat types that are also popular areas for human activities. This conflict between the needs of people and animals is a problem of human management.

Biologists expend a great deal of their efforts on habitat management. Chapter 15 describes some of the procedures and techniques that they use. If one were to measure the amount of time biologists expend on various conservation activities, the majority of that time would be associated with habitat management. This is especially true of terrestrial biologists. The current trend toward ecosystem and landscape management reinforces the precept that unless all habitat needs are met, populations of aquatic and terrestrial organisms cannot reach or approach their maximal levels. In the past, wildlife and fishery biologists have received criticism for managing for single species, usually game or commercially important organisms. To some extent this criticism is justified. However, a waterfowl biologist that manages or saves a wetland for waterfowl is in actuality also providing habitat for a large variety of other organisms. Even though such an area may be primarily managed for one species, many other species receive the benefit of that habitat. Conversely, other species may be diminished by a management activity directed at a single target species or group of species. In the future, habitat management strategies will take into account to a greater extent their effects on a variety of organisms.

SUMMARY

A broad form of terrestrial habitat classification is by biome or major ecosystem. Biomes are repeated on a global basis; a biome found in North America may also be found in Europe, Asia, or other regions of the world. Important North American biomes include tundra, northern coniferous forest, temperate forest, temperate grassland, desert, chaparral, Pacific coast coniferous forest, and tropical seasonal forest. Biomes are not monolithic habitat blocks, but contain various habitat types. The edges of biomes constitute ecotone areas.

Riparian zones, wet margins of lakes and ponds, and a variety of marshes, bogs, swamps, coastal wetlands, and beach-ocean interfaces represent transitional areas between terrestrial and aquatic habitats. These transitional areas occur across a continuum from almost totally terrestrial to almost totally aquatic.

Natural lakes, small impoundments, ponds, and reservoirs represent lentic aquatic environments. Rivers and streams represent lotic aquatic environments. Eutrophic aquatic systems are highly productive, while oligotrophic systems are relatively unproductive. The shallow water area of a lentic system constitutes the littoral zone; the limnetic and profundal zones are deeper water areas. Lentic water bodies can also be characterized by differences in water temperatures and oxygen content. Lotic systems are generally divided into riffle and pool habitats.

Estuaries, coastal wetlands, and beach–ocean interfaces represent important transitional zones. These highly productive areas contain a large number of organisms that are of great commercial and ecological importance.

While most habitat categories are general in nature, organisms also require more specific or functional habitat types. These specific habitat requirements can relate to such organismal needs as reproduction, food, shelter, concealment, and rest. Specific habitat needs for a particular species can vary depending on the age or sex of the organism, the time of year or time of day, or even the proximity of one type of habitat to another.

A large proportion of the efforts of a biologist involves habitat management. The trend in habitat management today is toward more holistic management rather than single-species management.

PRACTICE QUESTIONS

1. In which biome do you live? What are some of the characteristic plants and animals of that biome? What specific habitats do each use?

2. Why is Arctic tundra such a fragile biome?

3. For each North American biome, describe as many human activities as possible that have modified that biome over the last hundred years.

4. What are the primary differences among marshes, bogs, and swamps?

5. In what way does the issue of wetland loss affect the region where you live? Is it even an issue? Among whom? For what reasons?

6. Describe thermal stratification in lentic systems and describe what leads to fall overturn in these systems.

7. Describe a dozen different ways in which water current affects the lives of aquatic organisms in lotic systems.

8. Based on your past observations, choose six different bird species that occur in the area where you live, and describe the differences in where they nest and the types of nests they construct.

9. How can a reservoir affect aquatic and terrestrial organisms downstream from the reservoir?

10. What are estuaries, and why are they valuable to humans?

SELECTED READINGS

Bennett, G. W. 1970. *Management of lakes and ponds.* 2d ed. Van Nostrand Reinhold, New York.

Forman, R. T. T., and M. Godron. 1986. *Landscape ecology.* John Wiley & Sons, New York.

Gill, F. B. 1995. *Ornithology.* 2d ed. W. H. Freeman, New York.

Horne, A. J., and C. R. Goldman. 1994. *Limnology.* 2d ed. McGraw-Hill, New York.

Morrison, M. L., B. G. Marcot, and R. W. Mannan. 1992. *Wildlife-habitat relationships: Concepts and applications.* University of Wisconsin Press, Madison.

Smith, R. L. 1990. *Ecology and field biology.* 4th ed. HarperCollins, New York.

Vaughan, T. A. 1986. *Mammalogy.* 3d ed. Saunders College Publishing, Fort Worth, Tex.

LITERATURE CITED

Brown, P. W., and L. H. Fredrickson. 1986. Food habits of breeding white-winged scoters. *Canadian Journal of Zoology* 64:1652–1654.

Cowardin, L. M., V. Carter, F. C. Golet, and E. T. LaRoe. 1979. *Classification of wetlands and deepwater habitats of the United States.* FWS/OBS-79/31. U.S. Fish and Wildlife Service, Washington, D.C.

Crance, J. H. 1986. *Habitat suitability index models and instream flow suitability curves: Shortnose sturgeon.* Biological Report 82(10.129). U.S. Fish and Wildlife Service, Washington, D.C.

Fogarty, M. J., K. A. Arnold, L. McKibben, L. B. Pospichal, and R. J. Tully. 1977. Common snipe. Pages 188–209 *in* G. C. Sanderson, ed. *Management of migratory shore and upland game birds in North America.* The International Association of Fish and Wildlife Agencies, Washington, D.C.

McLaren, I. A. 1958. *The biology of the ringed seal (Phoca hispida Schreber) in the Eastern Canadian Arctic.* Bulletin 118. Fisheries Research Board of Canada.

Melquist, W. E., and A. E. Dronkert. 1987. River otter. Pages 626–641 *in* M. Novak, J. A. Baker, M. E. Obbard, and B. Malloch, eds. *Wild furbearer management and conservation in North America.* Ministry of Natural Resources, Ontario, Canada.

Needham, P. R., and A. C. Jones. 1959. Flow, temperature, solar radiation and ice in relation to activities of fishes in Sagehen Creek, California. *Ecology* 40:465–474.

O'Farrell, T. P. 1987. Kit fox. Pages 422–431 *in* M. Novak, J. A. Baker, M. E. Obbard, and B. Malloch, eds. *Wild furbearer management and conservation in North America.* Ministry of Natural Resources, Ontario, Canada.

Ploskey, G. R., L. R. Aggus, and J. M. Nestler. 1984. *Effects of water levels and hydrology on fisheries in hydropower storage, hydropower mainstream and flood control reservoirs.* Technical Report E-84-8. U.S. Army Corps of Engineers, Environmental and Water Quality Operational Studies, Washington, D.C.

Pulliam, H. R., and G. S. Mills. 1977. The use of space by wintering sparrows. *Ecology* 58:1393–1399.

Raleigh, R. F., L. D. Zuckerman, and P. C. Nelson. 1986. *Habitat suitability index models and instream flow suitability curves: Brown trout, revised.* Biological Report 82(10.124). U.S. Fish and Wildlife Service, Washington, D.C.

Stuber, R. J. 1982. *Habitat suitability index models: Black bullhead.* FWS/OBS-82/10.14. U.S. Fish and Wildlife Service, Washington, D.C.

13

HABITAT SAMPLING AND ASSESSMENT

Biologists must be able to assess the biota and habitat components of a fishery or a wildlife system. Chapter 9 described a variety of methods used in assessing populations of target or related organisms; this chapter builds on those previously described assessment skills, focusing on the assessment of habitat. Biota assessment is seldom done without some form of companion assessment of habitat.

Consider a situation in which population or community assessment has indicated that an area has a particular density or number of a certain target species or group of species. A biologist may want to increase or decrease those numbers, or to determine why they are increasing or decreasing. Habitat assessment is likely to be needed to answer a variety of questions, such as, "Is a particular habitat limiting?" "What characterizes that habitat?" "How much of that habitat is present?" Even after habitat assessment has led to the development of a conservation strategy, the habitat may need to be reassessed to determine the success or failure of any implemented management practices.

Wildlife and fishery biologists often identify habitat as the most important overall ingredient in managing terrestrial and aquatic animals. Because of the importance of habitat, biologists should be familiar with the methods used in its measurement and assessment. In practice, a wide variety of techniques are available for evaluating habitat. The purposes of this chapter are to introduce some of the more common procedures used for sampling and assessing habitat and to describe some of the uses of such information.

▼

13.1 Aquatic Habitats: Water Characteristics

A variety of parameters, such as water temperature, oxygen content, current velocity, pH, and nitrate levels, may need to be measured when assessing an aquatic habitat. Various methods and devices are available for measuring these water characteristics. The devices used range from precision electronic instruments to rather simple field kits or tools used for obtaining approximate measurements.

Water temperature

Water temperature has widespread influences on aquatic ecosystems. For example, water temperature is important in determining spawning chronology in fish communities. Fishery biologists may monitor water temperatures near the spawning season of particular species to predict optimal times for collection of eggs and sperm for aquaculture purposes. Water temperature also influences metabolic activity, and thus growth rates, in most aquatic organisms, as well as the dissolved oxygen-holding capacity of water. Even rates of development and lengths of hatching periods for fish and amphibian eggs are influenced by water temperature. Many other examples of the effects of water temperature could be provided.

Water temperatures often vary with depth. Temperature variation from the surface to the bottom of a water body can be determined by removing water samples from various depths and measuring the temperature of each sample with a hand-held thermometer or electronic temperature measuring device. Devices such as **Kemmerer** or **van Dorn water samplers** (fig. 13.1) can be remotely triggered to obtain water from a particular depth. Water temperatures can also be obtained with electronic probes that can be lowered to various depths. Obtaining temperatures at different depths is especially important in water bodies that thermally stratify. In lotic or shallow lentic water bodies where temperature stratification is lacking or poorly developed, near-surface water temperature information is usually adequate. However, in shallow water habitats, temperature fluctuation over a twenty-four-hour period is usually greater than in deeper water bodies.

A

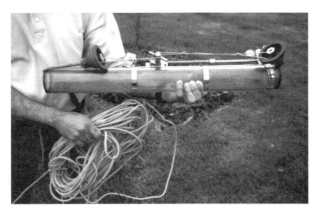

B

FIGURE 13.1 (A) The Kemmerer water sampler and (B) the van Dorn water sampler are used for sampling water at various depths.

Dissolved oxygen levels

Knowing the concentration of dissolved oxygen in a body of water is important because a lack of oxygen may result in stress or death for aquatic organisms. Even if low dissolved oxygen levels do not cause death, stress from **hypoxic** (low oxygen) conditions can reduce growth rates and increase the susceptibility of animals to diseases, pathogens, and parasites. Oxygen concentrations of approximately 3–5 milligrams per liter are commonly considered adequate for warmwater fishes, while levels of 6–9 milligrams per liter or above are usually adequate for coldwater fishes. Low oxygen levels do not necessarily cause fish kills. Meyer and Herman (1990) reported that dissolved oxygen levels of less than 2 milligrams per liter, but usually less than 1 milligram per liter, are indicative of potential fish kill conditions. Some fishes, however, such as northern pike and fathead minnows, often survive the winter in shallow, ice-covered northern water bodies where dissolved oxygen levels drop below 1 milligram per liter. If oxygen levels under ice indicate that a water body is likely to become oxygen deficient (see section 15.3), biologists have a number of management options. Liberalized fishing, that is, allowing anglers to harvest more fish, is one such strategy that may be implemented.

Dissolved oxygen can be measured in water samples either by titration or electronically. **Titration** is the measurement of the quantity of a given constituent of a solution—dissolved oxygen in this instance—through the addition of reagents of known concentration until a particular chemical reaction, often indicated by a color change, is completed. Measurements of dissolved oxygen can also be obtained by lowering a dissolved oxygen probe into the water; the probe is connected to an electronic meter.

Light penetration

The amount and kind of light that penetrates to various depths helps to determine the primary

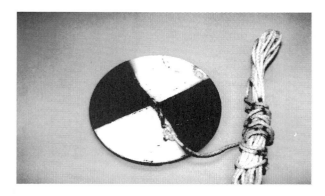

FIGURE 13.2 Water transparency is commonly approximated with a Secchi disk.

productivity of an aquatic system. Light penetration in aquatic systems varies with water depth, color, and the amounts and kinds of materials suspended in the water. Water transparency is often estimated with a **Secchi disk** (fig. 13.2). The Secchi disk is a metal, plastic, or wooden plate approximately 20 centimeters in diameter, which usually has white and black quarters. A line attached to the disk is marked in known length increments. The disk is lowered into the water until it disappears from view. It is then lowered another meter or so, then raised until it reappears. The average of the depths at which the disk disappears and reappears is recorded as the Secchi disk visibility. Secchi disk visibility is normally measured from the shaded side of a boat and in areas where aquatic vegetation will not hinder or obscure the disk. There are also a variety of photoelectric devices that measure light intensity when lowered to various depths.

Light penetration or water transparency information has many uses. For example, if turbidity is limiting primary production, management practices to reduce turbidity can be implemented (see section 15.3).

Current velocity

Current velocity is an important variable in lotic systems. It can be measured with current meters such as the **pygmy water meter** (fig. 13.3). Water current varies with depth and position in the stream channel, so it is usually necessary to measure current velocity at different depths and at different locations in relation to the stream bottom and shoreline. A simple measure of surface current speed can be obtained by timing the progress of a floating object along a known distance of stream. An object such as a lemon or orange that has a specific gravity similar to water is recommended. Rapid changes in current speed over a twenty-four-hour period are not unusual. Strip chart recorders that can be set and left unattended can obtain water current information over time.

Human construction activities can physically alter stream channel characteristics, current velocities, and current patterns. Changes in current and other flow characteristics affect aquatic organisms as well as terrestrial species associated with water bodies. Measures to protect streams may be necessary near construction sites. For example, the development of a highway along a river or stream may require stream channel rerouting (fig. 13.4). Channel rerouting often has substantial effects on stream currents, riffle and pool features, and other stream characteristics that influence aquatic organisms. To mitigate the effects of construction on lotic systems, the rerouted portions should be designed to provide riffles and pools of similar value to those destroyed. Fishery biologists should obtain at least a subjective classification of flow patterns and the relative proportions of habitat types, such as pools and riffles, prior to stream

FIGURE 13.3 The pygmy water meter is used to measure current velocity in lotic systems. This photograph shows only the portion of the meter that is submerged in the water.

FIGURE 13.4 Highway development can influence the channel characteristics and thus the fishery and wildlife values of lotic and associated riparian habitats. (*Photograph courtesy of Brigham Young University.*)

modification; this information can be used in assessment of postconstruction fish habitat and in mitigation decisions. Totally aquatic species are not the only organisms influenced by stream modification. Other organisms, such as river otters, American dippers, and riparian plants, are also important aspects of lotic ecosystems that are influenced by construction activities.

Other water characteristics

The hydrogen ion concentration in a solution is measured as **pH.** (see section 14.1). **Acidity** is the capacity to donate hydrogen ions. Acidic waters have a pH of less than 7. Acidic water bodies can occur naturally, such as in some areas of the Pacific Northwest and in northeastern North America, but water bodies can also become more acidic due to industrial activities, as discussed in section 14.1. Increased acidity in water bodies can limit or eliminate various invertebrate and vertebrate biota. Electronic meters, titration methods, and even pH-sensitive paper (turns a specific color depending on pH) can be used to measure pH.

Alkalinity (**basicity**) is a measure of carbonates and bicarbonates in water and often reflects the carbonate content of surrounding soils. Alkalinity is usually expressed in terms of calcium carbonate equivalents. Waters with higher levels of calcium and magnesium carbonates, within limits, are more biologically productive. Alkalinity is usually measured using titration techniques.

Hardness is best defined as the total concentration of calcium and magnesium ions in water. These two elements are generally the most abundant ions in fresh water. Hardness and alkalinity are closely related but are measured in different ways. Water bodies with minimal concentrations of calcium and magnesium are generally lower in productivity. Hardness can be measured by titration or with electronic meters.

Conductivity (**specific conductance**) is a measure of the ability of water to convey electrical current. In general, the greater the concentration of ions in water, the greater its conductivity. Conductivity can also be used as an indication of water hardness. Conductivity is usually measured with electronic meters.

Measurement of other water quality characteristics, such as nitrogen and phosphorous content, can also be important. Causes of high nitrogen and phosphorus levels in water bodies include runoff from fertilizers placed on croplands and lawns, natural waste from domesticated livestock, and incompletely treated human sewage. Domesticated livestock holding facilities near water bodies can be a substantial source of nitrogen. High nitrogen levels in the form of ammonium can be limiting to fishes and may even cause mortality in some fish populations. Both nitrogen and phosphorus, especially the latter, can contribute to high productivity and excessive aquatic vegetation in water bodies.

There are many other chemical characteristics of water that can be measured; chloride, chlorine, and carbon dioxide content are examples. The water characteristics measured depend on what information is needed in a particular situation.

Procedures for water quality analysis are quite specific because the interacting effects of turbidity, time, and temperature on water samples may influence various water quality characteristics. Specialized water quality analysis laboratories are the best source of accurate evaluation of water samples. These laboratories provide exact specifications for the collection, shipping, and care of water samples. The cost of accurate water analysis can be quite high and is often a major budgetary consideration.

13.2 Aquatic Habitats: Water Body Characteristics

Surface area

The surface area of a water body is an important descriptor in fishery and, to a lesser extent, wildlife work. In fishery evaluations, standing stock is measured in weight per unit of surface area, usually in kilograms per hectare. Numbers of fish stocked are usually determined in relation to surface area. Surface area, in conjunction with productivity, is related to potential production of fishes and thus the number of angler hours of fishing a water body can support. The use of water bodies by some wildlife species is also influenced by surface area. For example, molting ducks such as redheads and canvasbacks traditionally gather on large water bodies; these molt gatherings may number many thousands of birds. The water bodies on which these ducks molt must be shallow and produce adequate food, but just as important, must be large enough so that the birds can remain far from disturbances.

Maps of water bodies are available from a variety of sources. National Wetlands Inventory maps can be obtained through the U.S. Fish and Wildlife Service; these maps provide information on both the size (surface area) and classification of water bodies. Most natural resource agencies produce maps of specific water bodies that provide surface area information.

If appropriate maps giving surface area are not available, surface area can be determined from aerial photographs of known scale, such as those available in the United States through the Consolidated Farm Service Agency. If aerial photographs are not available, it may be necessary to make a map of the water body surface. Orth (1983) described a mapping procedure using an **alidade** that works well on small water bodies. The alidade is a straight-edged device that is used for sighting along a straight line; it is used on a plane table (fig. 13.5A). The alidade method involves the establishment of two survey points a known distance apart along the shoreline of the water body. The biologist then triangulates to stakes or known points around the water body (fig. 13.5B). Triangulations are sketched from each baseline point to mark shoreline points and to provide a scaled map of the water body. There are additional ways that outlines of water body surfaces can be obtained; the use of aerial photographs and the alidade method are just two examples.

Once the outline of a water body has been mapped, one of numerous methods by which surface area can be determined is by using a manual or electronic **polar planimeter** (fig. 13.6). When the outline of the water body is traced with this device, the planimeter reading indicates the area enclosed by the tracing. The scale size of the map or photograph is then used in conjunction with the planimeter reading to determine surface area.

Depth

Depth contours are another characteristic of water bodies that are often used for conservation purposes (fig. 13.7). Depth contours of water bodies can be determined by means of weighted cables or lines with depth markings, poles or boards with depth markings, or electronic echolocators such as fish finders. Depth contours can be mapped by taking depth measurements in a systematic fashion, usually along a series of predetermined lines (**transects**). Depth information can also be entered into a computer with a coordinate system and the contours mapped using computer software.

A

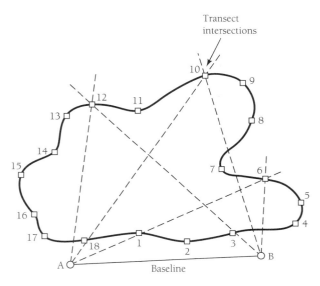

B

FIGURE 13.5 (*A*) The alidade, a straight-edge device with a sight, and the plane table are tools used for mapping surface areas. (*B*) Areas can be mapped by triangulating outlining features from two points on a baseline of known distance.

Contour maps depicting depth can be used to evaluate fish habitat or to select sites for fishing piers, scuba diving areas, boat docks, or fish habitat structures such as old Christmas trees or other artificial reefs. Water depth and contour information is also used extensively in management practices involving increases or decreases in water levels.

Water level fluctuations in a water body can affect populations of many aquatic and terrestrial species (see sections 15.3 and 15.4). While most water level fluctuations influence aquatic organisms, these are not the only biota affected. Water releases from some reservoirs, for example, can periodically flood sandbars, islands, and other low areas. In some areas, it has been necessary to reduce these fluctuations during spring and early summer to protect the nests and young of animals that use these areas (fig. 13.8).

Water depth and contour information also allow biologists to use water level increases or decreases as a management technique. Knowledge of depth contours can be used to answer many questions, such as, "How much water volume or surface area will be present in a water body if a certain amount of water is removed?" "What bottom areas will be exposed if a certain amount of water is removed?" "How much surface area or water volume will be added if a certain amount of water is added?" For example, walleye, white bass, and white crappie populations did not benefit from water level management plans in Kansas reservoirs unless at least 20% of the bottom area was exposed during planned water removals. Information on depth contours would indicate

FIGURE 13.6 Polar planimeters, such as this electronic version, can be used to estimate the area of an irregularly shaped surface. The scale size of the image being traced must be known.

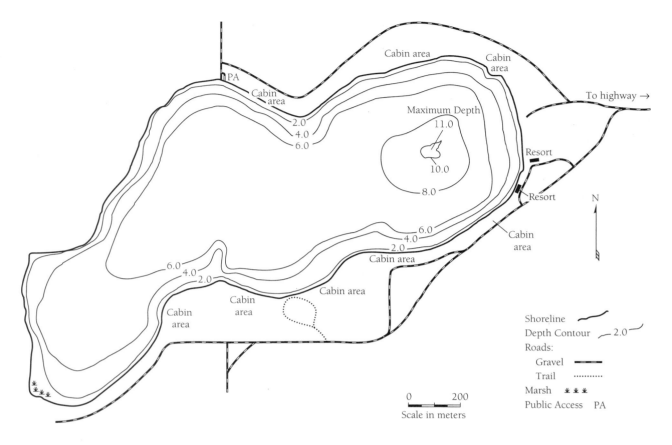

FIGURE 13.7 Depth contour maps of water bodies such as this lake are important in evaluating fish habitat. The contour lines in this example are at 2-meter depth intervals.

how much water would need to be removed to attain 20% exposure, and what areas would be exposed.

Volume

If the depth contours and area of a water body are known, water volume can be determined. In fishery work, water volume estimates are needed for chemical treatments. For example, biologists often need to determine the amount of fertilizer that should be added to increase primary productivity or the amount of piscicide needed to eliminate a fish species or community. Water volume informa-

tion can also be useful in wildlife work. Grizzly bears in Yellowstone National Park, for example, obtain fish primarily from small streams used by the fish for spawning purposes. The water volume characteristics of these streams can be used to identify important feeding sites for the bears.

Shape

Another characteristic that is often measured in lentic water bodies is **shoreline development**. Shoreline development is the ratio of shoreline length to the circumference of a circle with the same area as the water body. Thus, a round water

A

B

FIGURE 13.8 Piping plovers *(A)* often nest *(B)* on sites such as this sandbar along the upper Missouri River that are vulnerable to the fluctuating water levels common below many dams. *(Photograph courtesy of the National Biological Service.)*

body would have a low shoreline development value, while a water body with numerous bays and arms would have a higher shoreline development value. Shoreline development is basically a measure of shoreline irregularity and has applications in both fishery and wildlife work. The usefulness of ponds and other wetlands as breeding habitat for waterfowl, for example, can be evaluated in relation to both water body size and shoreline development. Water bodies with greater

shoreline development also have more areas that are protected from wind and waves and therefore provide more habitat for some fish species.

13.3 Aquatic Habitats: Plants and Other Substrates

Emergent and submergent aquatic plants can be identified and measured in a variety of ways. The percentage of the area of a water body that has submergent aquatic plants can be measured using echolocation. In clear, shallow water bodies, aerial photographs can reveal the extent of coverage of submergent vegetation; emergent vegetation can be photographed even if the water is turbid. Infrared photography can also be used for this purpose. By eliminating the blue colors and emphasizing greens, reds, and the infrared spectrum (contiguous to the red end of the visible spectrum), this technique highlights growing vegetation. Also in clear, shallow water, floating sampling quadrats (for example, those measuring 0.5 × 0.5 meters) can be dropped randomly or systematically in the water body. The amounts of aquatic vegetation enclosed by each quadrat can be used to estimate the percentage of the water surface overlying submergent or underlying emergent aquatic plants. If individual species of plants are identified, this method can also be used to estimate the relative species composition of aquatic vegetation. Aquatic plant species composition can also be assessed by setting up parallel transects in the water body and assessing the proportion of each transect intercepted by a particular plant species. Collection of aquatic plants for identification can be conducted by hand or by using simple devices such as rakes or grappling hooks. Collection of aquatic plants for determination of characteristics such as volume or weight can be done by randomly or systematically using sampling quadrats and clipping all vegetation enclosed by each quadrat.

FIGURE 13.9 Some species, such as the pied-billed grebe, build floating nests of dead aquatic vegetation.

The abundance, distribution, and types of aquatic plants in a water body can affect a variety of terrestrial species as well as fishes and other aquatic animals. For example, birds such as pied-billed and horned grebes use aquatic vegetation to form a floating nest (fig. 13.9). Yellow-headed blackbirds attach their over-water nests to emergent vegetation such as cattails. Several species of frogs and toads can lay their gelatinous strings of eggs on aquatic vegetation, as can some fishes such as yellow perch. Aquatic vegetation provides important escape cover and food production areas for fishes. Aquatic vegetation also provides substrate and food for a variety of invertebrates that are important in the aquatic food web. Either an overabundance or a lack of emergent and submergent aquatic plants can be cause for concern to the biologist.

Long-term data describing trends in aquatic plant abundance and species composition may be valuable to biologists. Changes in aquatic vegetation may be directly related to siltation, nutrient enrichment, changes in water levels, changes in animal species composition, or other potentially human-influenced factors. The disappearance of rooted aquatic plants in a water body, for example, may indicate severe sedimentation or other pollution problems. Losses of aquatic plants or changes in plant species composition can affect all biotic

components of an aquatic system. Such changes can also be important to human users, particularly as they influence water-related recreational activities. Long-term documentation of these changes is important in gaining the public support needed to correct or mitigate water quality problems.

Abiotic benthic substrates are most often categorized according to particle size. Particle sizes range from very fine silt to large boulders, with sand, gravel, and rocks of various sizes in between. Substrates consisting of small particles can be sampled with grabs (see section 7.8) or by hand. The particles can then be separated by size using a series of differently sized meshes or screens. Such measurements are not usually necessary to categorize large rocks and boulders.

Bottom particle size has a major effect on habitat suitability for many organisms. For example, silt is often a good location for aquatic vegetation, but a poor location for fish egg deposition. Large rocks serve as a good attachment substrate for invertebrates and as cover for fishes. In lotic habitats, fishes may seek reduced current velocities associated with large rocks. Many fish species need a specific size of gravel in which to successfully spawn. Chinook salmon, for example, tend to spawn on larger gravel than the other Pacific salmon (Scott and Crossman 1973). Most salmonids have a low tolerance for silt in intergravel spaces. Inverse relationships have been found between the quantity of sand and silt in a substrate and survival rates of emerging coho salmon and steelhead fry (Phillips et al. 1975) and pink salmon fry (Heard 1991).

Dead vegetation can also be important as an aquatic substrate. Dead trees, brush, and other nonliving vegetation can serve the same functions as abiotic substrates; they can serve as substrates for attachment or spawning, or they can moderate the effects of wind and wave action. These substrates can occur naturally or can be established by humans. In lotic systems, areas containing accumulations of dead trees and brush represent the best habitat for many fish species and other aquatic organisms. Wood ducks, green kingfishers, and various other bird species commonly perch on or feed around dead trees or brush protruding from

FIGURE 13.10 This fallen tree in a river channel provides hiding cover, invertebrate prey, and reduced current velocity for fishes as well as loafing, feeding, and perching sites for birds. Various mammals, amphibians, and reptiles also use such habitat. *(Photograph courtesy of the National Biological Service.)*

the stream surface (fig. 13.10). Such dead vegetation accumulations are sometimes quantified in terms of number per linear distance (number per kilometer) or number per unit of surface area (number per hectare). Dead vegetative and abiotic substrates in aquatic systems have utility to more than fishes and birds; a variety of mammals, amphibians, reptiles, and invertebrates also use such structures.

Live trees and brush are also beneficial substrates. Swamps dominated by bald cypress and southern coastal areas dominated by red mangrove, for example, provide abundant structural components that are beneficial to a variety of aquatic and terrestrial organisms.

13.4 Terrestrial Vegetation: Grasslands, Shrublands, and Other Nonforested Habitats

Several techniques exist for obtaining information concerning terrestrial vegetation. These techniques all involve taking random or systematic samples that accurately describe the habitat in question. In many studies of grassland or other low-growing vegetation, the percentage of the ground obscured by grasses, forbs, and other vegetation is recorded. The percentage of the ground obscured by plants as viewed from above is known as **coverage**. Daubenmire (1959) developed a method of estimating plant coverage using 20 × 50 centimeter sampling quadrats, often referred to as **Daubenmire quadrats** (fig. 13.11). The percentage of each quadrat covered by a certain plant species or group (such as forbs) is estimated and recorded in one of six coverage classes (0%–5%, 5%–25%, 25%–50%, 50%–75%, 75%–95%, and 95%–100%). For example, in an uncut alfalfa field in late spring, alfalfa would be likely to cover a high percentage of the ground within a 20 × 50 centimeter quadrat, and would probably be placed in the 75%–95% or 95%– 100% coverage class. Frequency of occurrence, the percentage of samples in which a particular plant species or group occurs, can also be estimated using these quadrats. Other types of sampling quadrats used in a similar manner may be square or even round.

FIGURE 13.11 Plant coverage within this 20 × 50 centimeter Daubenmire quadrat is recorded in one of six coverage classes (0%–5%, 5%–25%, 25%–50%, 50%–75%, 75%–95%, and 95%–100%). The quadrat is color-coded to allow more accurate estimation of coverage classes. The small accessory quadrat is used for measuring frequency of occurrence of grasses. *(Photograph courtesy of M. Rumble, U.S. Forest Service.)*

FIGURE 13.12 The line-intercept method works well for determining the coverage of plants such as this big sagebrush. *(Photograph courtesy of the Enviornmental Science and Research Foundation, Idaho Falls, Idaho.)*

Another common technique used in assessing coverage and species composition (the percentage of total coverage each plant species composes) of vegetation is the **line-intercept method** (Canfield 1941). In this method, a measuring tape is stretched for several meters between two stakes at a level above or near the top of the vegetation for which coverage is to be estimated. The percentage of the tape intercepting (passing over the top of or touching) a particular plant species represents the coverage for that species (fig. 13.12). If a biologist wished to estimate the coverage of big sagebrush in a high desert area in Idaho, for example, a 10-meter tape could be stretched between two stakes at a number of sample sites. The biologist would compute the percentage of the tape overlying or intercepting big sagebrush at each sample site (centimeters overlying or intercepting big sagebrush/total centimeters in the 10-meter tape × 100), and the average of all the sample sites would provide an indication of coverage of big sagebrush in the study area.

Point frames can be used to assess coverage of vegetation; both basal and canopy coverage can be determined in this manner (Higgins et al. 1994). A point frame is usually a metal frame with guided holes into which metal rods (usually 10 or more) can be inserted. The frame is set up above the vegetation, and the rods are lowered until they touch a stem or leaf (fig. 13.13). The first species of vegetation touched is recorded, as is bare ground if no plant is intercepted by the rod. When rods first touch vegetation they are recorded as canopy data points; plants touched below the canopy at the soil interface are usually recorded as basal data points. Not only can a single rod touch more than one plant species, it can also touch the same species more than once. The percentage of all rods first touching a particular species of vegetation provides the estimated canopy coverage for that species; for example, if three of ten rods in a point frame first touched western wheatgrass when lowered, then the estimated canopy coverage for that grass would be 30% (3/10 × 100). Actual estimates of coverage would be based on multiple point frame samples from

FIGURE 13.13 Point frames are time-consuming to use, but they provide considerable accuracy in estimating coverage and species composition of grassland plants. *(Photograph courtesy of the Department of Animal and Range Sciences, North Dakota State University.)*

FIGURE 13.14 A visual obstruction board provides a measure of vegetation density. *(Photograph courtesy of the U.S. Fish and Wildlife Service.)*

the area being studied. Basal vegetation data points can be used in a similar manner to estimate basal coverage.

Biologists are often interested in the effectiveness of vegetative cover in concealing nesting birds, deer fawns, or other wildlife. This concealment value of vegetation can be measured by determining the degree to which a marked board or rod is obscured; marked boards or rods used for this purpose are called **visual obstruction measurement devices (visual obscurity measurement devices)**. One commonly used visual obstruction measurement device is a board divided into 0.5-meter marked intervals. The board is usually 2.5 meters long, approximately 30 centimeters wide, and may be hinged in the middle so that it can be folded and carried in the field (Nudds 1977). This device is sometimes referred to as a **Nudds board**. The board is hand-held or held in an upright position by stakes at a nest site, bedding site, or other area where a biologist wishes to obtain a quantitative measure of cover (fig. 13.14). An estimate of the percentage of each 0.5-meter section that is obscured from vision is recorded. Usually five categories, 0%–20%, 21%–40%, 41%–60, 61%–80%, and 81%–100%, are used in estimating the coverage of each board section. The distance at which the board is viewed needs to be appropriate to the vegetation

type to assure maximum variation in readings between sites. For example, if the board were viewed from 50 meters away at a site with very thick vegetation, 81%–100% of all the board sections would probably be obscured from view regardless of vegetation differences among sites. If the board were viewed from only 2 meters away at such a site, a high percentage of the board would be visible despite considerable differences among microhabitats. Nudds boards are most commonly viewed from 15 meters. Another method of measuring visual obstruction involves some type of background marked like a large checkerboard that is photographed to estimate the percentage obscured.

The **Robel pole**, another type of visual obstruction measuring device, is a pole approximately 1.5 meters in height marked into 5-centimeter units which often alternate between light and dark (Robel et al. 1970). The observer views the pole from a 4-meter distance and at a height of 1 meter, and records the lowest 5-centimeter mark visible (fig. 13.15). The Robel pole is often read from four directions spaced 90° apart.

FIGURE 13.15 Readings obtained from a Robel pole provide information on the height and density of vegetation. *(Photograph courtesy of the U.S. Fish and Wildlife Service.)*

13.5 Terrestrial Vegetation: Forests

In forested areas, biologists can evaluate the understory for coverage, species composition, and visual obstruction using techniques similar to those discussed in section 13.4. However, additional measures, such as **canopy coverage (canopy closure)**, **canopy volume**, **diameter at breast height**, **basal area**, and tree height, are used to assess forest habitat. Forest canopy coverage is the percentage of the forest floor with tree cover directly overhead. Canopy volume is the space occupied by tree crowns (leaves and branches) per unit of area or per tree. Diameter at breast height, often referred to as DBH, is the diameter of the trunk (**bole**) of a tree at approximately 1.4 meters (approximately human breast height). Basal area is the cross-sectional area of standing tree trunks per hectare as measured at breast height.

Readings of canopy coverage can be obtained using a hand-held **spherical densiometer** (fig. 13.16). This device consists of a concave, mirrored surface marked with a grid. The canopy is reflected in the mirrored surface, and the overhead area not covered by canopy is estimated. Biologists can also make estimates of canopy coverage visually, without any specific devices to aid them. For example, a biologist may simply estimate canopy coverage in broad categories such as 0% to less than 40%, 40% to less than 70%, and 70% to 100%.

Estimates of canopy coverage are useful to wildlife biologists in describing and predicting the potential wildlife values of timber stands. Areas with a high percentage (over 70%) of coniferous canopy coverage in cold regions are often important for wintering deer, elk, and a variety of smaller animals. Such areas provide thermal cover that reduces the loss of radiant heat to the atmosphere by wintering animals. For Merriam's wild turkeys in coniferous forest habitat, areas with a high

FIGURE 13.16 Spherical densiometers are used to estimate canopy coverage in a forest. (*Photograph courtesy of the U.S. Forest Service.*)

percentage of canopy coverage may provide sites where snow depth is reduced and pine seeds are available. Dense canopy coverage in the summer may provide cool areas for animals. Thus, canopy coverage has a moderating effect on temperature extremes and snow cover, especially in coniferous forest areas. Canopy coverage along streams has an influence on water temperature and even on siltation, and thus may influence the suitability of a stream for species such as rainbow trout, blacknose dace, or bridgelip suckers (fig. 13.17).

Diameter at breast height is measured using a DBH tape to obtain a direct reading of tree diameter by fitting the tape around the bole of the tree. DBH measurements can be used to determine basal area. The basal area of trees per unit of area is sometimes used by wildlife biologists to describe cover value in forested areas. Wild turkey roosting sites, for example, have commonly been described in terms of basal area of trees per hectare.

FIGURE 13.17 Canopy coverage over streams can strongly influence water temperature and thus the suitability of a stream for various fishes.

The **fixed radius plot method** can be used for estimating tree density and basal area per hectare. A sample plot is described by a circle drawn at a measured distance from the plot center, and all trees occurring within the plot are counted. Diameter at breast height can be directly measured for all trees in the plot. Multiple sample plots can then be used to estimate the average number of trees (by species and total) and basal area per hectare. Fixed radius plots provide accurate data, but this method is labor-intensive.

The basal area and density of tree species are often estimated using a **variable radius method** that is less labor-intensive than the fixed radius plot method. The variable radius method is plotless and requires the use of a prism or angle gauge at random sampling sites in a forest to determine which trees will be included in the inventory. The observer stands at a sampling site and views the surrounding tree boles (at breast height) through the prism (fig. 13.18A) or angle gauge (fig. 13.18B). The observer counts the number of boles that appear wider than the view through the angle gauge, or that are displaced to the right or left when viewed through the prism. The number of trees counted depends on the distance of the trees from the observer and the DBH of the trees. This method has also been modified for determination of canopy coverage of shrubs. The operating principles and basic mathematics behind this method are explained in Grosenbaugh (1952) and in other forest measurement texts.

A

B

FIGURE 13.18 *(A)* A prism (shown in use) or *(B)* an angle gauge can be used in a quick plotless method for estimating the basal area of trees. *(Part A photograph courtesy of the U.S. Forest Service.)*

In areas where tree distribution is random, a plotless method known as the **point-centered-quarter method** is commonly used (Cottam and Curtis 1956). The sampling points can be randomly located or placed along transects. At each sampling point the distance to the nearest tree in each of four quarters is measured, and the DBH and species of the nearest tree are recorded for each quarter. Computations provide estimates of tree density per hectare by species as well as average and total basal areas. Additional discussion and references on this method are included in Higgins et al. (1994).

Forest composition (number of trees of a particular species per hectare ÷ total trees per hectare × 100) can be determined by recording tree species while estimating densities by plot or plotless methods. The line-intercept method can also be utilized to estimate forest composition. Forest composition is given in terms of numbers of trees, usually above some minimum size, rather than coverage of the ground, as was described earlier for grasses, forbs, and shrubs.

Tree height and height of the base of the crown can be estimated to the nearest meter using a **ranging optometer**. Field models of the ranging optometer are usually about the size of a small camera and are thus easy to transport. Most ranging optometers provide distance readings in meters. Because the uppermost branches of many trees cannot be viewed from directly below, the observer must move away from the tree until the top is visible; at this point the distance to the top of the tree in meters is recorded from the optometer, as is the horizontal distance to the base of the tree. The height of the tree is then determined by solving for the vertical side of a right triangle. Tree heights can be more accurately determined with a **clinometer** (fig. 13.19), an instrument with a scale inside the viewing area that is calibrated to read zero when the device is exactly horizontal. Each clinometer is scaled for use at a specific baseline distance from the tree to be measured, such as 20.1 or 30.5 meters. An observer using a clinometer scaled for 30.5 meters would stand that distance from the bole of the tree being measured

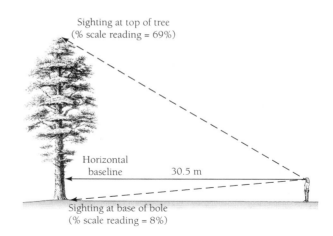

FIGURE 13.19 Biologists can use a clinometer to estimate tree height. In this example the observer is using a clinometer scaled for a 30.5-meter baseline.

and sight the top of the tree and subsequently the bole of the tree at the base. A biologist can use a clinometer to obtain a percent scale reading of the angle from the horizontal line to the top and base of the tree. If, for example, the percent scale reading was 69% for the top and 8% for the base, the height of the tree would be estimated by adding these two percentages, converting them to a decimal, and multiplying by the baseline distance. The height in this example would be 0.77 × 30.5 = 23.5 meters. Tree height, along with height to the base of the crown and crown width, are used in estimating canopy volume.

Tree height measurements could be used for such assessments as evaluating potential bald eagle nesting sites or determining whether adequate numbers of roosting trees are available for a wild turkey population. Many birds segregate horizontally in the vegetative community; that is, different species use different portions of the forest canopy and understory. Some bird species are highly specific to a particular horizontal stratum. Thus, measurements of tree height and the availability of horizontal strata could be of primary interest to a wildlife biologist.

13.6 Habitat Relationships to Microclimate and Weather

In addition to sampling and assessing vegetation and other landscape features, biologists may want to evaluate the secondary effects of those features on microclimate. **Microclimate** is the climate on a very local scale and differs from the general climate of the region (Smith 1990). The climate near the ground in tall-grass prairie, the climate in a snake denning area, and the climate under a fallen tree are all examples of microclimates. Some habitat features can moderate temperatures, wind speeds, precipitation, or even humidity. Wildlife may select habitats, at least in part, based on such microclimate characteristics.

Selection of a wintering microhabitat with reduced winds or higher temperatures, for example, allows an animal to maintain its body temperature with minimal use of additional energy to produce heat (see section 5.6). Ring-necked pheasants in the northern plains and intermountain region of North America are subject to stress during cold and windy periods, particularly if they cannot find protective cover and nearby food sources. Biologists have found that dense stands of cattails can reduce wind speed by as much as 95%; these areas provide protective cover that greatly reduces the wind chill factor for many wintering wildlife species. These animals, when located in good winter cover that protects them from the wind, can survive prolonged periods of cold that could otherwise cause mortality. Cool temperatures and high winds can also place unusual stress on young waterfowl when they are on bodies of water that lack protective emergent vegetation. Sitka black-tailed deer in Alaskan coastal areas seek old-growth forests for wintering because their high canopy coverage and volume reduce snow accumulation on the forest floor (Bloom 1978). Gray wolf dens and warm rocks used by reptiles are both exam-

ples of microclimates influenced by habitat, primarily by abiotic components in these cases.

Animals may likewise seek microclimates associated with landscape features. The leeward side of a hill or a forest patch, for example, can provide protection from wind. Caribou or bison may select windswept hillsides to reduce problems with biting insects. Therefore, the secondary influence of landscape features on microclimate may need to be assessed as part of terrestrial habitat sampling and assessment.

Temperatures in the terrestrial environment can be measured directly using outdoor thermometers, minimum-maximum thermometers, or continuous recording devices, or they can be obtained from local weather station records. **Anemometers** or other wind measuring devices can provide information on wind speed. Wind speeds can also be continuously or periodically recorded with electronic equipment and later downloaded to a computer. Biologists often obtain precipitation information from local weather stations. However, precipitation gauges can be placed in study areas at little cost. Snow accumulation can be measured at various sites using poles with length markings.

In some cases, annual variations in precipitation or other weather patterns have an overriding influence on habitat. In such cases, measurements of weather variables may provide a low-cost means of predicting habitat conditions and thus survival or reproductive rates for certain wildlife populations. High reproductive success in Gambel's quail, California quail, and other quail species in the Southwest, for example, is associated with above normal winter-spring rainfall, high soil moisture, and a resultant increase in new-growth vegetation. Conversely, unusually dry conditions can lead to extremely low reproductive success in these species. Therefore, biologists can predict habitat conditions and reproductive success in these quail species based on rainfall and soil moisture conditions. Snow goose reproductive success in the Arctic can be predicted based on satellite photographs of snow cover taken during May, when nesting begins. If snow cover is extensive in late May and early June, the habitat is suboptimal for nesting, and many pairs may forgo nesting until the following year.

13.7 Herbage, Browse, Mast, and Other Plant Foods for Wildlife

Biologists may wish to obtain information on the availability of **herbage** (grasses and forbs), **browse** (leaves, shoots, and stems of shrubs and trees), **mast**, and other plant foods utilized by wildlife species. Mast includes various fruits of trees and shrubs, especially nuts such as acorns and fleshy fruits such as blackberries and wild plums.

To measure herbage biomass, plant material can be clipped from small sample plots or quadrats in grass-forb habitat and weighed. Biomass of grasses and forbs can also be estimated using an electronic device called a **radiometer** that measures reflected light from the vegetation. Dry weight data from clipped sample plots in different areas can be compared to assess food availability and the relative productivity of potential food plants (fig. 13.20). For example, herbage biomass at forest sites where ponderosa pines have been thinned to a low density and low basal area can be compared with that at denser and higher basal area sites to determine the influence of logging practices on herbage availability and production. Likewise, sites protected from domesticated and wild ungulate feeding by exclosures can provide information on the combined effects on vegetation of **grazing** (feeding on herbage) and **browsing** (feeding on browse) (fig. 13.21). Mast can be sampled on trees by direct observation and counts, passively by the use of various traps to collect falling mast, or by means of sample plots under

FIGURE 13.20 Clippers and a circular sample plot can be used to collect herbage samples in a forest understory. Subsequent determination of the dry and wet weight of these samples can be used to evaluate the influence of different forestry practices on understory vegetation. (*Photograph courtesy of M. Rumble, U.S. Forest Service.*)

trees for fallen mast. Mast samples can be used for a variety of purposes; for example, they may indicate winter food availability for species such as wild turkeys and white-tailed deer.

In addition to availability information, biologists are likely to need information on the nutritional value of herbage, browse, mast, and other plant foods. Measurements of the crude protein, digestible protein, and digestible energy content of plant foods may prove valuable to a wildlife biologist. It should be remembered that digestible protein and digestible energy must be determined in reference to different animal groups, such as seed-eating birds or ruminants. Crude protein measurements are usually directly related to the value of plant foods to animals. However, if plants contain oils or other compounds, such as tannin, that inhibit digestion, they can be relatively high in crude protein and still be of little value to a

A   B

FIGURE 13.21 A comparison of vegetation inside these exclosures with that outside primarily reflects (A) grazing by brant near the Arctic coast in Alaska and (B) the influence of browsing by wintering mule deer and domesticated livestock in California. (*Part B photograph courtesy of K. Jensen.*)

particular animal species. Digestible protein and digestible energy provide more accurate estimates of the actual nutritional value of plant food sources, but involve more expensive laboratory analyses than does the measurement of crude protein. The moisture content of vegetation is also important because higher moisture content is usually associated with increased digestibility.

Wildlife biologists are interested in knowing which plants provide the highest digestible protein and digestible energy levels for wild animals; they are also interested in how these nutritional levels change with time of year and plant growth stage. Migrations of large ungulates and other animals may be directly related to the availability of high-protein forage in a particular area or habitat. The spring migration of elk to higher altitudes in the Rocky Mountains, for example, is related to the high-protein forage available at these higher altitudes. Soils may have a strong influence on the nutrient levels and growth rates of various plant

foods. Fire, by releasing nutrients from organic matter, may increase digestible protein values as well as forage growth rates on a burned area. Some wildlife species, through their grazing activities, stimulate new vegetative growth that is higher in digestible protein than ungrazed vegetation. Brant, for example, graze on grasses and sedges in their Arctic nesting and brood-rearing grounds; their grazing promotes renewed vegetative growth and sustains the high protein levels that are valuable to these geese.

Habitats with plentiful, high-quality forage or other plant foods are able to support a more abundant and diverse wildlife assemblage than are habitats that have similar potential but are in poorer condition. For example, if a winter range for bighorn sheep is overgrazed and overbrowsed by domesticated livestock, its value to the bighorn sheep, as well as to other wintering wildlife and domesticated livestock, is reduced. In such a situation, biologists would probably find reductions in the quantity and quality (digestible protein and digestible energy) of the herbage and browse compared with sites of similar potential that had not been mismanaged.

13.8 Prey Populations

In fisheries, prey animals such as invertebrates, frogs, or small fishes are considered to be a portion of the biota; prey is not considered to be a part of the habitat (see section 1.2). While the same is generally true in wildlife, some wildlife biologists include prey animals as a portion of the habitat. If a biologist were describing or evaluating brood habitat for wood ducks, for example, information on invertebrate populations, the primary food (prey) source for the young birds, would be critical. Even for more typical predators, such as kit foxes or diamondback water snakes, an assessment of small mammals, fishes, amphibians, and other potential prey could be included in habitat assessment.

Prey availability is an important factor whether or not it is considered to be a part of or separate from habitat. A wide variety of techniques are available for sampling and assessing prey populations. Techniques applicable to the sampling and assessment of prey populations are presented in chapters 7 and 9.

13.9 Landscape-Level Assessment

The U.S. Fish and Wildlife Service has developed a series of **habitat suitability index models** for both fish and wildlife species (Schamberger et al.

1982). These models, generally referred to as HSI models, are based on suitability indices formulated from variables that affect the life cycle and survival rate of each species. Information from literature sources on species-habitat relationships has been consolidated to create these indices. The models are valuable to biologists both as a habitat assessment tool and for predicting the likely effects of human alterations of habitat.

Assume, for example, that an urban fishery biologist is directing a project to improve a previously polluted stream within the city of Richmond, Virginia. The biologist is interested in determining whether the stream habitat is now appropriate for smallmouth bass, a fish species native to that area. Consulting the HSI model for smallmouth bass (Edwards et al. 1983) would allow the biologist to determine whether the stream was now suitable for that species. The habitat variables in the model include substrate type, percentage of pools, percentage of cover (stumps, boulders, crevices), depth of pools, pH, dissolved oxygen, turbidity, temperature, water level fluctuations, and gradient. These habitat variables are measured and used in a mathematical equation that will indicate the likely population abundance of smallmouth bass. The biologist must also be aware that the status of each of these stream variables is linked to the surrounding landscape.

Habitat evaluation for many species can be enhanced by or may even require a landscape-level approach (see section 2.11). For example, forest species such as the worm-eating warbler, painted-bunting, and pileated woodpecker are negatively affected by habitat fragmentation, a landscape characteristic. Extensive roadless areas and intermixed meadows are important components of elk habitat. Most studies of wildlife habitat can be greatly enhanced with information on habitat characteristics at a landscape level.

There are several methods of evaluating habitat features at a landscape scale. Broad-scale habitat patterns can be mapped with high-altitude photography and infrared imagery. Satellite imagery can also provide broad-scale information on landscapes, but may lack sufficient resolution for

Total Basins / 10 sq-miles

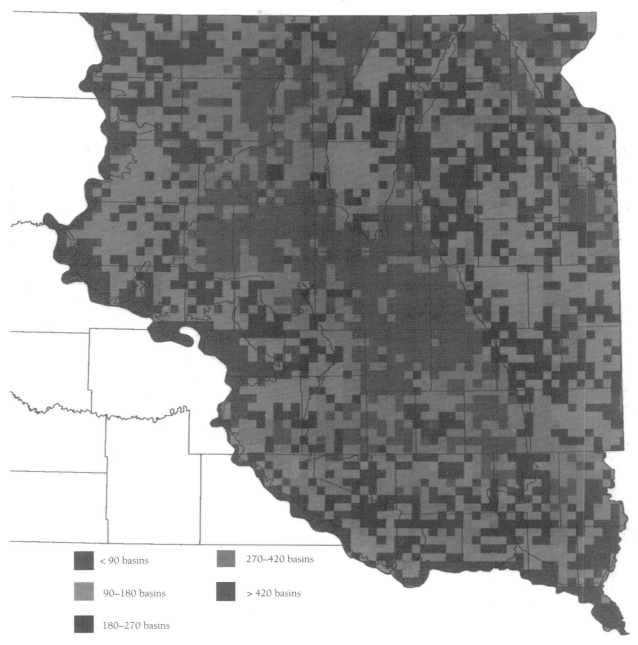

< 90 basins

90–180 basins

180–270 basins

270–420 basins

> 420 basins

many purposes; the practical minimum area that can be resolved with technology available to the public is about 0.4 hectares. Information from the landscape-level maps produced by these methods can be digitized and entered on computers using appropriate software. Such a system for the computerized mapping and analysis of geographic features using digitized data is termed a **geographic information system** (**GIS**). Computers and landscape-level GIS provide the capacity for quick eval-

◀ **FIGURE 13.22** (*opposite*) This GIS map codes hectares of wetland basins per unit of area in eastern South Dakota to provide a landscape-level view of wetland distribution.

uations of habitat patches, corridors, wetlands, and other habitat features in terms of area and perimeter (fig. 13.22). GIS technology also allows for the updating of habitat maps through time and the tracking of changes over time. GIS is being widely used by many natural resource agencies for a variety of applications.

Measures of the horizontal diversity of habitats, such as edge (see section 2.11), interspersion, and juxtaposition, can also be determined using GIS. **Interspersion** is a measure of the intermixing of different habitat types. For example, a full section of land (1.6 kilometers × 1.6 kilometers) consisting of two large crop fields, corn and soybeans, with no noncropland between the fields has very low interspersion. An area of the same size containing a small willow- and cattail-lined stream, a patch of native prairie, two shelterbelts, three ponds, a fenceline with shrub and tall grass cover, and separate fields of sorghum, soybeans, and corn has a much higher interspersion. **Juxtaposition** is the proximity of habitats needed by an organism. For example, a common kingsnake may require sunning sites, winter denning sites, sites for depositing eggs, sites for hunting rodents, and several other habitat types within a certain area. If these needed habitats are not available within a reasonable distance of each other, the area may support no or few common kingsnakes. Techniques for quantifying interspersion and juxtaposition are described in Anderson and Gutzwiller (1994).

Information on geographic features such as habitat types, soils, roads, and land ownership can be stored in GIS data bases. These features can then be viewed simultaneously (superimposed) and their relationships explored. For example, a biologist can use GIS to superimpose roads, habitat types, soils, and other characteristics on the home ranges of species such as mule deer or red foxes. This capability allows the biologist to evaluate the home ranges of animals or groups of animals in relation to habitat features. Selection of certain habitats and avoidance of other features by the animals could thus be detected by the biologist. Avoidance (or lack of avoidance) of areas near roads by radio-collared elk, for example, could be detected using GIS.

Another important application of GIS is called **gap analysis** because it identifies gaps in the protection of species and species diversity (Scott et al. 1993). Using this technique, biologists can evaluate a large area in terms of species diversity to determine which sites should receive priority in efforts to conserve the greatest variety of species. Such a study could involve one group, such as birds, or it could involve a broad array of organisms. For instance, an area of intermixed national forest and private lands containing various habitat types could be mapped on GIS; then the distribution of bird species in that area could be superimposed. Certain areas would probably show the greatest overlap in bird species and thus might be considered as unusually important for maintaining bird species diversity, particularly if such habitats were limited in area and subject to potential alteration or elimination. By superimposing land ownership patterns, biologists could then determine which or how many of these important areas were currently in public ownership. Such areas might be identified for protection from activities that could destroy their value to birds; for example, they might be designated for no logging or for some other specified logging treatment that would maintain their particular habitat.

In this manner, biologists can identify gaps in species or species diversity protection within existing habitat and land ownership patterns. If an area contains endangered or threatened species, these gaps in protection may receive special attention. Because gap analysis may identify privately owned areas in need of long-term protection from development, its possible implications in terms of public relations and landowner rights must be carefully considered. Communication with all parties involved is critical if special easements, purchases, or other long-term arrangements to provide habitat protection in these gaps are to be attempted.

▼ SUMMARY

Habitat is often considered to be the most important overall ingredient in conserving populations of wild organisms. Biologists use a variety of techniques in sampling and assessing aquatic habitats. Water characteristics such as temperature, oxygen content, pH, hardness, and nutrient levels influence the growth and survival of aquatic organisms, and measurements of these characteristics provide important information to biologists. Aquatic habitats can also be characterized by size, depth, volume, and shape. Maps depicting the shapes and depth contours of water bodies are often available from natural resource agencies; they can also be created using various mapping techniques.

Aquatic vegetation is a major habitat component for fishes and aquatic invertebrates as well as many aquatic-oriented terrestrial species. Biologists can estimate submergent and emergent vegetation abundance using transects, quadrats, or even aerial photography. Long-term trends in aquatic vegetation can be an indicator of water quality changes. Abiotic substrates in water bodies are categorized on the basis of particle size. Live or dead trees and other woody vegetation in water bodies have value to both aquatic and terrestrial organisms.

In terrestrial ecosystems, much habitat analysis involves the sampling and assessment of vegetation. The coverage and species composition of vegetation is often estimated using quadrats or other types of sample plots. Plant coverage and composition can also be estimated with the line-intercept method or with a point frame. The density of vegetation as viewed horizontally can be measured with various visual obstruction measurement devices.

In forested habitats, additional characteristics such as canopy coverage, canopy volume, diameter at breast height, basal area, and tree height are often assessed. Canopy coverage is commonly measured with a spherical densiometer. Tree density and basal area can be sampled in circular sample plots with fixed boundaries. Variable radius methods requiring the use of a prism or angle gauge can also be used to estimate tree densities and basal areas in forested habitats; variable radius methods are less labor-intensive than are fixed radius plots. The point-centered-quarter method is another plotless method commonly used to estimate tree density.

Vegetation, rocks, and other habitat features can have a strong influence on microclimate characteristics such as wind speed, temperature, and humidity. Because of the importance of microclimates to wild animals, biologists may want to assess the influence of vegetation and landscape features on microclimate. Weather conditions in a region can sometimes be a good predictor of broad-scale habitat conditions.

Plant foods such as herbage, browse, and mast are often assessed as a measure of habitat condition. A variety of sampling methods, from traps that collect falling mast to clipping herbage, are used by biologists. Biologists may monitor the nutritional values of vegetation by assessing crude protein, digestible protein, and digestible energy.

Prey populations are not considered to be a part of the habitat component by fishery biologists. Some wildlife biologists consider prey to be a part of the habitat component, but others consider it to be part of the biota component of a wildlife system.

Landscape-level evaluation of habitat characteristics is often accomplished using aerial photography or computerized geographic information systems (GIS). Gap analysis is greatly facilitated by the availability of GIS.

PRACTICE QUESTIONS

1. Assume that a fishery biologist is collecting monthly information on temperature, dissolved oxygen concentration, and pH in a small lake. For each characteristic, list two possible reasons for collecting this information.

2. Describe the use of a Secchi disk.

3. Why might information on water volume in a water body be of value to a fishery biologist?

4. What is shoreline development, and why is it an important characteristic of water bodies?

5. List three reasons why the assessment of aquatic plant abundance may be important.

6. Why would a biologist need information on the substrate composition of a water body? Provide four reasons.

7. What is a point frame, and how is it used?

8. Describe the use of a visual obstruction measurement device. Describe two situations in which such a tool might provide important information for a biologist.

9. Define the following terms: canopy coverage, diameter at breast height, and basal area.

10. What are two instruments that can be used in estimating tree height?

11. What is microclimate, and why is it important in animal habitat selection?

12. Name three measures of the nutritional quality of herbage, browse, and mast for wild animals.

13. What is a GIS? Provide two examples of ways in which GIS could be used by a biologist.

14. What is gap analysis?

15. Define interspersion, and relate it to the current farming practices in which field sizes are increasing.

SELECTED READINGS

Anderson, S. H., and K. J. Gutzwiller. 1994. Habitat evaluation methods. Pages 592–606 *in* T. A. Bookhout, ed. *Research and management techniques for wildlife and habitats.* The Wildlife Society, Bethesda, Md.

Bisson, P. A., J. L. Nielsen, R. A. Palmason, and L. E. Grove. 1982. A system of naming habitat types in small streams, with examples of habitat utilization by salmonids during low streamflow. Pages 62–73 *in* N. B. Armantrout, ed. *Acquisition and utilization of aquatic habitat inventory information.* Western Division, American Fisheries Society, Bethesda, Md.

Clesceri, L. S., A. E. Greenberg, and R. R. Trussell, eds. 1989. *Standard methods for the examination of water and wastewater.* 17th ed. American Public Health Association, Washington, D.C.

Cooperrider, A. Y., R. J. Boyd, and H. R. Stuart, eds. 1986. *Inventory and monitoring of wildlife habitat.* U.S. Department of the Interior, Bureau of Land Management, Denver, Colo.

Higgins, K. F., J. L. Oldemeyer, K. J. Jenkins, G. K. Clambey, and R. F. Harlow. 1994. Vegetation sampling and measurement. Pages 567–591 *in* T. A. Bookhout, ed. *Research and management techniques for wildlife and habitats.* The Wildlife Society, Bethesda, Md.

Koeln, G. T., L. M. Cowardin, and L. L. Strong. 1994. Geographic information systems. Pages 540–566 *in* T. A. Bookhout, ed. *Research and management techniques for wildlife and habitats.* The Wildlife Society, Bethesda, Md.

Morrison, M. L., B. G. Marcot, and R. W. Mannan. 1992. *Wildlife-habitat relationships: Concepts and applications.* University of Wisconsin Press, Madison.

Orth, D. J. 1983. Aquatic habitat measurements. Pages 61–84 *in* L. A. Nielsen and D. L. Johnson, eds. *Fisheries techniques.* American Fisheries Society, Bethesda, Md.

LITERATURE CITED

Anderson, S. H., and K. J. Gutzwiller. 1994. Habitat evaluation methods. Pages 592–606 *in* T. A. Bookhout, ed. *Research and management techniques for wildlife and habitats.* The Wildlife Society, Bethesda, Md.

Bloom, A. M. 1978. Sitka black-tailed deer winter range in the Kadashan Bay area, Southeast Alaska. *The Journal of Wildlife Management* 42:108–112.

Canfield, R. H. 1941. Application of the line interception method in sampling range vegetation. *Journal of Forestry* 39:388–394.

Cottam, G., and J. T. Curtis. 1956. The use of distance measures in phytosociological sampling. *Ecology* 37:451–460.

Daubenmire, R. 1959. A canopy-coverage method of vegetational analysis. *Northwest Science* 33:43–64.

Edwards, E. A., G. Gebhart, and O. E. Maughan. 1983. *Habitat suitability information: Smallmouth bass.* FWS/OBS-82/10.36. U.S. Fish and Wildlife Service, Washington, D.C.

Grosenbaugh, L. R. 1952. Plotless timber estimates: New, fast, easy. *Journal of Forestry* 50:32–37.

Heard, W. R. 1991. Life history of pink salmon (*Oncor-hynchus gorbuscha*). Pages 119–230 *in* C. Groot and L. Margolis, eds. *Pacific salmon life histories.* University of British Columbia Press, Vancouver.

Higgins, K. F., J. L. Oldemeyer, K. J. Jenkins, G. K. Clambey, and R. F. Harlow. 1994. Vegetation sampling and measurement. Pages 567–591 *in* T. A. Bookhout, ed. *Research and management techniques for wildlife and habitats.* The Wildlife Society, Bethesda, Md.

Meyer, F. P., and R. L. Herman. 1990. Interpreting the scene. Pages 10–18 *in* F. P. Meyer and L. A. Barclay, eds. *Field manual for the investigation of fish kills.* Resource Publication 177. U.S. Department of the Interior, Fish and Wildlife Service, Washington, D.C.

Nudds, T. D. 1977. Quantifying the vegetative structure of wildlife cover. *Wildlife Society Bulletin* 5:113–117.

Orth, D. J. 1983. Aquatic habitat measurements. Pages 61–84 *in* L. A. Nielsen and D. L. Johnson, eds. *Fisheries techniques.* American Fisheries Society, Bethesda, Md.

Owensby, C. E. 1973. Modified step-point system for botanical composition and basal cover estimates. *Journal of Range Management* 26:302–303.

Phillips, R. W., R. L. Lantz, E. W. Claire, and J. R. Moring. 1975. Some effects of gravel mixtures on emergence of coho salmon and steelhead trout fry. *Transactions of the American Fisheries Society* 104:461–466.

Robel, R. J., J. N. Briggs, A. D. Dayton, and L. C. Hulbert. 1970. Relationships between visual obstruction measurements and weight of grassland vegetation. *Journal of Range Management* 23:295–297.

Schamberger, M., A. H. Farmer, and J. W. Terrell. 1982. *Habitat suitability index models: Introduction.* FWS/OBS-82/10. U.S. Fish and Wildlife Service, Washington, D.C.

Scott, J. M., F. Davis, B. Csuti, R. Noss, B. Butterfield, C. Groves, H. Anderson, S. Caicco, F. D'erchia, T. C. Edwards Jr., J. Ulliman, and R. G. Wright. 1993. *GAP analysis: A geographic approach to protection of biological diversity.* Wildlife Monographs 123.

Scott, W. B., and E. J. Crossman. 1973. *Freshwater fishes of Canada.* Bulletin 184. Fisheries Research Board of Canada, Ottawa.

Smith, R. L. 1990. *Ecology and field biology.* 4th ed. HarperCollins, New York.

14

HABITAT
DEGRADATION

Habitat can be altered by numerous natural and artificial means. Human-influenced alterations of habitat, both intentional and unintentional, can have far-reaching effects on ecosystems. Habitat degradation or destruction is not as simple to define as it may appear; as in art and most other endeavors, beauty is in the eye of the beholder. The plowing of a section of grassland and its replacement with a wheat field would be viewed as habitat degradation by a biologist interested in grassland organisms; it would not be viewed as habitat degradation by a wheat farmer. To an upland bird biologist, the flooding of a riparian zone due to the construction of an impoundment would obviously be destruction of that habitat for terrestrial wildlife purposes, but it would not necessarily be viewed as habitat degradation by a fishery biologist. The construction of the impoundment would probably not be viewed as habitat degradation by a water-skier, but would be by someone who had enjoyed canoeing that river stretch before its impoundment.

Sometimes multiple changes in habitat can occur over time. A forest habitat could be changed into an agricultural crop field, eventually becoming an abandoned field of forbs, grasses, and shrubs, and then be turned into a parking lot. With each change, different people would view the resulting habitat differently.

Unfortunately, some of our most productive natural habitats, such as wetlands and riparian zones, have been and often still are viewed as "wasteland" by many people. When a "positive use" (positive usually meaning some direct, immediate, and obvious economic benefit to humans) is found for such "wasteland," the change is viewed by many people as beneficial.

Some habitat degradation or destruction is purposeful. For example, when humans disturb an area of Arctic tundra for petroleum production, they know that the area will be degraded. There may be long discussions concerning how much and what kind of damage will occur, but the activity is carried out with full knowledge that some level of degradation is going to result. Some habitat degradation or destruction is not done purposefully. For example, increased aquatic acidity resulting from the combustion of fossil fuels was not a result of human actions taken with the knowledge that this combustion would destroy aquatic life in some water bodies; aquatic degradation was an unforeseen by-product of human activity.

Habitat degradation changes the biotic community that is present. When habitat changes, one type of habitat is destroyed, but another is created. Thus some organisms are harmed, while others benefit. By the 1960s, Lake Erie had been degraded by humans to a point where the native biota of the lake had been greatly modified. Years of municipal, agricultural, and industrial pollution had resulted in dramatic shifts in the plant and animal communities of the lake. Damage to the native biota was so severe that many people pronounced Lake Erie "dead." The lake was not really "dead"; in fact, it probably had more life than ever, just not the flora and fauna present before degradation. Efforts to reduce pollution were successful, and by the late 1980s the lake had recovered sufficiently to allow some of the native biota to rebound. Once again, Lake Erie is "alive." The lake now faces another human-induced threat in the form of the accidentally introduced zebra mussel. Zebra mussels, by multiplying rapidly, utilizing available nutrients and oxygen, and physically covering submerged surface areas, will affect other organisms in the lake. Only time will tell what effects this exotic species will have on the native biota of the lake.

Degradation of habitat is not necessarily a process accomplished only by humans. When a beaver dams a stream to ensure itself a home, when a tree produces chemicals that inhibit the growth of nearby plants, or when a white-tailed prairie dog clips grass around its burrow so that it will be less vulnerable to predation, those organisms are modifying the habitat to their benefit. Each modification degrades that habitat for some species and improves it for others. It is impossible for organisms, humans included, to exist without in some way changing or degrading the habitat in which they live. Humans are different in that they can make conscious choices. Even if the final choice is to degrade an area, humans often have some idea of the effects they will have. The degree to which destruction or degradation occurs, however, can often be reduced through correct management.

This chapter describes some of the ways in which habitats are degraded or destroyed by humans. Habitat degradation or habitat destruction will be interpreted in this text as human-induced changes that most biologists consider to be negative in relation to native or desirable plant and animal populations and communities. It should be noted that the majority of this chapter addresses the habitat-degrading practices involved in human activities such as urbanization, agriculture, and logging. Management practices associated with these activities can also be beneficial to wild organisms, but the primary focus of this chapter is on the degradation they can cause. Chapter 15 will also address habitat alteration, but it will address purposeful alterations made by biologists specifically to affect particular populations or communities.

14.1 Air Pollution

Air pollution in its various forms has many implications with regard to habitat degradation. While many facets of air pollution could be discussed, three particular problems will be highlighted here: acid deposition, global climate change, and ozone depletion.

Acid deposition, which is often inappropriately called **acid rain** or **acid precipitation**, results primarily from fossil fuel combustion. When rain, snow, fog, dew, or solid airborne particles contain high levels of acidity, acid deposition occurs. Rain is naturally somewhat acidic; however, human activities can increase its acidity. As fossil

fuels such as coal and oil are burned, sulfur oxides and nitrogen oxides are released into the atmosphere. When combined with water they form sulfuric acid and nitric acid. When deposited on land or in water, these acids are harmful to many organisms.

The level of acidity or basicity is usually expressed as pH, a measure of hydrogen ion concentration (fig. 14.1). A neutral solution has a pH of 7. If a solution has a pH above 7, it is basic, or alkaline, and if it has a pH below 7, it is acidic. Each whole number on the pH scale from 1 to 14

pH	Substance
14	← NaOH solution (lye), 4%
	← Oven cleaner
13	← Limewater
	← Hair remover
12	← Household ammonia
	← Washing soda (about 1%)
11	
10	← Soap solutions
9	← Baking soda
8	← Bicarbonate of soda (about 1%)
	← Blood
7	← Pure water
	← Cow's milk
6	
	← Unpolluted rainwater
5	← Squash, pumpkin
4	← Tomatoes
	← Oranges
3	← Vinegar; soft drinks
2	← Limes
	← Normal stomach acidity (pH 1.0-3.0)
1	← Acid stomach

Basic / Neutral / Acidic

FIGURE 14.1 The pH scale. Unpolluted rainwater is already acidic, but acid deposition events generally involve much lower pH values. The pH scale runs from 1 to 14, with 7.0 being neutral, pH below 7 acidic, and pH above 7 alkaline.

represents a tenfold difference in hydrogen ion concentration. For example, a solution with a pH of 6 is ten times more acidic than a pH 7 solution, and a pH 4 solution is one hundred times more acidic than a pH 6 solution.

The environments most immediately affected by acid deposition are those with low alkalinity levels. For example, an area with soil pH of 7.2 would be more susceptible to acid deposition problems than an area with soil pH of 8.1. The more basic an environment, the more hydroxide ions it has; these hydroxide ions neutralize hydrogen ions. **Buffering capacity** is the ability of water or soil to react with and neutralize acids. Areas with good buffering capacity are not immune from acid deposition problems, but they are not as likely to exhibit immediate negative effects.

With respect to habitat of wild organisms, the effects of acid deposition can be manifested in numerous ways. By harming particular plants and animals, acid deposition can substantially alter habitats and affect the organisms dependent on those habitats. In aquatic systems, acid deposition can result in mortality among fishes, aquatic plants, and microorganisms (Howells 1990). Acid deposition may also occur at levels that do not kill organisms, but instead reduce their ability to reproduce or to survive early developmental stages. Adult brook trout, for example, can survive acidity levels that their eggs and young cannot tolerate. In addition, aquatic invertebrate numbers are reduced by acid deposition. Such a reduction affects fishes and birds, such as waterfowl, that depend on aquatic invertebrates as a food source (McAuley and Longcore 1988). Birds such as common loons and common mergansers that depend on fishes as a food source are also affected (McNicol et al. 1987). Acid deposition can also negatively affect the cycling of elements such as nitrogen, or act to release elements such as aluminum, lead, mercury, and cadmium that can be toxic to aquatic organisms.

Acid deposition also has the capacity to affect terrestrial plants. However, there are many unanswered questions regarding the effects of this pollution source on terrestrial plants. One well-established effect is that of acid deposition on the chemistry of soils and the concomitant effects on terrestrial plant roots (Raven et al. 1993). The development of roots, as well as their uptake of nutrients and water from the soil, are affected. Essential minerals, such as calcium and magnesium, readily wash out of acidic soils, while heavy metals, such as manganese and aluminum, dissolve in acidic soils and become available for absorption by roots in toxic amounts.

Forest decline is common in the portions of the world most severely affected by acid deposition. It is difficult to trace declines in a system as complex as a forest ecosystem to a single factor. Acid deposition, however, is a probable contributing factor, as are insects, diseases, other forms of human-induced pollution, and weather (drought or severe winters). Forest decline is probably a result of numerous factors acting in combination.

To date, parts of the northeastern United States and southeastern Canada have been the areas most severely affected by acid deposition in North America (fig. 14.2). Large areas of land in other parts of the world have also been negatively affected by acid deposition, and still other areas throughout the world are at risk. There is little question that acid deposition has had and will continue to have major negative effects on aquatic and terrestrial resources. In addition to its negative effects on fishery and wildlife resources, its economic costs resulting from the destruction of timber, decreased agricultural productivity, damage to human structures, and damage to domesticated livestock are large. Actual dollar figures for this damage are difficult to calculate, but in the United States, the figure is estimated to be $5 billion per year (Miller 1990). Until the acid deposition problem is adequately addressed, environmental degradation from this pollution source will continue.

Global climate change is another potential result of air pollution that has important implications for fishery and wildlife resources. Increases in **greenhouse gases** that result from human activities are thought to be contributing to global warming. Greenhouse gases include carbon dioxide, water vapor, ozone, methane, nitrous oxide, and chlorofluorocarbons. These gases, when re-

FIGURE 14.2 Soil sensitivity to acid deposition varies depending on the natural buffering capacity of an area. A large part of North America is sensitive to acid deposition.

leased into the atmosphere, may act somewhat like the glass in a greenhouse. The glass lets in visible light from the sun, but keeps some of the resulting infrared radiation (heat) from escaping; the resulting heat buildup raises the ambient temperature in the greenhouse. Human activities, such as fossil fuel combustion, release large amounts of carbon dioxide (CO_2) into the atmosphere (see section 2.2). Deforestation, especially in tropical rain forest areas of the world, also contributes to the greenhouse effect because when vegetation is removed, plants that would use CO_2 are eliminated.

A rapid change in global climate resulting from greenhouse warming would have severe ecological ramifications. Such ecological changes would obviously affect wildlife and fishery resources and would have many other far-reaching economic and social implications. World climate change is not new; climatic conditions on the earth have changed often through geologic time.

These past changes, however, usually occurred at a relatively slow pace and took place prior to the development of current human population levels. Gradual climatic modifications that affect ecosystems and their biota allow time for many organisms to either adapt or move. Rapid, human-induced climatic change may not allow the time necessary for organisms to adapt to the new conditions or disperse to areas with more appropriate climates. Habitat fragmentation (see section 14.7) adds to this problem by restricting the movements of some animals. Even if such human-induced changes were gradual enough so that most organisms could adapt to them, there would still be substantial negative economic and demographic effects on humans. For example, think of the effects of the melting of polar ice and the resulting higher sea levels that would result from global warming. Many coastal cities would be under water. Also think of the effects global warming would have on agricultural crops by changing precipitation levels and patterns.

Another result of air pollution that has important implications for fishery and wildlife resources is the depletion of the atmospheric ozone layer that protects the earth from ultraviolet radiation. Most of the potential negative effects of ozone depletion are usually mentioned in relation to human problems such as skin cancer, eye cataracts, severe sunburn, suppression of the immune system, or reductions in agricultural production. However, ozone depletion would affect many non-human organisms as well. Ozone depletion can result in negative occurrences such as retardation of the growth of plants, the death of larval organisms such as shrimp and crabs, and the depletion of plankton. Think back to the discussion in section 2.3 of trophic levels, food webs, and pyramids of energy and biomass. A reduction in plankton would reduce the bases of aquatic food webs and therefore negatively affect all higher trophic levels, including humans.

Ozone depletion has been associated with the use of chlorofluorocarbons (CFCs). Natural processes, such as volcanic eruptions, also affect ozone levels; however, there is considerable evidence that CFCs are the primary cause of ozone depletion. As CFCs are released from air conditioners and refrigerators (where they function as coolants), aerosol cans (where they function as propellants), and other sources, they enter the atmosphere. The CFCs move upward to the stratosphere, where, under the influence of ultraviolet radiation, they release chlorine atoms, which speed the breakdown of ozone (O_3) into oxygen (O_2). With less ozone in the stratosphere, more ultraviolet radiation can reach the surface of the earth.

Aside from their potential for direct damage to plants and animals, decreased ozone levels can also add to the global warming problem already mentioned. Currently, ozone depletion is primarily a problem in polar regions, especially in the Southern Hemisphere. While CFC usage is being reduced, this form of pollution is still a major concern.

The thin envelope of air surrounding our planet is essential to all living organisms. While various forms of air pollution can harm fishery and wildlife resources, the broader ways in which such pollution affects all organisms, including humans, should be of paramount concern.

14.2 Water Pollution

There are numerous materials that pollute water, some of which have more negative effects than others. Important aquatic pollutants include oxygen-demanding wastes, sediments, organic chemicals, inorganic chemicals, and heat. Aquatic pollutants can affect not only those organisms that live in water, but also those organisms that use water for drinking or other purposes. Thus, both aquatic and terrestrial species are affected by water pollution. The effects of pollution on surface water will be emphasized in this text, but pollutants also damage groundwater.

Oxygen-demanding wastes include domestic sewage, domesticated livestock wastes, and a host of other biodegradable organic wastes that cause water to be depleted of dissolved oxygen. Because the majority of aquatic organisms obtain their oxygen from that dissolved in water, such depletions can have severe negative effects. When aquatic systems receive large amounts of oxygen-demanding wastes, the decomposers that utilize these wastes increase in numbers. Many of these organisms (**aerobic decomposers**) use dissolved oxygen during their activities. Other decomposers (**anaerobic decomposers**) do not use dissolved oxygen, but can produce toxic substances, such as ammonia, through their anaerobic processes. As dissolved oxygen decreases or toxic substances increase, aquatic organisms are affected. These effects can be lethal, or they can act to reduce growth or reproduction, or cause organisms to be more susceptible to parasites, pathogens, and diseases. There is also a potential for the direct spread of diseases by this pollution source.

Sediments are water-insoluble particles, primarily soil, which erode into surface waters from surrounding land. Some sediments come from natural erosional processes, while others result from human activities such as agriculture, mining, logging, or construction. Large sediment particles settle to the bottom of a water body. Smaller sediment particles can remain suspended in the water column for longer periods of time. As sediments settle to the bottom, they can smother fish eggs or young, invertebrates, aquatic plants, and a variety of other organisms. They also, over time, reduce the depth of a water body. This "shallowing" effect can have substantial ecological ramifications. The smaller sediment particles that remain suspended can reduce the efficiency of sight-feeding organisms, can clog the gills of fishes or the filtering systems of filter feeders such as many mollusks, and, by lowering light levels, can reduce the primary production of an aquatic ecosystem. Sediments can also act as vehicles for the transport of nutrients, toxic metals, pesticides, herbicides, and other substances from terrestrial to aquatic systems.

Organic chemicals, such as oil, pesticides, detergents, and a wide array of other contaminants, can also have negative effects on aquatic biota. Some are cancer- or tumor-causing in humans and other organisms, some result in added nutrient input, others are toxic, and some cause damage by collecting on the outer surfaces of organisms. All can reduce the utility of water bodies as habitat for plants and animals. For example, federal, state, provincial, and territorial governments often issue advisories concerning water quality. These may take the form of fish consumption advisories, or even advisories against water contact or use.

Oil contamination originates from various sources, ranging from oil drained from a lawn mower to massive marine oil spills. The Exxon Valdez spill off the coast of Alaska illustrated the extensive damage and costs that can result from such contamination. Included in this source of pollution is all of the gasoline and similar hydrocarbons that evaporate into the atmosphere and eventually make their way into aquatic systems. Oil pollution from a variety of sources continues to be a major problem, and much more could be said about the negative effects of this form of contamination on aquatic and terrestrial systems.

Inorganic chemicals and nutrients such as toxic metals, acids, salts, nitrogen, and phosphorous can also be damaging to aquatic habitats. Acids released by mining activities, industry, or acid deposition can degrade habitat. Salts from irrigation runoff, urban storm drains, or industry can negatively affect aquatic systems. Toxic substances such as mercury, cyanide, arsenic, and chromium can toxify aquatic ecosystems. Nitrogen and phosphorous can be discharged into aquatic systems from a variety of sources. Excess amounts of these nutrients result in excessive algal growth or dense stands of other unwanted aquatic vegetation (fig. 14.3). Such excessive aquatic plant growth can block light needed by more desirable vegetation, can cause noxious odors, and can reduce the aesthetics of aquatic systems.

FIGURE 14.3 Excessive plant growth, such as this dense stand of water-fern in a Texas water body, can not only result in oxygen depletion, but can also produce toxic substances and reduce the recreational utility of a water body. (*Photograph courtesy of B. Blackwell.*)

Such growth can also interfere with a variety of recreational activities. In addition, when the excess vegetation dies, its decomposition can lead to decreases in dissolved oxygen and increases in toxic substances.

Heat can also be a water pollutant when water is removed from a water body and used to cool power plants or other industrial facilities. When the heated water is returned to the water body, it changes the aquatic habitat. Native plants and animals adapted to lower water temperatures may be unable to exist at the new higher water temperatures.

In addition, water released from a reservoir that is either warmer or colder than that normally found in a river or stream can degrade aquatic habitats (see section 12.3). This water may also lack oxygen if it is released from an oxygen-poor hypolimnetic zone. Such water releases can have ecological effects far downstream from the release point.

14.3 Urbanization

As the human population increases, the rate of urbanization also increases. Even if human population levels were stable, urbanization might continue. The growth of cities and suburbs has many ramifications for fishery and wildlife resources. Its most obvious effect is habitat loss.

As land use patterns change, the types of organisms present also change. Those species having specific habitat needs not met in human-developed areas will experience population reductions, while other species that can use these habitats will increase. The species aided may be desirable or undesirable ones. For example, house sparrows do well in urban areas, as do Canada geese. Most people consider Canada geese (in acceptable numbers) to be desirable and house sparrows to be undesirable. White-tailed deer and raccoons can also thrive in areas with large human populations. Many other species are generally intolerant of either the habitat changes resulting from urbanization or the increased presence of humans. The increased number of domesticated cats in urban areas, for example, has a negative effect on wildlife, especially on songbird numbers. Even if natural habitats remain within urban areas, these areas often exist as habitat fragments within the urban setting; the utility of these small islands of habitat to many terrestrial organisms is marginal (see section 14.7).

The effects of urbanization on aquatic organisms are usually even more profound. Natural and artificial water bodies are often drained and used as construction sites when urbanization occurs. Even if they are not totally drained, they are often reduced in size or in some other way negatively modified. If they cannot be drained, concrete-lined "rivers" are often the result; such "rivers" are the ultimate in habitat degradation. They have little utility for aquatic or riparian life forms. However, there is a current positive trend toward protection and creation of aquatic habitats in urban areas.

Another aspect of urbanization that should be mentioned is its effect on climate. Because of the heat-absorbing properties of structures such as buildings, roads, and parking lots, large urban areas tend to become islands of heat on a landscape. This heat helps to create environmental changes in urban areas, such as more rainfall, warmer temperatures in winter and summer, and cloudier conditions. These changes affect not only the urban environment but also the surrounding areas. If more rain falls on a large urban area, for example, will less be available for nonurban areas? Little is known about the ramifications of such climatic changes on the biota of these areas.

An additional effect of urbanization is increased interaction between humans and wild animals, which often leads to a variety of problems. Conflicts between people and animals are increasing as humans encroach on areas traditionally used by wildlife. For example, the building of a housing development in an area historically used by deer as a winter range can result in the deer causing damage to landscape plantings; it can also lead to reduced survival and poor health for the deer. To further compound the problem, mountain lions can follow deer into such areas, resulting in lion attacks on pets and people. In many areas, increases in Canada goose populations have led to problems for airports, golf courses, and a variety of other human enterprises. Houses, sheds, and other structures can provide excellent habitat for animals such as striped skunks, and the edible refuse associated with human habitations is an attractant to bears and raccoons. Many species, however, cannot exist in habitats intermixed with lawns or parking lots; the habitat needed by these species is completely lost when urbanization occurs. Try to envision a grizzly bear living in a human-inhabited urban setting.

While contact with wild organisms, both plant and animal, is desired by many people, problems do arise. By their nature, most nondomesticated creatures are wild, and they do not react to humans in the ways depicted in cartoon fantasies. Human tolerance tends to wane when a fox squirrel gets into an attic and shreds for a nest the quilt Great-Grandmother hauled across the Oregon Trail, or when deer eat prize roses, or when an American alligator eats the family dog for lunch. Additionally, some animals may lose their innate wildness in urban settings, becoming caricatures of their wild counterparts; the way we perceive them also changes. The wild and wily mallard, white-tailed deer, or yellow-bellied marmot may become a popcorn-eating creature that is quite different from its conspecifics in the wild. In many people's view this tends to degrade the organisms. Some animals, such as coyotes, have been able to adapt to new surroundings while still remaining essentially in a wild state. In many urban areas, coyote populations have been able to expand, but differ from their rural counterparts in being rather nonvocal in their behavior. Similarly, the American crow was in the past a crafty bird of rural areas that disdained places of human activity; it is now commonly found in urban areas but is still a wary creature.

While urbanization does have benefits for some aquatic and terrestrial species, its general results are usually habitat degradation and reduced wildlife and fishery resources. The field of urban wildlife and fisheries is an emerging and rapidly growing area of study. Based upon world human population growth projections, we can expect continuing urbanization and increased human-wildlife interactions.

14.4 Agriculture

Of all human activities, none has directly modified large areas of natural landscapes more than agriculture. Vast expanses of land, especially in the temperate grassland and temperate forest biomes, have been changed by this activity. As human populations have increased, their need for food and natural fibers has increased as well. As agricultural technology has changed, the effects of agriculture on habitat have also changed. Some of these effects have been essentially unavoidable as the need for land devoted to agriculture has increased. Some agricultural activities, however, have had unnecessarily negative effects because of the manner in which they have been and are being conducted.

North American agriculture, more so today than ever, is influenced by a wide array of economic and social forces. Capital outlay has increased as land prices have increased; as equipment has gotten larger, more complicated, and more costly; and as herbicide, pesticide, and fertilizer use and costs have increased. It has become increasingly difficult for the agricultural community to spend money or forgo income to benefit wildlife and fishery resources. However, practices such as lax pollution controls and agricultural subsidies, which shift costs from agricultural producers to the general public, are being scrutinized more now than at any time in the past. The costs of agricultural activities that degrade the environment, which were once passed on to the commons and to other private property, are now more often being viewed as costs to those who cause the degradation. The **commons** are those resources such as air, water, wildlife, fish, and publicly owned land that are owned by everyone but by no individual in particular. More will be said about the commons in section 14.10.

Agricultural activities have changed terrestrial and aquatic habitats in a variety of ways. Some of these changes can be viewed as beneficial, but many have been primarily negative. Generally positive changes include pond construction and the production of food for wildlife. Irrigated farmland in arid regions has provided abundant food and water for a variety of wildlife species. Some domesticated animal grazing systems benefit both wildlife and domesticated livestock. Agricultural practices that clear woodlands for small fields have been beneficial to such species as the northern bobwhite and eastern cottontail, although

they have not been beneficial to other species, such as the American redstart and black bear. Some of these habitat changes were and are essentially unavoidable as humans in their use of land, change landscapes.

Some other agricultural activities have been unnecessarily damaging. Some land use practices promote accelerated soil erosion, which clogs water bodies with sediments (fig. 14.4). Increasingly efficient methods of draining lowlands, including the use of dikes, have encouraged the spread of agricultural crops into marginal areas. Increased use of fertilizers and pesticides has also had negative effects on terrestrial and aquatic wildlife. The change from smaller to larger field sizes has negatively affected wildlife by reducing interspersion and edge habitat. The removal of obstructions such as trees or fencelines, which hinder the movement of irrigation equipment or increasingly large tractors and farm implements, has also had nega-

FIGURE 14.4 Planting row crops too close to a stream with no buffer zone leads to excess siltation. Note the corn at the stream margin and the corn that has fallen as a result of erosion. (*Photograph courtesy of C. Berry.*)

FIGURE 14.5 Little habitat for wild organisms remains in areas where agriculture of this intensity is practiced. Species diversity is a casualty in such areas. *(Photograph courtesy of R. Meeks.)*

tive effects. Fence-to-fence agriculture, and in some cases the total removal of fences and their associated habitat, has been harmful (fig. 14.5). The use of irrigation in agriculture has also had some widespread negative effects on both terrestrial and aquatic systems. Irrigation has allowed the movement of agriculture into areas previously too dry for such pursuits, with concomitant negative effects on most of the native biota of such areas.

Another negative aspect of agriculture has been its tendency toward monoculture systems. **Monoculture** is the growing of a single species, such as corn or wheat, over a large area. In addition to reducing habitat diversity, a monoculture provides excellent habitat for certain plant pathogens, and undesirable plant species that thrive in the monocultured crop. The damage caused by these organisms elicits an agricultural response in the form of herbicides and pesticides that can cause additional environmental damage. Monoculture is a good example of how changing a habitat reduces some species but aids others. What more could a corn rootworm, a pest on that crop, want than field after field planted with corn? Species diversity is an early casualty of monoculture.

Agricultural practices that involve domesticated livestock also have caused some unnecessary problems. When livestock grazing is properly managed, it can benefit vegetation and wild animals. However, incorrect grazing practices can be damaging to vegetation as well as to wildlife and fishery resources, and will eventually be costly to the landowner as well. If too many cattle, sheep, or goats are fenced into too small an area for too long a time, habitat degradation occurs. Domesticated livestock grazing activities can reduce the numbers and biomass of desirable grasses (called **decreaser species** because they decrease under grazing pressure) and increase the numbers of less desirable grasses. These less desirable grasses can be **increaser species** (which increase under grazing pressure) or **invader species** (which invade from other areas into heavily grazed areas). These changes can have profound effects on wildlife using the area. Mechanical damage to streambeds and riparian zones is another common result of domesticated livestock activity. Riparian zones are attractive feeding and watering places for domesticated livestock, and this often leads to overuse of such areas by the animals (fig. 14.6A and B).

In the United States today, there are a number of contentious issues concerning domesticated livestock grazing, as well as mining and other consumptive pursuits, on federally owned land by private citizens and companies. There are questions about whether current levels of grazing can be sustained on a long-term basis and how much of a negative effect grazing has on wild animals and their habitat. There is also concern about whether other uses of federal land, such as camping, fishing, hiking, and wildlife viewing, are receiving adequate attention and prioritization. To further fuel this issue, the grazing fees currently charged on federally owned land are below the fees charged on similar private land. Federal laws require that federally owned lands be managed for multiple purposes (see section 18.8), but there are questions concerning the effectiveness of such laws. Because federal lands cover such a large area, especially in Western states, the way grazing and other activities are managed on these lands is of great importance.

Changes are occurring in agriculture, as evidenced by increases in minimum-till and no-till

A

B

FIGURE 14.6 Riparian zone degradation due to domesticated livestock grazing affects both terrestrial and aquatic organisms. The damage can be very obvious (A) or more subtle (B). The scene depicted in (B) may look fairly normal, but the degradation reflected by sloughing stream banks, a lack of shoreline vegetation and undercut banks, and other problems indicates poor grazing management. (*Photographs courtesy of W. Platts.*)

farming, more efficient use of fertilizers and pesticides, removal of some marginal land from crop production, and other improvements that are reducing environmental damage. Regardless of the environmental effects of agriculture, one aspect cannot be changed: agriculture will always change landscapes, with some species being benefited and others being harmed.

14.5 Logging

Products made from wood have long been used by humans in a variety of forms, as has wood as a fuel source. To obtain wood, trees must be cut. In some situations the cutting of trees has minimal

or positive effects on wild animal populations, while in other situations such cutting can be devastating. The effects are dictated by the scope and kind of logging practiced and the wild organisms in question.

When a tree is cut, individual organisms using that specific tree are affected. While individual organisms will always be affected in this way by logging, its most negative effects occur when entire populations or communities are adversely influenced. For example, while the removal of an old tree serving as the home of a cavity-nesting organism, such as a southern flying squirrel, is important to that individual organism, logging has a greater effect when so many old trees are removed that the population of that cavity nester is substantially reduced.

In most forest management activities, humans tend to replace multispecies forests with monoculture forests. This type of ecosystem simplifica-

FIGURE 14.7 Clear cutting of large areas can reduce the utility of forested areas to both terrestrial and aquatic organisms. Severe erosion is often associated with this forest management practice. *(Photograph courtesy of K. Jenkins.)*

tion causes problems similar to those described for agricultural crops by reducing habitat diversity and in turn the species diversity of organisms that inhabit the forest. Mature natural forests generally have several layers of understory and overstory, including shrubs, smaller tree species, larger tree species, and other layers. They also tend to have an interspersion of differently aged vegetation. Monoculture reduces these layers by encouraging not only a single species of tree but also a single year class of trees. As tree diversity is reduced, so are food supplies, nesting areas, and other niche characteristics used by various animals. Such monoculture practices also promote larger-scale problems with diseases, parasites, pathogens, and undesirable species. The natural system of checks and balances inherent with species diversity is lost.

When extensive logging occurs, habitat fragmentation (see section 14.7) occurs. Extensive logging has also been shown to increase the vulnerability of animals such as elk to hunting and human disturbance. Logging, particularly where **clear cutting** (removal of all trees) occurs (fig. 14.7), can also lead to accelerated soil erosion and stream warming that negatively affects organisms living in water bodies within the watershed. Erosion and warming affects downstream areas as well. Pacific salmon fisheries, for example, are affected by timber management practices not only in the streams where the fishes reproduce, but also in downstream areas through which they migrate.

One of the primary characteristics of logging activities is the road network needed for log removal. Such roads usually damage stream habitat through the resulting erosion and vehicular traffic. Additionally, a newly constructed logging road network opens an area to increased human use for other activities. Although this can be beneficial for some recreational activities, areas previously accessible only to foot traffic are opened to a variety of motorized vehicles, even after logging operations are completed. Areas that were once remote and received little human disturbance become more accessible; this increased accessibility can have negative effects on some species. A road placed too close to a bald eagle or northern goshawk nest, for example, may cause the birds to abandon the nest or to avoid further use of the nest site.

Destructive forest wildfires throughout North America in the late 1800s and early 1900s were followed by increased fire suppression activities. The natural burning patterns of forests were disrupted by fire suppression, and this resulted in habitat changes, often to the detriment of species diversity. Forest fires that occur in areas subjected to frequent burning are less destructive than fires that occur in unburned forests with a great deal of fire fuel buildup. However, logging does not always lead to fire suppression; the burning of unusable parts of trees from logging operations during droughts has led to wildfires and loss of human life (Biswell 1989).

Not all logging activities need be destructive. **Selective cutting** of specific trees or **block cutting** (clear cutting of small areas) can usually be accomplished with less damage, especially if care is taken in log removal and road construction. In some cases block cutting can even serve as a partial substitute for fire. Some wildlife species can be benefited in numerous ways by correct logging practices. Ruffed grouse, for example, use aspen trees of a variety of sizes. Managing aspen stands so as to maintain a combination of recently cut areas, young-tree areas, and mature-tree areas will fill a variety of ruffed grouse habitat needs. Similarly, wild turkeys can benefit from the creation of forest openings, as they use such sites for feeding purposes. However, these forest openings may also increase populations of other less desirable species, such as the parasitic brown-headed cowbird. This species, which lays its eggs in the nests of other species, prefers forests with openings. By creating habitat for the brown-headed cowbird, logging can reduce the nesting success of other birds.

The cutting of dead trees can sometimes be beneficial. In some areas, massive outbreaks of insects, such as western pine beetles, have produced large blocks of forest in which over 50% of the trees are dead. Salvage logging of these areas reduces fire danger, provides timber, and promotes forest regeneration. It also reduces the habitat available to a variety of species such as cavity-nesters.

In the last thirty to forty years, many areas of public land in the western United States have been logged even when this resulted in below-cost timber sales (a form of subsidized logging). The usual arguments in favor of this practice concern jobs and the control of construction material costs. This practice has, however, disrupted much forest habitat in the western United States. Logging practices on federally owned land will continue to receive much attention from a variety of groups, such as logging companies, politicians, advocacy groups representing a variety of interests, biologists, people who make their living by logging, and others.

Humans will continue to use wood for a variety of purposes. When logging is correctly managed, the damage it causes can be minimized. The potential for short-term and long-term damage to aquatic and terrestrial populations, like the value of the timber itself, is an economic factor that should be considered when logging is planned. All too often in the past, only the obvious monetary value of the timber removed was considered in economic analyses. When decisions are made concerning the location, extent, or kind of logging to be carried out, they should be based on full economic considerations. The timber is not the only resource involved.

14.6 Wetland Loss

Wetlands have often been viewed by many people as "wasteland." Coastal wetlands, floodplain forests, bogs, marshes, and other wetland types have generally been viewed as places to drain, fill, or in some other way make dry. Over the last two hundred years, housing developments, harbor building, agricultural encroachment, the diking of rivers, and a variety of other human activities have led to extensive losses of wetlands (fig. 14.8). California, for example, has lost 91% of its wetlands, Ohio 90%, Iowa 89%, Indiana 87%, and Missouri 87%. Twenty-two states have lost over one-half of their wetlands (Dahl and Johnson 1991).

The value of wetlands to both terrestrial and aquatic organisms, including humans, is great. Wetlands represent a habitat type that is exceptionally productive. A large percentage of commercially important marine species live a portion of their lives in coastal wetlands and estuaries. North American waterfowl populations are dependent on wetlands, as are vast numbers of non-game organisms, ranging from salamanders to

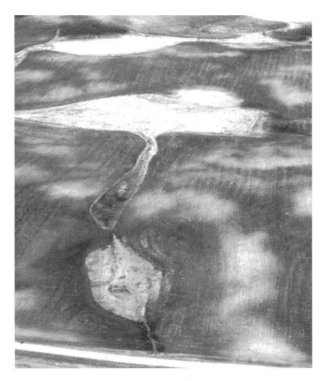

FIGURE 14.8 Wetland drainage not only results in decreased wildlife and fishery resources, but also negatively affects many of the other benefits of wetlands to humans. The drainage ditches shown in this photograph demonstrate how wetlands are often drained into road ditches and other wetlands. *(Photograph courtesy of the U.S. Fish and Wildlife Service.)*

butterflies. Wetlands also contain large numbers of plant species dependent on this type of habitat.

Wetland inhabitants are adversely affected by wetland loss in numerous ways in addition to the actual loss of their habitat. For example, as wetland area declines, waterfowl and shorebirds are forced into smaller and smaller areas. Such concentrations increase the likelihood of disease outbreaks. Great numbers of ducks and geese concentrate in the late winter and early spring in the Rainwater Basin area in south central Nebraska. This is an area where over 90% of the natural wetlands have been drained. Thousands of waterfowl often die during this concentration period due to avian cholera.

Wetlands serve as natural water purifying systems, and as those systems are reduced, water pollution problems are aggravated. The microclimate of an area is modified when water is artificially removed through wetland loss. This climate modification affects all organisms, including humans and their agricultural crops and domesticated livestock. Numerous other direct and indirect negative consequences can be attributed to wetland loss.

Wetlands are currently one of the focal points in an often heated discussion concerning habitat maintenance versus personal property rights (see section 14.10). The benefits of wetlands extend to all people, but wetlands are mainly the property of individuals. This debate is an excellent example of the way biological, social, economic, and other issues can interact to create a complex problem. It is sometimes argued that wetlands are only good for ducks, mosquitoes, snakes, or whatever other wild organisms directly use the area. Those who support wetland drainage argue that these areas are being made "useful" to humans when they are drained and used for direct human purposes such as agricultural crops or buildings. A person making such a statement is demonstrating a lack of knowledge concerning the negative ecological and economic ramifications of wetland loss to humans, as well as the negative effects of that loss on wild organisms. Humans are a part of the ecosystems in which they live, not separate from or above them.

14.7 Habitat Fragmentation

Clear cut forests, power line rights-of-way, series of dams and reservoirs on rivers, agricultural fields, highways, and urban sprawl all have two things in common: all are the result of human activities and all result in habitat fragmentation (see section 2.11). The isolated bits of habitat resulting from these activities do not function ecologically in the same way as comparably sized undivided habitat blocks. Such altered landscapes can affect some animal migration patterns, nesting or roosting patterns, feeding behavior, and a myriad of other activities pursued by organisms. In addition, some species simply need undisturbed areas. Roads and accompanying traffic, activities associated with power lines, and recreational activity on water bodies may prevent some animal species from using an area.

There are some wildlife benefits from habitat fragmentation, such as increased ecotone areas and the creation of openings in large blocks of timberland, but these benefits must be weighed against drawbacks. One of the most pressing problems in both wildlife and fishery work will be finding ways to ameliorate the negative effects of continued habitat fragmentation.

14.8 Channelization

Stream channelization and associated activities, such as dredging (primarily for deepening water bodies) and snag clearing (removal of dead trees), are forms of habitat degradation that affect both aquatic and terrestrial biota. In our efforts to improve shipping channels or to move water more rapidly out of an area by straightening or deepening streams and removing flow obstructions, humans change the characteristics of lotic systems and the biota that can inhabit them. The ability of lotic systems to release energy is modified, and this energy is often expended in ways that harm the general environment and the works of humans. In many situations our human purposes seem contradictory. On the one hand we view water as an important economic resource to be saved, protected, and stockpiled, while on the other hand we try to move it out of an area as rapidly as possible so that we can plant crops (which need water) or construct buildings in floodplains.

The initial problem with channelization is that stream or river length is reduced. Mark Twain, in his book *Life on the Mississippi*, described such a loss of river length on the lower Mississippi River between Cairo, Illinois, and New Orleans, Louisiana. In his time as a riverboat pilot on the Mississippi in the mid-1800s, the distance between the two towns decreased by 67 miles because of cutoffs that changed the river channel. These human-influenced cutoffs decreased river length from 1,040 miles to 973 miles. He calculated that the river was being shortened by 1.3 miles per year, and that at that rate of loss, in 742 years the Mississippi River between Cairo and New Orleans would be just 1.75 miles long. The point is that as a river is straightened, river length (and therefore habitat) is lost, if perhaps not to the extent that Mark Twain facetiously predicted.

Even after rivers are shortened, further degradation occurs. Dredging to deepen channels and snag clearing to remove obstructions, two other methods of speeding water flow, are also destructive to habitat. Dredging disturbs the development of benthic communities. In addition, materials dredged from such areas need to be placed somewhere; this placement can cause further habitat damage. Snag clearing removes one of the most important habitat features of a lotic system. The areas around snags have been shown to be one of the most attractive locations for fishes and other aquatic organisms. Even organisms such as birds, reptiles, and amphibians utilize these snags.

Most human efforts to move water more rapidly out of an area cause further problems. That water must go somewhere; usually it becomes a problem for other humans downstream.

Another form of river and stream degradation that should be mentioned is dam construction. The impoundment of a free-flowing river changes that water body from a lotic into a primarily lentic environment, thereby destroying lotic habitat (see section 12.3). The construction of impoundments also establishes barriers to the movement of organisms, especially fishes. These barriers have their greatest effects on migratory species. Many efforts, usually only partially successful, have been made to assist various fish species around or over these obstructions. The series of fish ladders and other structures built into dams on Pacific Northwest rivers to assist Pacific salmon are an example.

14.9 Introduced Organisms and Competition

The previous sections of this chapter have highlighted habitat degradation or destruction resulting from direct physical changes in habitat caused by humans. Biological changes that result from competition with or activities of other organisms can be just as destructive as the previously mentioned forms of degradation. Some of these changes occur through natural expansion in the geographic range of an organism, while others are obviously induced by humans through the accidental or purposeful introduction of organisms. An example of the latter would be habitat degradation resulting from the grazing of domesticated livestock (see section 14.4).

Introduced organisms other than domesticated livestock can also result in degraded habitats. The exotic common carp, for example, has caused widespread aquatic habitat damage that has contributed to the reduction of numerous native organisms in North America. The activity of this species, among other effects, increases water turbidity and decreases the abundance of rooted aquatic plants. Common carp have negatively affected populations of terrestrial organisms associated with water bodies as well as native aquatic organisms. The exotic house sparrow competes for nest sites with native birds such as the purple martin and eastern bluebird. Kudzu is an introduced plant that has degraded large terrestrial areas in the southeastern United States. Purple loosestrife has had and is having negative effects on native plant species in wetland areas. There

are many other examples of introduced forms that have caused habitat degradation. Introductions sometimes seem like simple solutions to problems, or at worst, actions that will have little or no effect. However, one added species or one subtracted species can make the difference between a correctly functioning ecosystem and one that is degraded and unstable.

14.10 Habitat Degradation, the Commons, and the Takings Issue

Pollution and other forms of degradation result in habitat modification. The commons, as already described, represents resources such as air, water, wildlife, fish, and publicly owned land that are owned by everyone but by no specific individual. The takings issue involves a response by some people to protect what they believe to be their rights as private property owners. They believe that they should not be inhibited from pursuing activities involving their private property. If regulations or laws intended to protect the commons and other private property reduce their ability to use their property in some way, they believe that this represents a taking of their property rights. In some cases when they believe that takings have occurred, these private property owners believe that they should be monetarily compensated. In the United States, such compensation has been sought through legal action for a variety of perceived takings, including civil rights laws, public health and safety laws, conservation measures, and others.

The takings issue has far-reaching implications for natural resource conservation with regard to a variety of regulatory activities such as clean water and endangered species legislation and the decisions of land-use planning boards and commissions. Following are sample scenarios of situations in which takings would be considered by some people to be occurring. A property owner wants to construct an apartment building, but cannot because the area is zoned for single-family dwellings. This would represent a taking of the ability of that property owner to collect rent from multiple apartment units rather than from one house. A land developer owns a piece of property that has on it one of the few remaining populations of a particular plant species. If the property owner is inhibited from building an amusement park on that piece of land because of that plant, this would represent a taking. A farmer wants to plant an agricultural crop within a few meters of a lake, but law or regulation prohibits farming within 30 meters of that lake. Those 30 meters would represent a taking, and compensation would be due for the loss of income from the crops that would have been planted on that strip of land.

To better understand the relationships among pollution, habitat degradation, the commons, private property, and the takings issue, one must first briefly review the history of pollution control in the United States. Initially, pollution control was directed at **point source pollution**. Point source pollution emanates from single, identifiable, concentrated sources, such as factory smokestacks or drainpipes or municipal sewage effluents. These sources release pollutants into the environment at a specific location and represent the most easily identifiable pollution sources. Laws and regulations were promulgated that required some point polluters to reduce pollution from these sources. One of the rationales underlying such regulation or law involves the costs that were being passed on by polluters to the commons and to other private property owners. For example, when pollution emanating from a sewage effluent pipe degrades a resource such as a river or lake, that degradation reduces the value of that resource. Someone will eventually have to pay to correct the action of the polluter. It might be a local government that incurs costs to improve the water body for recreation, it might be another industry that

incurs costs to make the polluted water usable in its industrial process, or it might be an individual landowner who wants to use the water for some purpose but cannot until it is purified. Controls on point source pollution shifted the cost burden, at least partially, from other private property owners and the commons to the polluter. Initially there was resistance to this shift in costs, but the shift is now generally well established in the mind of the public. However, there are some people who would like to see such laws and regulations reduced or eliminated; to them, such laws and regulations represent takings and a threat to economic development.

The next phase in pollution control, which is at least partially responsible for fueling the takings issue, involves **nonpoint source pollution**. This term refers to more diffuse and dispersed forms of pollution that are not released at a single identifiable point. For example, fertilizer runoff from a dozen farms is nonpoint source pollution. It does not enter a pipe and then flow into a lake or onto a neighbor's land at one point; it enters at multiple points. Efforts to control nonpoint source pollution and to partially shift its costs to the nonpoint polluters have met with resistance. This resistance is just one facet of the takings issue response by some people to increased regulation of activities conducted on private property.

The concept of the right of an individual to do what he or she wants to with property he or she owns may seem like a cornerstone of the free enterprise system. Usually property rights and takings arguments are based on the Fifth Amendment of the U.S. Constitution, which states that private property "shall not be taken for public use without just compensation." However, there is a long legal and court history that addresses what represents a taking, and there are many laws and regulations that limit the rights of property owners. For example, there are laws and regulations that prohibit one property owner from draining water from his or her land onto the property of another person. There are rules that keep someone from opening a manufacturing plant or a cocktail lounge next to your home. There are regulations that prohibit people from having noxious plants, such as Canada thistle, on their land. A listing of all the controls on what individuals can or cannot do on or to their own property would be lengthy. All are legal safeguards that protect the personal property rights of others, the health and safety of the public, the civil rights of all people, and the commons. For example, all of the above scenarios that limit the property rights of some protect the rights of others. The rules or regulations mentioned above, and many others, do represent the limitation of some property rights, but such limitations are generally not supported in court cases as infringements of Fifth Amendment rights. Such rules and regulations are generally accepted by most people, even most advocates of the takings issue.

Most of the takings issue problems involving habitat degradation occur in the course of attempts either to shift some pollution costs back to the degraders or to modify land use practices away from activities that property owners have traditionally pursued with little or no regulatory interference. Unfortunately, the effect of what someone does on or to their own property does not stop at their property line, nor does that effect remain a problem only for the commons. Pollution and habitat degradation caused by airborne soil particles resulting from improper agricultural practices can be used as an example. Consider a few of the locations where these soil particles come to rest. They enter the lungs and eyes of people, domesticated livestock, and wild animals. They end up in automobile, tractor, and airplane engines. They fill lakes, ponds, and drainage ditches. They invade homes, factories, restaurants, and grocery stores. It is obvious that these airborne soil particles affect many people in many ways. In addition, the loss of topsoil eventually results in losses to the polluting landowner through reduced crop production.

The question is, at what point do the rights of the private property owner supersede the rights of other people? In the United States, two hundred years of laws and court decisions have upheld the tenet that no one has the absolute right to use property in a way that may harm the public health or welfare or that may damage the quality of life for neighboring property owners or for the community as a whole. However, a continuous stream of national and state legislation that addresses the takings issue is currently being introduced—and in many cases passed. Such legislation either provides for monetary compensation for perceived takings, or imposes increased paperwork requirements on agencies or boards that serve to impede perceived takings (red-tape legislation).

The takings issue will continue to be a divisive one. Because of its complexity, it will not be easily resolved. It represents an area in which continued efforts at conflict resolution are needed. Conflict resolution with regard to the takings issue has and probably will continue to be primarily political and legal in nature. Section 20.5 further addresses the topic of conflict resolution.

SUMMARY

Habitat can be degraded by numerous natural and artificial means. Human-influenced degradation can be intentional or unintentional. As habitats are degraded, they become more usable to some organisms and less usable to others.

Three types of air pollution that have substantial effects on wild organisms are acid deposition, potential global climate change, and ozone depletion. Environmental changes in pH affect both aquatic and terrestrial organisms. Global climate changes may result from increased greenhouse gases in the atmosphere. Ozone depletion results primarily from the use of chlorofluorocarbons.

Important water pollutants include oxygen-demanding wastes, sediments, organic chemicals, inorganic chemicals, and heat. Oxygen-demanding wastes reduce habitat usability for organisms that obtain their oxygen from that dissolved in water.

Urbanization results in habitat loss and increased interactions between people and wildlife. The field of urban wildlife and fisheries is an emerging and growing area of study.

Agricultural activities modify large areas of natural landscapes. The increased use of herbicides, pesticides, and fertilizers has had dramatic effects on wildlife and fishery resources, as have increased field size and monoculture. Domesticated livestock can also contribute to habitat degradation.

Timber management practices vary in their effects on wildlife and fishery resources. Not only does the cutting of trees represent an agent of habitat change, but so do other aspects of forestry practices such as monoculture and road construction. All too often, economic analyses of logging revolve around timber value alone. The timber is not the only resource involved.

Wetlands are all too often viewed as "wastelands" that need to be used for the obvious, immediate economic benefit of humans. The economic benefits of wetlands go far beyond their values associated with conversion to agricultural crops or housing developments, or even as wildlife and fishery resources. They are an integral part of the landscape that contributes many economic, ecological, and aesthetic benefits.

Habitat fragmentation due to activities such as forest clear cutting, power line rights-of-way, and highway construction reduces the utility of habitat to most wild organisms. Isolated bits of habitat do not function ecologically in the same manner as comparably sized unfragmented habitat blocks.

Channelization and associated activities, such as dredging and snag clearing, are forms of habitat degradation that affect lotic systems. River length is shortened and specific habitats are reduced. Dams serve as barriers to migratory animals and also change lotic water bodies into lentic habitats.

Purposefully and accidentally introduced organisms can degrade physical habitat or compete with native organisms. Introductions often seem to be a simple solution to problems, but they can easily degrade or destabilize ecosystems.

Pollution, habitat degradation, the commons, private property, and the takings issue are all interrelated. The takings issue is primarily the result of attempts to have degraders of natural resources pay more of the costs of their polluting activities and to modify habitat-degrading land use practices that property owners have traditionally pursued with little or no regulatory interference.

PRACTICE QUESTIONS

1. Describe the process that results in acid deposition.

2. List some human activities that result in the release of greenhouse gases and describe how these gases may lead to global climate changes.

3. How do oxygen-demanding wastes lead to aquatic habitat degradation? What are some of the sources of oxygen-demanding wastes?

4. What causes urban heat islands? Select five species of wild animals with which you are familiar and describe how this increased heat could affect them.

5. List a dozen different ways in which agriculture can negatively affect wild organisms.

6. Define decreaser, increaser, and invader species, and describe some of the ecological and economic ramifications associated with decreased populations of decreaser species.

7. What is meant by forest clear cutting, and what other less ecologically destructive forest management practices can be used?

8. List some benefits of wetlands other than their benefits to wildlife and fishery resources.

9. What is habitat fragmentation? List a number of human activities taking place where you live that contribute to habitat fragmentation.

10. What are some of the purposes of channelization, and how might changes resulting from channelization affect terrestrial wildlife?

11. What is involved in the takings issue? Describe four situations in the area where you live in which takings could be an issue.

SELECTED READINGS

Burger, G. V. 1978. Agriculture and wildlife. Pages 89–107 *in* H. P. Brokaw, ed. *Wildlife and America.* Council on Environmental Quality, Washington, D.C.

Cairns, J. Jr. 1978. The modification of inland waters. Pages 146–162 *in* H. P. Brokaw, ed. *Wildlife and America.* Council on Environmental Quality, Washington, D.C.

Cairns, J. Jr. 1986. Restoration, reclamation, and regeneration of degraded or destroyed ecosystems. Pages 465–484 *in* M. E. Soulé, ed. *Conservation biology: The science of scarcity and diversity,* Sinauer Associates, Sunderland, Mass.

Clark, E. H. 1985. *Eroding soils: The off-farm impacts.* The Conservation Foundation, Washington, D.C.

Commoner, B. 1990. *Making peace with the planet.* Pantheon Books, New York.

Ehrlich, P. R., and A. H. Ehrlich. 1991. *Healing the planet: Strategies for resolving the environmental crisis.* Addison-Wesley, Reading, Mass.

Everhart, W. H., A. W. Eipper, and W. D. Youngs. 1975. *Principles of fishery science.* Cornell University Press, Ithaca, N.Y.

Hedgpeth, J. W. 1978. Man on the seashore. Pages 163–195 *in* H. P. Brokaw, ed. *Wildlife and America.* Council on Environmental Quality, Washington, D.C.

Leopold, A. S. 1978. Wildlife and forest practices. Pages 108–120 *in* H. P. Brokaw, ed. *Wildlife and America.* Council on Environmental Quality, Washington, D.C.

Mason, C. F. 1991. *Biology of freshwater pollution.* 2d ed. John Wiley & Sons, New York.

Wagner, F. H. 1978. Livestock grazing and the livestock industry. Pages 121–145 *in* H. P. Brokaw, ed.

Wildlife and America. Council on Environmental Quality, Washington, D.C.

LITERATURE CITED

Biswell, H. H. 1989. *Prescribed burning in California wildlands vegetation management.* University of California Press, Berkeley.

Dahl, T. E., and C. E. Johnson. 1991. *Status and trends of wetlands in the conterminous United States, mid-1970s to mid-1980s.* U.S. Department of the Interior, Fish and Wildlife Service, Washington, D.C.

Howells, G. P. 1990. *Acid rain and acid waters.* Ellis Horwood, New York.

McAuley, D. G., and J. R. Longcore. 1988. Food of juvenile ring-necked ducks: Relationship to wetland pH. *The Journal of Wildlife Management* 52:177–185.

McNicol, D. K., B. E. Bendell, and D. G. McAuley. 1987. Avian trophic relationships and wetland acidity. *Transactions of the Fifty-second North American Wildlife and Natural Resources Conference* 52: 619–627.

Miller, G. T. Jr. 1990. *Resource conservation and management.* Wadsworth Publishing, Belmont, Calif.

Raven, P. H., L. R. Berg, and G. B. Johnson. 1993. *Environment.* Saunders College Publishing, Fort Worth, Tex.

15

HABITAT MANAGEMENT

A large proportion of the strategies used to conserve wildlife and fishery resources involve the management of habitat. Habitat is often manipulated to improve or degrade the ecological characteristics required by a particular species or group of species. Such habitat manipulations also have positive and negative effects on nontarget organisms, and these effects should be taken into consideration when implementing any habitat management technique. A species or community may simply be more or less desirable to humans in that particular place or at that particular time. The desirability of a species or community depends on the management objectives on that specific situation. For example, improving habitat for Canada geese may be an objective in a natural setting, but degrading habitat for that species may be an objective where the geese are a problem, in, for example, an urban setting.

In some instances, the management strategy implemented may address multiple needs of a population or community. For example, controlled burning of an area may be done to improve food supply, escape cover, nesting cover, and other habitat characteristics for grassland songbirds. The management practice implemented may also address a single need. For example, the construction and placement of American kestrel nesting boxes is directed only at the reproductive requirements of that species.

Habitat management can take many forms. It can involve the addition of natural structures, such as gravel and rock to encourage walleye spawning. It can involve the removal of natural structures, such as trees if grassland species are the target organisms. It can involve the addition of artificial objects, such as eastern bluebird nesting boxes, or the removal of artificial objects, such as old buildings to reduce striped skunk populations. Habitat management can also be aimed at modifying the ecological succession that would naturally occur if no action were taken. For example, the water level in some wetlands can be raised to inhibit cattail growth.

Most habitat management activities in the past have been directed primarily at single species of sport or commercial importance. There is, however, an increasing tendency to direct habitat - management efforts at multispecies combinations. In addition, more effort is being directed at habitat management for native and nongame species now than in the past.

Habitat management has always helped more species than just the target organisms; most habitat management efforts are beneficial to a variety of species. However, the recent trend toward the promotion of species diversity requires that biologists take more of an ecosystem approach to habitat manipulation, rather than just directing activities at a single species. In reality, such an ecosystem approach can usually benefit sport or commercial species as well as the other organisms present.

The purpose of this chapter is to provide an overview of some of the most commonly used habitat management techniques. This discussion will follow the order of the habitat types presented in chapter 12: terrestrial habitats, terrestrial and aquatic transitional habitats, standing water habitats, flowing water habitats, and estuarine habitats.

15.1 Management of Terrestrial Habitats

The management techniques employed in terrestrial habitats are highly diverse. Common techniques include efforts at improving food production, cover, and water availability. As previously mentioned, some techniques are intended to benefit particular species or communities, while other techniques are primarily intended to reduce the utility of (degrade) habitat for some species or communities.

Food Production

Food production techniques in terrestrial habitats typically concentrate on mast, browse, or grasses and forbs, including agricultural crops. As mentioned in section 1.3, most managed wildlife species are primarily herbivorous; therefore, this discussion primarily involves vegetation management. Even if carnivorous wildlife species, such as mountain lions or northern harriers, are the target species, most of the habitat management techniques implemented still involve vegetation management. Predator population abundance is related to food availability, and predators often prey on herbivorous species. If vegetation is managed for herbivorous prey, populations of carnivores will be affected. If insectivorous birds, such as Brewer's sparrows or Blackburnian warblers, are the target species, the protection of food supply is often a concern. In such situations, biologists may monitor pesticide use and runoff from agricultural and urban areas to ensure that insect populations are not depleted. Because these insect prey species are often herbivores, habitat management is still involved.

Three primary methods used to improve food production are propagation, release, and protection. **Propagation** involves the planting of seeds or the transplanting of young trees or shrubs. Shelterbelt planting, for example, is a type of propagation. Plant species such as Russian olive, red cedar, and wild plum are often included in shelterbelt plantings because the mast they produce attracts game and nongame wildlife species. Propagation is often used in urban settings to provide specific types of foods for songbirds. Much information is available on habitats, such as mast-producing plants, selected by various songbirds (for example, Cody 1985).

Browse is typically used as food by a variety of animals, including mule deer, black-tailed jack rabbits, porcupines, and ruffed grouse. As with mast, browse can be propagated. Planting is a common method of increasing browse.

There are very few wildlife species that do not include in their diets at least some grasses or forbs, or their seeds, or the insects that feed on these plants. A common propagation practice that allows the establishment of grasses and forbs is range reseeding. In general, sowing seeds, either by hand or mechanically, without seedbed preparation is unlikely to be successful. In most cases, it is best to prepare a seedbed of exposed soil so that the newly planted seeds can germinate and grow with reduced competition. The species planted will vary by geographic location. Similarly, plantings of cereal crops in food plots for wildlife can provide both food and cover for many game and nongame species. The crops planted vary by region, but such plots most commonly contain corn, millet, sorghum, or a combination of these plants. Plots of tall varieties of sorghum, sometimes mixed with corn or shorter sorghum varieties, can provide important winter cover and food in northern areas. Such plots are particularly important in areas of intensive agriculture. They need to be large enough so that drifting snow will not destroy their cover and food values in severe winters. In urban settings where growing plants is less feasible, feeding stations with different types of seeds or other foods can be provided for a variety of animals as a substitute for propagation.

Mention should also be made here of the ever-increasing efforts to restore communities of native grasses and forbs by reseeding. At one time, the cost of this practice was prohibitive because seeds for these native plants were at a premium. Seed costs have been reduced as demand has resulted in increased seed production. Increased native plant reseeding has resulted in an economically important form of alternative agriculture.

Release refers to the reduction of competition from other, less desired plant species. If competing plants are removed from the area surrounding a mast-producing tree, for example, more mast production can be expected. Conversely, release can also occur after a reduction in the target plant species. Release is an important food management tool, and can be accomplished by mechanical methods, chemical methods, or burning.

Mechanical release methods can include pulling a chain or cable through vegetation to break and uproot undesirable plants, using conventional agricultural tillage equipment such as disks or plows, or even hand pulling. Another mechanical release technique is the use of domesticated livestock, whose grazing and trampling can stimulate or depress the growth of various plants. In urban settings, mechanical methods such as mowing or pruning can result in the release of desired plants.

Timber management can also be used by biologists for improving food production. Thinning of tree stands allows light penetration and increased production in the understory. Clear cutting can also be used as a management tool. Clear cutting has a bad reputation because of the environmental damage caused by clear cuts over huge tracts of land (see section 14.5). However, timber stand improvement using small clear cuts or block cuts can provide open areas in which grasses, forbs, and young trees can sprout, providing much wildlife food. Selective cuts can also promote forest rejuvenation.

Management for uneven-aged timber stands can be especially important for some animals. For example, biologists manage ruffed grouse in northern states such as Minnesota and Michigan by managing aspen stands. Different portions of the stands are cut at approximately ten-year intervals such that four stages of aspen are always available, from newly sprouting aspens in recently cut areas to mature aspens that have not been cut for thirty or more years. The first-stage areas (most recently cut) are important feeding areas for developing young. Adults feed and roost most often in third-stage areas (cut twenty years ago). The second (cut ten years ago) and fourth (cut thirty years ago) stages are used by the adults during breeding. This type of habitat management is also useful in urban environments, where the promotion of different layers in a forested area can be used to promote species diversity. This practice can be applied to everything from small backyard areas to larger, more natural areas in the urban landscape.

Chemical control of competing plants can also be used for improving food produciton, but many biologists are concerned about the long-term and sublethal effects of chemicals. Perhaps the best advice, if chemical control is deemed to be the most effective release method in a particular situation, is to consider the use of small hand-held sprayers that will allow highly selective placement of herbicides.

Fires, whether intentionally set or naturally occurring, have influenced both forest and grassland habitats in North America for thousands of years. Fire as a release mechanism can remove dead vegetation that hinders new growth, free nutrients to enrich the soil, reduce invader species and encourage native species, and in general create and maintain habitats attractive to wildlife. Fire can also result in earlier vegetative growth in the spring. Fire is a natural force in many ecosystems. For example, estimates of historic lightning-set fires in the northern Great Plains range from 6 to 92 fires per 10,000 square kilometers per year (Higgins 1984). Fires deliberately set by native peoples were apparently even more frequent on North American prairies than lightning-set fires (Higgins et al. 1989).

A **prescribed burn** (**controlled burn**) is a human-induced fire that mimics the course of nature. Prescribed burns can be categorized as **reclamation burns**, which are used on mistreated and unmanaged lands, or **maintenance burns**, which are set regularly on a particular area as a management practice. **Backfires**, which are set upwind from a barrier such as a road or a cleared section of land, burn against the wind. Such burns tend to move slowly and burn thoroughly, so they are often used for reclamation burns. **Headfires**, those set to move with the wind, tend to burn more rapidly and less thoroughly and are more often used for maintenance burns. A backfire may also be set in an area where a biologist wants a headfire to stop burning.

Prescribed burns generally should be undertaken before the growing season, and certainly before seed heads of desired grasses begin to develop. The actual timing of prescribed burns varies with latitude. Prescribed burns may be done in February or March in Texas or Florida grasslands, but not until April or May in North Dakota grasslands. The timing of a prescribed burn also varies with the life cycle of the plant species being managed. For example, biologists in northern regions who wanted a prescribed burn to favor a cool-season grass, such as tall or intermediate wheatgrass, would burn between mid-April and mid-May. However, if they wanted the prescribed burn to favor a warm-season grass, such as big bluestem or Indian grass, the burn would be delayed until early June. Prescribed burns can also be done during the fall, again with the actual date varying by latitude.

Woodlands can also be burned to improve food production. Reclamation burns are likely to be hot and intense because unmanaged woodlands often contain a substantial amount of fuel. Maintenance burns are higher in frequency and lower in intensity, as the usual management objective is to decrease litter and allow understory plants such as grasses and forbs to develop. Thus, maintenance burns are generally set during cooler times of the day when humidity levels are high.

Protection, as the word implies, refers to attempts to protect plants until they mature and produce the desired food. Protection may involve screening or fencing the plants. Like propagation, protection is a common habitat management strategy in urban settings. Mast, browse, grasses, and forbs can be enhanced by protection.

Cover Development

Cover development is an important habitat management tool because wildlife species depend on a variety of cover types. Often it is difficult to distinguish between management for cover and management for food because both aspects of a habitat may be improved by the same management techniques.

Cover can be provided by such features as hedgerows, dense timber stands, shelterbelts, brush piles, and natural or artificial roosts or nests. Maintaining undisturbed cover along fences, and in railroad rights-of-way, roadsides, and backyards can encourage many wildlife species. An effective management strategy for species that nest along roads is to delay roadside mowing until most nesting has been completed. In some areas, such roadside habitat management is a policy. Dense nesting cover, often a mixture of grasses and legumes such as alfalfa, can be planted to enhance nesting success, while in other situations, native grasses are used for such cover development.

Some wildlife species, such as many of the squirrels and woodpeckers, rely on den or nesting trees. Protection of such habitat is important for these wildlife species. The U.S. Forest Service identifies and clearly marks den and nesting trees on some of its lands. These trees are protected, even in areas where logging is occurring. Artificial nest structures are built for many wildlife species. Some waterfowl species, such as Canada geese and mallards, will nest in artificial structures such as nest baskets (fig. 15.1). Nest boxes can be used for animals such as squirrels, American kestrels, bluebirds, and wood ducks (fig. 15.2). An interesting development in urban and even rural settings is the placement of bat boxes that bats can use for roosting.

The reproduction and survival of some species can be promoted through other cover-development techniques. The building of islands in water bodies, for example, can provide areas safe from nonaerial predators. Islands can be purposefully built during the construction of artificial water bodies, formed by excavating material from the neck of a peninsula on an existing water body, or added to existing water bodies by placing materials in the water at particular locations (fig. 15.3).

Nesting by undesirable species can be discouraged with cover management techniques. Areas on buildings that are attractive to house sparrows or rock doves for nesting and roosting can be reduced by eliminating ledges and small projections from building designs. Squirrels can be discouraged from nesting in house attics by effectively screening small openings. Spotted and striped skunks and rodents can be discouraged from

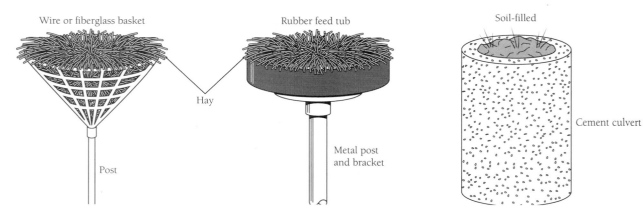

FIGURE 15.1 Baskets, feed tubs, and cement culverts can all be used to create nesting sites for waterfowl.

using the undersides of buildings by adequately sealing such areas to prevent their entry.

A variety of other devices and techniques can be used to control animal access to areas needing protection. For example, areas can be fenced to prevent or control access by domesticated livestock and allow the development of better wildlife cover. Interstate highway fences can be constructed in urban areas to prevent deer or other large ungulates from gaining access to areas of high traffic. In western areas of North America, consideration must be given to the migrations of species such as mule deer. Highway underpasses are sometimes constructed to allow the movement of these animals. Orchards may need to be fenced to prevent access by deer, which feed on fruit. Nesting baskets can be placed over water to reduce predation by terrestrial animals. Metal collars can be placed around trees with natural nest sites, on trees and poles supporting artificial nesting structures (see fig. 15.2), or on birdhouses and bird feeders in backyard urban settings to reduce predator access. The sheet metal around the bases of telephone poles seen in many areas of the eastern and central United States is used to reduce fox squirrel and raccoon access; power outages can occur if these animals gain access to electrical wires. Electric barriers are used to reduce nonaerial predator access to

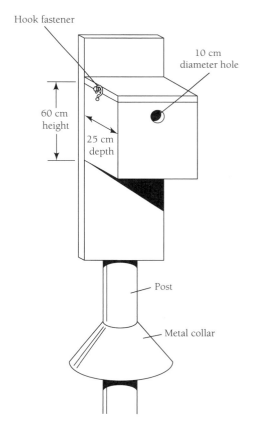

FIGURE 15.2 A nest box constructed to encourage wood duck nesting. Several inches of wood shavings are placed in the box, which should be cleaned annually. The box can be mounted on a post or attached to a tree at a height of 3–10 meters above the ground or water.

FIGURE 15.3 Waterfowl nesting islands can be created by excavating earth across a peninsula to form an island or by hauling sufficient fill to an area to create an island. A nesting structure has also been placed on this island. *(Photograph courtesy of S. Vaa.)*

numerous ecotone areas. The northern spotted owl, which uses old-growth forests in the Pacific Northwest, is an example of an interior species. Similarly, many songbirds in the eastern United States are dependent on large patches of woodland habitat. For example, Robbins et al. (1989) found that the likelihood of occurrence for Neotropical migrants such as red-eyed vireos, wood thrushes, scarlet tanagers, and ovenbirds increased as the area of forest increased to 100 hectares or more.

peninsulas as a means of protecting nesting waterfowl and other ground-nesting birds (see fig. 10.1).

Many research studies involve the placement of exclosures that prevent animal access to a certain area (see fig. 13.21). These "control" areas are then compared with areas where access is allowed. For example, fences that prevent elk access to riparian habitat in wintering areas can provide information on the effects of elk on willows and other vegetation.

Biologists must be aware of habitat needs on a broad landscape level, particularly with the increasing emphasis on ecosystem management. The juxtaposition of habitat types is important to both terrestrial and aquatic organisms. Northern bobwhites, for example, require food, cover, and other habitat types in close proximity so that the needs of a covey or pair can be met within a relatively small home range. Species such as the northern bobwhite that require a variety of habitats in close proximity are sometimes referred to as **edge species**.

Other species, referred to as **interior species**, require large patches of a relatively homogeneous habitat type, such as tall-grass prairie or temperate forest. These species may be negatively affected by management practices that fragment larger patches of habitat into smaller patches with

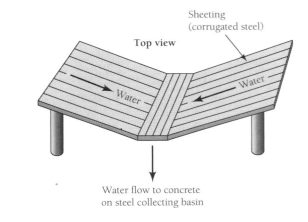

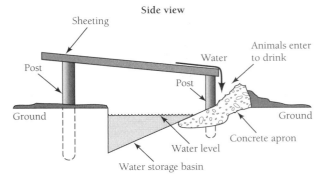

FIGURE 15.4 Guzzlers are often constructed in arid regions to provide a water supply for wildlife. The top view shows the corrugated metal collar that collects water and directs it into an underground storage basin. The side view shows how the water is caught by a concrete apron below the collar. The water then flows into the storage basin. Animals can walk down the concrete apron to reach the water.

FIGURE 15.5 The depression in this rock used in an urban landscape holds water that is used by many birds.

Water Development

The need for development of water sources varies depending on the abundance of water in a given area and the drinking needs of target organisms. Water development might include the improvement of natural water sources such as springs, seeps, and water holes, the construction of **guzzlers**, or the placement of backyard birdbaths. Guzzlers, also known as gallinaceous guzzlers because they are particularly useful to upland game birds, are often constructed in arid regions to provide water to wildlife (fig. 15.4). Guzzlers typically are watertight tanks set into the ground and filled by means of a rain-collecting collar. The rainwater collected is stored underground to reduce evaporation. Birds and mammals walk down a sloping ramp to the water. There are numerous guzzler designs.

In some situations, water development for terrestrial species may involve the construction of artificial water bodies. Small farm and ranch ponds, for example, are particularly useful in some regions where natural watering places for wildlife are a limiting factor.

In urban areas, natural water bodies are often degraded, with resulting decreases in water availability. In these circumstances, something as simple as a ceramic birdbath or a rock with a depression that will hold water (fig. 15.5) can be useful to wildlife.

15.2 Management of Terrestrial and Aquatic Transitional Habitats

Many of the habitat management techniques discussed in section 15.1 also apply to terrestrial and aquatic transitional habitats. Such management techniques have utility not only for terrestrial species, but also for aquatic and semi-aquatic organisms.

A variety of habitat management techniques have been specifically developed and designed for wetland management. Wetland areas are important to many animals and often serve as important nursery and food production areas. Wetlands can be natural or artificial. Natural wetlands are often managed to maintain a desirable interspersion of open water and emergent aquatic vegetation. A wetland that is half open and half plant-covered will generally be used by more bird species and other wildlife than a wetland covered with emergent vegetation such as cattails, especially if the plant-covered and open-water areas are interspersed about the wetland. However, while extensive and dense growth of emergent vegetation can limit the usefulness of wetlands to some species, it can be beneficial to others. In areas with severe winters, such vegetation can provide important winter cover for animals such as sharp-tailed grouse, ring-necked pheasants, white-tailed deer, and a variety of nongame species. Wetland vegetation can be manipulated by mechanical or chemical means, or controlled biologically, as with muskrats. If too few muskrats are present, cattails and other emergent vegetation may completely cover a shallow wetland. If muskrats are overly abundant, they may reduce the vegetation far more than would be desirable. Domesticated livestock can also be used to reduce emergent vegetation in wetland habitats through grazing and trampling. Burning can also be used as a management tool for wetlands. Vegetation can also be

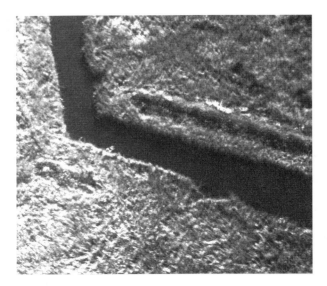

FIGURE 15.6 The level ditch shown in this aerial photograph was created by excavation. In this wetland, the ditch holds water even during dry periods when the remainder of the marsh is dry. During wet periods, the ditch provides an area of deeper water. *(Photograph courtesy of M. Kjellsen.)*

burned over ice or over water to remove dead plant material and increase nutrient cycling.

If a natural wetland is too uniformly shallow, techniques are available to develop deeper areas that are more likely to provide open water or to provide some water in drought years. These deeper areas can be produced in a variety of ways. Earth-moving equipment can be used to excavate some areas. This excavation is usually done in the form of **level ditching** (fig. 15.6), in which ditches are dug in the wetland, often in a zig-zag pattern. Another means by which deeper areas can be formed in wetlands is blasting with ammonium nitrate, dynamite, or other explosives (fig. 15.7). Care must be taken during blasting or level ditching to ensure that too large or deep an area is not made, because smaller wetlands may be dewatered.

Any type of excavation or artificial manipulation of natural wetlands should be subjected to careful scrutiny because of its possible influences on the appearance and function of the wetland. However, in many areas where wetlands have become choked with dense emergent vegetation, the

wetlands have already been greatly modified by agricultural and other human land use practices.

Water level manipulation is another management tool for both natural and artificial wetlands. Natural wetlands can sometimes be enhanced by installing water level management devices such as outlet control structures. If wetland areas are interconnected, various portions of these wetlands can be kept in different successional stages with such control structures (fig. 15.8). Some natural wetland areas, such as floodplain riparian forests in the southern United States, can be managed by water level manipulation during summer and winter for a variety of wildlife species. Wintering mallards and breeding and wintering wood ducks are often targeted in the management of these riparian wetlands. Water level determines the availability of invertebrates as well as mast, such as acorns, needed by waterfowl. Management for hydric soils, hydric plants and their seeds, detritus, and the resulting invertebrate production in

FIGURE 15.7 Explosives can be used to create new wetlands or modify existing wetlands.

FIGURE 15.8 The water level in this restored wetland is maintained at the height of the water control structure. (*Photograph courtesy of R. Meeks.*)

these riparian forests can be important to a variety of terrestrial and aquatic organisms.

There are efforts under way to produce totally artificial wetlands to mitigate losses of natural wetlands. This may seem like a simple procedure; flood an area and it becomes a wetland. In reality, it is not so simple. The aquatic plants and animals normally found in a wetland may need to be introduced. At this time, little is known about starting a wetland from "scratch."

Artificial wetlands can be created in many areas, generally by building a dam or by intentionally flooding an area using dike systems. Water levels can be manipulated in these areas, especially if sufficient water is available for pumping and subsequent flooding of wetland areas at various times of the year. Many biologists flood artificial wetlands during spring and fall migration periods for shorebirds and waterfowl. During midsummer, the water levels can be lowered, and either natural hydric plants or planted crops can be allowed to grow in the basin. Reflooding will then provide highly attractive habitat for various wildlife species.

There is also growing interest in restoring wetland areas, such as marshes and swamps, that were drained at some time in the past. Often, all that is required is the filling or plugging of the drainage ditches initially dug to drain the wetland (fig. 15.8; fig. 20.1). However, if drainage tile was put in place to drain an area, the tiles may need to be removed or broken. In some cases, even if

drained for many years, reflooded wetlands still contain a bank of wetland plant seeds just awaiting the correct conditions to germinate and grow. In other cases, reflooding alone may not result in the reestablishment of the plants needed for complete restoration.

Coastal wetlands in North America include tidal salt marshes, tidal freshwater marshes, and mangrove wetlands. Such wetlands are particularly prone to damage caused by urban development and pollution because of the high human population densities in many of these areas. Losses of the protective islands and shallow water areas that tend to reduce wave action in coastal wetlands are also reasons why these habitats have been degraded. Current management techniques for these wetlands include laws and regulations intended to prevent their degradation.

Beach-ocean interfaces are also often protected by laws and regulations that prevent their degradation. In addition, beach stabilization structures, the addition of sand, and other measures are employed to manage this habitat.

15.3 Management of Standing Water Habitats

The manageability of lentic habitats varies by water body type. For example, smaller water bodies are easier to manipulate than larger water bodies. This section is divided into subsections concerning the management of dissolved oxygen levels, water levels, water access, aquatic vegetation, terrestrial landscaping, fertilization, and artificial structures.

Management of Dissolved Oxygen Levels

Some lentic systems can undergo **winterkill** if dissolved oxygen is depleted under the ice.

FIGURE 15.9 The results of a winterkill are highly visible just as the ice leaves a water body in the spring. The risk of winterkill in shallow water bodies increases with nutrient loading, thickness of ice, depth of snow that covers the ice, and duration of ice cover.

Additionally, naturally produced toxic substances, such as hydrogen sulfide and ammonia, can increase under ice cover. Both low dissolved oxygen levels and toxic substances can negatively affect fishes and other aquatic organisms (fig. 15.9). The loss of oxygen occurs when respiration (use of oxygen) by plants, animals, and bacteria exceeds the production of oxygen by plants during photosynthesis. Winterkill is therefore more likely to occur if a water body is shallow and contains large amounts of decaying organic material, especially algae or other aquatic plants. Such conditions are more likely in eutrophic waters. The probability of winterkill also increases with duration of ice cover, thickness of the ice, depth of snow on the ice, and ice opaqueness. Clear ice allows for good light penetration, and photosyn-thesis can occur under such ice. However, snow on the ice will block light penetration. Under winterkill situations, water bodies first lose their oxygen near the bottom, and oxygen depletion then progresses upward toward the ice.

Water bodies that are prone to winterkill are generally managed in one of three ways: the entire system of water and surrounding land (the watershed) can be managed to reduce nutrient and sediment inputs, aeration systems can be installed, or such waters can simply be managed as "marginal" waters.

Water bodies can be prone to winterkill because of incorrect land management in the drainage basin. Reducing the inflow of nutrients, organic material, and sediments into a water body can help to reduce or avoid winterkill. Such inflows can come from industrial, residential, agricultural, or other sources. Controlling these inflows requires control of land use, which often involves private property, so a variety of social, legal, and economic issues involving property rights must be considered (see section 14.10). This is the primary reason why stopping or reducing nutrient and sediment inflow is difficult to accomplish. Thus, biologists are often forced to treat symptoms rather than causes of this problem.

Aeration systems prevent winterkill by maintaining dissolved oxygen levels and providing an open water area for the dispersion of potentially harmful toxic gases. One type of aeration system is illustrated in figure 15.10. In this system, an air compressor is placed on the shore and an air line is extended into the water body. As bubbles rise from the air line, a current is created that carries the warmer (typically near 4°C) water from near the bottom of the water body to the surface, and an opening is maintained in the ice. Because these systems are called aeration systems, most people assume that the air bubbles produced are the primary source of oxygen input into the water body. In reality, little oxygen transfer occurs by this means. Most of the added oxygen is produced by phytoplankton during photosynthesis because the open area in the ice allows sunlight penetration. In addition, just maintaining an exposed water sur-

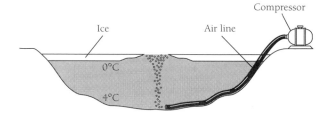

FIGURE 15.10 An aeration system that uses a compressor on shore to pump air to the lake bottom. The rising bubbles carry the warmer (4°C) water at the bottom to the surface, and a hole remains open in the ice.

face allows for a small amount of oxygen addition by surface mixing.

If aeration systems are not correctly managed and maintained, aquatic organism death can be caused by the system itself. For example, if the system is not started until oxygen levels are depressed and toxic gases such as hydrogen sulfide are present, it can cause warmer unoxygenated water from the bottom of the water body to mix with oxygenated water above it. As a result, the average oxygen availability is reduced and winterkill may occur because the toxic gases are also present.

The cost-effectiveness of using an aeration system needs to be closely evaluated. There are costs involved in the construction, maintenance, and operation of such systems (for example, electricity to run the compressor). A conservation agency would be more likely to invest money in an aeration system in an area that had few fishing waters. If other, higher-quality water bodies were present in the same locale, aeration probably would not be undertaken. Aeration systems are very popular with lake associations and other citizen groups and private individuals, and it is not uncommon for such people or groups to install systems on public or private water bodies in which they wish to prevent winterkill.

In much of North and South Dakota, Minnesota, Wisconsin, Michigan, and southern Canada, numerous shallow, eutrophic water bodies are naturally prone to winterkill. Conservation agencies in many of these areas manage these

water bodies as "marginal" waters, meaning that they are managed for short-term production of sport fishes, recognizing that the frequency of winterkill will be high. For example, perhaps winterkill can be expected once every three years. Such marginal water bodies can still provide many hours of sport angling. They tend to be quite productive, and fishes that are stocked after a winterkill often have rapid growth rates. In North and South Dakota and Canada, fishes such as the northern pike, yellow perch, and even rainbow trout are commonly stocked into these waters. Some of these marginal waters can also be used for aquaculture purposes. Bait fishes such as fathead minnows can be raised, and sport fishes such as walleyes and yellow perch can be reared to advanced sizes, captured, and then stocked elsewhere.

Summerkill, like winterkill, is a result of oxygen deprivation. Summerkill generally occurs in eutrophic waters. As with winterkill, summerkill can be associated with incorrect land use practices in the watershed. Consider one of the marginal, eutrophic water bodies discussed in the previous paragraphs on winterkill, and the following conditions. During midsummer, hot and sunny weather dominates for a two-week period, and a large **algal bloom** develops. Bloom is a commonly used term that denotes a rapid increase in algal populations. The weather then changes, and a five-day period of cool, cloudy, calm weather occurs. The clouds block some sunlight, some of the algae die, and the calm weather allows for little replenishment of oxygen by wind action. Just before sunrise, after several days of this weather pattern, respiration by plants and animals, combined with the extra respiration caused by increased numbers of bacteria feeding on the decaying algae, may reach a point at which oxygen supplies are depleted. Summerkill occurs if the dissolved oxygen level drops too low to support some or all of the fishes present.

Summerkill can also occur in a stratified water body. In eutrophic waters, the hypolimnion often loses its oxygen. The hypolimnion is the zone of decomposition, and its oxygen cannot be replenished because photosynthesis does not oc-cur at that depth. Hypolimnetic water cannot mix with better-oxygenated water in the epilimnion because of water temperature differences (see section 12.3). High winds can cause destratification, however, allowing complete mixing, and the combination of low oxygen levels, warm water temperatures, and the presence of other dissolved gases such as hydrogen sulfide can lead to a partial or complete summerkill in the entire water body. Even without total mixing, a partial summerkill can occur; fishes restricted to the hypolimnion by temperature requirements may be the only ones to die if that layer loses its oxygen.

Aeration systems can be used to decrease the likelihood of summerkill. In small water bodies that can become stratified, aeration systems are used to help prevent stratification. Thus, the entire water body circulates and oxygen is available throughout. In addition, summer aeration is at times used in oligotrophic water bodies where the hypolimnion can act as a nutrient trap. Destratification will allow the nutrients to cycle throughout the water body for use by plants.

Fishes vary in their sensitivity to low dissolved oxygen levels and thus in their susceptibility to winterkill or summerkill. Most fish species can live at lowered oxygen levels if they are slowly acclimated to reduced concentrations. However, many sport fishes, such as walleyes, trouts, and basses, are the first to die when dissolved oxygen levels are reduced. Fishes such as northern pike, bullheads, and common carp are more tolerant of low dissolved oxygen levels. While many fishes require at least 4−5 milligrams of dissolved oxygen per liter of water, these tolerant fish species can exist at oxygen levels below 1 milligram per liter. Partial kills can be a problem for fishery biologists. Biologists responsible for waters at risk of winterkill or summerkill would usually prefer a complete kill rather than a partial kill that allows less desirable fishes to survive and reproduce the next spring. However, complete kills seldom occur. Little is known about the tolerances and the effects of winterkill and summerkill on most nongame fish species and other aquatic organisms.

Water Level Manipulation

Planned water level manipulation can be a useful tool for biologists who manage water bodies. A newly constructed water body generally has an initial period of high productivity because as it fills, nutrients are continuously being added to the system. Water level manipulation is sometimes used to mimic this initial period of high productivity.

Decreases in water level (**drawdowns**) concentrate predators and prey; thus, many biologists use drawdowns to force members of a stunted species, such as bluegills, out of cover so that predators, such as largemouth bass, can reduce their populations. This management practice improves habitat for the predator species while degrading habitat for other species, in this case the stunted panfishes. Increased mortality of the small panfishes will result in faster growth rates and improved size structures for both predator and prey species. Drawdowns expose the bottom of a water body to oxidation, and nutrients are thus released from bottom soils. Drawdowns also allow for compaction of silt, resulting in improved water clarity after reflooding. Drawdowns may also affect nontarget organisms, either positively or negatively.

Drawdowns allow the growth of terrestrial vegetation on exposed shoreline areas. Such vegetation may be natural (for example, smartweed and many sedges will commonly grow in such environments), or may result from human seeding of plants such as millet. When reflooded, this vegetation results in increased populations of bacteria, phytoplankton, zooplankton, and aquatic insects. Obviously, this increased overall food supply results in increased productivity for fishes; wildlife that feed on invertebrates, such as shorebirds and many duck species, also benefit. Some fishes may also increase recruitment as a result of the flooded vegetation. Largemouth bass recruitment, for example, may be low in turbid water bodies that lack submerged aquatic plants. Flooding of terrestrial vegetation results in improved habitat for largemouth bass, and their recruitment temporarily increases. Flooded terrestrial vegetation is also attractive to waterfow because it provides seeds as a food source.

Flooded vegetation can help increase water transparency. Some waters remain turbid because they contain suspended clay particles. These particles have a negative charge, and each particle repels others. The material remains in suspension because the particles cannot clump and settle to the bottom. Decaying aquatic vegetation provides an attachment site for these clay particles; thus, biologists can expect improved water transparency in such waters after terrestrial vegetation has been flooded. Increased water clarity is beneficial to some species and reduces populations of other species. Sight-feeding predators, such as horned grebes and various aquatic organisms, may become more effective in clearer water, while prey species may be reduced in numbers because of increased vulnerability to predation. Increased water clarity can also lead to increased productivity due to the increased photosynthetic activity that results from greater light penetration.

Water level management does not always include the addition or removal of water; it may instead involve the modification of the bottom of a water body. Dredging to deepen a water body functions in much the same manner as adding water. Dredging is a costly procedure in existing water bodies and is much more likely to be economically feasible if the excavation occurs when artificial water bodies are being constructed.

While water levels can be increased or decreased by pumping or siphoning, water level management is most often done in water bodies that have a preexisting means of water retention and release, such as small impoundments or reservoirs. Many such water bodies are multiple-use impoundments, meaning that they were constructed for water supplies, irrigation, flood control, electricity generation, recreation, navigation, and other purposes. Because these multiple uses often conflict, biologists find it difficult to implement a water level manipulation plan that will satisfy all users. Conflicts among multiple uses, even just recreational uses, can complicate water level

management. Most recreational facilities are designed for a certain water level. Water levels above this elevation can cause flooding and shoreline erosion from waves, and water levels below this elevation can result in the facilities being difficult to use; for example, boat ramps may be out of the water, or camping and picnicking areas may be too far from the water. Water level manipulation intended to improve a fishery can have substantial negative effects on other organisms. For example, island-nesting birds can be reduced if the bottom of the water body around the island is exposed. Despite these problems, water level management can be one of the most successful tools for reservoir and small impoundment fishery management.

Management of Water Access

Aquatic habitat management in urban settings is often less habitat management than access management. The management emphasis is less upon habitat and more upon making space available for large numbers of people. Urban water bodies are often made more accessible to anglers and other users by providing access along open shorelines, fishing piers, and rock jetties. In most of these situations, angling effort is so high that habitat management within the water body could not possibly provide fish populations of sufficient density, so stocking is sometimes a common practice (see section 10.7). At the extreme end of the urban habitat spectrum are very artificial fisheries, often placed in permanent or temporary water bodies such as swimming pools, which primarily serve youngsters, the elderly, and the physically disadvantaged. Such fisheries are simply based on water-holding mechanisms to support put-and-take fisheries, and little habitat management is involved.

Water access management is also important in nonurban settings. Road networks, boat launching areas, and parking lots are important means of access. In addition, amenities such as fish cleaning stations, rest rooms, picnic areas, and even food, bait, or fuel concessions may also be needed. Access management is usually based on the amount of fishing or on other public uses at a particular water body.

Management of Aquatic Vegetation

Fishery biologists are commonly called upon to address problems associated with aquatic vegetation in water bodies. Most human users tend to view aquatic vegetation as a problem that needs control. It can make fishing, as well as water skiing and swimming, more difficult and can impede the use of outboard motors. Overly abundant vegetation can allow some panfish species to avoid predators and overpopulate, and it can increase the likelihood of winterkill or summerkill. Some aquatic vegetation, primarily some blue-green algae species, also produces substances that can be toxic to aquatic and terrestrial animals.

Although excess vegetation can be a problem, aquatic vegetation has many positive attributes. Aquatic vegetation produces tremendous amounts of food for aquatic and terrestrial organisms. Farm ponds constructed in grassland ecosystems have proved particularly valuable for nesting ducks. Nesting hens and their ducklings find an abundant supply of aquatic invertebrates associated with the submergent vegetation in these ponds. Further, emergent vegetation is important escape cover for ducklings and escape and nesting cover for a variety of other animals.

Many aquatic stages of insects cannot exist without vegetation. Vegetation buffers the effects of wind action and therefore serves to make water clearer and to reduce wave height in large water bodies. Vegetation provides cover for young fishes, both for escape from predators and for protection from wind and wave action. In reality, most aquatic communities benefit far more from aquatic vegetation than anglers, boaters, swimmers, and other recreational resource users realize. For example, various studies have indicated that largemouth bass/bluegill communities function best when submerged aquatic plants cover 20%–40% of the water body surface area. Therefore, if vegetation is a problem, the best means of vegetation management may be a combination of controlled and uncontrolled areas.

Aquatic vegetation control methods can be categorized as mechanical, chemical, or biological. Mechanical control methods can range from simply using a rake to keep a fishing lane or swimming area open to expensive, boat-mounted mechanical vegetation harvesters that cut and remove vegetation. Dredging of shallow water areas can result in deeper water and reduced vegetation, but this method is expensive. Sometimes opaque plastic sheeting used to cover small areas of bottom can provide an effective control. Drawdowns compact bottom soils and generally result in reduced rooted aquatic plant abundance for a few years. A drawdown over the winter can be especially effective because it ensures freezing and desiccation of soils and plant material as well as compaction of soils.

Chemical control involves the use of herbicides to reduce vegetation. It is important that a biologist using chemical control methods be able to identify the aquatic plant species causing the problem because many chemicals are selective. Chemicals that are effective on algae, for example, are often ineffective on vascular aquatic plants. In the United States, both the Environmental Protection Agency and the Food and Drug Administration control the types of herbicides that can be used in aquatic habitats. The effects of chemical control tend to be temporary; plants typically regrow quickly after the herbicide has been applied. Thus, application at regular intervals may be needed, which can become quite expensive. In addition, many chemicals are designed to eliminate plants completely, which can be quite detrimental to aquatic communities, especially if the entire water body is treated.

A common biological control method is the introduction of grass carp. This exotic species, a native of Asia, is herbivorous as an adult. Its introduction into North America was quite controversial, given the problems created by its close relative, the common carp, and other exotic introductions. However, grass carp have been widely introduced, and unfortunately are reproducing in some of the large river systems of the United States. Triploid grass carp, which are sterile, can be used to avoid this problem. Grass carp commonly eliminate nearly all vegetation in a water body; it is difficult to obtain partial vegetation control with this method. In addition, grass carp are long-lived and can affect a water body into which they have been introduced for ten or more years. Fishery biologists are often unwilling to use this control method in water bodies where the fish species present rely on aquatic vegetation or food organisms associated with it because complete vegetation removal could harm the structure and dynamics of fish populations (Bettoli et al. 1993). Similarly, wildlife biologists are concerned with grass carp effects on vegetation needed by species such as canvasbacks.

It should be obvious that the various mechanical, chemical, and biological methods for controlling excess aquatic vegetation most often treat symptoms rather than causes. Large expenditures can be made with very transitory results. The best method of controlling excess vegetation is avoiding the causes. Correct planning and construction of new artificial water bodies can reduce the problem; for example, large, shallow expanses of water should be avoided. For already constructed artificial water bodies or natural water bodies, correct watershed management will reduce the problem. By controlling erosion and nutrient input, excess vegetation problems can usually be minimized.

Terrestrial Landscaping

Aquatic habitats, both lentic and lotic, can often benefit from terrestrial landscaping. Properly placed trees and shrubs can reduce the effects of wind on water bodies, help to reduce water temperatures, and provide an aesthetically pleasing setting. They also can produce quiet water areas for the enjoyment of bird watchers, anglers, swimmers, and other human users. In addition, maintaining shoreline areas in trees, grasses, or forbs improves water quality by reducing erosion. Such terrestrial landscaping can have negative

effects by providing raptor perches and increasing predation.

Fertilization

Fertilization is sometimes used as an aquatic habitat management tool, most often to increase the carrying capacity for fishes in a water body. This tool should be used only in areas that have infertile soils and resulting infertile waters, as commonly occurs in the southeastern United States, where soils are highly leached. An infertile 2-hectare pond in Alabama, for example, may contain a standing stock of only 15 kilograms per hectare of largemouth bass and 40 kilograms per hectare of bluegills, as compared with a typical 2-hectare pond in the Midwest, which may contain 40 kilograms per hectare of largemouth bass and 150 kilograms per hectare of bluegills. More fertile waters and water bodies in the northern part of North America should not be fertilized because of the increased risk of winterkill. In established waters, phosphorus is typically the limiting nutrient. Thus, phosphate fertilizers often are the only needed form of fertilization.

While not a form of fertilization, the addition of lime is sometimes used as a means of counteracting environmental damage resulting from acid deposition or acid mine drainage. Liming neutralizes acidity directly, and it also increases the buffering capacity of a water body. Liming is usually not a cost-effective habitat management tool except under certain conditions.

Artificial Structures

Fishery biologists sometimes place artificial structures in lentic water bodies to attract fishes. A variety of materials can be used, including brush piles or Christmas trees weighted by cement blocks (fig. 15.11), automobile tires bundled into pyramids, concrete blocks and rubble piles, and old car or bus bodies (primarily used in the ocean). In some marine situations, old oil well drilling rigs and old

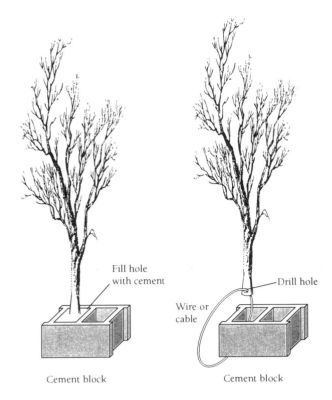

FIGURE 15.11 Fish attractors can be constructed from trees or brush and cement blocks.

ships are sunk to serve as artificial reefs. Most biologists use artificial structures as fish concentrators, and do not expect an increase in fish productivity (Pardue and Nielsen 1979). However, extremely large artificial reefs can apparently result in increased productivity (Prince et al. 1979).

Biologists involved in pre-construction planning for a new impoundment should attempt to retain as much of the standing timber and natural structure as possible. The concentrating effects of standing timber have been established for many fish species (Willis and Jones 1986), and it is much easier to retain the timber than to add artificial structures at a later date. Unfortunately, after a number of years, flooded timber begins to decompose, and its usefulness declines. It is important to note that the decomposition of organic materials in water bodies has at times created concerns over water quality, especially in human water supply impoundments. For example, some

regulatory agencies are concerned that the decomposition of standing timber in reservoirs or brush piles added to a reservoir as fish concentrators will cause increased levels of trihalomethane (a suspected human carcinogen). However, Layher (1984) reported that such practices contribute little to the degradation of water quality compared with inputs from the remainder of the watershed. In the unique case of some large, shallow hydropower impoundments in northern Canada, decaying vegetation has caused methylation of mercury, resulting in contamination of aquatic life. This contamination is a risk to humans when fishes and waterfowl from such water bodies are used as human food.

15.4 Management of Flowing Water Habitats

The manageability of lotic habitats varies with the size of the water body. Management techniques used in large rivers are quite different from those appropriate for small streams.

Stream Habitat Management

Stream habitat management is a common practice in coldwater streams; warmwater and coolwater streams are less often managed. Much of this habitat management is undertaken to improve degraded sections of streams. Stream habitat improvement has been an important part of coldwater stream management in the Rocky Mountains, Appalachian Mountains, and upper Midwest (Michigan, Minnesota, and Wisconsin) for many years. Stream management in warmwater and coolwater streams has primarily developed in the last two decades, but is increasing.

Human activities such as domesticated livestock grazing (see fig. 14.6) and other agricultural practices (see fig. 14.4), building and road construction (see fig. 13.4), forest cutting (see fig. 14.7), and urbanization can result in losses of streamside vegetation, losses of overhanging banks, and a wider and shallower stream basin. A shallower stream basin without overhanging vegetation can lead to increased water temperatures, which can result in the loss of coldwater fish habitat if the water warms too much. Various habitat management practices can be used to alleviate such problems.

One of the best methods of stream habitat improvement is the preservation of natural conditions. Therefore, fencing can be an important tool. Simply by fencing a stream to exclude domesticated livestock or vehicles, biologists can allow the stream to recover its original vegetation and channel. Fencing can be expensive, however (Platts and Wagstaff 1984), and the cost-effectiveness of such a strategy must be considered.

Many artificial stream habitat improvement structures can be built in the streambed to improve the quality of habitat. **Current deflectors** and **bank covers** (fig. 15.12) are used to narrow streams and provide overhead cover along the erosional zone or straight reaches. Narrowing a stream increases its current velocity, which can result in lower water temperatures and increased scouring effects of the current on the streambed. Overhead cover provides shade and can also result in lower water temperatures. Many fishes seek the cover provided by a simulated overhanging bank.

Current deflectors can be constructed from large rocks or logs. If logs are used, they must be dug into the stream bank to prevent their loss during times of high water flow. Deflectors can also be constructed with rock-filled wire baskets called **gabions**. Often, current deflectors are used to direct stream current into a stream bank or tree roots, creating a natural overhanging bank (fig. 15.13).

The type of bank cover used depends on the stream bottom (Hunt 1988). In streams with sandy bottoms, pilings can be driven into the stream bottom to support the overhead cover. Where bot-

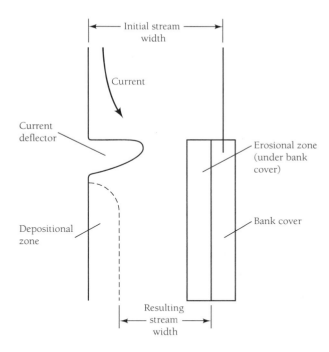

FIGURE 15.12 Current deflectors (such as rocks or logs) can be used to direct current. In this case, the current is directed under an artificial bank cover (see fig. 15.14). The increased current velocity will prevent silt from accumulating under the structure, resulting in an overhanging bank. The area of the bank behind the current deflector will become a depositional zone. Thus, the stream width will be narrowed and current velocity increased by use of these habitat improvement techniques.

toms are hard, biologists often use "skyhook" or "lunker" structures (fig. 15.14). Wire, rocks, and boards can all be used to create these structures. After soil is placed on the top and vegetation allowed to regrow, users such as anglers and picnickers can rarely tell that the structures are anything other than natural.

Brush bundles can be placed on the depositional zones found on inside bends of streams (fig. 15.15), where they collect silt and eventually help to narrow the streambed. **Half-logs** (fig. 15.16) can be used to provide overhead cover in the middle of a stream. Although half-logs are occasionally used in coldwater streams, they are more commonly used in warmer streams that contain smallmouth bass.

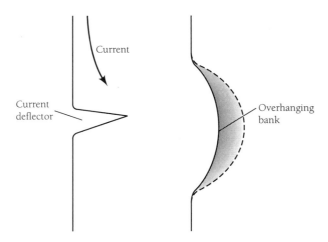

FIGURE 15.13 A current deflector being used to create a natural overhanging bank. This method can be especially effective when the current is deflected toward the roots of a large tree.

Dams, generally small ones, are also used to improve stream habitat. Generally, a deeper-water area, or plunge pool, forms below the dam. Dams are typically used only on steep-gradient streams. In shallow-gradient streams, a dam slows water flow, opens the overhead canopy, and can allow too much stream warming. Similarly, removal of vegetation along small coldwater streams, for example, at high altitudes in the Rocky Mountains, allows greater sunlight penetration and warmer waters and results in increased productivity. However, in some coldwater streams, temperatures during midsummer are often very near the upper limit tolerated by coldwater fishes. Shade from vegetation is essential for keeping water temperatures cool in such streams in the upper Midwest and in low-elevation mountain streams. Stream habitat improvement experts always stress that the proper techniques must be decided on a case-by-case basis. There is no standard procedure that can be applied to all stream types.

Small impoundment construction can also be used in areas where streams are naturally intermittent. If water is collected in a small impoundment during high runoff periods, a minimum stream flow can be provided over the entire year by means of planned water releases. A stream that has little utility for constant aquatic life can thus be transformed into one with continuous populations of aquatic organisms.

Poor land use practices in the watershed of a stream can lead to high sedimentation rates. Such sediments can fill interstitial spaces in gravel that may be important for aquatic invertebrate production or as spawning habitat for fishes. Habitat improvement structures that narrow a stream and increase its velocity can sometimes clean sediment

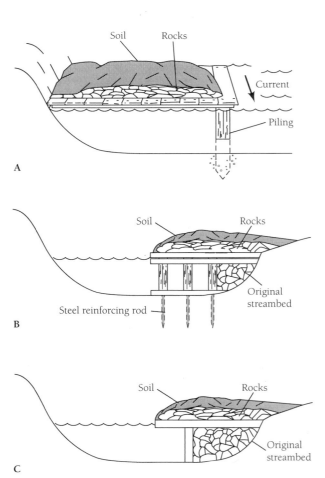

FIGURE 15.14 Various types of bank covers can be used to simulate overhanging banks. (*A*) A structure built directly on pilings and covered with rocks and soil can be used on a sandy bottom. (*B*) A "lunker" structure; steel reinforcing rods can be used if the stream bottom is sufficiently soft. (*C*) A "skyhook" structure held in place with rocks and soil can be used on a rock-bottomed stream.

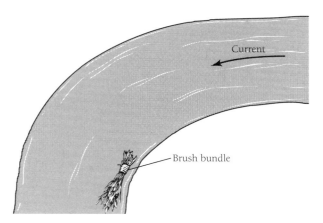

FIGURE 15.15 A brush bundle can be staked on the inside bend of a stream. The bundle can simply be a number of tree branches tied together. Silt will collect around the bundle, resulting in a narrowing of the streambed.

from a stream reach. In extreme cases, biologists can add sediment traps to a stream, or even enter the streambed and clean the gravel mechanically by dislodging the sediment and actually "vacuuming" it out of the water (Sigler and Sigler 1990).

Special note should be made here of the challenges of stream habitat management in urban environments. Some of this habitat management involves the construction of habitats that closely resemble those found in more natural settings. For example, Rapid Creek flows through Rapid City, South Dakota. By placing large numbers of stream habitat improvement structures in this stream, biologists have been able to maintain a large population of wild brown trout. Population quality of wild fishes in an urban environment

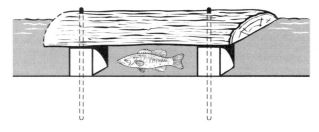

FIGURE 15.16 Half-logs can be suspended near the surface in the middle of a stream to provide overhead cover for fishes.

often can be maintained only through the use of harvest regulations (see chapter 17).

The maintenance of **minimum instream flows** is another stream habitat management practice that is becoming increasingly important. To maintain populations of aquatic organisms, every stream must have some minimum amount of water flow, and at least that minimum must be maintained throughout the year, not just for eight months, ten months, or eleven and one-half months. Maintaining such minimum flows is usually a matter of human user management. Individuals, companies, and government agencies may all have the legal right to remove water from streams. Usually such removal is for industrial, agricultural, or municipal purposes. It should also be noted that removal of groundwater can affect surface water levels. Only relatively recently has the need for "water rights" for aquatic organisms become an important issue.

The problem of maintaining minimum instream flows is most prevalent in western North America where water is scarce. The problem is often exacerbated by increasing human populations. Attempts to maintain minimum instream flows in arid areas can result in conflict among the various users of stream water. A farmer may need water to irrigate orchards or crops, a city may need water for domestic use, a factory may need water for its manufacturing process, and the aquatic organisms in the stream also need water. Satisfying all users can be a problem, especially if all of these water needs added together exceed the total flow of a river or stream; this situation is not uncommon. Efforts to provide minimum instream flows are being made in many areas, and often conflict resolution must occur where the issue is contentious. Much more will be said about conflict resolution in section 20.5.

River Habitat Management

Most artificial structures placed in large rivers are there for reasons other than fish habitat improvement. For example, the channelized sections of rivers such as the Missouri and Mississippi are used for barge traffic. Numerous bank stabilization structures have been built along these rivers. **Wing dikes** are bank stabilization structures that are usually covered with rip-rap (large rocks) and are used to concentrate the current in the middle of the river channel (fig. 15.17), thereby keeping the channel sufficiently deep for barge passage. While such structures can be quite attractive to many aquatic species, they do not compensate for the loss of habitat diversity resulting from the channelization (see section 14.8).

One river habitat management practice that benefits fisheries is the protection of natural snags. A more common procedure has been to remove the snags to allow for more rapid water flow. Protection of these naturally occurring structures can provide habitat for a variety of aquatic organisms.

Fishways are artificial structures that are designed to allow fish passage over or around a barrier. Natural barriers in a river, for example, might be modified by using explosives to blast a series of pools or steps in rock. Fish ladders and other mechanisms to allow fish passage are often built over or around dams. Such devices are commonly used on dams constructed on rivers that serve as migratory routes for anadromous species such as the Pacific salmon on the west coast of North America and the Atlantic salmon on the east coast, often with limited success.

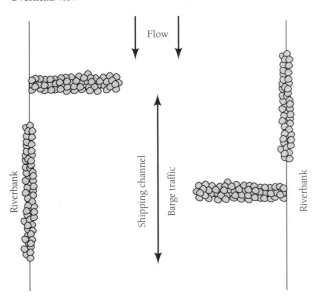

Overhead view

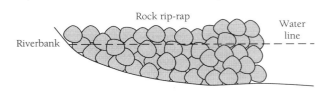

Side view

FIGURE 15.17 Wing dikes, covered with large rock rip-rap, are often constructed in large rivers. The primary function of these devices is to concentrate river flow in a center channel to provide deep water for barges. Because of the turbulence caused by the wing dikes, the opposite shore is typically protected with rip-rap as well.

15.5 Management of Estuarine Habitats

Much of the management of estuarine habitat is focused on the protection of this very productive habitat type. The rate of loss of estuarine habitat is high along both the Atlantic and Pacific coasts of North America and along the Gulf of Mexico. Estuaries are essential reproductive and nursery habitat for many marine fishes, crustaceans, mol-

lusks, and other aquatic species. In addition, many birds and mammals use this habitat type for all or part of their lives.

Deepening channels for navigation and filling wet areas to form solid land causes substantial losses of estuaries and estuarine wetlands in the United States. In addition, wave action from ship traffic causes erosion of estuary margins. The National Marine Fisheries Service estimated that destruction of estuaries has cost the fishing industry in the United States $200 million per year since 1954 (Owen and Chiras 1990). Contamination of estuaries with toxic substances has also affected fishery and wildlife resources, and has affected humans that are part of the food webs involving organisms from contaminated areas.

The value of estuarine habitat, both in economic and aesthetic terms, is finally being recognized in the United States, and protection and restoration efforts are being made. For example, the U.S. Army Corps of Engineers plans to divert fresh water from the Mississippi River into Louisiana coastal marshes. Intrusion of salt water has greatly altered these wetlands, and the return of fresh water, which in the past was provided by spring floods, will facilitate recovery of this habitat.

SUMMARY

Terrestrial habitat management is typically focused on food production, cover development, and water availability. Most sport and commercial wildlife species are primarily herbivorous, and even most managed carnivores rely on herbivorous prey; therefore the management of plants is an important part of wildlife conservation. Food production and cover development can be managed through propagation, release, or protection.

Fire is an important terrestrial management tool that is used for both grasslands and woodlands. Fire removes dead vegetation that hinders new growth and releases nutrients to enrich the soil. Prescribed burning is an artificial technique that mimics the course of nature.

Cover management is important to many terrestrial animals. Many wild animals nest on the ground and can be managed by techniques such as delaying the mowing of roadside ditches or planting dense nesting cover. Some wildlife species rely on den or nesting trees, so such trees are often protected from disturbance during timber harvest in public woodlands. Nest structures or nesting islands are built for many wildlife species. Use of particular areas by certain animals can also be prevented with habitat management techniques.

Many of the habitat management techniques used for terrestrial habitats can also be used for aquatic and terrestrial transitional habitats. Natural wetlands are often managed to produce an interspersion of open water and emergent aquatic vegetation. Vegetation can be manipulated through mechanical removal, chemical control, or biological control. If natural wetlands are too uniform in depth, level ditches can be excavated throughout the wetland, or explosives can be used to form deeper areas. Water level manipulation in natural or artificial wetlands can also be used to maintain wetland vegetation in the desired successional stage.

Management of lentic water habitats for fisheries often requires consideration of oxygen levels. Eutrophic waters at northern latitudes are susceptible to winterkill caused by under-ice oxygen depletion. Aeration systems can be placed in such waters to prevent winterkill, but the cost-effectiveness of such a strategy must be assessed. Eutrophic waters at all latitudes are susceptible to summerkill. Oxygen levels may drop below those needed to support fishes and other aquatic life, especially just prior to dawn.

Planned water level manipulation is an important habitat management tool for small impoundments and reservoirs. Newly flooded small impoundments and reservoirs typically have an initial flush of productivity. Many water level management plans are designed to mimic this period of high initial productivity.

Control of overly abundant aquatic vegetation is another habitat management technique used in lentic waters. Control can be accomplished by mechanical, chemical, or biological methods. Biologists are cautious about vegetation control because removal of too much vegetation can harm aquatic communities.

Fertilization is a management technique that can be applied to infertile water bodies, such as those found in the southeastern United States, to increase their carrying capacity.

Artificial structures are often added to lentic water habitats. Where natural structure is limited, structures such as brush piles will attract many fish species. Generally, concentration of fishes for sport angler harvest is the goal of biologists that install these artificial structures.

Habitat management in lotic waters is most commonly attempted in smaller coldwater streams. Much of this management is undertaken to improve degraded sections of streams. Sometimes habitat improvement simply involves the reestablishment of natural conditions. In other cases, in-stream structures are constructed. Experts stress that stream habitat improvement techniques must be chosen on a case-by-case basis. Another major concern with regard to stream habitat management is maintenance of minimum in-stream flows.

PRACTICE QUESTIONS

1. Why is it difficult to distinguish between the use of fire as a habitat management technique to improve food production and its use to improve cover? Provide an example of a prescribed burn applied to grasslands and a burn applied to woodlands that would improve both food and cover.

2. In the future, clean air standards may reduce the ability of wildlife biologists to undertake prescribed burns in some areas. Name some alternative techniques that could be used instead of burning.

3. In which of the North American biomes would it be worthwhile to construct a guzzler? What animal species might benefit from guzzlers in each of these biomes?

4. What is dense nesting cover? List some animal species that might benefit from the use of this habitat management technique. How would they benefit?

5. What are several differences between a restored wetland and an artificial wetland?

6. How does an aeration system maintain an opening in the ice in a lentic water body?

7. How does planned water level manipulation in an older reservoir mimic the conditions that occur in a newly constructed reservoir? Is this desirable? Why?

8. Aquatic vegetation control is a two-edged sword. Too much vegetation can impair the functioning of a fish community, but so can too little vegetation. What problems are associated with each of these two scenarios?

9. Small dams are sometimes constructed as habitat improvement structures in small streams. However, they can be appropriate for some environmental conditions and inappropriate for others. Assume that you are managing for brook trout. Would you use a small dam on a high-elevation stream in the Rocky Mountains in which water temperatures typically did not exceed 13°C during midsummer? Would you use a small dam on a low-elevation stream in Minnesota or Wisconsin in which summer water temperatures sometimes approached 21°C? In both cases, why or why not?

10. What is the status of estuarine habitats in the United States?

11. Are minimum instream flows a concern in the area where you live? If so, describe the various human users involved.

SELECTED READINGS

Fredrickson, L. H., and M. K. Laubhan. 1994. Managing wetlands for wildlife. Pages 623–647 *in* T. A. Bookhout, ed. *Research and management techniques for wildlife and habitats*. The Wildlife Society, Bethesda, Md.

Herke, W. H., and B. D. Rogers. 1994. Maintenance of the estuarine environment. Pages 263–283 *in* C. C. Kohler and W. A. Hubert, eds. *Inland fisheries management in North America*. American Fisheries Society, Bethesda, Md.

Johnson, D. L., and R. A. Stein, eds. 1979. *Response of fish to habitat structure in standing water*. Special Publication no. 6. North Central Division, American Fisheries Society, Bethesda, Md.

Kie, J. G., V. C. Bleich, A. L. Medina, J. D. Yoakum, and J. W. Thomas. 1994. Managing rangelands for wildlife. Pages 663–688 *in* T. A. Bookhout, ed. *Research and management techniques for wildlife and habitats*. The Wildlife Society, Bethesda, Md.

Mannan, R. W., R. N. Conner, B. Marcot, and J. M. Peek. 1994. Managing forestlands for wildlife. Pages 689–721 *in* T. A. Bookhout, ed. *Research and management techniques for wildlife and habitats*. The Wildlife Society, Bethesda, Md.

Orth, D. J., and R. J. White. 1994. Stream habitat management. Pages 205–230 *in* C. C. Kohler and W. A. Hubert, eds. *Inland fisheries management in North America*. American Fisheries Society, Bethesda, Md.

Ploskey, G. R. 1986. Effects of water-level changes on reservoir ecosystems, with implications for fisheries management. Pages 86–97 *in* G. E. Hall and M. J. Van Den Avyle, eds. *Reservoir fisheries management: Strategies for the 80s*. Reservoir Committee, Southern Division, American Fisheries Society, Bethesda, Md.

Rodiek, J. E., and E. G. Bolen. 1991. *Wildlife and habitats in managed landscapes.* Island Press, Washington, D.C.

Schneberger, E., ed. 1970. *A symposium on the management of Midwestern winterkill lakes.* North Central Division, American Fisheries Society, Bethesda, Md.

Summerfelt, R. C. 1994. Lake and reservoir habitat management. Pages 231–261 *in* C. C. Kohler and W. A. Hubert, eds. *Inland fisheries management in North America.* American Fisheries Society, Bethesda, Md.

VanDruff, L. W., E. G. Bolen, and G. J. San Julian. 1994. Management of urban wildlife. Pages 507–530 *in* T. A. Bookhout, ed. *Research and management techniques for wildlife and habitats.* The Wildlife Society, Bethesda, Md.

Warner, R. E., and S. J. Brady. 1994. Managing farmland for wildlife. Pages 648–662 *in* T. A. Bookhout, ed. *Research and management techniques for wildlife and habitats.* The Wildlife Society, Bethesda, Md.

Wesche, T. A. 1993. Watershed management and land-use practices. Pages 181–203 *in* C. C. Kohler and W. A. Hubert, eds. *Inland fisheries management in North America.* American Fisheries Society, Bethesda, Md.

LITERATURE CITED

Bettoli, P. W., M. J. Maceina, R. L. Noble, and R. K. Betsill. 1993. Response of a reservoir fish community to aquatic vegetation removal. *North American Journal of Fisheries Management* 13:110–124.

Cody, M. L., ed. 1985. *Habitat selection in birds.* Academic Press, San Diego, Calif.

Higgins, K. F. 1984. Lightning fires in North Dakota grasslands and in pine-savanna lands of South Dakota and Montana. *Journal of Range Management* 37:100–103.

Higgins, K. F., A. D. Crossway, and J. L. Piehl. 1989. *Prescribed burning guidelines in the northern Great Plains.* EC 760. South Dakota Cooperative Extension Service, Brookings.

Hunt, R. L. 1988. *A compendium of 45 trout stream habitat development evaluations in Wisconsin during 1953–1985.* Technical Bulletin no. 162. Wisconsin Department of Natural Resources, Madison.

Layher, W. G. 1984. Compatibility of multiple uses: Potable water supplies and fisheries. *Fisheries* (Bethesda) 9 (6): 2–11.

Owen, O. S., and D. D. Chiras. 1990. *Natural resource conservation: An ecological approach.* 5th ed. Macmillan, New York.

Pardue, G. B., and L. A. Nielsen. 1979. Invertebrate biomass and fish production in ponds with added attachment surface. Pages 34–37 *in* D. L. Johnson and R. A. Stein, eds. *Response of fish to habitat structure in standing water.* Special Publication no. 6. North Central Division, American Fisheries Society, Bethesda, Md.

Platts, W. S., and F. J. Wagstaff. 1984. Fencing to control livestock grazing on riparian habitats along streams: Is it a viable alternative? *North American Journal of Fisheries Management* 4:266–272.

Prince, E. D., O. E. Maughan, D. H. Bennett, G. M. Simmons Jr., J. Stauffer Jr., and R. J. Strange. 1979. Trophic dynamics of a freshwater artificial tire reef. Pages 459–473 *in* H. Clepper, ed. *Predator-prey systems in fisheries management.* Sport Fishing Institute, Washington, D.C.

Robbins, C. S., D. K. Dawson, and B. A. Dowell. 1989. *Habitat area requirements of breeding forest birds of the middle Atlantic states.* Wildlife Monographs 103.

Sigler, W. F., and J. W. Sigler. 1990. *Recreational fisheries: Management, theory, and application.* University of Nevada Press, Reno.

Willis, D. W., and L. D. Jones. 1986. Fish standing crops in wooded and nonwooded coves of Kansas reservoirs. *North American Journal of Fisheries Management* 6:105–108.

Part IV

THE HUMAN USERS

16

ASSESSMENT
OF HUMAN
USERS

To thoroughly understand a fishery or a wildlife system, it is essential to understand the human user component. Biologists can obtain a wealth of information from and about the individuals and groups who directly and indirectly use wildlife and fishery resources. Currently, more information is available on consumptive resource users than on nonconsumptive users; however, assessment of nonconsumptive users is increasing, and this trend is likely to continue.

Human dimension assessment may involve a variety of information collected in numerous ways. Some of the information may be biological in nature, such as species taken, species seen, or the size of organisms taken. Other information may be demographic in nature, such as the age or sex of the people who utilize the resource. Still other information may concern human attitudes on a wide array of natural resource issues.

Economic value is another type of information that is often needed concerning wildlife and fishery resources. Economic theory and practice are human endeavors; thus, economic assessment is included in this chapter. The economic values of natural resources can be difficult to determine accurately. For example, the economic value of taking a fishing trip or viewing songbirds is not measured solely by the amount of money spent in either of these activities. They provide satisfaction beyond the items that require monetary expenditures. In addition, because such assessment involves human ethics and values, the economic value placed on a natural resource can vary from one individual or group to the next. Like economic values, human ethics and values can be difficult to determine.

The initial portion of this chapter highlights some of the ways in which natural resource information is gathered from humans and the reasons why such information is collected. Following this discussion of survey and census information gathering, a brief discussion of natural resource economics is presented. Human ethics and values are then addressed because they are logical outgrowths of any discussion of human dimension assessment and natural resource economics.

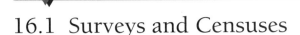

16.1 Surveys and Censuses

This discussion pertains primarily to the collection of biological and demographic data from resource users. Survey and census information gathering for the purpose of wild organism population assessment is described in section 9.2. This chapter builds on that previous information concerning censuses and surveys.

Surveys and censuses can be conducted on-site (in the field) as natural resource agency personnel meet directly with the public; they can involve observation of users by resource personnel; or they can involve telephone or mail data collection. At times combinations of formats are used. Common examples of on-site data collection formats are fishery creel surveys and wildlife roadside bag checks (fig. 16.1). Data collection can also involve information on nonconsumptive users such as hikers, campers, or bird-watchers. The information gathered can include a wide variety of data, depending on the purposes of the survey or census.

By conducting creel surveys or censuses, fishery personnel can obtain biological information such as the size, species, and number of fishes being caught. If those data are obtained in conjunction with information on time spent fishing, harvest and success rate estimates can be calculated. Determination of hours spent angling is one of the more common and important objectives of such creel surveys or censuses. Natural resource agencies often set management goals for particular water bodies in terms of the expected catch per angler-hour of fishing effort for a particular species (see section 20.6). Determination of the success or failure of management strategies is often impossible without data on catch (or harvest) and effort (angler-hours). Such information can also be used to formulate new regulations or other corrective measures if management objectives are not being met.

Biological data can also be obtained through bag surveys or censuses of wildlife species. Road-

counts. When surveys are performed, specific sampling techniques must be used so that the data obtained can be expanded to an estimate of the whole (see section 9.2). The specifics of how, when, and where surveys need to be conducted are beyond the scope of this text, but an example will demonstrate the difficulties that can be associated with obtaining reliable survey data.

Assume that angler-hours spent per year (annual total fishing effort) must be determined in order to assess resource use on a particular water body. One cannot sample every day, so which days are chosen? Think of how the data would be biased if only weekends or holidays were sampled. How are night anglers surveyed compared with day anglers, or boat anglers compared with shore anglers? If ice fishing occurs in enclosed fish houses, how are people and numbers of lines being used counted? These and other variables combine to make accurate survey data difficult to obtain.

Some of the information obtained may not be biological in nature, but instead is intended to profile the users. Demographic data such as user age, sex, distance traveled to utilize a resource, personal income level, education level, and methods used to take organisms are common components of surveys or censuses. This information helps biologists to determine who is using a resource, how, and possibly why. Such information may affect the way the resource is conserved.

Some data can best be obtained by face-to-face contact with users. In fishery work, resource personnel can move from one user or group of users to another in order to obtain information. One can also wait at an access point, such as a boat dock or fishing pier, for users. For wildlife bag surveys or censuses, biologists can use primary roads that funnel users to particular locations, or travel around an area checking hunters that they meet. For surveys of campers, various campsites can be visited. For hikers, resource personnel can gather information at various points on a trail system.

There are a number of distinct advantages to direct personal contact. First, the information

FIGURE 16.1. A great deal of useful information can be obtained in face-to-face contact with resource users, as in this bag check of big game hunters. *(Photograph courtesy of L. Parker, Wyoming Game and Fish.)*

side bag surveys may be used to determine the age, sex, and number of deer or other game species harvested by hunters. This information can be used to assess the success or failure of management objectives set in terms of hunter success rates, male-female ratios, or adult-young ratios. Face-to-face meetings between biologists and fishery and wildlife resource users also have obvious utility for law enforcement and public relations purposes.

Censuses and surveys need not target only game organisms. For example, the National Audubon Society annually conducts a Christmastime bird count. This effort is national in scope, is carried out by thousands of volunteers, and provides numbers and distributions for all birds observed. This information is useful in determining long-term trends in bird species distribution and abundance.

Most biological data are collected from users by means of surveys (partial counts); censuses are seldom done because they require complete

received is more likely to be accurate. For example, which length measurement of a fish do you think would probably be accurate, one made by a biologist or one made by an angler? Wildlife resource personnel at a roadside bag check can see a doe or a fawn deer that might have been reported as an antlered buck if some other type of survey or census had been used. An estimate of time spent bird-watching is likely to be more accurate if resource personnel can discuss starting and finishing times with a bird-watcher who is unsure of when he or she started, ate lunch, or finished the activity.

Another advantage of personal contact is the public relations work that can be accomplished. Users often have questions about laws or regulations, where and how to be successful in their recreational activity, and other information. Personal contact allows them access to natural resource personnel who can answer such questions.

There are also drawbacks to personal contact, especially if it is conducted improperly. Personal contact is expensive, and most natural resource agencies have limited funds. Another drawback of personal contact is the possibility of inaccurate results if the questions and information obtained are not standardized; the information obtained should be the same regardless of who performs the interviews. An additional drawback is that resource users who are particularly satisfied or dissatisfied are often more talkative and cooperative than users who are more neutral in their success or opinion.

Personal interviews involve interactions between people, and thus require special care. The interviewer may be interrupting the recreational activity of the user, or the user may be finished and want to return home. In addition, the interviewer represents authority, and different people react differently to authority. These and other factors, some of them psychological in nature, combine to result in a need for proper interviewing techniques. The interviewer should be courteous and informative, and should attempt to develop trust. Resource personnel should also convey to the user the reasons for the survey or census so

that he or she understands that the information requested is important enough to justify the interruption.

Gathering on-site field information does not always require actual personal contact with users. For example, aerial surveys or censuses can be conducted from a variety of types of aircraft, vehicle counting devices can be placed at resource use access points, or a biologist can obtain estimates of use just by viewing an area.

There are other means of obtaining information from users without face-to-face contact. Survey or census instruments, such as questionnaires, can be provided at particular locations. They can be completed by the user and left at that location or returned by mail. In some situations, specific users may be surveyed or censused by mail. It is a common practice for many natural resource agencies to obtain data by requesting that users provide information concerning catch or harvest success, area used, time spent on the recreational activity, and other data. There is a growing tendency to request that detailed activity logs be kept by some users; this is usually done on a voluntary basis. The U.S. Fish and Wildlife Service annually requests information from waterfowl hunters by having them report the birds taken by species, date, and location. Wings and tail feathers from the birds taken are also requested so that information on species and age structure can be obtained. In some fishery surveys, anglers are asked to provide a variety of detailed information as well as scales from fishes caught.

Telephone surveys or censuses are also used to gather information. The use of telephones is intermediate between mail surveys and face-to-face interviews in the degree of personal contact. It allows for an exchange of questions and answers, which can increase both accuracy and public relations benefits.

Most people will readily cooperate with resource personnel in data collection because they are interested in natural resources and their successful conservation. They are often interested in the results obtained and will usually respond positively when asked if they would like information on the results of the survey or census. Supplying such information provides educational benefits and is also a useful public relations tool. It is of great importance to remember, however, that data gathering should be done with specific purposes in mind. It is too costly to collect data without a good reason, and users' time is valuable to them and should not be wasted.

The user census and survey information available from a variety of federal, state, provincial, territorial, and local sources is voluminous. In the United States, the most extensive general survey of natural resource users is the *National Survey of Hunting, Fishing, and Wildlife Associated Recreation* (U.S. Department of the Interior 1993). This survey, conducted every five years, contains a wealth of information. The following examples represent just a few of the types of information that can be gleaned from this survey. All represent 1991 data for people sixteen years of age or older. Of all the big game hunter-days expended in the United States, 88% were expended on deer hunting. Almost 17 million people photographed wildlife in the state in which they resided. There were approximately 4.5 million people in the United States who considered themselves to be fly anglers, and they spent almost 28 million days a year fly fishing. Approximately 93% of the residents of Alaska participated in wildlife-associated recreation. Approximately 29% of the residents of Vermont hunted or fished. Approximately 45% of the residents of Tennessee were participants in nonconsumptive wildlife and fishery activities. More specific surveys conducted by a variety of natural resource agencies provide even more detailed information.

16.2 Attitude Assessment

Wildlife and fishery biologists have always been adept at obtaining biological information from resource users, but they have often been much less effective in assessing user attitudes. As mentioned in section 1.2 of this text, the biota and habitat aspects of a fishery or a wildlife system have received the most attention from biologists; the human user components of these three-part systems, especially attitudes, has received much less attention. Many biologists, while well educated in the biota and habitat aspects of a fishery or a wildlife system, lack sufficient educational depth in the human dimension aspect. The current trend in the higher education of future fishery and wildlife biologists is to correct this deficiency. While all biologists cannot become human dimension experts, they should be knowledgeable about basic aspects of the human dimension. They should also be able to work effectively with people specifically educated in the human dimension area. This increased attention to the human dimension, especially the attitudes of resource users and how such attitudes can be assessed, should result in the development of more appropriate natural resource conservation planning.

Attitude assessment can include direct, indirect, consumptive, and nonconsumptive users of a resource and also segments of the general public not included in those categories. Because wildlife and fishery resources in the United States belong to the public as common resources, all people have some stake in how those resources are managed. In addition, attitude assessment goes far beyond whether someone is satisfied with a particular law or regulation, believes that boat access areas are adequate, has been successful in bagging a trophy deer, or has added three new species to the list of birds observed. Each fishery or wildlife resource use is a multiple-satisfaction activity with a whole series of potential recreational outcomes that can accrue to users. All of these outcomes must be understood if accurate attitude assessments are to be obtained.

Biologists need to understand how to measure, evaluate, and integrate the socioeconomic, cultural, and political elements of the conservation environment (Krueger et al. 1986). Typically, attitude assessment relates to the beliefs, values, behaviors, and socioeconomic and demographic characteristics of user groups or publics and how those characteristics can be incorporated into an overall natural resource conservation scheme (Gigliotti and Decker 1992).

Most human dimension theorists agree that attitudes can be divided into two separate components: affect, the emotional component, and belief, the perception (cognitive) component. ("Affect" is used as a noun by human dimension specialists, and should not be confused with the words affect [verb] and effect [noun] as they are commonly used.) Many also add a behavioral (intentions and actions) component (conation) to the model (Fishbein and Ajzen 1975). If one is to understand public preferences and attitudes toward natural resource conservation actions and policies, then it is necessary to examine user beliefs.

Beliefs represent the information that a person has about an object and what links that object has to some attribute. For example, the belief that "beavers are abundant" links the object "beavers" to the attribute "abundant." The object of a belief may be a person, group, behavior, policy, or event, and the associated attribute may be any object, trait, property, quality, characteristic, outcome, or event. Beliefs can vary in strength from person to person; that is, people usually differ in terms of the ways in which an object is associated with the attribute in question. Beliefs about an object can be positive or negative, and the attitude of a person is viewed as corresponding to the total affect associated with his or her belief. For example, the evaluation of beavers by a business executive will be positive or negative based on her belief concerning a statement such as "beavers are abundant" and the sum of other beliefs she holds about beavers. If she values beavers, then the belief that "beavers are abundant" will be a positive belief.

Pierce (1979) defined values as "preferred outcomes or preferred ways of achieving those outcomes." In the above example, if the business executive values "the presence of beavers," then her holding that value would probably lead to a positive rating for the belief that "beavers are abundant." Conversely, if a person such as a farmer has contact with beavers only in contexts such as a beaver dam flooding his land, plugging irrigation ditches, or the animals eating his agricultural crop, the belief that "beavers are abundant" would probably receive a negative rating. Understanding the attitude a person has toward something requires learning about the belief and value systems of that person. Understanding the diverse reasons for the attitudes of people who participate in wildlife- and fishery-related activities, what they expect from these activities, and their sources of satisfaction and dissatisfaction can allow a biologist to better incorporate the human dimension component into conservation decisions. As mentioned, the diverse reasons for the attitudes of consumptive, nonconsumptive, indirect, and direct users all require assessment.

The most common method currently used to obtain attitude information is surveying people by means of questionnaires. When such surveys are properly conducted, a representative sample of people and attitudes can be obtained. In a representative sample, all are represented, not just those with a strong interest one way or the other.

Public input into natural resource decisions has often been dominated by vocal minorities ("squeaky wheels"). Assessment based on just those individuals is usually inaccurate. For example, voluntary response questionnaires, such as tear-out questionnaires found in magazines or those left on location, although popular methods of collecting public input, are not valid survey methods and often obtain the opinions of only the vocal minority. Opinion call-ins are not valid survey methods, nor are compendia of letters of complaint or support, nor are surveys performed at public meetings. Such surveys lack control over and knowledge about the sample.

A disadvantage of even a valid questionnaire format used as a public participation tool is that it does not allow for two-way communication. This disadvantage can be overcome by linking questionnaires with other public involvement strategies, such as meetings or interviews, that involve two-way communication.

The process of developing an attitude survey instrument or questionnaire does not begin with writing questions. The initial step is to develop specific goals, objectives, and strategies with regard to information needed. Section 20.6 describes the planning process, including the development of goals, objectives, and strategies. The question, "What management decisions do we intend to make?" must be answered first. The answer to that question is used to develop goal statements. The question, "What information is needed to make these management decisions?" leads to development of objectives. The question, "How, specifically, can we obtain this information?" leads to strategy statements. Specific questions, such as whom to sample and how many people to sample, must also be answered, and other aspects of the survey strategy determined. Only after all of these issues are settled can question writing begin. The development of survey goals, objectives, and strategies—indeed, the whole planning process—works most effectively if a team of people is involved.

Question wording is crucial to obtaining valid information. If questions are vague, incomplete, ambiguous, or open to different interpretations by respondents, the survey results will be of little use. Consider the following pitfalls in question writing.

Avoid imprecise questions. The question, "Do you want a quality songbird management program for this area?" is imprecise. A songbird biologist or any number of bird-watchers would interpret the term "quality" quite differently. Another example of an imprecise question would be, "Did you fish during the last year?" Does this mean during the last 365 days, this calendar year, last calendar year, or during the year shown on your fishing license?

Avoid double-barreled questions. These are questions with two or more parts; for example, "Were you satisfied with your ring-necked pheasant harvest and pheasant hunting regulations this year?" Such questions can confuse respondents because they may have a different answer for each part. It is safest to avoid lengthy and complex questions. Use the simplest wording possible that conveys the intended meaning.

Avoid irrelevant questions, questions that the surveyed people are not really in a position to answer. For example, one would not ask a hunter, "Do you think the deer harvest this year exceeded management objectives of harvesting 50,000 deer for the state?" Few hunters could give a meaningful response.

Another serious mistake is the use of loaded questions. These are questions that channel the response in a particular direction. "Don't you think the mourning dove season should be eliminated?" is an example. Such wording implies that the season should indeed be eliminated. "What is your opinion about the mourning dove season?" or "Do you support or oppose the mourning dove season?" are more appropriately worded questions. Another type of loaded question results from using an unequal set of response choices. Giving the respondent a choice among "superior," "excellent," and "good" does not provide sufficient choices. There should be a full range of choices, and none should be obviously different from the rest, as, for example, "How would you rate the Wildlife Department—superior, excellent, good, fair, poor, not worth a cent?" The last response is very different from the others, and the question is loaded by its implication that the Wildlife Department is so bad that a special response category is needed.

Most biologists are not well qualified to write survey questions. Assistance from experienced human dimension professionals is needed; that is why the team approach produces excellent results. The biologist knows what information is desired, and the human dimension expert knows how to word questions to obtain accurate information.

Biological data tend to be quantitative in nature; for example, how many of a particular species were taken or seen, whether they were male or female, or what their lengths were. Attitude information tends to be qualitative; rather than "yes" or "no" or a specific number, an attitude assessment question has a range of potential responses. For

Here are some statements that deal with your feelings about catching trout. Please indicate the extent to which you agree or disagree with each of the following statements. Please circle one number for each item.

	Strongly disagree	Slightly disagree	Neutral	Slightly agree	Strongly agree
1. A fishing trip can be successful to me even if I don't catch trout.	1	2	3	4	5
2. The bigger the trout I catch, the better the fishing trip.	1	2	3	4	5
3. When I go fishing, I am satisfied only when I catch some trout.	1	2	3	4	5
4. Catching a "trophy" trout is the biggest reward for me.	1	2	3	4	5
5. It does not matter to me what type of trout I catch.	1	2	3	4	5
6. How I catch a trout is as important to me as actually catching one.	1	2	3	4	5
7. If I thought I would not catch trout, I would not go fishing.	1	2	3	4	5
8. The more trout I catch the happier I am.	1	2	3	4	5

FIGURE 16.2. The Likert scale method is an attitude survey method that provides information on both direction and intensity of affect. (*Courtesy of L. Gigliotti.*)

this reason, specific instruments for attitude measurement have been developed. Their intent is to obtain qualitative information that can be put into a quantifiable format.

Attitudes can be measured with a survey device called the **Likert scale method** (fig. 16.2). Attitudes have both direction of affect (positive, negative) and intensity of affect; a good attitude scale will measure both of these components. Attitude statements are developed that relate to the object of concern, and respondents are asked to respond to the statements on a scale ranging from "strongly agree" to "strongly disagree." The survey instrument depicted in figure 16.2 measures different aspects of the importance of catching trout to trout anglers. The total of all responses results in an index measuring the importance of catching trout, while the individual items explore the component parts of this concept.

Another means of measuring attitudes is the **semantic differential method**. This method rates an object or concept on a scale with each endpoint anchored by an adjective or phrase. Figure 16.3 is an example of the use of this method to evaluate the experience of a person being checked by a conservation officer. Note that it also measures both direction and intensity of affect. The **checklist scale method** (fig. 16.4) provides a list of items and asks the respondent to indicate whether each applies in a certain context. The **frequency scale method** (fig. 16.5) asks a question about the likelihood of a certain event or action. In the **ranking scale method** (fig. 16.6), the respondent is asked to order a list of items according to personal preference or some other standard. With the **magnitude scale method**, an attempt is made to build a scale (fig. 16.7). The respondent is given a short list of items and asked to allocate a given number of points among them to represent their importance or value.

Forcing respondents to indicate the presence of an attitude when none exists can result in inaccurate information. To avoid this problem, "no opinion," "do not know," or other neutral responses can be added to the response categories. These are valid responses, and including them can have an important effect on attitude assessment accuracy. Which of these neutral response categories to

How would you describe your treatment by the conservation officer for your most recent contact?

1.	Fair	1	2	3	4	5	Unfair
2.	Rough	1	2	3	4	5	Kind
3.	Professional	1	2	3	4	5	Unprofessional
4.	Good	1	2	3	4	5	Bad
5.	Unfriendly	1	2	3	4	5	Friendly

FIGURE 16.3. The semantic differential method of attitude assessment uses a scale with each endpoint anchored by an adjective or phrase. (*Courtesy of L. Gigliotti.*)

include and when to include them depends largely on the specific situation.

The attitude assessment methods in figures 16.2–16.7 are all examples of closed-ended formats; that is, respondents choose one or more appropriate responses from a given set of alternatives. The advantages of closed-ended questions are that they are quick and easy for the respondent to use and that they readily lend themselves to quantitative analysis. A disadvantage of closed-ended formats is that the validity of the results depends on the adequacy of the given set of alternative responses as well as on the wording of the question. Therefore, designing closed-ended questions requires in-depth knowledge of the survey topic.

By contrast, open-ended questions, such as "What would you like done in this area that would improve your whale-watching experience?" do not provide alternative responses from which to choose, but instead provide for a free and potentially more informative response. Open-ended questions are best used to generate ideas and to explore issues; they are also used when adequate knowledge for the development of a list of appropriate response categories is lacking. They are more difficult for the respondent to answer, and the responses are more difficult to quantify.

Please rate from zero (0) to nine (9) the importance of *each* reason for why you fish. Please circle *one* number for each reason.

		Not a reason									A very important reason
1.	To catch fish to eat	0	1	2	3	4	5	6	7	8	9
2.	To catch fish for fun and excitement	0	1	2	3	4	5	6	7	8	9
3.	For companionship (friends and/or family)	0	1	2	3	4	5	6	7	8	9
4.	To get away and relax	0	1	2	3	4	5	6	7	8	9
5.	To enjoy nature	0	1	2	3	4	5	6	7	8	9
6.	To enjoy my fishing equipment	0	1	2	3	4	5	6	7	8	9
7.	To catch trophy-sized fish	0	1	2	3	4	5	6	7	8	9

FIGURE 16.4. The checklist scale method of attitude assessment provides a list of items and asks the respondent to indicate whether each applies in a certain context. (*Courtesy of L. Gigliotti.*)

You find a raptor with a broken wing. How likely might *you* be to contact a raptor rehabilitation center concerning the bird? Please circle only one number for your response.

Never   Very likely
1   2   3   4   5   6   7   8   9

FIGURE 16.5. The frequency scale method requires the respondent to indicate the likelihood of an event. (*Courtesy of L. Gigliotti.*)

Open-ended questions work best in personal interviews. In mailed surveys, open-ended questions are often left blank or lead to an increase in nonresponses to the survey.

These and any other attitude assessment tools can reveal much information about user attitudes including those that conflict with what the biologist believes is biologically correct and what different portions of the public may want; all of these preferences must be considered. What should a biologist do, for example, if he or she is involved in an effort to restore an extirpated native species, such as the gray wolf, to an area and assessment of the human dimension indicates that this is the course of action desired by the general public, but not by the people in the area of the restoration?

A study conducted by Fazio (1987) can be used as a brief example of the information that can be obtained through attitude assessment. In that study, members of The Wildlife Society were surveyed with regard to their opinion on the wildlife resource values that are most important to communicate to the public. An initial open-ended opinion questionnaire was followed by a second questionnaire that asked respondents to assess 28 value statements that had been developed from the first survey. The value statements were assessed on a five-point scale that ranked them from extremely important to less important. Fazio found that "protection of habitat" was regarded as the most important value to be communicated. A few of the other 27 values were ranked as follows (the number in parentheses after the statement represents its order of importance): that all species are important to diversity (2); that wildlife is a reflection of the overall health of our environment (3); that wildlife plays a role in enriching the quality of human life (4); that wildlife is a renewable resource if managed (5); the negative effects of the human population on wildlife (10); the importance of restoration of native vegetation (16); and the food value of wildlife for humans (25).

Please rank (using the numbers 1 through 6) the following methods of sewage sludge management according to your overall preference. Use 1 for your first choice.

_____ Burying in landfills

_____ Incineration (burning)

_____ Application to agricultural land for production of human food

_____ Application to agricultural land for production of animal feeds

_____ Use as lawn and garden fertilizer or soil conditioner

_____ Application to forest lands

FIGURE 16.6. The ranking scale method requires the respondent to rank a list of items. (*Courtesy of L. Gigliotti.*)

To show how much consideration you believe should be given to each of these categories by planners when considering development in your area, *divide 100 points among the four categories below.* The higher the number of points given to a category, the more importance you feel it should have compared with the other categories. Remember that the total must equal 100.

_____ Human health   _____ Environmental quality

_____ Economics (costs)   _____ Aesthetics (beauty) of the area

Total = 100

FIGURE 16.7. In the magnitude scaling method, the respondent is provided with a list of items and asked to allocate points among them. (*Courtesy of L. Gigliotti.*)

While such information helped to answer the question, "What values need to be communicated?" a second question, "How can we successfully communicate these values?" is also important. Answering that question would require additional attitude surveying.

16.3 Economic Assessment

Natural resource economics is a field of study in itself, and is much too involved to address here except in a cursory fashion. **Natural resources** are things such as land, water, and wildlife that are useful and valuable in the condition in which they are found. **Economics** is the study of the allocation of resources. The human social aspects of natural resource use constitute an important element of resource value. Assessments of the economic values of wildlife and fishery resources may take into consideration a variety of economic activities associated with these resources.

It is important to have information on the economic values associated with particular natural resources, and such information will become even more critical in the future. Without a knowledge of their economic value, it is difficult to assess the importance of natural resources compared with

other economic factors. For example, it is likely that for wetland conservation to be ultimately successful, the total economic values of wetlands will need to be judged against the economic values of competing activities, such as agricultural pursuits and commercial and residential land development. When economics is the prime criterion for decisions on whether a particular area or habitat type is worth conserving instead of developing, adequate and correct economic information needs to be available.

Unfortunately, it is difficult to obtain adequate economic information concerning natural resources such as a fishery or a wildlife system. They are inherently much more difficult to assess than some other human undertakings, such as agriculture, forest crops, or shopping malls, because a portion of their value is attitudinal, and attitudes vary from person to person. Economic assessment of wildlife and fishery resources may involve direct, indirect, consumptive, or nonconsumptive uses or users, or any combination thereof. The best current economic data available relate to publicly owned resources used in a consumptive manner. It should be noted, however, that the nonconsumptive use of natural resources is a rapidly growing area; it is also an area in which better economic information is needed.

Economic assessment of wildlife and fishery resources may include the economics of hunting, fishing, trapping, and other consumptive wildlife and fishery activities. It may also include activities such as game farming, game ranching, and aquaculture. It can also include **fee hunting** or **fee fishing**, which involves charging fees for the right to trespass and take either publicly or privately owned organisms on private land. **Ecotourism** is an economic activity associated primarily with the nonconsumptive use of natural resources. Ecotourism is tourism primarily directed at viewing specific habitats or organisms in their wild state. Examples include bald eagle watching tours, trips to view tropical rain forests, desert wildflower viewing trips, tours to view autumn foliage in temperate forests, and a wide variety of other similar activities.

What is the economic value of a particular natural resource, what are the problems associated with its assessment, and how is an assessment made? A good method of starting to answer such questions is to provide some examples of the problems involved in the economic assessment of fishery and wildlife resources. Many large reservoirs in North America are managed by the U.S. Army Corps of Engineers (COE). When each reservoir was constructed, the COE assigned particular economic values to a variety of uses of that reservoir. These economic values to a large degree dictated how each reservoir was managed. By definition, multiple-use reservoirs have multiple functions. Power generation, irrigation, flood control, navigation, and recreation are just some of their uses. However, the economic values of these uses, often determined many years ago when the reservoirs were constructed, may differ from their current economic values; these values may also be quite different in the future. Conflicts have arisen because the original economic assessments placed more emphasis on the reservoirs' nonrecreational values than on their recreational values. The original emphases were also generally on consumptive uses, such as irrigation and power generation, rather than nonconsumptive uses, such as bird-watching, boating, and hiking. Use patterns have changed, and so have the economics of the situation, but the management of these reservoirs has been slow to change. The COE is now in the position of trying to address the multiple and often conflicting and changing uses of the reservoir system. The conflicts should be obvious. At what reservoir level is power production greatest? At what reservoir level is flood control most effective? How is downstream navigation best protected? At what water level are the fishery resources best served? When do water discharges need to be reduced or stabilized to protect downstream nests of endangered or threatened bird species? How much water can be appropriated for irrigation, and when and where? How much concern should there be for nonconsumptive uses? These issues are essentially economic in nature; however, biological, social, and political issues must be considered as well.

Another example of how natural resource economics and biological, social, and political considerations interact is the issue of old-growth forests in the Pacific Northwest. The lightning rod of this issue has been the effect of timber harvest on the northern spotted owl, but the economic and other issues go much further. What of the economic, aesthetic, and biological values of the myriad of other animal and plant species dependent on those old-growth forests? What economic and social factors affect the people and towns dependent on that diminishing forest resource? What are the economic and social effects on the commercial and sport salmon fisheries that are dependent on salmonid spawning habitat protected by old-growth forests? How does one modify the time frame (one year, two years, thirty years) in which the economic assessments are made? These and other issues require, among other things, adequate economic assessment before enlightened conservation decisions can be made.

Economic issues are directly related to the human dimension because economics is a human undertaking. Identifiable economic benefits from wildlife and fishery resources accrue in numerous ways. For example, the **gross economic value** (benefits) of a fishery or a wildlife system is composed of (1) the value of the federal, state, provincial, territorial, and local services that contribute to the natural resource; (2) user expenditures (license fees, equipment, travel, etc.); (3) the value beyond actual expenditures that users would be willing to pay to use the resource; and (4) the additional amounts nonusers would be willing to pay to conserve the particular fishery or wildlife system.

The first economic component, federal, state, provincial, territorial, and local services that contribute to the value of a natural resource, may not contribute directly to user benefits, but does contribute to the value of the resource. The biologists who manage the natural resource, the businesses that invest in services specifically for users, such as boat dealers and sporting goods stores, and a variety of other services that contribute to or support use of the natural resource, such as motels, restaurants, and service stations, all have input into the economic value of the resource. However, some assessments of these economic benefits can be misleading. Cleanup costs for the *Exxon Valdez* oil spill, for example, exceeded a billion dollars. Technically, that expenditure added to the economic benefits of Alaska and the United States; however, few people would view that situation as an economic benefit of any kind.

User expenditures are the second component of the economic value of a natural resource. The amount that users expend to pursue an activity constitutes a major portion of the economic value of a natural resource. For example, users may spend money for fuel, food, and lodging in the course of a particular trip to use a natural resource. Costs for licenses, boats, motors, binoculars, bird seed, guns, fishing equipment, special clothing, and other items also add economic value. As mentioned earlier in this chapter, the U.S. federal government conducts a survey of recreational activity associated with wildlife, fishery, and related resources (U.S. Department of the Interior 1993). Among the information provided by this survey are estimates of such monetary expenditures by users. This information is available not only for hunting and fishing, but also for a variety of nonconsumptive activities such as observing, feeding, and photographing fishes and wildlife, maintaining plantings that benefit wildlife, and park visitations primarily for wildlife viewing purposes.

The third component of the economic value of a wildlife or fishery resource is the value beyond actual user expenditures that users would be willing to pay to use the resource. This value is called **consumers' surplus**. It is a theoretical value that would be collectable only if the user could be induced to pay this excess amount. Because many wildlife and fishery resources are not privately owned, but instead are a portion of the commons, and because valuations differ from one person to the next, this value can be difficult to assess. The basis for determining consumers' surplus is a simulation of the demand schedule for a natural resource. A demand schedule is simply a curve that indicates the quantities of a good or service that consumers are willing and able to purchase at various prices (fig. 16.8). Demand curves generally slope downward, indicating that decreased amounts of goods or services are desired at higher prices. If, for example, *v* number of visits to use a natural resource actually occur at *c* cost per visit, actual user expenditures can be calculated (Dwyer et al. 1977; Loomis 1993). The difference between the demand curve and the actual expenditures can then be used as an

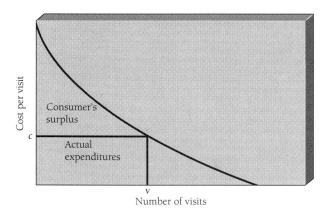

FIGURE 16.8. A demand schedule depicting the relationship between actual expenditures and consumers' surplus. *v* represents the actual number of visits; *c* represents the actual cost per visit.

estimate of the consumers' surplus, that is, the amount users would pay in excess of actual expenditures. Rockland (1985, 1986) provided excellent guides to the economic benefits of and economic models related to fishery resources.

The fourth item considered in determining the value of a fishery or a wildlife system, the value of the natural resource to nonusers, does not directly contribute to user benefits, but does represent a contribution to the economic value of the resource. It is a difficult value to determine, but it does exist. For example, someone in New York may never see a grizzly bear in Alaska, a snail darter in Tennessee, or a tropical rain forest in Central America, but these organisms and ecosystems may have value to that person. The value is in knowing that the natural resource exists. Obviously such values vary from person to person; some may value the existence of an organism or ecosystem highly, others not at all.

An additional economic value that is often overlooked is the value of the fishery and wildlife resources that are harvested by consumptive users such as hunters and anglers. These organisms have value to the user, but again, such values can be difficult to determine. A pen-raised ring-necked pheasant in a grocery store may cost $7–$8. A dinner of ring-necked pheasant in a restaurant may cost anywhere from $10 to much more in very fashionable restaurants. Some species used for sport purposes are not commercially raised; how can their economic value be estimated? The consumptive user who harvests such an organism and consumes it is obviously obtaining a benefit. Its cost may be great, based on the expenditures made to harvest the organism, but some value is being obtained. There are no specific mechanisms available to make such value estimates, but such values can be determined in other ways. For example, the American Fisheries Society (1992) produced a publication intended to assist in assessing the values of various fish species killed accidentally or purposefully; this listing can also be used for other economic assessment purposes. It provides values of fishes by length and weight for a variety of species. This is just one tool of many available that can be used to assess the economic value of individual organisms.

A problem involved with the economic assessment of natural resources is the fact that estimates are based only on current values and knowledge. What is the economic value of some undescribed or nondescript insect before and after it is learned that the insect contains the key ingredient needed to cure arthritis? What is an Atlantic salmon worth today compared with its economic value in the year 2096? In most cases, we do not know the answers to such questions, but we still try to place an economic value on an item. Changes with time and knowledge are factors that can cause a natural resource to be overvalued or undervalued when assessments are projected into the future.

It is important to reiterate the difficulty of obtaining a total assessment of the monetary value of a fishery or a wildlife system. There are many benefits that are difficult to include in economic models. Activities such as fishing, bird-watching, visits to national parks, and other recreational pursuits that utilize fishery and wildlife resources are multiple-satisfaction activities. What is the value of the social well-being one obtains from a natural resource? What are the physiological benefits? What are the psychological benefits? What are the cultural benefits? These hard-to-measure benefits make it difficult to put dollar values on such activities.

16.4 Human Ethics and Values

While not a form of resource user assessment, consideration of human ethics and values can be important in managing a fishery or a wildlife system. Ethics and values are directly related to human attitudes and to numerous other aspects of natural resource valuation. The ethics and values of a person or group can greatly influence the worth they place on a fishery or a wildlife system. Many topical areas could be covered in this portion of the text, but comments will be restricted to three areas: ethics and values in general, conservation ethics and values, and animal rights and animal welfare ethics and values. Different belief systems are represented in each of these areas. A further discussion of ethics and values is presented in section 20.5. That discussion revolves around biopolitics and conflict resolution.

Ethics (standards of conduct and moral philosophy) and **values** (the qualities of a thing according to which it is thought of as being more or less desirable, useful, or important) are obviously dependent on the individual. While the ethics and values of some people would allow them to follow wildlife and fishery regulations minimally, others would go further. For example, while regulations might allow the harvest of a hen northern pintail, some people would not do so if they believed that the birds were at low population levels. While some people think nothing of damaging fragile habitat with an all-terrain vehicle, others would not consider such an activity.

Even when trying to be ethical, humans can cause problems. A few years ago, during a study of piping plovers (endangered in some parts of North America and threatened in others) and least terns (an endangered species), a sandbar area on the Missouri River was posted as a no-recreational-use zone because the plovers and terns were nesting there. River sandbars in the region are popular recreational spots. Some users entered the restricted area, removed all the eggs from the nests, played volleyball, and then returned the eggs to the nests. They were trying to be ethical as they viewed it, but their actions were still harmful to the birds. In other cases, people entered such restricted areas and purposefully destroyed nests. Those people may even have been attracted by the signs; this would be indicative of a totally different set of ethics and values.

Ethics and values need not be viewed only as a consumptive use issue. Examples like that of the volleyball players above demonstrate that non-consumptive users are also involved in this issue. Even indirect users are involved. The recycling of paper, aluminum, and other recyclables, for example, is increasing for more than just economic reasons. Recycling is seen as ethical by many people because of the trees and habitat it saves from cutting and the reduction in environmental damage due to mining and manufacturing. The increased interest of the general public in protecting organisms and habitat is a result of decisions based on ethics and values by those who support conservation; those who do not support such conservation have also made a decision based on ethics and values. Two people can have diametrically opposed opinions and both consider their particular viewpoint to be ethical.

Another example of the ambiguity of what is or is not ethical is the debate over saving individual organisms harmed by oil spills. Large amounts of time and money are spent in trying to save a relatively few animals each time an oil spill coats the bodies of birds or mammals. Many animals are damaged and killed as a result of this pollution source (fig. 16.9). Some people would contend that it is ethical to save these few animals; others would contend that it is unethical to expend all of that effort on a few when similar efforts or expenditures directed at total damage control would be much more valuable. The initial cleanup of a few birds and mammals after the *Exxon Valdez* oil spill in Alaska has been variously described as a positive, well-directed effort at reducing damage and as a double disaster. Those who consider it a double disaster view both the spill and the large initial expenditure on cleanup of a few individual animals as disasters.

Human-induced problems do not provide the only examples of ethical conflicts. A few years ago, a massive effort was made to free some icebound whales. While this effort made for good news coverage and seemed ethical to some, others felt that it was unethical to expend more money on a few

FIGURE 16.9. The ethics of saving a few individual animals harmed by an oil spill is often questioned when similar time and monetary expenditures could assist greater numbers of animals in other ways. It is probable that this American coot could have been saved if found alive and washed thoroughly to remove the oil that killed it.

individual whales than is annually expended for research for the benefit of all whales.

There are ongoing debates concerning the ethics of sport hunting and fishing and indeed the use of any organism by humans for any purpose, even as pets. These are individual decisions with no answers that are universally right or wrong. What may seem perfectly right and ethical to one person, even what is allowable by law, may seem unethical or wrong to another. This dichotomy of opinion on issues such as hunting, fishing, trapping, animal experimentation, keeping animals as pets, and many other similar activities will never be resolved until all people believe the same thing, and the chance of that ever occurring is remote.

Whatever action is taken by a person or group, it is usually done with the belief that their position is more ethical than that of their opposition. Many solutions to problems in a democratic society are decided by compromise, but in some situations compromise is not considered to be an option because of ethical considerations.

16.5 Conservation Ethics and Values

There is a wide range in belief systems that relate to how the environment and the fishery and wildlife resources it supports should be used or protected. The term conservation, for example, has been defined in many ways. One simply stated definition is "the wise management and use of natural resources." However, the concept of "wise management and use" varies depending on the person or group of people concerned. To some, conservation is akin to **preservation**, which is leaving natural systems the way they are. For example, those who advocate the preservation position would support the total closing of many areas to human activity. To others, conservation is management and use for the benefit of both humans and wild organisms. In this latter group, management and use can be directed primarily at benefiting either humans or wild organisms, or at some point in between.

Because belief systems with regard to conservation vary to such an extent, confusion often enters the picture. For example, a group that identifies itself as the **wise-use movement** advocates resource management and use that primarily favors grazing, logging, and mineral extraction on federally owned land over other uses. While it is the ethical belief of the wise-use supporters that their position represents the best form of conservation, many people with a different set of conservation ethics and values disagree with them. Groups such as the wise-use movement generally support resource management and use for short-term economic benefits at the expense of the long-term viability and economic benefits of the resources. They seldom question whether the short-term utilization they support is sustainable over a longer time frame. Those who disagree with the wise-use movement support a more balanced, long-term approach to resource management and use, an approach that more equitably considers both economics and the wild organisms involved.

16.6 Animal Rights and Animal Welfare

The last segment of this chapter concerns two other belief systems, **animal rights** and **animal welfare**. These belief systems are also rooted in ethics and values. This topic is so contentious that not all people would even agree on the following definitions and descriptions.

Animal rights involves a belief that other animals have rights equal to those of humans. While these rights are extended primarily to other mammals, advocates of this philosophy often extend such rights to other animals as well. They believe that any form of management or use by humans interferes with these rights. Animal rights advocates are not necessarily supportive of animal welfare. For example, if a deer population were starving or dying from some disease, animal rights supporters would advocate letting the deer fend for themselves. They believe that humans should not intervene. Some people in this group might recommend providing food to the deer, but this would not be a true animal-rights response. Obviously, any consumptive use, experimentation, or other similar activity using animals would be opposed by this group.

Animal welfare is a totally different ethical belief system. Correct handling and humane treatment of animals by humans is the primary goal of animal welfare advocates. Within the animal welfare philosophy, activities such as consumptive use and experimentation are allowable if they are conducted in as humane a manner as possible and if they do not threaten populations or species. The majority of biologists believe in animal welfare, and thus are primarily concerned with the well-being of populations and species.

There are a variety of federal, state, provincial, territorial, and local laws and regulations that protect animals. Many of these are quite specific as to what is and is not acceptable in relation to what humans can do to other animals. Wildlife and fishery biologists have developed additional guidelines for the handling and use of the organisms with which they work. Some organizations involved with the development of such guidelines are the American Society of Ichthyologists and Herpetologists, the American Fisheries Society, the American Institute of Fisheries Research Biologists, the American Society of Mammalogists, the American Ornithologists' Union, the Cooper Ornithological Society, the Wilson Ornithological Society, the Association for the Study of Animal Behaviour, and the Animal Behavior Society. There are also other groups that have developed additional guidelines for the care and treatment of animals. However, because biologists are also humans, there is individual variation in their commitment to animal welfare and their adherence to these guidelines.

Animal welfare supporters do not just include biologists. Consumptive and direct users, such as trappers, hunters, anglers, agriculturists, and pet owners, should all be concerned with the humane treatment of the animals that they use or keep. Numerous precautions, such as shooting only at animals that are within range, checking trap lines often, using proper fish release methods, keeping farm animals in well-ventilated buildings, and remembering to feed and water the dog or cat, contribute to proper concern for and care of animals. Even nonconsumptive and indirect users are involved with, and should be concerned about, the welfare of animals. For example, concerned bird-watchers do not disturb nesting birds, concerned snowmobilers do not chase deer, and concerned farmers may delay hay cutting during the height of the ring-necked pheasant nesting season. In most cases, the humane treatment of animals primarily requires a feeling of concern and the use of common sense.

▼ SUMMARY

Surveys and censuses are means of gathering biological information. They can be conducted on-site, through the observations of biologists, or through telephone or mail contact. Data on species, sizes, numbers, sex ratios, and other biological information can be collected. Demographic information on users can also be gathered. Personal contact during surveys and censuses can result in more accurate information and also serves a variety of public relations and law enforcement functions. Telephone and mail censuses and surveys lack the degree of personal contact available in face-to-face encounters but still have much utility.

Attitude assessment planning usually requires the interaction of biologists and human dimension experts. Assessment methods usually measure both the direction and intensity of attitudes. Proper question wording is critical if accurate information is to be obtained. A variety of methods are used to measure attitudes. Some involve closed-ended questions, while others involve open-ended questions. Closed-ended questions allow for the collection of data that are more easily quantified.

The total economic values of wildlife and fishery resources can be difficult to determine. Individuals or groups of people can place different values on outdoor activities, and such activities provide multiple satisfactions. Gross economic value comprises the economic benefits of the federal, state, provincial, territorial, and local services that contribute to a resource; user expenditures; consumers' surplus; and additional amounts nonusers would pay to preserve the resource. An additional benefit is the value of the fishery and wildlife resources that are harvested by consumptive users.

Ethics and values vary depending on the individual or group involved. What is ethical to some may be unethical to others. Ethics and values concern consumptive, nonconsumptive, direct, and indirect users. Ethics and values vary depending on whether an individual or group defines conservation as management and use directed primarily at the well-being of the biota and habitat, or management and use directed primarily at the economic benefits of that biota or habitat to humans. Positions on animal rights and animal welfare also vary with individual or group ethics and values.

PRACTICE QUESTIONS

1. List ten different types of biological information that can be obtained from fishery creel surveys and wildlife bag checks.

2. List six different types of demographic data that can be obtained in wildlife and fishery surveys and censuses.

3. What are some advantages of direct personal contact with users while performing a survey or census?

4. Human attitudes can generally be divided into two separate components. What are they?

5. Why are public meetings, tear-out questionnaires, and opinion call-ins of little value as accurate survey methods?

6. What is wrong with the following attitude assessment question? "The Wildlife Division is an excellent organization that is interested in direct user needs and indirect users as well. Agree or disagree!"

7. What is consumers' surplus?

8. Select a natural resource-related activity in which you participate, and list all the satisfactions that you receive from that activity.

9. Define conservation. Define preservation.

10. Describe the major differences between the animal rights and animal welfare philosophies.

SUGGESTED READINGS

Ad hoc Committee on Acceptable Field Methods in Mammalogy. 1987. Acceptable field methods in mammalogy: Preliminary guidelines approved by the American Society of Mammalogists. *Journal of Mammalogy,* Supplement to 68 (4): 1–18.

American Society of Ichthyologists and Herpetologists, American Fisheries Society, and the American Institute of Fisheries Research Biologists. n.d. *Guidelines for use of fishes in field research.*

Anderson, S. H. 1991. *Managing our wildlife resources.* 2d ed. Prentice Hall, Englewood Cliffs, N.J.

Bolen, E. G., and W. L. Robinson. 1995. *Wildlife ecology and management*, 3d ed. Macmillian, New York.

Decker, D. J., and G. R. Goff, eds. 1987. *Valuing wildlife: Economic and social perspectives*. Westview Press, Boulder, Colo.

Loomis, J. B. 1993. *Integrated public lands management*. Columbia University Press, New York.

Oring, L. W., K. P. Able, D. W. Anderson, L. F. Baptista, J. C. Barlow, A. S. Gaunt, F. B. Gill, and J. C. Wingfield. 1988. Guidelines for use of wild birds in research. *Auk,* supplement to 105:1A–41A.

Pollock, K. H., C. M. Jones, and T. L. Brown. 1994. *Angler survey methods and their applications in fisheries management*. Special Publication no. 25. American Fisheries Society, Bethesda, Md.

Prochaska, F. J., and J. C. Cato. 1983. Economic considerations for fishery management. Pages 447–456 in L. A. Nielsen and D. L. Johnson, eds. *Fisheries techniques*. American Fisheries Society, Bethesda, Md.

Randall, A. 1987. *Resource economics: An economic approach to natural resource and environmental policy*. 2d ed. John Wiley & Sons, New York.

Smith, C. L. 1983. Evaluating human factors. Pages 431–445 in L. A. Nielsen and D. L. Johnson, eds. *Fisheries techniques*. American Fisheries Society, Bethesda, Md.

Weithman, A. S. 1993. Socioeconomic benefits of fisheries. Pages 159–177 in C. C. Kohler and W. A. Hubert, eds. *Inland fisheries management in North America*. American Fisheries Society, Bethesda, Md.

LITERATURE CITED

American Fisheries Society. 1992. *Investigation and valuation of fish kills*. Special Publication no. 24. American Fisheries Society, Bethesda, Md.

Dwyer, J. F., J. R. Kelly, and M. D. Bowes. 1977. *Improved procedures for valuation of the contribution of recreation to national economic development*. Report 128. University of Illinois Water Resources Center, Urbana.

Fazio, J. R. 1987. Priority needs for communication of wildlife values. Pages 296–304 in D. J. Decker and G. R. Goff, eds. *Valuing wildlife: Economic and social perspectives*. Westview Press, Boulder, Colo.

Fishbein, M., and I. Ajzen. 1975. *Belief, attitude, intention and behavior: An introduction to theory and research*. Addison-Wesley, Reading, Mass.

Gigliotti, L. M., and D. J. Decker. 1992. Human dimensions in wildlife management education: Pre-service opportunities and in-service needs. *Wildlife Society Bulletin* 20:8–14.

Krueger, C. C., D. J. Decker, and T. A. Gavin. 1986. A concept of natural resource management: An application to unicorns. *Transactions of the Northeast Section of The Wildlife Society* 43:50–56.

Loomis, J. B. 1993. *Integrated public lands management*. Columbia University Press, New York.

Pierce, J. C. 1979. Water resource preservation: Personal values and public support. *Environment and Behavior* 11:147–161.

Rockland, D. B. 1985. *The economic benefits of a fishery resource: A practical guide*. Technical Report 1. Sport Fishing Institute, Washington, D.C.

Rockland, D. B. 1986. *Economic models: Black boxes or helpful tools? A reference guide*. Technical Report 2. Sport Fishing Institute, Washington, D.C.

U.S. Department of the Interior, Fish and Wildlife Service, and U. S. Department of Commerce, Bureau of the Census. 1993. *1991 national survey of fishing, hunting, and wildlife-associated recreation*. U.S. Government Printing Office, Washington, D.C.

17

MANAGEMENT OF HUMAN USERS

Most individuals interested in wildlife and fisheries choose this field of study because they delight in being outdoors, utilize wild organisms for recreational purposes such as hunting or fishing, obtain satisfaction from working with animals, and have a concern for these resources. However, the world of the wildlife or fishery biologist is not simply one of biota and habitat. In fact, many of the activities of a fishery or wildlife biologist revolve around managing the people who use these resources. Biologists often need to manage human activities in order to conserve the animal communities for which they are responsible.

Most wildlife and fishery regulations concern direct consumptive users of a resource, such as individuals who hunt, fish, or trap. However, nonconsumptive users are also regulated. For example, people wishing to view bald eagles may need to be regulated by location, number, and activity so as not to unduly disturb the birds, especially if they are nesting. Even indirect users are regulated. For example, regulations commonly limit the amount of water that can be removed from a stream by industry or agriculture.

This chapter will primarily address direct consumptive users because they are the ones most closely regulated. Regulation of consumptive users tends to be more complicated and addresses more conservation issues than does regulation of nonconsumptive users. For example, consider the variety of regulations that could be directed at two fish species, the largemouth bass and the riffle sculpin. Because the largemouth bass is pursued over a wide geographic area for sport purposes, a long list of regulations, such as how many can be taken and what size they must be to be harvestable, is promulgated to control its use. The riffle sculpin is not a sport species. Few people have heard of or seen this small stream fish, which is found primarily in coastal streams of Washington and Oregon. Regulation pertaining to this fish may consist of a protective rule that simply states that the species cannot be captured. Regulation may also be absent because no one uses this species consumptively. The riffle sculpin is obviously not a fish that would attract limits on size, number, or

methods of capture. Rules that protect the riffle sculpin may be more commonly found in the regulation of indirect users, often through environmental protection laws such as clean water legislation. The degree of regulation or nonregulation is not necessarily indicative of the biological importance of a species, but instead is based on the organism in question.

Regulations generally address one or more of the following aspects of consumptive use: "who" could pertain to items such as user age and residence, "what" to species or size, "when" to time of year or time of day, "where" to protected areas or portions of those areas, and "how" to type of bait or harvest method.

Management of human users is essential to the proper conservation of natural resources. People can often be the overriding factor in conservation decisions and in the likelihood of their subsequent success or failure. The purpose of this chapter is to discuss some aspects of human user management in relation to wildlife and fishery resources.

17.1 Purposes of Regulations

Controlling **overexploitation** (**overharvest**) is a common purpose of regulations. Overexploitation, however, can be a rather nebulous concept. Taken literally, overexploitation might be defined as the point at which harvest mortality causes undesirable reductions in fish, wildlife, or other populations (see section 3.3). However, no single definition of overexploitation exists. Overexploitation sometimes suggests such extensive harvest that insufficient adult organisms remain to replace themselves through reproduction. In other situations, the term is simply used to infer that the quality of the population has been impaired by some level of harvest. The distinction in terms of quality is similar to that made in comparing maximum and optimum sustained yields (see section 3.6).

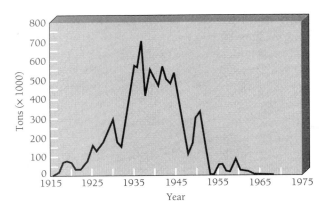

FIGURE 17.1 Catches of Pacific sardines in the Pacific Ocean off California, 1917–1968. Harvest increased gradually, and then dramatically, until the stock was overexploited. *(From Talbot 1973.)*

Commercial fisheries provide numerous examples of overexploitation. Commonly, a new resource is found, harvest increases over time, overexploitation eventually occurs, the adult stock rapidly declines, and commercial harvest plummets. A typical example was described by Talbot (1973) for the Pacific sardine fishery in the Pacific Ocean off the California coast (fig. 17.1). Overexploitation in this context does not mean that the resource or species is extirpated; it is simply reduced below the level needed to support a commercially viable venture.

In sport fisheries, there are no documented cases in which regulated sport angler harvest has been the primary factor in reducing an adult stock to the point at which it could not replace itself. The reasons for this are quite understandable. Most biologists believe that sport fishing becomes increasingly poor as the adult stock is reduced, and that most anglers quit fishing before the adult stock is reduced to the point at which insufficient adults remain for reproductive purposes. However, overexploitation of a sport fishery can result in a decrease in the quality or size structure of the fish population. For example, consider a largemouth bass population in a water body with a surface area of approximately 60 hectares that is located near a city with a population of 100,000 people. Angling effort at such a water body typically could be about 500 hours per hectare per year. Given this scenario and no regulation of the largemouth bass harvest other than a five fish per day **creel limit** (the number of fish that one person can harvest and remove in a day), one would expect to find few bass longer than 30 centimeters in the water body. There would be enough adult largemouth bass to reproduce and replace themselves, but the level of harvest would preclude many fish from reaching larger sizes.

Another reason for regulation may be to achieve a particular management goal. For example, the Colorado Division of Wildlife might wish to manage a specific elk population for the quality of harvestable bulls. The management objective might be to produce an elk population in which 40% of the bulls have four or more points on each antler. Such a population generally cannot exist without special harvest regulations.

Regulations may also be implemented to make a resource available to more users. For example, while a daily creel limit or **bag limit** (the number of individual wildlife organisms and sometimes fish that one person can harvest and remove in a day) or **possession limit** (the total number of individual organisms that can be possessed by one person at any one time) can be implemented primarily to attain a biological management objective, it can also be intended to ensure that a publicly owned resource is distributed among as many users as possible. Without limits, someone who is very adept at catching a steelhead or shooting a ruffed grouse may be able to take more than "their share" of these resources.

The purposes of regulation can be deeply rooted in psychology. Some individuals who use a resource may not be satisfied with a recreational experience unless they "get the limit." While some would consider such an attitude to be unethical, it can still be a factor in recreational outcomes. If the limit is six and that is what some users get, they are satisfied; if the limit is four and that is what they get, they are satisfied; if the limit is six and they get four, they may be dissatisfied. The particular number can be less important than the goal. The psychology of such situations can create unanticipated results. For example, consider a situation in which biologists wish to reduce harvest and therefore reduce the creel limit on rainbow trout from eight to four. Because of the psychology of limits, and therefore human goals, it is possible that harvest could actually increase with such a creel limit reduction. If the limit is eight and an angler has three fish, the angler may stop fishing; if the limit is four and the angler has three fish, that person may continue fishing until the fourth fish is caught.

Regulation may also be applied for reasons that are primarily social or political in nature. For example, while a management objective for watchable wildlife might call for increased populations of Canada geese in a metropolitan area, golf course groundskeepers, airport superintendents, or the general public might not desire such increases; in fact, they might even want decreases. The length of the hunting season for ring-necked pheasants in Midwestern states is another example. Biologically, a male-only season could be opened during the fall and run through late winter (perhaps mid-October through February) with no negative effects on the ring-necked pheasant population. Ring-necked pheasants are polygynous, and require as few as one male to every ten females for successful reproduction. Even with an extremely long season, the males would not be harvested to a sufficient extent to reach such a skewed sex ratio. Why, then, do many natural resource agencies set season lengths of one or two months? The answer may vary from state to state, but such season length limitations are often the result of social concerns. A commonly cited reason for shorter seasons is landowner tolerance. Even though wildlife species are a part of the commons, much hunting necessarily occurs on privately owned land. The commissions or boards that preside over most natural resource agency activities (see section 20.4) often express concern that landowners will not tolerate the bother and interruptions that would accompany a longer hunting season.

Are socially and politically based regulations wrong? There is no definitive answer to this question; the answer varies among individuals. An enthusiastic biologist may insist that resource decisions, on both regulations and other conservation strategies, should be based only on biology. Another person, such as an agency administrator, may have a different viewpoint. Very often administrators have more experience in biopolitical activities and recognize that a political dispute over a minor regulation proposal may incur a cost that carries over to larger, perhaps more important, matters. To administrators, a socially based regulation may be highly appropriate. Who is right? We certainly cannot say. However, it is important that people be aware that day-to-day decisions made by natural resource agencies are influenced by social and political considerations as well as by biological ones (see section 20.5).

17.2 History of Harvest Regulations

The earliest fish harvest regulations implemented in the United States eliminated the use of fish traps, nets, and seines, particularly near the mouths of streams and during spawning seasons (Redmond 1986). By the early 1900s, increasingly restrictive regulations had progressed to minimum length limits and closed spring seasons to allow fish to spawn at least once before being caught. Minimum length limits were often very low, such as 15 centimeters for trout species and 20 centimeters for largemouth or smallmouth bass.

During the first half of the twentieth century, maximum sustained yield (MSY) (see section 3.6) was the common management strategy. Biologists strove to maximize fish harvest without harming subsequent generations of a fish population. MSY was given impetus by liberalization of regulations after studies in the 1940s indicated that fishing could be allowed in the spring without affecting reproduction or numbers of fish harvested in following years. Studies in the 1940s and 1950s continued to support increased sport fish harvest. By the 1960s, however, investigations indicated that, in some small water bodies, predator fishes could be overharvested, upsetting the "balance" in the fish community and seriously affecting future fishing quality among both predators and panfishes. The imposition of effective **size limits** (the size of organisms that can legally be harvested) and **harvest quotas** (the specific number or total weight of organisms that can be harvested) prevented predator overharvest and maintained balanced fish populations that provided good fishing. With the advent of refined fish tagging and recovery methods, biologists demonstrated that high exploitation rates of **black basses** (largemouth, smallmouth, and spotted bass and other members of the genus *Micropterus*) and white crappies also occurred in large water bodies. Harvest restrictions applied to large reservoirs in the 1970s demonstrated that practical, properly applied, and well-enforced size limits could maintain sport fishing and a harvest of desirable quality. These management practices reflected an optimum sustained yield (OSY) strategy.

Redmond (1986) concluded that an important challenge for the future is to increase public acceptance of harvest restrictions to achieve the necessary compliance. Compliance will be necessary to maintain high-quality fishing for a growing number of more knowledgeable and better-equipped sport anglers. Compliance is typically accomplished through both enforcement and education.

Shaw (1985) summarized the history of wildlife regulation in the United States. The "Era of Abundance" lasted from 1600 through 1849. Wildlife was abundant, and human effects on wildlife populations were minimal. The "Era of Overexploitation" began in 1850 and lasted until 1899. Shaw used the destruction of the vast herds of bison as the best example of this overexploitation. This decline in wildlife abundance was accelerated by settlement of wild lands and by technological improvements that allowed increased harvest without any accompanying increase in conservation measures. During both of the above periods, regulation was infrequent. The "Era of Protection," 1900–1929, was the immediate response to the Era of Overexploitation. Formal regulation of wildlife harvest became well established during this time period.

The wildlife profession evolved during the "Era of Game Management," 1930–1965. While legal protection was an important part of wildlife conservation, it was apparent that biological information was also needed so that biologists could apply conservation strategies that would be effective for protecting particular species. The biology of the organism had to be understood before proper regulations could be applied. Wildlife conservation efforts during this era were largely limited to game species.

Finally, we have reached the "Era of Environmental Management." Conservation measures have spread beyond game species to reach all types of organisms and their accompanying habitats, with concomitant increases in the complexity of laws and regulations.

17.3 Contrast between Wildlife and Fishery Regulations

Fishery harvest regulations commonly include both number and size (length or weight) limitations. Harvest regulations for sport fisheries are based on the fact that many fish species can be caught and released with a reasonable expectation of survival. Hooking mortality does occur, and is likely to be highest when live bait is used or when water temperatures are relatively warm. Where substantial, hooking mortality can be minimized by artificial-lure-only requirements and by careful handling techniques. Wildlife regulations, in contrast, most often limit the number of organisms harvested; size regulations are less frequently used. Obviously, it is not possible to release a common snipe or a mountain goat once it has been harvested. This single difference greatly influences how a fishery can be regulated as compared with most wildlife resources.

The numbers of harvestable species and individuals (density) are usually lower for wildlife in a particular area than for fishes in a water body; this difference also contributes to the differences between fishery and wildlife regulations. Another difference relates to differing growth patterns. Fishes, because they exhibit indeterminate growth, have populations that can generally be characterized by a wider variety in individual fish sizes.

In fisheries, it is possible for nearby waters to be managed differently. Two water bodies within close proximity can be managed for entirely different sizes of fishes, or for entirely different fish species. Such water-body-specific conservation has few equivalents in wildlife. Larger management units are typical in wildlife because most wildlife species are less limited in their movements than fishes.

For nongame animals, there are also differences among the regulations aimed at protecting fishes, mammals, and birds. Most fish and mammal nongame species are either not protected by law or are protected by not being included on lists of organisms that can be harvested. Exceptions typically occur when a species is threatened or endangered. In contrast, most nongame bird species are protected because they are included in international agreements on migratory species and because the public has requested their protection.

17.4 Regulating "Who"

Harvest of wildlife and fishery organisms can be regulated simply by determining who can legally take these resources. For example, a state, province, or territory that has a limited allowable harvest for a particular wildlife species may set a reg-

ulation decreeing that only residents can hunt that species; nonresident hunting may not be allowed. Similarly, if a state, province, or territory allows a limited harvest of a fish species that is not particularly abundant, such as coho salmon, nonresident anglers may be prohibited from fishing.

In the United States, the user's level of skill, interest, and knowledge can also be used to limit participation in a recreational activity. A common example of this strategy is the hunter education programs that are mandatory in most states. Until individuals have demonstrated an appropriate level of knowledge by passing an examination, they are prohibited from purchasing a hunting license. There are calls for even more rigorous testing in some circumstances. For example, some people believe that proficiency should be tested before individuals are allowed to purchase archery licenses for big game species.

Most natural resource agencies allow minors to hunt, fish, or trap at reduced license fees. This is usually done to encourage the entry of new participants into these activities, and with the belief that these individuals are inexperienced and therefore less likely to be proficient. At times, reduced fee structures are provided for senior citizens. This is often considered justifiable because senior citizens may be on a fixed income.

Many natural resource agencies provide special considerations for landowners. Landowners often can obtain licenses at no or reduced cost, and may receive a substantial portion of the licenses when a lottery or drawing system is used. Other benefits for landowners include exemptions from license requirements for hunting or fishing for some species on their own land. The general philosophy underlying such benefits is that landowners are providing homes for organisms and thus are already contributing to the maintenance of the resource.

Individuals that have a physical disability, former prisoners of war, veterans, military personnel not living in their normal state of residence, and nonresident college and university students often receive special consideration for obtaining licenses. People with a physical disability may be allowed to harvest wildlife with equipment that cannot be used by the general public. Nonresidents in the military and attending colleges and universities often are allowed to purchase lower-cost resident licenses rather than more expensive nonresident licenses.

Regulating who can harvest a resource can also be accomplished by limiting the number of individuals that can participate. For example, limiting the number of licenses or permits can limit the number of participants. Limited permits and license drawings will be discussed further in section 17.5.

17.5 Regulating "What"

Regulating what can be harvested most often involves regulation by species, number, sex, or size. Each of these will be discussed separately.

Regulation by Species

In any particular area, there are many species that cannot be harvested. Some may be endangered or threatened, some may exist in low numbers in that particular area, while others may not be legally classified as game or sport species. Regulations on the harvest or nonharvest of particular species can be highly variable from location to location. For example, the mourning dove is classified as a game bird in some areas, but as a songbird in others. In some states, provinces, and territories, fish species are classified as game or nongame based on the taxonomic group, such as family, to which they belong. Thus, in some areas, darters, a type of fish too small to be a sport fish, are legally classified as game species and subject to protection because they belong to the same family (Percidae) as walleyes. In other areas, it is legal to take darters because they are not specifically listed as game species. Game species are often regulated in nu-

merous ways; nongame species are often regulated only with regard to prohibition of harvest.

Regulation by Number

The number of individuals to be harvested can be managed in a variety of ways. Numbers of permits to be sold for big game species are often set by geographic unit. When demand is greater than the number of permits available for a unit, some type of lottery or drawing is generally used to determine the successful applicants.

Quotas that allow the harvest of a certain proportion of a population are a theoretically attractive means of preventing overharvest of some species. Quotas can be set for wildlife, fishes, or any other species if there is sufficient knowledge of the numbers of that particular species and how many are needed to maintain viable and quality populations. Determining such quotas and monitoring the harvest, however, can entail substantial economic costs. Consider the largemouth bass as an example. Graham (1974) found that harvest of more than 40% of the largemouth bass in Missouri ponds resulted in a decreased size structure and less effective predation on panfishes. Based on this finding, a biologist in charge of a 20-hectare impoundment might decide to set a quota under which 40% of the largemouth bass in that population could be harvested. When that quota was reached, harvest would stop. However, it would be expensive and time-consuming to obtain a population estimate for largemouth bass in that impoundment, and even after a reliable estimate was made, a thorough, expensive creel survey or census would still be needed to monitor the amount of harvest. The quota system could certainly work, but it would be difficult to manage very many water bodies using this technique because of economic limitations. Quotas are commonly used for commercial fisheries, and are also used to manage paddlefish in some Missouri River states.

Quotas are sometimes applied to the harvest of wildlife species. For example, geese that congregate on waterfowl refuges during migration or wintering may be subject to high levels of exploitation on the surrounding lands. Biologists may attempt to manage specific populations of geese by setting quotas on the number that can be harvested. These quotas are designed to assure that breeding stock from a particular area is not depleted. Consider a hypothetical population of Canada geese that nests near Hudson Bay and spends much of late October and early November on a state waterfowl refuge in Minnesota. The Canada geese rest on large water bodies within the refuge and feed primarily on surrounding private lands. While moving to feeding areas or returning to the refuge, the Canada geese are hunted, both on portions of the refuge open to goose hunting and on private lands. Assume that breeding ground surveys and fall migration counts indicate that the fall population of these Canada geese has declined to 80,000 from about 140,000 over ten years, and that the management objective for this population has been set at 135,000 geese. Band returns indicate that 60% of all hunting mortality in this population is occurring on or near the state refuge in Minnesota. Biologists may place harvest quotas on this population to reduce mortality and allow the population to reach management objectives. When the harvest quota is reached, the season on Canada geese in that area is closed, at least until that population has migrated from the refuge. If the daily harvest is high, the season may last less than a week; if harvest is low, the season may be allowed to continue much longer. Such regulatory practices sound ideal, but they can be difficult in terms of public relations. Consider the pressures brought upon those making the decision to close a season early because quotas have been reached.

Quotas are actually quite common in wildlife regulation, although the term is seldom used. For example, when the number of big game licenses to be sold is determined for a geographic unit, that number is a reflection of the estimated number of animals present and the population level that biologists wish to maintain in that area. Knowing these numbers and having a historical indication of what hunter success rates will be allows biologists to determine the number of animals that can be taken, and therefore the number of licenses issued.

Creel limits in fisheries and bag limits for wildlife also regulate what is taken. If a fish population is overharvested, common sense seems to dictate a reduction in the number of fish that each angler can harvest. Despite the apparent attractiveness of this response, fishery biologists have found that creel limits are usually an ineffective tool for preventing overharvest. To effectively control overharvest, creel limits would typically have to be so low that most anglers would consider them unacceptable. Consider the following case history. During 1982, there was no creel limit for white crappies in Kansas, and anglers were allowed to keep all the crappies they caught. At about this time, there were concerns that overharvest might be negatively affecting some white crappie populations in the state. The solution proposed by most anglers was to enact a statewide creel limit for crappies. When questioned about the appropriate number, most people recommended a limit somewhere between 25 and 50. During 1982, a creel survey was undertaken at Clinton Reservoir in the northeastern part of the state. Based on the creel survey results, estimates were made of the reduction in harvest that would have occurred had various creel limits been in effect (fig. 17.2). A creel limit of 30 would have reduced total harvest by 1%, and a creel limit of 10 (much too restrictive for most anglers) would have reduced harvest by only 11%. It would have taken a creel limit of 5 to result in a harvest reduction of 26%, which may or may not have been sufficient to improve the quality of the white crappie population. Where overharvest truly is a problem, creel limits generally have to be very restrictive to reduce harvest sufficiently.

As a result, creel or bag limits are often applied for social or political, rather than biological, reasons. For example, at one time, one southern state had a creel limit of 37 for crappies. This unusual number was the result of a compromise between two commissioners; one wanted a creel

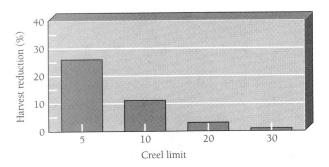

FIGURE 17.2 Reductions in the white crappie harvest that would have resulted if various creel limits had been in effect for Clinton Reservoir, Kansas, during the 1982 fishing season. No creel limit was actually in effect that year.

limit of 50, and the other thought that 25 would be more appropriate. As another example, the Kansas Department of Wildlife and Parks has instituted a statewide creel limit for crappies since the time of the study described in figure 17.2. This regulation was not intended to prevent crappie overharvest. Rather, it was intended to send a message to anglers that crappies are a limited resource and that they do have value. Daily bag limits on many wildlife species are implemented for similar reasons.

Migratory ducks are sometimes managed with a point system. Under such a system, various species of ducks are assigned different point values, and when the last bird a hunter harvests puts the total over a certain number of points, such as 100, the hunter is done for the day. This technique encourages knowledgeable hunters to harvest more abundant, lower-point species and avoid high-point species that are found in lower numbers. The point system has also been used to encourage harvest of drakes instead of hens where males outnumber females, as occurs in most duck species. For example, mallard hens typically have a higher point value, while the more abundant and distinctly colored drakes generally have a lower point value.

Some unique species are managed with a lifetime limit. For example, a limited number of permits are available each year for bighorn sheep in the Badlands of North Dakota. Applicants can ob-

tain only one license in a lifetime. The license is good for one hunting season, and the permittee cannot apply for another license in subsequent years.

A creel or bag limit may serve to distribute harvest more equitably among a variety of users over a longer period of time. In goose hunting areas with quotas, for example, low daily bag limits allow more hunters to pursue geese over a longer period of time. Restrictive creel or bag limits also help to reduce the waste of harvested animals that may occur when anglers or hunters harvest a large number of a species in a single day. At the least, a user arriving one day and not catching many fish or seeing much game does not hear stories about the massive numbers of animals that were taken and removed the previous day.

Wild animals are limited resources, and creel and bag limits add perceived value to them. If a natural resource agency sets no creel or bag limit for a species, users may believe that it has little value. The same philosophy applies to possession limits. Possession limits are often set at two to five times the daily bag or creel limit. Possession limits are difficult to enforce. A search warrant would have to be obtained to check home freezers and commercial freezer lockers to determine whether an individual had violated the possession limit.

Regulation by Sex

Regulation by sex of organisms harvested can also be used to modify harvest. This is a common practice in wildlife, but is seldom used in fisheries. Sexual dimorphism or dichromatism in the animal to be regulated is an obvious prerequisite. Polygynous species are often regulated in this manner. Ring-necked pheasant harvest, for example, can be regulated by allowing harvest of males only, as described in section 17.1. The presence of fewer males will have no effect on reproductive success the following spring. There is even some question as to whether, biologically, a bag limit on males is needed at all. Few hunters would still be afield in an area where males composed only 10% of a population. Remember, however, that there are reasons for regulations other than biological ones.

Similar arguments can be made for some mammalian species in which sex can be readily determined, such as white-tailed deer. Allowing harvest of antlered deer only is one way to maintain large harvests without negatively affecting population numbers. In the case of deer, there is, however, another consideration. When hunters harvest male ring-necked pheasants, they do not generally differentiate between young and old males; this cannot be said for harvest of large ungulates, in which antler or horn size is a consideration. Unlimited buck harvest can lead to a reduction in population age structure (see section 3.7). Many other species that can be readily sexed before harvest are regulated by sex. Examples include bull versus cow and calf elk, gobbler versus hen wild turkeys, and ram versus ewe and lamb bighorn sheep.

Monogamous species or species in which the sexes cannot be readily differentiated before harvest must be managed with more care to ensure that population numbers are not, over time, adversely affected by harvest. Examples of such species are northern bobwhite, American woodcock, and white-winged dove.

Regulation by Size

Size limits are commonly used as a fishery management technique, and are sometimes used in wildlife management. Wildlife size limits usually pertain to structures such as antlers or horns, rather than body size, and are commonly termed appearance restrictions. Such a regulation might restrict harvest of mule deer or elk to those individuals with four or more antler points per side. Similarly, bighorn sheep might be regulated with a requirement for a full horn curl, or perhaps a 3/4 curl (fig. 17.3). Whether or not such appearance regulations actually increase the proportion of large or trophy animals in a population is debatable (Strickland et al. 1994). For example, regulation of elk harvest using point restrictions may re-

FIGURE 17.3 Some big game species are managed with appearance regulations. This bighorn ram has horns that exceed a half curl, but have not yet reached a three-quarter curl. Full curl is attained when the tip of the horn reaches the base of the horn, above the eye. (*Photograph courtesy of South Dakota Tourism.*)

sult in increased kill of illegal (having insufficient points) bulls and decreased overall yield (Boyd and Lipscomb 1976). The genetic implications of such a strategy are also not clear. Thelen (1991) suggested that harvest of only bulls with high numbers of antler points could be genetically counterproductive.

While size implies both length and weight, the majority of size regulations in fishery work are based on organism length (fig. 17.4). Length is easy to measure, and enforcement on this basis is easier. A fish can be gutted but transported with the head intact to allow measurement by enforcement officers. Common length limit regulations include **minimum length limits**, under which all fish less than a specified length must be released, and fish longer than the specified length may be kept; **protected slot length limits**, under which all fish within a specified length range must be released, while those outside the slot length range may be kept; **reverse slot length limits**, under which all fish within a specified length range may be kept, while those outside the slot length range must be released; and **maximum length limits**, under which fish up to a maximum length may be kept, but all longer fish must be released. We will further distinguish these types of length limits

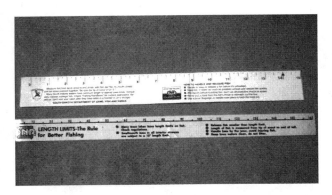

FIGURE 17.4 Many natural resource agencies provide rulers for measuring fish length. A common technique is to produce "bumper stickers" such as these that can be attached directly to boats.

using largemouth bass examples for minimum and slot length limits, and a cutthroat trout example for maximum length limits.

Consider a fishery biologist responsible for a 30-hectare impoundment in the midwestern United States that has appropriate habitat for largemouth bass. Water transparency in the impoundment exceeds 45 centimeters for most of the year, and submerged aquatic vegetation covers approximately 25% of the surface area during midsummer. A sample of largemouth bass is obtained by electrofishing at night during the spring. The size structure of 20-centimeter and longer largemouth bass in the population sample is depicted in figure 17.5. It should be immediately apparent that the quality of the population is low. Of the 80 fish (20 centimeters and longer) collected, only 9 (12.7%) were 30 centimeters or longer. An appropriate management objective in this case might be 40%–70% of these fish being 30 centimeters or longer. Before a strategy can be implemented to attain this management objective, the cause of the poor size structure must be understood.

Poor size structure in largemouth bass in small impoundments can result from either overpopulation or overharvest. Overpopulation is characterized by high density, slow growth, and poor body condition. As a result, few largemouth bass grow to sizes longer than 30 centimeters. Over-harvested populations are characterized by lower density, faster growth, and good body condition, but few fish exceed 30 centimeters because most are harvested by the time they reach this length. Once a biologist has determined which of these problems is occurring, the appropriate length limit type can be selected.

The size structure of an overharvested largemouth bass population that has fish with moderate to fast growth rates, and that has excessive angling mortality, can be improved with a minimum length limit. If such a description fit the population depicted in figure 17.5, one would expect to see a substantial increase in the number of fish that exceeded 30 centimeters within one or two years after the imposition of a 38-centimeter minimum length limit.

Conversely, if the largemouth bass population depicted in figure 17.5 was near carrying capacity and exhibited slow growth and poor body condition, then a protected slot length limit could be used to improve its size structure. A slot length limit commonly used in Midwestern states protects largemouth bass from 30 to 38 centimeters long. Harvest of smaller fish is allowed, and is necessary, to reduce population density. If density is reduced, growth rates will increase, and fish will begin to grow past 30 centimeters. The fish

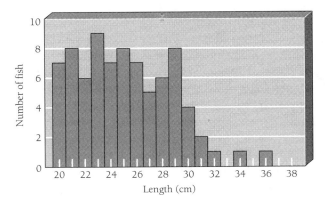

FIGURE 17.5 Size structure for a sample of 20-centimeter (cm) and longer largemouth bass collected by night electrofishing during the spring in a small Midwestern impoundment.

are then protected until they reach 38 centimeters, and fish above the slot length range can again be harvested by anglers. A 30–38-centimeter protected slot length limit was imposed for largemouth bass in Murdo Lake, a 19-hectare impoundment in central South Dakota (fig. 17.6). In the 1989 spring electrofishing sample, only 22% of the 20-centimeter and longer largemouth bass exceeded 30 centimeters. The slot length limit was imposed that spring. By 1991, nearly 60% of the 20-centimeter and longer largemouth bass exceeded 30 centimeters, and over 25% exceeded 38 centimeters.

Some natural resource agencies have quite specific missions. Consider a fishery biologist working for the National Park Service. The National Park Service is primarily geared toward protection of the resource. Harvest is a secondary consideration, and is generally not allowed unless it will not affect the success of the primary mission. Gresswell and Varley (1988) used a 33-centimeter maximum length limit to manage the Yellowstone cutthroat trout population in Yellowstone Lake in Yellowstone National Park. The lake received high levels of angling effort. The maximum length limit allowed fish less than 33 centimeters in length to be harvested, but all longer fish had to be immediately released. After this regulation was implemented in conjunction with an artificial-lures-only requirement, 5- to 10-year-old fish reappeared in the lake, 45-centimeter fish were abundant, 51-centimeter fish were present, and anglers caught and released five large (illegal size) fish for every legal-length fish they caught. In reality, this regulation returned the Yellowstone cutthroat trout population to a size structure that was more characteristic of the lake before human impacts. The spawning runs of these large Yellowstone cutthroat trout out of Yellowstone Lake then became energetically profitable for grizzly bears to

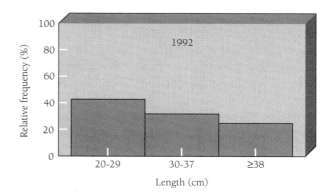

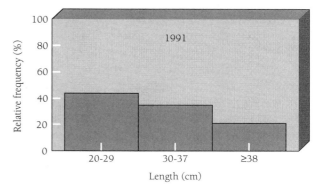

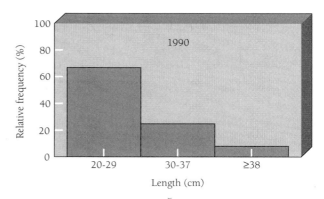

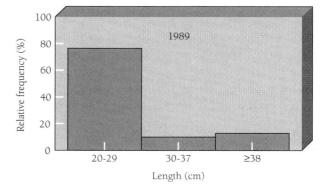

FIGURE 17.6 Size structure for 20-centimeter (cm) and longer largemouth bass collected by night electrofishing during the spring from 1989 to 1992 in Murdo Lake, South Dakota. A 30–38-centimeter protected slot length limit was imposed in 1989.

fish. Thus, the food base for a threatened species, the grizzly bear, was increased.

Most fishery biologists find length limits to be the best way to manage fish populations when overharvest is a problem. Given sufficient enforcement efforts to ensure compliance, length limits decrease harvest, but not fishing opportunity. When angling effort is sufficiently high, few fish will exist in a population that are longer than the minimum length limit or the upper end of a protected slot length limit. However, that does not mean that fishing must cease. Anglers have the opportunity to continue fishing, although they rarely have the opportunity to legally harvest a fish.

At the far end of the spectrum of length limit regulations are total **catch-and-release fisheries**. Under catch-and-release regulations, it is legal to capture a species, but any fish caught must be released. Some areas are managed as trophy fisheries with these regulations. Brown, cutthroat, and rainbow trout under catch-and-release regulations commonly reach large sizes. However, such changes occur only where habitat is appropriate. Such regulations are not a panacea that guarantees the production of large fish. Catch-and-release fishing is growing in popularity with anglers, even where it is not a regulation.

17.6 Regulating "When"

Regulations specifying when organisms can be taken usually involve either time of year or time of day. Time restrictions were some of the earliest harvest regulations used in the United States. Many early regulations attempted to limit harvest to a single time of the year and prohibited harvest during breeding seasons. Limiting season length can regulate the amount of harvest that actually occurs. Especially in hunting, seasons are generally open during the time of year when most populations are at peak density and a harvestable surplus typically is available. Most bird and mammal hunting seasons, for example, are in the fall in the Northern Hemisphere. Exceptions do occur in which hunting is allowed at other times of year, such as spring wild turkey seasons, but in this case, only males are legal game.

Another tool of the biologist is manipulation of the time of day when a species can be taken. Half-day hunting can be used to reduce harvest, or can simply be a tool to reduce harassment of migratory birds and thus the likelihood of premature migration. Duck harvest can be limited by opening hunting time at or after sunrise rather than before sunrise. The movement and vulnerability to hunting of most ducks is greatest in the crepuscular periods prior to sunrise and after sunset. Ducks are so vulnerable to hunting after sunset that a sunset daily closure is mandatory in the United States.

Most hunting is a daylight activity. Exceptions, such as nighttime raccoon hunting, are rare. Fishing can be much more of a nighttime activity than hunting, but night fishing is prohibited in some situations by some natural resource agencies.

The use of closed seasons often presents a dilemma for fishery biologists. Most fishing season closures occur during the late winter and early spring. The idea of disallowing sport harvest during a time period that includes the spawning season for many species is attractive to anglers. If overharvest is suspected of reducing the quality of a fish population, people often are quick to request a closed season. However, if a fish population truly is being overharvested, then the season must be closed at the time of year when most harvest typically occurs if sufficient harvest reduction is to be accomplished. For example, Redmond (1974) summarized case histories for largemouth bass harvest in five Missouri impoundments that ranged in size from 9 to 85 hectares. In the first 4 days after the impoundments were opened to fishing, 40%–69% of the largemouth bass were harvested by anglers. To prevent overharvest of largemouth bass in these impoundments, a closed season would obviously have to last 11 months and 26 days, or perhaps longer. Therefore, fishery biologists generally prefer to find alternatives to closed seasons to prevent overharvest in situations in which exploitation is substantial.

We certainly do not mean to infer that there are not social and biological justifications for closed seasons in fisheries. Much tradition often surrounds opening day, and excitement can build as that day approaches. Closed fishing seasons are still used in some northern states, but are uncommon in southern states. Kiefer et al. (1995) reported that even catch-and-release fishing of male smallmouth bass during the spawning season may negatively affect reproductive success, which could justify a closed season at this time of year.

17.7 Regulating "Where"

The locations where hunting or fishing is allowed are commonly regulated. Refuges and sanctuaries are provided for some wildlife species, with a common example being waterfowl refuges. Waterfowl refuges may provide a sanctuary from hunting in some cases, while in others they may attract waterfowl to an undisturbed area but actually increase hunter harvest. For example, as in the case described in section 17.5, a refuge may allow geese to rest undisturbed between feeding flights to lands surrounding the refuge. However, the birds may be subject to hunting off the refuge, and may even be overexploited without some type of control on total harvest.

Sanctuaries have rarely been used as a freshwater fishery management tool, although they are now common in commercial marine fisheries. A sanctuary of sufficient size could be used to prevent overexploitation of a freshwater fish species. Overcoming the potential opposition to such a closure could be difficult, however, especially if

the area had traditionally been a popular fishing location. There are cases where sanctuaries have been provided in freshwater fisheries. For example, the Wisconsin Department of Natural Resources maintains sanctuaries for lake trout in the Wisconsin waters of Lake Superior. These sanctuaries typically are primary spawning grounds for the lake trout, and no sport or commercial harvest is allowed there.

Geographic hunting units are often delineated by natural resource agencies, especially where licenses are limited in number and granted by lottery. Thus, harvest can be regulated according to these units. Harvest can be higher in areas where population density of a species is high than in areas with lower population densities. Although they are less common, similar differentiations can be made in different fishing waters.

Another important factor determining where a person can hunt, fish, watch birds, hike, canoe, or pursue any other recreational activity involves trespass laws. Landowners, in conjunction with trespass laws, determine how access is obtained. This issue may initially seem to be a simple one—that is, it might be assumed that if land is privately owned, access is under the control of the landowner. In reality the situation is not so simple. In some areas, for example, hunting is allowed on private land unless that land has been posted against hunting or trespassing. Another example involves laws designating waterway ownership. Some streams and lakes are publicly owned, while others are in private ownership. In some situations, a stream or river is open to public access, but the banks are not. In other cases, a stream or river is open to public access, but users cannot touch the banks or the bottom of the watercourse. Where a person can go and what they can do varies a great deal depending on regional laws.

There are also a wide variety of other laws and regulations that determine where activities can be pursued. Laws and regulations controlling activities in road rights-of-way, near dwellings, and in proximity to domesticated livestock all control "where" aspects of resource use.

17.8 Regulating "How"

Regulating the methods that can be used to take organisms certainly influences the eventual harvest. For example, restricting anglers to shore-only fishing and prohibiting angling from boats would reduce harvest of fishes commonly found in limnetic areas, such as walleyes, white bass, and striped bass. A commonly used fishery regulation is a restriction on the type of **terminal tackle** (**terminal gear**) that can be used. Terminal tackle is the fishing equipment at the end of a fishing line, such as live bait or artificial lures. The characteristics of the angling public and the amount of harvest can be influenced by whether anglers are restricted to flies only or artificial lures only or whether live bait is allowed. When the Colorado Division of Wildlife restricted certain portions of the Cache la Poudre River to fly fishing only, angler use dropped to 10% of its former levels (Klein 1974). As often happens, however, some of the results of this regulation were unexpected. Even though use dropped substantially, the people most likely to fish in the special regulations areas were highly skilled anglers, and they effectively caught the fish in those river sections.

There are a number of other laws and restrictions regulating how fishes and other aquatic organisms can be taken. Common restrictions control the number of lines and hooks that can be used, whether fishes can be snagged (hooked somewhere in the body other than the mouth), whether animals can be speared, whether lines must be tended, whether artificial light can be used, and whether and what kinds of traps, nets, or seines can be used. In most cases, the use of electricity, piscicides, or explosives is illegal except with special permits.

Similarly, wildlife biologists can regulate hunting weapons to influence harvest. Big game hunters using rifles, slugs in shotguns, primitive firearms (black powder), and archery all have dif-

ferent efficiencies. Even within these groups there can be further differentiation; for example, rifle caliber or power, the minimum draw weight of a bow, or the number of shells or cartridges that a weapon can hold or a hunter can carry while hunting can be regulated. How organisms are taken can also relate to nonhunting mortality. For example, lead shot can no longer be used for hunting waterfowl because of its toxicity (Bellrose 1982). Nontoxic shot, such as steel, must be used. There are many other potential restrictions on how wildlife can be taken, including those on the use of dogs, artificial light, baits, calls, and a wide variety of other techniques or devices.

Even regulating how access can be gained can influence hunter or angler use and the characteristics of the users that are present. Access by walking only, for example, will certainly draw a different clientele than access by motorized vehicle. In addition, there is likely to be less overall effort and harvest where motorized vehicles are not allowed.

17.9 Regulations for Nonharvested Species

Many people enjoy viewing terrestrial and aquatic organisms; such nonconsumptive use may still require regulation. For example, if sufficient numbers of bird-watchers seek a rare bird species such as the California clapper rail, even simply for viewing, they may damage salt marsh habitat. Large numbers of people wishing to view breeding and nesting ospreys or bald eagles can disturb the birds to the point at which they are not successful in their reproductive attempts. Protected (no trespassing) areas are sometimes delineated around bald eagle or osprey nests in the lower forty-eight states to prevent viewers and others from disturbing the breeding and nesting process. Posting of no trespassing signs on islands and sandbars has been used to protect nesting piping plovers and least terns from nonconsumptive disturbance (see section 16.4). At times, protection may be needed for sensitive habitats, such as sand dunes or wildflower areas.

Many fishes are also considered watchable wildlife in appropriate habitat conditions, such as in freshwater and marine environments where water clarity is high. Swimmers, snorkelers, and scuba divers desiring to view the underwater world are increasing in numbers. In some areas, numbers have increased to the point where environmental damage was occurring, and diver numbers had to be limited. Similarly, people commonly gather to watch the spectacle of migrating salmonids as they leap to pass over barriers.

Species that are considered game animals may be important watchable organisms in some locations. Regulations that prohibit angling or hunting may be implemented to maintain high population abundance for those species in those locales. The organisms may be in a refuge situation, or they may be in a location that for human safety purposes precludes harvest. Waterfowl refuges attract many people who wish to view waterfowl; often such users view the animals before, during, and after hunting seasons.

17.10 Public Relations and Regulation

In the past, biologists often determined what techniques would be applied to address a conservation problem, then informed the public of what they intended to do. Those days are gone, and for more than one reason. Today, public input into management decisions is essential. Natural resource agencies are cognizant of this change; in fact, many are hiring human dimension specialists to address this facet of resource conservation. In addition, conservation programs today need to be geared toward satisfying the needs of many user groups, not just the anglers and hunters who were traditionally interested in natural resources. For agencies to be successful in maintaining wild populations of plants and animals, the support of the general public is needed. Most biologists are aware that the future welfare of many wild organisms depends on broad-based public support.

Public involvement in the past was usually in the form of public meetings. These were often attended by very few people who had rather narrow interests. Those who attended were also usually willing to accept only those options provided by wildlife and fishery professionals. Biologists can no longer present a short, selected list of already-planned conservation alternatives. There are many more people and groups interested in a much wider range of conservation options. It is also important to schedule public meetings sufficiently early in the process so that public input can be obtained on various potential strategies. Such meetings also have to be made available to a broad spectrum of the public.

As wildlife and fishery professionals have evolved in their knowledge of the human dimension, it has become apparent that public meetings are not the only means through which public opinion should be obtained. In fact, a few loud voices, often termed "squeaky wheels" (see section 16.2), at a public meeting can actually have far more influence than their opinions should carry. Thus, many natural resource agencies are beginning to collect broader data on public opinions and attitudes. Instead of simply holding a public meeting and gaining the views of fifty people concerned with an issue, an agency might mail a questionnaire to all individuals who use a resource (or a statistically valid sample), as well as to other citizens who do not use the resource. Certainly, more valid public opinion information is collected by such means. Many natural resource agencies survey citizens at regular intervals to obtain information on their preferences and attitudes. Such data can then be used to direct the future efforts of the agency.

17.11 The Regulation Process

The regulation process is becoming increasingly associated with and influenced by politics (see section 20.5). Final decisions on regulations (and other conservation decisions as well) are usually not made by professional biologists.

When a state resource in the United States, such as a nonmigratory bird species, is involved, the citizens of the state control regulation. However, such decisions can still be questioned. Legal action may be taken by nonresidents to try to obtain some measure of input into state decisions. Federal agencies may become involved in interstate aspects of regulations affecting a variety of aquatic and terrestrial organisms. Regulations affecting endangered and threatened species, migratory birds, and marine fisheries require federal action, often in conjunction with other countries.

As the science of wildlife and fisheries progresses and improves, biologists have more and better information that can influence and improve their decisions. Very often, this knowledge has led to specific strategies for fishes that vary by water body. Similarly, wildlife regulations commonly vary by geographic unit. The negative aspect of this improved management has been an increase in the complexity of regulations. Some wildlife and fishery professionals are concerned that complex regulations could prevent participation in recreational activities by new and marginally interested individuals. Participation by such individuals is desirable.

17.12 Information and Education

Regulations are not the only means by which biologists can modify human behavior and attitudes. Increasingly, biologists are educating the public through dissemination of information. The formats used are highly diverse: articles in popular and natural resource agency magazines, pamphlets to be distributed to the public (fig. 17.7), news releases, public meetings, call-in talk shows, and public service announcements on radio and television have all been used.

Most natural resource agencies have an information and education staff to help with information dissemination. The primary responsibilities of these people lie in providing as much information as possible to the widest possible audience. Their education and background can be invaluable during the production of informational leaflets, booklets, and other communication forms. For example, fishery biologists working for a natural resource agency have the technical expertise to provide information for a booklet that describes fish management in private ponds. However, the information and education personnel have the ex-

FIGURE 17.7 Many informational publications are produced for individuals interested in natural resources.

pertise to ensure that the appropriate writing style and presentation format will be used for the booklet.

Professional societies, such as The Wildlife Society and the American Fisheries Society, are also involved in information and education efforts. Their activities range from work at the local level to work at the federal and even international levels. For example, the American Fisheries Society was highly involved in the educational effort that resulted in the Wallop-Breaux Amendment to the Dingell-Johnson Act (see section 18.5).

Much education can occur through informal contacts as well. A biologist who encounters anglers at a boat ramp or hunters at a bag check station can provide them with valuable information not only on local fish and wildlife communities, but also on larger issues such as the Clean Water Act.

Education cannot take the place of regulations and enforcement of regulations; it is simply an additional tool. Enforcement of regulations is essential for their success. However, public relations and information dissemination techniques can be used to educate the users who must comply with the regulations. If the majority of users understand and accept a regulation, it is more likely to succeed, and enforcement will certainly be easier to accomplish.

Education can also be used to modify human behavior without regulation. Many natural resource agencies devote substantial amounts of time and money to maintaining good relationships between resource users and landowners. In some areas, public service announcements remind users that they are guests on private lands and should act responsibly and courteously. Many agencies are encouraging users to be more nonconsumptive, for example, by promoting catch-and-release fishing without specific regulation. Fishery and wildlife resources are renewable, but should be considered limited. Educational efforts can be directed at users to convince them that there are many recreational benefits to be obtained from each natural resource-related activity, and that

reaching a goal such as a bag or creel limit is not the only benchmark for a successful recreational experience.

▼ SUMMARY

Biologists need to manage human activities in order to conserve the natural resources for which they are responsible. Management of the human user component of fisheries and wildlife systems, while always important, has received ever-increasing attention within the profession. Regulations on human activities generally affect who can use resources, what species or sizes can be taken, when the activities can occur, where the activities are allowed, and how the activities can be accomplished.

Regulations are commonly imposed to prevent overexploitation of a resource; however, this is not the only justification for regulation. Specific regulations may be used to achieve a specific management goal. Regulations may also be implemented to divide a resource more equitably among users. Finally, regulations may be applied for reasons that are primarily social or political.

Both wildlife and fishery regulations and laws have progressed through a series of changes over time in North America. Early settlers found abundant resources, which were exploited without regulation. Through time, as fishery and wildlife resources declined, conservation philosophies changed, and regulations and laws were altered.

The primary difference between wildlife and fishery regulations lies in the fact that anglers often are able to release the fish that they catch. Hunters, in contrast, cannot release a wildlife species once it has been harvested. In addition, fish movement is limited by the boundaries of a water body; wildlife species are more mobile.

Regulating "who" can use wildlife and fishery resources can simply involve a determination of who can legally harvest organisms. Regulation of "what" can be harvested generally involves restrictions on the species, number, sex, or size of organisms that can be taken. Regulation of "when" use can occur generally involves either time of year or time of day. The location "where" recreational activities can occur is commonly regulated as well. Finally, natural resource users are often restricted in "how" they can legally pursue a recreational activity.

Regulations are also needed to manage nongame or nonharvested species. The disturbance caused when a large number of wildlife or fishery watchers converge on a site can negatively affect the organisms. Thus, disturbance may need to be eliminated by means of no trespassing regulations, or regulation of the number of people allowed to participate.

In the past, biologists often determined appropriate conservation strategies and informed the public of their plans. Today, public input into such decisions is essential. The support of the general public who do not consumptively use resources is especially important.

Enforcement of regulations is an essential component of wildlife and fishery management, but information and education programs can be equally important. An informed public is likely to be more supportive of regulations and less likely to violate them.

PRACTICE QUESTIONS

1. What are the three components of a fishery or a wildlife system? If necessary, refer back to section 1.2.

2. Why is human user management such an important component of wildlife and fishery conservation?

3. Describe a situation where you live in which a wildlife or fishery regulation was imposed primarily for social or political, rather than biological, reasons.

4. What roles do law enforcement and education programs play in determining the success of regulations?

5. If your objective for a mule deer population was to produce a buck harvest that included a substantial proportion of individuals with four or more points, what regulations might you use to reach that objective?

6. Your objective for largemouth bass in a 30-hectare impoundment is to produce a population in which 40%–70% of the 20-centimeter and longer bass are also longer than 30 centimeters. Currently, only 5% of the fish exceed 30 centimeters. What information is needed to select the appropriate length limit to attain your objective?

7. Make a list of some bird and mammal species that can be managed with separate regulations for each sex.

What characteristics of each species make such regulations possible?

8. Choose a nongame species in your area, and describe potential human activities that might affect this species and require regulation.

9. Why do fishery biologists consider daily creel limits to be relatively ineffective for prevention of fish population overharvest?

10. Discuss a situation in which establishment of a sanctuary might be an appropriate conservation strategy for a wildlife population and for a fish population. Include the management objective for each population in your discussion.

SELECTED READINGS

Barnhart, R. A., and T. D. Roelofs, eds. 1987. *Catch-and-release fishing: A decade of experience.* California Cooperative Fishery Research Unit, Humboldt State University, Arcata.

Denney, R. N. 1978. Managing the harvest. Pages 395–408 *in* J. L. Schmidt and D. L. Gilbert, eds. *Big game of North America: Ecology and management.* Stackpole Books, Harrisburg, Pa.

Hawkins, A. S., R. C. Hanson, H. K. Nelson, and H. M. Reeves, eds. 1984. *Flyways: Pioneering waterfowl management in North America.* U.S. Fish and Wildlife Service 1984-667–157. U.S. Government Printing Office, Washington, D.C.

Noble, R. L., and T. W. Jones. 1993. Managing fisheries with regulations. Pages 383–402 *in* C. C. Kohler and W. A. Hubert, eds. *Inland fisheries management in North America.* American Fisheries Society, Bethesda, Md.

Richardson, F., and R. H. Hamre, eds. 1990. *Wild trout IV: Proceedings of the symposium, Yellowstone National Park, September 18–19, 1989.* 1990-774–173/25037. U.S. Government Printing Office, Washington, D.C.

Strickland, M. D., H. J. Harju, K. R. McCaffery, H. M. Miller, L. M. Smith, and R. J. Stoll. 1994. Harvest management. Pages 445–473 *in* T. A. Bookhout, ed. *Research and management techniques for wildlife and habitats.* The Wildlife Society, Bethesda, Md.

Wise, J. P. 1991. *Federal conservation and management of marine fisheries in the United States.* Center for Marine Conservation, Washington, D.C.

LITERATURE CITED

Bellrose, F. 1982. Impact of ingested lead pellets on waterfowl. Pages 633–641 *in* J. T. Ratti, L. D. Flake, and W. A. Wentz, eds. *Waterfowl ecology and management: Selected readings.* The Wildlife Society, Bethesda, Md.

Boyd, R. J., and J. F. Lipscomb. 1976. An evaluation of yearling bull elk hunting restrictions in Colorado. *Wildlife Society Bulletin* 4:3–10.

Graham, L. K. 1974. Effects of four harvest rates on pond fish populations. Pages 29–38 *in* J. L. Funk, ed. *Symposium on overharvest and management of largemouth bass in small impoundments.* Special Publication no. 3. North Central Division, American Fisheries Society, Bethesda, Md.

Gresswell, R. E., and J. D. Varley. 1988. Effects of a century of human influence on the cutthroat trout of Yellowstone Lake. Pages 45–52 *in* R. E. Gresswell, ed. *Status and management of interior stocks of cutthroat trout.* Symposium 4. American Fisheries Society, Bethesda, Md.

Kiefer, J. D., M. R. Kubacki, F. J. S. Phelan, D. P. Philipp, and B. L. Tufts. 1995. Effects of catch-and-release angling on nesting male smallmouth bass. *Transactions of the American Fisheries Society* 124:70–76.

Klein, W. D. 1974. *Special regulations and elimination of stocking: Influence on fishermen and the trout population at the Cache la Poudre River, Colorado.* Technical Publication no. 30. Colorado Division of Wildlife, Denver.

Redmond, L. C. 1974. Prevention of overharvest of largemouth bass in Missouri impoundments. Pages 54–68 *in* J. L. Funk, ed. *Symposium on overharvest and management of largemouth bass in small impoundments.* Special Publication no. 3. North Central Division, American Fisheries Society, Bethesda, Md.

Redmond, L. C. 1986. The history and development of warmwater fish harvest regulations. Pages 186–195 *in* G. E. Hall and M. J. Van Den Avyle, eds. *Reservoir fisheries management: Strategies for the 80s.* Reservoir Committee, Southern Division, American Fisheries Society, Bethesda, Md.

Shaw, J. H. 1985. *Introduction to wildlife management.* McGraw-Hill, New York.

Strickland, M. D., H. J. Harju, K. R. McCaffery, H. M. Miller, L. M. Smith, and R. J. Stoll. 1994. Harvest management. Pages 445–473 *in* T. A. Bookhout, ed. *Research and management techniques for wildlife and habitats.* The Wildlife Society, Bethesda, Md.

Talbot, G. B. 1973. The California sardine-anchovy fisheries. *Transactions of the American Fisheries Society* 102:178–187.

Thelen, T. H. 1991. Effects of harvest on antlers of simulated populations of elk. *The Journal of Wildlife Management* 55:243–249.

18

WILDLIFE AND FISHERY LEGISLATION

Wildlife and fishery legislation in the United States and Canada reflects over a century of efforts by individuals and groups to conserve natural resources for the good of all citizens. Early colonists and other settlers in North America found an abundance of wildlife and fishery resources that seemed beyond depletion. With the settling of North America and the associated increases in human population and industrialization, the negative effects of uncontrolled use of common resources soon became evident. To some, the future lay in dividing and depleting wildlife and fishery resources with no thought to their future availability; to others with greater vision, the potential for the management of wildlife and fishery resources as renewable resources was evident.

The wisdom of those who advocated conservation of our natural resources led eventually to the establishment of a North American heritage in wildlife and fishery legislation. The common historical backgrounds, stable governments, and democratic natures of the United States and Canada were important factors in the evolution of the existing framework of wildlife and fishery legislation in these two societies. Other nations with stable governments have also developed important foundations in wildlife and fishery legislation, but none more strongly reflects the values of democratic societies than the legislation found in North America.

Extensive wildlife and fishery legislation exists within individual states, provinces, and territories and within the federal governments of the United States and Canada. It would be impossible to review all such legislation in this text. The primary emphasis of this chapter is on U.S. federal legislation as it affects natural resources on a broad geographic scale. Some comparisons with Canadian legislation are also provided. An attempt is made to interweave the state, provincial, territorial, and federal perspectives with regard to jurisdiction over wildlife and fishery resources.

18.1 States' Rights Doctrine

Wildlife and fishery resources in any nation are dependent on legislation as well as on the traditions that led to that legislation. In Canada and the United States, immigrants found vast expanses of land with abundant wildlife, fishery, and other natural resources. Most early settlers in North America were from the British Isles, France, and other European countries, and had lived under laws and traditions that largely reserved hunting and fishing rights for royalty, landowners, and the privileged few. These immigrants shared in the abundant wealth of fishery and wildlife resources in North America and rejected the European system. Early tradition, as well as court decisions in the United States, confirmed that the colonies and then the states, representing the people, assumed ownership of and responsibility for wildlife and fishery resources. This meant that these resources belonged to everyone, not just the privileged class. In the United States, the right of the individual states to control wild animal populations within their borders falls under the realm of the **states' rights doctrine**.

The states' rights doctrine, as it relates to fishery and wildlife resources remained largely intact through the 1800s. The U.S. Supreme Court acknowledged state ownership of wildlife and fishery resources in two particularly important cases. The first case, McCready v. Virginia in 1876, confirmed the right of the Commonwealth of Virginia to limit access to oysters within state waters. In the case of Geer v. Connecticut in 1896, Edgar M. Geer had attempted to transport American woodcocks, ruffed grouse, and northern bobwhite, legally taken in Connecticut, across the state line. The court ruled that Connecticut had the right to limit transport of legally taken game from the state for sale or use outside state borders. The

Geer decision is viewed as the landmark states' rights decision and is commonly referred to in discussions and court decisions involving state versus federal jurisdiction over fishery and wildlife resources.

In the United States, any federal involvement in the conservation of wildlife and fishery resources or their control through the legal system was considered as the usurping of rights that constitutionally belonged to the states. These rights are based on the Tenth Amendment to the U.S. Constitution, which states, "The powers not delegated to the United States by the Constitution nor prohibited by it to the states, are reserved to the states respectively, or to the people."

Actual colonial and state legislation limiting the taking of wildlife and fishery resources by restricting the time of year harvest could occur, requiring licenses, and even closing seasons occurred early in the history of the United States. Prior to independence, twelve of the thirteen colonies had already enacted closed seasons on deer, and Massachusetts actually closed deer hunting for several years starting in 1718 (Leopold 1933). New York required the first hunting license in 1864. By 1880, all of the states had game laws, although enforcement was often limited and sporadic. The colony of Massachusetts closed fishing during the spawning period for species such as Atlantic cod, haddock, pollock, and Atlantic mackerel as early as 1652. By 1877, twenty-seven states had outlawed nets and seines for capturing fishes (Redmond 1986). Additional information concerning the history of fishery and wildlife regulations appeared in section 17.2.

The Canadian Constitution Act (1867) provided for federal regulation of "sea coast and inland fisheries." The system that currently exists, as in the United States, recognizes provincial and territorial responsibilities as well as selected federal (migratory birds and coastal fisheries) responsibilities for wildlife and fishery resources. Manitoba, Saskatchewan, and Alberta, the prairie provinces, did not receive authority over and ownership of most wildlife and fishery resources

until 1930, under the Natural Resources Transfer Agreement with the Government of Canada. Crown lands (public lands) in Canada are under provincial rather than federal control, a major difference from the United States.

18.2 The Federal Legislative Process in the United States

Prior to discussing wildlife- and fishery-related legislation, a brief description of the U.S. federal legislative process is appropriate. Legislative proposals can originate in a variety of ways—with individuals, groups, members of Congress, or the executive branch of the federal government—but all proposed legislation must be introduced by a member of Congress. Several members of the House of Representatives and Senate can co-sponsor the same bill. Each year many bills are introduced into Congress, but few become law. The typical route by which proposed legislation becomes federal law in the United States is illustrated in figure 18.1.

Federal legislation can have strong influences on wildlife and fishery resources, and continuing changes in this area can be expected. Important legislation, such as the Endangered Species Act or Clean Water Act, must periodically be reauthorized by Congress. Thus, passage of legislation and years of successful implementation do not guarantee the future status of that legislation. Wildlife and fishery biologists, conservation groups, other advocacy groups, and the general public can have an important influence on the reauthorization of legislation, the development of amendments to legislation, the introduction of new legislation, and opposition to legislation they are against. The legislative process involves a good deal of biopolitical interaction; section 20.5 contains a discussion of biopolitics.

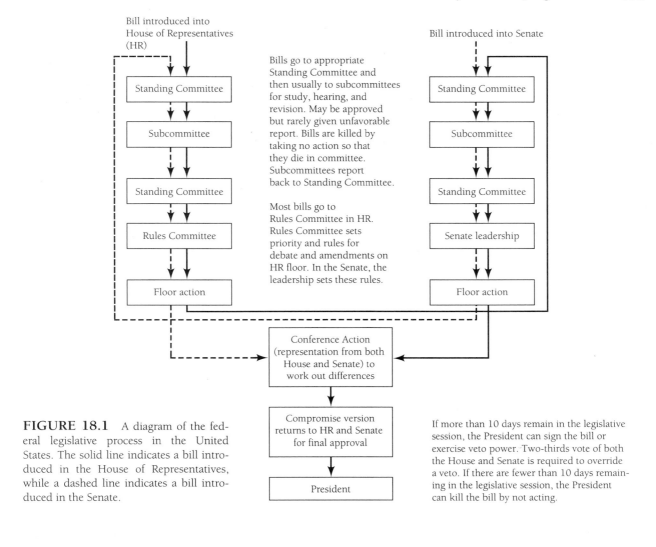

Bill introduced into
House of Representatives
(HR)

Bill introduced into Senate

Standing Committee

Subcommittee

Standing Committee

Rules Committee

Floor action

Standing Committee

Subcommittee

Standing Committee

Senate leadership

Floor action

Bills go to appropriate
Standing Committee and
then usually to subcommittees
for study, hearing, and
revision. May be approved
but rarely given unfavorable
report. Bills are killed by
taking no action so that
they die in committee.
Subcommittees report
back to Standing Committee.

Most bills go to
Rules Committee in HR.
Rules Committee sets
priority and rules for
debate and amendments on
HR floor. In the Senate, the
leadership sets these rules.

Conference Action
(representation from both
House and Senate) to
work out differences

Compromise version
returns to HR and Senate
for final approval

President

FIGURE 18.1 A diagram of the federal legislative process in the United States. The solid line indicates a bill introduced in the House of Representatives, while a dashed line indicates a bill introduced in the Senate.

If more than 10 days remain in the legislative session, the President can sign the bill or exercise veto power. Two-thirds vote of both the House and Senate is required to override a veto. If there are fewer than 10 days remaining in the legislative session, the President can kill the bill by not acting.

18.3 Early Federal Legislation in the United States: The Lacey Act

The U.S. federal government became directly involved in wildlife legislation with the passage of the Lacey Act, which is also called The Game and Wild Bird Preservation Act of 1900. The Lacey Act, most importantly, regulates the interstate shipment of illegally killed animals. The Lacey Act was viewed as helping state law enforcement and conservation efforts and for that reason was not considered to be an infringement on the states' rights doctrine. The U.S. government is empowered by the Constitution to regulate interstate commerce, and thus its involvement in wildlife and fishery conservation through the Lacey Act is constitutionally protected. The Lacey Act and most other U.S. federal legislation described in this chapter are listed in table 18.1. The table contains the title of each act, its date of enactment, and a brief description of its importance.

A piece of legislation similar to the Lacey Act, the Black Bass Act of 1926, was originally passed to control the interstate shipment of illegally taken largemouth bass. This legislation was later amended to include other freshwater and saltwater fishes and even shellfishes. In 1981, amendments to the Lacey Act repealed the Black Bass Act and provided for a single comprehensive law. Wildlife, fishes, shellfishes, some plants, and other protected organisms are now covered under the Lacey Act, as are parts or eggs of wildlife, fishes, and other protected organisms. If someone were to illegally kill a sage grouse in eastern Oregon and transport that bird, or even its leg or some other body part, across the state line to Idaho, they would be in violation of the Lacey Act. Likewise, an angler from Alabama fishing illegally in Mississippi and transporting a catch of channel catfish and largemouth bass across the border to Alabama would be in violation of the Lacey Act. Protected animals killed illegally on tribal reservations by nontribal members and transported any distance, even without crossing reservation lines, also constitute a violation of the Lacey Act.

The Lacey Act also regulates international commerce in protected organisms. Anyone who transports, ships, or sells protected species contrary to foreign law or U.S. federal statutes, as well as state law, is in violation of the Lacey Act. Thus, if the horn of the protected black rhinoceros from eastern Africa or illegally taken eggs of greater or lesser rheas from Argentina were shipped into the United States, those shipments would constitute a Lacey Act violation. As another example, assume that someone was illegally netting white sturgeons in the Snake River in Washington and then selling the eggs to a second party, who transported them to eastern Europe for sale as caviar. Both the people illegally capturing the fish and those purchasing and transporting the illegally taken fish or eggs would be in violation of the Lacey Act. If a third party, who knew that the eggs were illegally taken, was storing them prior to shipment, they too would be in violation.

The Lacey Act also prohibits the importation of certain species into portions of the United States. The importation of any species of mongoose, for example, is prohibited by the Lacey Act. The prohibition of the introduction of injurious species was originally a feature added in the interest of agriculture and horticulture (Bean 1983). This prohibition has been broadened to incorporate a wide array of organisms, including some plants, because of their possible negative effects on natural ecosystems. For example, the introduction of peacock cichlid, a fish from South America, is prohibited because of its probable competition with native species.

In Canada, the Wild Animal and Plant Protection and Regulation of International and Interprovincial Trade Act was passed in 1992; this act has many similarities to the Lacey Act. It prohibits interprovincial, interterritorial, and international commerce in illegally taken wild organisms or their parts, and also prohibits the introduction of species potentially harmful to ecosystems within Canadian federal, provincial, or territorial borders.

18.4 Migratory Bird Treaties and Legislation

Early in the 1900s, Canada and the United States recognized that migratory birds, such as waterfowl, needed protection within a national and international framework. Without national and international legislation and cooperation, migratory birds could not be scientifically conserved. In 1913, the U.S. Congress passed the Migratory Bird Act, which declared that migratory and insectivorous birds were under the custody and protection of the federal government. This 1913 legislation was challenged in two federal district courts and was judged invalid on the basis of the earlier Geer decision. Even though the U.S. federal government recognized the opposition of many states to any federal regulation of wildlife, the federal government exercised its constitutional right

TABLE 18.1 A chronological listing of some important wildlife, fishery, and environmental legislation at the federal level in the United States.

Legislation (year)	Importance
Lacey Act (1900)	Placed federal control on interstate or foreign commerce (import or export) in illegally taken wildlife, fishes, plants, or other organisms. The Black Bass Act of 1926 was incorporated into the Lacey Act through amendments in 1981 to form a single, comprehensive legislative act.
Migratory Bird Treaty Act (1918)	Gave the federal government legal jurisdiction over migratory birds. Provided for the setting of seasons and regulation of harvest of migratory game birds. Provided for the coordination of conservation efforts for migratory birds between the United States and Canada.
Migratory Bird Conservation Act (1929)	Established the Migratory Bird Conservation Commission and allowed for purchase of migratory bird refuges.
Migratory Bird Hunting Stamp Act (1934)	Required the purchase of a federal duck stamp to hunt waterfowl and some other migratory birds. Funds generated were directed primarily toward purchases of waterfowl habitat or protection of such habitat with easements.
Fish and Wildlife Coordination Act (1934)	Required federal agencies to coordinate activities that might affect wildlife or fisheries. Fishery and wildlife resources influenced by federally funded projects, such as dams, were to be given equal consideration with other project benefits. This legislation was amended in 1946 and 1958.
Taylor Grazing Act (1934)	The beginning of federal control of domesticated livestock grazing on Bureau of Land Management lands.
Pittman-Robertson Act or Federal Aid in Wildlife Restoration Act (1937)	Initiated a federal excise tax on arms and ammunition, the income from which is distributed to states for use in a variety of wildlife-related projects. Hunting license income in states receiving these funds must remain with the state wildlife and fishery agency.
Bald Eagle Protection Act (1940)	Protected the bald eagle. Amended in 1962 to protect the golden eagle.

(*Table 18.1 continued on following page*)

Legislation (year)	Importance
Dingell-Johnson Act (1950) or Federal Aid in Sport Fish Restoration Act (1950)	Modeled after the Pittman-Robertson Act. Initiated a federal excise tax on fishing tackle that is used for restoration and management of state fishery resources. States are eligible for this funding only if fishing license fees are restricted for use to the state fishery and wildlife agency. The Wallop-Breaux Amendment to this legislation in 1984 extended the federal excise tax to other fishing and boating gear and more than tripled this source of funding.
Multiple Use-Sustained Yield Act (1960)	The legal basis for multiple-use policies on national forest lands. The U.S. Forest Service must consider fisheries and wildlife with other uses of national forest lands.
Wilderness Act (1964)	Protected areas of public land given wilderness status from most development and damaging forms of use.
Classification and Multiple Use Act (1964)	The legal basis for multiple-use policies on Bureau of Land Management lands.
National Environmental Policy Act (NEPA) (1969)	Established the Council on Environmental Quality and the Environmental Protection Agency. This legislation also requires environmental impact statements for projects that receive federal funding and affect the quality of human life.
Wild and Free-Roaming Horses and Burros Act (1971)	Required feral horses and burros to be treated as an integral part of the natural ecosystem on Bureau of Land Management and national forest lands.
Marine Mammal Protection Act (1972)	Removed the states' authority over management of marine mammals, except through federally approved management plans. Harvest of selected marine mammals is allowed.
Federal Environmental Pesticide Control Act (FEPCA) (1972)	Grouped pesticides into general and restricted use categories and required certified applicators. The intent was to stop applications that cause unreasonable adverse effects on the environment.
Endangered Species Act (ESA) (1973)	Provided for the listing of species as endangered or threatened. Provided for species recovery plans and the acquisition and protection of critical habitat (see chapter 11).
National Forest Management Act (NFMA) (1976)	Directed the U. S. Forest Service to maintain viable and well-distributed populations of all native vertebrate species in national

(Table 18.1 continued on following page)

Legislation (year)	Importance
	forests. The U. S. Forest Service must comply with the National Environmental Policy Act in its forest management plans. ·
Federal Land Policy and Management Act (FLPMA) (1976)	Contemporary with and similar to the National Forest Management Act but directed at Bureau of Land Management lands. Provided for protection of areas of critical concern and coordinated management planning with the U. S. Forest Service.
Toxic Substances Control Act (1976)	Extended federal protection of the environment to substances other than pesticides.
Clean Water Act (1977)	Required the Environmental Protection Agency to set standards to limit pollutants released into surface waters by point sources such as factories and sewage treatment plants. Set goals to keep water safe for fishing and swimming.
Public Rangeland Improvement Act (1978)	Provided further support for multiple use by requiring improvement of some federal lands that had been degraded by domesticated livestock.
Fish and Wildlife Conservation Act (1980)	Provided a framework for conservation of nongame wildlife and fish species.
Alaska Lands Act (1980)	Established extensive tracts of national wilderness lands in Alaska.
North American Wetlands Conservation Act (1989)	Made conservation of wetlands a federal priority and authorized funding for that purpose. Authorized the use of appropriated funds to protect wetlands in Canada and Mexico as well as in the United States.

to sign treaties with foreign nations by signing a treaty with Canada (with Great Britain signing) in 1916 on migratory birds. Canada, in supporting and signing the Migratory Bird Treaty with the United States, likewise acknowledged the need for a national and international approach to the conservation of migratory birds.

Following the signing of the Migratory Bird Treaty, the Migratory Bird Treaty Act of 1918 was enacted to implement the treaty in the United States; in Canada the treaty was implemented by the Migratory Bird Convention Act of 1917 (Gilbert and Dodds 1992). In the United States, the passage of the Migratory Bird Treaty Act meant that the Supreme Court no longer needed to decide the constitutionality of the 1913 Migratory Bird Act. The Migratory Bird Treaty Act in the United States and the Migratory Bird Convention Act in Canada made it illegal in both countries, except as permitted though regulations set by federal authority, to kill, possess, sell, purchase, or otherwise deal in migratory birds or their parts, eggs, or nests. States, provinces, and territories may make and enforce migratory bird laws and regulations if they are consistent with or more restrictive than those allowed under federal jurisdiction. Section 20.5 delineates the process of working within this legislation to set waterfowl hunting regulations in the United States.

The constitutionality of the Migratory Bird Treaty Act was challenged in the U.S. Supreme Court in 1920 in the case of Missouri v. Holland. Ray Holland was a U.S. game warden enforcing the Migratory Bird Treaty Act in Missouri. The state of Missouri contended that enforcement of the Migratory Bird Treaty Act was a violation of constitutional rights reserved to the states under the Tenth Amendment. The authority of the U.S. government to use police powers to arrest persons violating the Migratory Bird Treaty Act was upheld in this important decision. The power of the U.S. government to enforce this legislation is derived from its constitutional powers to make and uphold treaties with foreign nations. Interestingly and wisely, the Supreme Court noted that migratory birds are transient between states and nations and

that without the treaty, there might soon be no migratory birds left for any power to regulate. The upholding of the Migratory Bird Treaty Act was seen as a major defeat to the states' rights view of wildlife control.

In the United States, the Migratory Bird Treaty Act did not authorize the acquisition of migratory bird habitat. The passage of the Migratory Bird Conservation Act (1929) resulted in the formation of the Migratory Bird Conservation Commission and allowed the Secretary of the Interior to purchase and lease migratory bird refuges. The Migratory Bird Hunting Stamp Act (1934) provided funding for the purchase and administration of **Waterfowl Production Areas** and the purchase of **wetland easements** (in later years). Under this act, hunters in the United States pursuing waterfowl and some other migratory birds, such as sandhill cranes, must purchase a Migratory Bird Hunting and Conservation Stamp (fig. 18.2). This legislation is now more appropriately titled the Migratory Bird Hunting and Conservation Stamp Act, and the stamp itself is commonly referred to as a "duck stamp."

Canada passed a Migratory Bird Hunting Permit Act in 1966 that requires waterfowl hunters to purchase a federal migratory bird hunting license. Funds from the purchase of this federal license go into the general fund, with no stipulation on their use. However, in 1984, Canada implemented a Wildlife Habitat Conservation Stamp Program to accompany the federal migratory bird hunting license as a vehicle for raising money for habitat conservation projects throughout the nation. Funds from the purchase of the Wildlife Habitat Conservation Stamp, primarily by hunters, are transferred to and administered by Wildlife Habitat Canada, a nonprofit private corporation. Although waterfowl and wetland projects have been supported by these funds since 1984, the funds are not exclusively dedicated to this purpose.

Funds from the sale of duck stamps in the United States have provided substantial benefits in terms of migratory bird habitat purchases and easements. The preservation of these habitats has

A

B

FIGURE 18.2 Funds from the sale of the Migratory Bird Hunting and Conservation Stamp (A) in the United States are primarily used to purchase or otherwise protect wetlands and associated upland habitat, such as this Waterfowl Production Area (B). The vegetation seen in the background serves as upland nesting cover for waterfowl and habitat for many other organisms.

also benefited many other nonmigratory species of animals as well as plants. Funds from the stamp were inadequate for acquisition of sufficient wetlands by fee title and permanent easements, and it was necessary to borrow against future "duck stamp" income. Under the Accelerated Wetlands Acquisition Act (1961), additional federal funds were made available for the purchase of wetlands or wetland easements over a seven-year period. Funding for the Accelerated Wetlands Acquisition Act was later advanced for additional years and has recently been based on one-year extensions. These borrowed funds were to be repaid by income from the Migratory Bird Hunting and Con-

servation Stamp Act. However, the funding advanced for wetland acquisition has in large part been forgiven by Congress and does not need to be paid back from duck stamp funds. Payment of part of the cost of preserving wetlands from the general fund (funds provided by all taxpayers) seems reasonable considering the variety of benefits to society resulting from wetland preservation.

In 1936, a treaty for the conservation of migratory birds (and migratory game mammals) was signed by the United States and Mexico. Another treaty was signed by the United States and Japan in 1972 to cover birds migrating between the United States, Japan, and their territories. Canada does not have migratory bird treaties with countries other than the United States.

In 1986, Canada and the United States jointly signed the North American Waterfowl Management Plan (NAWMP), an ambitious plan to provide for the conservation of waterfowl habitat in North America; in 1994, Mexico also became a partner in NAWMP (Baldassarre and Bolen 1994). The goals of NAWMP involve the protection and enhancement of waterfowl habitat throughout North America, with the objective of restoring ducks and geese to 1970s population levels.

18.5 Funding: A Carrot-and-Stick Approach

In 1937 the Pittman-Robertson Act, or Federal Aid in Wildlife Restoration Act, was passed in the United States. This important legislation enacted an 11% tax on arms and ammunition and provided that the proceeds be held in the federal treasury for use in wildlife restoration and management by the states. The act takes a "carrot-and-stick" approach by encouraging states to reinvest their hunting license income in wildlife resources. The funds available to states through the Pittman-Robertson Act are commonly referred to as **PR**

funds or **federal aid funds**, and represent "the carrot." These funds are available to a state only if its income from the sale of hunting licenses goes directly to the state agency responsible for wildlife and fishery resources. If license money is taken away from that agency, the state loses its PR funds; this represents "the stick." Wise federal legislators saw that income from hunting licenses would be a great temptation to state governments that might want to divert it to a variety of non-wildlife-related projects and expenses. Efforts are constantly made at both the federal and state levels to divert PR funds to purposes other than those intended by the Pittman-Robertson Act. Only a watchful and well-informed public interested in wildlife resources can ensure that these user fee monies continue to be directed to their intended purposes.

PR funds are apportioned to states on the basis of total state area and number of paid hunting license holders, each composing 50% of the apportioning formula. No state is to receive less than 0.5% and no state more than 5% of the total annual PR funds available. PR funds pay up to 75% of the cost of state projects that are submitted to the U.S. Fish and Wildlife Service and accepted for federal aid participation. The state must provide 25% in nonfederal matching funds. State management and research projects involving the acquisition, restoration, rehabilitation, and improvement of wildlife habitat are eligible for PR funding.

In 1970 an amendment to the Pittman-Robertson Act directed that monies from an existing federal excise tax on pistols and revolvers be placed in the PR fund. Since 1975, half of the federal tax on bows and arrows also goes into the PR fund. Part of the annual revenue from these taxes is divided among the states for hunter safety programs, including construction, operation, and maintenance of outdoor target ranges.

The Dingell-Johnson Act (1950), or Federal Aid in Sport Fish Restoration Act, was modeled after the Pittman-Robertson Act and contains similar benefits, except that they apply to fishery resources. The Dingell-Johnson Act provides for the proceeds of a tax on fishing tackle to be held in the federal treasury and granted to the states for restoration and management of state fishery resources. Funds from the Dingell-Johnson Act are often referred to as **DJ funds** or federal aid funds. As with PR funds, DJ funds are available to states only if their income from fishing licenses is restricted to use by the state agency responsible for fishery and wildlife resources. Also, as with PR funds, constant attempts are made at both the state and federal levels to divert such funding away from fisheries. Distribution of DJ funds to states is based 40% on land area and 60% on fishing licenses sold. As with the Pittman-Robertson Act, states must contribute a 25% nonfederal match in order to use DJ funds.

In 1984 the **Wallop-Breaux Amendment** to the Federal Aid in Sport Fish Restoration Act extended the federal excise tax to include additional fishing gear. An excise tax on electric motors and sonar devices was also added, as was a federal tax on fuel for motor boats. The Wallop-Breaux Amendment also added import duties on foreign-manufactured fishing tackle and boats. The Wallop-Breaux Amendment has more than tripled the federal aid funds available to states for fishery work.

The Pittman-Robertson and Dingell-Johnson acts have greatly benefited wildlife and fishery resources in the United States. We are aware of no other country with similar legislation. The wording of these legislative acts has also successfully protected state hunting and fishing license fees from diversion into other programs.

18.6 Other Federal Wildlife and Fishery Legislation

A variety of other U.S. federal legislation affects wildlife and fishery resources; a few important examples follow. The Fish and Wildlife Coordination

Act of 1934 (amended in 1946 and 1958) signaled the intention of the U.S. government to coordinate all the activities of federal agencies that affect wildlife and fishery resources. Under this act, federally funded projects influencing fishery or wildlife resources must be coordinated with the U.S. Fish and Wildlife Service (USFWS). For example, if a major water development project using federal funds, such as a dam with hydroelectric facilities, were planned, the USFWS would have to be consulted and the effects of the project on fishery and wildlife resources given equal consideration (1958 amendment) with its other aspects. Equal consideration for fishery and wildlife resources may require **mitigation** for lost habitat. For example, the influence of a dam on wildlife is likely to be negative because of the loss of riparian and other bottomland habitat associated with the original river or stream. Mitigation might involve the purchase of private land with riparian and bottomland habitat values in another part of the state in which the project is developed. Even if such areas are purchased and put under public ownership and management, it is still difficult to mitigate for the extensive riparian habitat destruction caused by many water projects. In general, the mitigation process has not always been successful. The U.S. Congress usually appropriates funding for mitigation only after a project is completed, and such funding is often not provided at all.

The Bald Eagle Protection Act (1940), amended in 1962 to include the golden eagle, provides for complete protection for both species in the United States. The U.S. Endangered Species Act (1973) was discussed in chapter 11 and thus is not included here. It should be noted, however, that the bald eagle, because of its status as an endangered or threatened species in various portions of its range, is also protected by the Endangered Species Act. In Canada, endangered species such as the muskox and bald eagle also receive special protection.

The Marine Mammal Protection Act (1972) in the United States addressed a variety of marine mammal issues. States lost their authority to manage marine mammals through this legislation, but can regain in some of that authority by instituting conservation programs that meet federal standards. The Marine Mammal Protection Act places restrictions on the harvest of marine mammals. Limited exceptions are allowed for native peoples and for incidental kills of fishes and other aquatic animals. The act also allows for harvest of selected marine mammals if populations are kept at optimum carrying capacity, a term not clearly defined in the legislation. The act must be considered in international treaties and agreements, as in the Northern Fur Seal Treaty between the United States, Japan, Russia, and Canada. The Canadian federal government also has jurisdiction over marine mammals in its national waters.

The Fish and Wildlife Conservation Act (1980) provided a framework for the conservation of the many nongame species in the United States. These species are often overlooked unless they are endangered or threatened. Unfortunately, Congress did not provide an adequate funding mechanism for the Fish and Wildlife Conservation Act. However, Congress did require the U.S. Fish and Wildlife Service to conduct a study on potential funding sources. One current possibility, supported by natural resource agencies and many private conservation groups, is the "fish and wildlife diversity funding initiative." This initiative, if it becomes law, would place a surcharge on a variety of outdoor recreation products, such as backpacks, tents, binoculars, birdseed, bird feeders, and recreational vehicles, that are not currently taxed under Dingell-Johnson or Pittman-Robertson legislation. The proceeds would be restricted to use in various nongame conservation programs and would be distributed to state natural resource agencies in a manner similar to Pittman-Robertson and Dingell-Johnson funds. The "fish and wildlife diversity funding initiative" would raise an estimated $350 million per year for fishery and wildlife conservation, outdoor recreation, and environmental education. Even in times of fiscal conservatism, such legislation can pass because it is viewed as a "user pay" tax rather than a general tax increase.

The U.S. government recognized the importance of international wetland protection for conserving migratory birds and other species in the North American Wetlands Conservation Act of 1989. Under this legislation, funds can be spent for wetland acquisition and protection in Canada and Mexico as well as in the United States. Funding for wetland protection in Canada and Mexico is particularly important to the success of the North American Waterfowl Management Plan. Interest earned on Pittman-Robertson funds prior to their distribution to the states is now reserved for the funding of wetland acquisition and improvement under this legislation. These interest-generated funds previously went into the federal general fund and had long been targeted by conservation interests; their use in helping to fund the North American Wetlands Conservation Act seems highly appropriate. Additional funds come from fines collected by the Department of Justice for violations of the Migratory Bird Treaty Act and from federal appropriations.

18.7 General Environmental Legislation

While many of the following pieces of legislation are usually described as environmental legislation, they go far beyond that. All are important to public health, safety, and welfare as well as to wildlife and fisheries resources.

The National Environmental Policy Act (1969) in the United States, popularly known as NEPA, is broad in its implications because it established the Council on Environmental Quality as well as the Environmental Protection Agency. In addition, through NEPA, **environmental impact statements** (EIS) are required for all projects receiving federal funds that substantially affect the quality of human life. The construction of a dam or a highway using federal funds, for example,

would require preparation of an EIS that includes the possible effects of the project on wildlife and fishery resources. The Council on Environmental Quality must assess the EIS before allowing a project to proceed. Environmental impact statements have been required for many activities, even including actions by the U.S. Fish and Wildlife Service to set annual waterfowl seasons; in that case, parties interested in stopping waterfowl hunting attempted to do so through the EIS requirement. Many fishery and wildlife professionals become involved in writing environmental impact statements, either as employees of natural resource agencies or as employees of private consulting firms that contract for such work.

The Council on Environmental Quality was mandated by executive order in 1979 to evaluate the effects of major federal actions abroad. In addition, the United States signed a Convention on Environmental Impact Assessment with several European countries in 1991 requiring the signatories to consult one another regarding transboundary environmental impacts.

Environmental impact assessment in Canada has been mandated through a cabinet decision or directive (not a law or act) called the Federal Environmental Assessment in Planning Decision, or FEARO (1978). Under this decision, Canadian federal guidelines for environmental impact assessment were established. Projects using federal funds, on Crown lands, or being undertaken by federal agencies require environmental impact assessment under this decision. The assessment is conducted by a federally appointed committee.

Regulation of various toxic substances released into the environment can be extremely important to wildlife and fishery resources. By their nature, pesticides are designed to kill certain organisms or groups of organisms that damage crops, threaten human health, or in some other way cause economic damage. Pesticide control in the United States was initially under the Federal Insecticide, Fungicide, and Rodenticide Act of 1947 (FIFRA), which was later amended and replaced by the Federal Environmental Pesticide Control Act of 1972 (FEPCA). FEPCA provides for federal control of the sale of pesticides and their use when inconsistent with labeling. Under FEPCA, pesticides were placed into two categories, general use and restricted use. Restricted use pesticides require certified applicators. The intent of the legislation is to stop applications that involve unreasonable adverse effects on the environment. Human health is the primary concern, with wildlife and fishery considerations secondary. However, prohibition of use due to wildlife and fishery concerns can occur, as in the cases of strychnine and compound 1080 (sodium monofluoroacetate), removed from use as coyote poisons in 1972. The Toxic Substances Control Act of 1976 extended control to substances other than pesticides.

Despite pesticide control legislation, numerous instances of direct mortality and secondary effects on wild organisms due to pesticides still occur. Some of these incidents cause considerable mortality in some species, particularly in the young. In addition, pesticide effects that do not kill organisms outright often remain undetected (fig. 18.3). For example, sublethal effects, such as reduced growth rates due to reductions in the invertebrate foods available to young ducks and gal-

linaceous birds, are particularly difficult to identify (Grue et al. 1988).

Water quality affects aquatic plants and animals as well as terrestrial animals associated with aquatic systems. Thus, the U.S. Clean Water Act of 1977 is extremely important to wildlife and fishery resources. This legislation directs the Environmental Protection Agency to set standards that limit pollution in lakes, streams, and other surface waters.

18.8 Legislation Directed at Management of Federal Lands

Legislation influencing the management of federal lands in the United States is particularly important to wildlife and fishery resources, especially in the western half of the United States where these lands cover an extensive area. The Taylor Grazing Act of 1934 was the beginning of control over grazing by domesticated livestock on vacant, unappropriated, and unreserved federal lands under the jurisdiction of the Bureau of Land Management (BLM). The Taylor Grazing Act clearly states that nothing in the act should be construed so as to restrict hunting and fishing on these lands. It also notes that grazing permittees have no right to interfere with hunting, fishing, and many other recreational activities on these lands. Without this stipulation, public recreation on BLM lands could have been excluded by grazing or other commercial interests.

The Multiple Use-Sustained Yield Act (1960) is the legal basis for multiple use of national forests in the United States. The Classification and Multiple Use Act of 1964 provided a similar policy on BLM lands. These legislative acts state that these federal lands will be managed on a multiple use basis, with wildlife and fisheries given equal consideration with other uses (fig. 18.4). In spite

FIGURE 18.3 Agricultural pesticides sometimes kill invertebrates in wetlands, resulting in reduced growth rates or even death for ducklings such as these blue-winged teal. Some pesticides may also directly kill birds through contact or ingestion.

FIGURE 18.4 The Chugach National Forest in Alaska must be managed for multiple uses, including wildlife and fishery resources, under the Multiple Use-Sustained Yield Act. *(Photograph courtesy of K. Jensen.)*

of these legislative actions, timber, livestock, and mining interests have generally received higher priority than wildlife and fishery resources on federal lands. The two multiple use acts described above are still extremely important because they give legal impetus to a more equal prioritization of the different uses of and resources on federal land.

The Public Rangeland Improvement Act of 1978 provided further support for the multiple use concept by requiring improvement of federal lands that have been degraded by domesticated livestock grazing. This legislation, which imposed higher grazing fees and grazing restrictions, provoked a response on the part of some ranchers who grazed domesticated livestock on federal land. This movement was popularly called the "**sagebrush rebellion**," and it set forth strong demands for the transfer of federal lands to private ownership. Many ranchers, however, recognized that federal land, if sold, would go to the highest bidder, and that they could lose the grazing leases they controlled while the land was in federal ownership. In addition, if grazing lessees bought the land, they would also need to begin paying state property taxes on land they formerly leased at low rates without paying taxes. Economic considerations prevailed, and the demand for a massive sale of federal lands quieted. However, there is always

a threat of the loss of federal lands to private ownership. The potential ceding of federal lands to individual states is also a concern, as states could subsequently sell the land to private owners or in other ways exclude the public. Additionally, states might also be more likely to favor timber, mining, and grazing interests. It is clear from past and current efforts of some special interest groups that the existence of federal land in the United States and the right of public access to this land cannot be taken for granted.

The National Forest Management Act (NFMA) of 1976 required federal forest management plans to comply with the National Environmental Policy Act. This legislation specifically directs the U. S. Forest Service to maintain the biological diversity of native vertebrate species within national forests. The contemporary Federal Land Policy and Management Act of 1976 (FLPMA) had many similarities to NFMA, but was directed at BLM lands. Both pieces of legislation enhance the multiple use concept on federal lands.

The Wilderness Act of 1964 mandated the preservation in a natural state of large tracts of national forests, national parks, and other federal lands in remote areas. Areas are set aside in which roads, timber harvests, commercial development, and buildings are prohibited. Prior mining laws allow mining activity in areas where valid claims were filed before 1983. Where such mining is initiated, reasonable road access to the mine must be provided by the federal government. Additional lands were added through amendments to the Wilderness Act such as the Endangered American Wilderness Act of 1978 and the Alaska Lands Act of 1980. The Alaska Lands Act established large tracts of wilderness in Alaska; these areas constitute over two-thirds of the national wilderness system in the United States. The wilderness designation of extensive areas in Alaska has sometimes been controversial even among wildlife and fishery biologists. However, this act has been important in preventing logging, road development, and other economic activities in extensive areas of Alaska and can help to preserve these pristine ecosystems.

18.9 Legislation for Retirement of Agricultural Lands

Some of the legislation that most influenced the environment and public health, safety, and welfare, involved U.S. Department of Agriculture (USDA) programs. In 1956, Congress passed the Soil Bank Act, which diverted some land from crop production and placed it in protective vegetative cover for a period of three to ten years. In 1960 and 1961, the Soil Bank reached its peak, with over 12 million hectares enrolled. The highest enrollments in the Soil Bank program were in the Plains states from Texas to North Dakota. The Soil Bank provided habitat for a wide variety of wildlife species.

More recently, the **Conservation Reserve Program** (CRP) of the 1985 Food Security Act (Farm Bill) and the conservation compliance elements of the **sodbuster** and **swampbuster provisions** of this 1985 act have provided important benefits to the environment and to a wide variety of wild organisms. The swampbuster provisions include the loss of payments to farmers receiving federal crop subsidies through USDA if they produce specified crops (commodity crops such as corn or soybeans) on wetlands drained or otherwise converted to farmland after December 1985. After the passage of the 1990 Farm Bill, any wetland conversion, even without planting a crop, became a violation subject to loss of USDA subsidies. The sodbuster provisions require landowners plowing grassland to implement a soil conservation plan to keep soil erosion within limits specified.

The Conservation Reserve Program provides payments to landowners who take highly erodible cropland out of production. Under the CRP, extensive amounts of highly erodible cropland have been idled and seeded to grass or grass-legume cover for a period of ten years. The primary goals of the original CRP legislation were to reduce farm surpluses, control erosion, and improve water quality. The amount of land enrolled in the CRP exceeded 13 million hectares by 1993. In the northern Plains states, where much of the CRP land is located, these lands provide extensive areas of excellent nesting and other habitat for a variety of wild animals. Increased numbers of species such as bobolinks, grasshopper sparrows, sharp-tailed grouse, greater prairie-chickens, ring-necked pheasants, mallards, white-tailed deer, and others that benefit from grassland cover have generally resulted from this program. Fishery resources have also benefited from the CRP through the reduction of siltation and chemical pollution of aquatic habitats.

While payments to landowners for idling CRP lands are substantial, they are comparable to or lower than other payments that were made through USDA for other programs, such as crop subsidies, crop deficiency payments, annual set-aside payments, and disaster payments, prior to the implementation of the CRP. Thus, the CRP provides both producer benefits and conservation benefits at a lower price. CRP lands are not supposed to be grazed or mowed during the ten-year enrollment period, unless mowing is used for weed control. However, pressure from CRP enrollees, especially during drought when there is a high demand and good prices for hay, has often led to the allowance of substantial amounts of mowing and grazing on CRP land.

Unfortunately, land retirement programs such as CRP sometimes penalize those who have practiced good land stewardship by not plowing marginal ground. Many landowners who plowed areas of short-grass or mixed-grass prairie to plant wheat in areas such as eastern Colorado were rewarded with CRP contracts. Those landowners who had continued to graze livestock rather than plowing these lands, despite the higher immediate economic returns from wheat, were unjustly penalized by not receiving CRP contracts. Anticipation of soil retirement programs can even encourage plowing of marginal areas so that the landowner can qualify for the program. This penalization of good land stewardship needs to be corrected in future land retirement programs.

The Soil Bank and the CRP, while accomplishing many goals and providing many benefits to wildlife and fishery resources, were both initiated as temporary or short-term programs. Land retirement programs should be of a longer-term nature, with permanent vegetative cover required and with mowing, haying or grazing, and fire used sparingly and only to enhance the quality and wildlife value of these lands.

One longer-term conservation program for wetlands and farmland in the United States is the Water Bank Program, initiated in 1972 (Water Bank Act 1970). This USDA program provides for ten-year agreements protecting wetlands from drainage and placing protective cover on previously tilled uplands near these wetlands. Although the agreements are of a ten-year duration, they can be renewed. The Water Bank Program has remained active since 1972, but it involves a much smaller land area than either the Soil Bank or the CRP. The Water Bank Program provides multiple benefits, including flood control, improved water quality, water conservation, reduced siltation and wind erosion, and wildlife habitat. Funding for this program has been limited, however, and its future status and funding is uncertain. Canada, though facing crop surpluses, erosion, and other problems similar to those of the United States, has not enacted land retirement programs similar to the Soil Bank, CRP, or Water Bank.

18.10 Treaties, Tribes, and Wildlife and Fishery Resources

In the United States, the right of indigenous peoples to hunt and fish on their own tribal lands, and to set their own seasons and manage their own wildlife and fishery resources, is generally not challenged by the states. However, off-reservation tribal hunting and fishing rights are often contested. Some aspects of the hunting and fishing rights of indigenous peoples have been highly controversial and are still being litigated; differing interpretations of treaty wording are often the cause of such litigation.

The U.S. government signed numerous treaties with indigenous tribes during the period when territories were being settled. The specifications in these treaties are as legal as those in treaties signed with foreign nations. The treaties conferred many advantages on settlers, military establishments, railroads, and other nonindigenous users of land, and indigenous tribes gave up substantial areas of land in most treaties. In exchange for giving up land, the tribes were sometimes promised, in very specific terms, certain rights to fish and hunt. These treaties often may seem to conflict with the tenet that states have the authority to regulate wildlife and fishery resources within their borders. However, rights guaranteed to tribes by federal treaty supersede states' rights to wildlife and fishery resources.

The rights of indigenous peoples to hunt or fish, especially off reservations, under some treaties have been a source of much controversy. In parts of Wisconsin, for example, spring spearing of walleyes by tribal members on traditional but off-reservation sites has caused much controversy. Similar treaty rights are being exercised in the hunting of large ungulates on some off-reservation sites in states such as Idaho and Washington. The U.S. government attempts to enforce the hunting and fishing rights specified in these treaties because they are legal and lawful agreements.

Indigenous peoples exercising off-reservation fishing rights that are guaranteed by treaty cannot be required to purchase a state license, as their rights are more extensive than those of other citizens of the state (United States v. Winnans, 1905) (Sigler 1995). However, the state may restrict fishing by indigenous peoples to the extent that it is necessary to conserve a particular stock or run of fish (Puyallup Tribe, Inc. v. Washington

Department of Game, 1968). Where a stock of fish is being depleted, the state must first restrict anglers and commercial operations not covered by treaty rights.

Under treaty rights, tribes in the state of Washington may take salmon "at usual and accustomed places." An important District Court decision in the state of Washington in 1975, popularly called the Boldt Decision (United States v. Washington), provided a clear interpretation of the quotas of anadromous fishes allowed to indigenous peoples. Judge Boldt declared that tribes could take up to 50% of the annual harvest at traditional fishing sites (Gilbert and Dodds 1992). The total tribal and nontribal harvest must allow for adequate escapement of spawning fishes for conservation purposes. The Supreme Court upheld the Boldt Decision in Washington v. Washington State Commercial Passenger Fishing Vessel Association (1979), but did stipulate that fishes taken on reservations be counted toward the 50% of the harvest allotted to the tribes (Bean 1983). The Boldt Decision had important implications in the allocation of fishery resources in the northwestern United States, especially Pacific salmon and steelheads.

Reservation authorities have the right to allow or refuse hunting or fishing rights to people who are not tribal members. They can also require and enforce the purchase of a tribal hunting or fishing license by nonmembers. In some cases, state natural resource agencies also require a state hunting or fishing license for people who are not tribal members but who hunt or fish on reservation lands. The matter of state jurisdiction over hunting and fishing on reservation lands by nontribal members is still unresolved in many situations and has been the subject of several court cases. Aspects such as the degree to which a state is involved in fishery or wildlife conservation on the reservation—for example, in the stocking of fish or the management of migratory wild ungulates—are important in this jurisdictional question (Sigler 1995). Jurisdiction over wildlife and fishery resources on nontribal land (owned by nonindigenous peoples) within historical reservation

boundaries has also been controversial in some states.

Most larger reservations now have their own natural resource agencies, sell licenses to nonmembers and members of the tribe, and in many cases employ law enforcement personnel. Some of these tribes, such as the Yakama Indian Nation in Washington, the White Mountain Apaches in Arizona, and the Mescalero Apaches in New Mexico, also employ university-educated biologists to assist in managing their wildlife and fishery resources. Such tribes have had much success in natural resource management. Funds from the sale of licenses, including fees of up to $10,000 for a trophy bull elk, can result in substantial income for the tribe and for further conservation activities.

Many aspects of the hunting and fishing rights of indigenous peoples in the United States still remain unclear. The rights of the states, the tribes, and all citizens involved will eventually be clarified. Unfortunately, much time and money is being expended in litigation over these rights. In the meantime, many cases of cooperation among states, tribes, and the federal government are providing hope for the resource, the average citizen, and the tribes. Wildlife and fishery conservation on tribal lands has improved greatly in the past decade, and new natural resource opportunities for the public are developing. Tribal resources can be an important part of the natural resource base of a state, attracting hunters, anglers, and others, and benefiting the overall economy (fig. 18.5).

In Canada, the rights of indigenous or aboriginal people appear to be broader than those in the United States. Certainly, indigenous people in western Canada have greater rights based on Section 12 of the Resources Transfer Agreement. Basically, indigenous peoples under federal treaty may hunt for food during all seasons of the year on all unoccupied Crown lands, such as the Northern Provincial Forest, and on any lands to which they have "acquired a right of access." The courts have determined that all wildlife management units, wildlife refuges, and parks where an open sport hunting season has been held are open to year-round hunting by indigenous people;

FIGURE 18.5 Some tribal reservations have excellent wildlife and fishery resources and are managing those resources to provide income for the tribe and recreational opportunities for tribal members and nonmembers. Here, a tribal biologist and tribal personnel are conducting a stream survey on the Ft. Belknap Reservation in Montana. *(Photograph courtesy of W. Stancill, U.S. Fish and Wildlife Service.)*

the fact that a season was opened provides the right of access. Private land may also be hunted outside of the regulated game season if the indigenous people have been granted permission by the landowner.

The recent Sparrow decision in Canada has had a major effect on resource use. Basically, the decision states that in the case of federally governed species, such as waterfowl and some fishes, the sport hunting and fishing and commercial fishing rights of nonindigenous people are secondary to the food-gathering rights of indigenous people. If harvest restrictions are required, they are to be applied first to the nonindigenous commercial and sport users.

18.11 Jurisdiction on Federal Lands

As noted earlier, provincial governments in Canada have jurisdiction over most Crown lands.

The Canadian federal government maintains jurisdiction over migratory birds, marine mammals, and all animals within national parks, wildlife refuges, and the Yukon and Northwest territories. It also has jurisdiction over mammals such as caribou that migrate over provincial-territorial boundaries (Gilbert and Dodds 1992).

In the United States, federal holdings, such as national forests and BLM lands, remain under the jurisdiction of the federal government. The states, however, have much control over wildlife and fishery resources on those federal lands (see section 20.4). For example, elk hunters in Idaho or Montana hunt primarily on federal lands. The states are responsible for elk hunting regulations, hunting licenses, and law enforcement, and their jurisdiction over elk has for the most part been unchallenged. Likewise, fishing on most federal lands is regulated by the individual states.

Special federal fees for hunters, anglers, and other recreational users of federally owned lands have been suggested (Thomas 1984). Under such a plan, recreational users of federal lands would pay an access fee. For example, an elk hunter in a national forest might be required to purchase a $10.00 or $20.00 access permit along with his or her state hunting license. Such fees could provide considerable economic support for wildlife and fishery management efforts on federal lands. Such federal fees are generally viewed as a threat to the states' rights doctrine. However, the **Property Clause** of the U.S. Constitution allowed the federal government broad jurisdiction over federal lands. The Property Clause essentially states that the federal government has jurisdiction over its own property. The extent of such jurisdiction over wildlife and fishery resources in relation to state responsibilities has been defined largely by Supreme Court decisions relating to this possible area of controversy (Bean 1983).

Only rarely has the U.S. federal government attempted to unilaterally allow removal of resident wild animals on federal lands. The Secretary of Agriculture allowed the removal by shooting of large numbers of deer in the Kaibab National Forest and the Grand Canyon National Game

Preserve to protect those lands from substantial and long-term damage due to excessive numbers of deer (Hunt v. United States, 1928). The Supreme Court upheld the right of the Secretary of Agriculture to order this deer kill to protect its property.

In New Mexico State Game Commission v. Udall (1969), action was brought against the Secretary of the Interior for allowing the killing of deer for research purposes within Carlsbad Cavern National Park without state authorization (Bean 1983). The Secretary did not claim that actual damage was occurring to the park, but that research was necessary to prevent future damage. The U.S. Court of Appeals ruled that the Secretary had the right to allow removal of deer for research purposes, thus allowing the park to evaluate deer populations and prevent future damage to federal property. New Mexico and several other states were concerned that the deer removals could signal a federal attempt to usurp state responsibility for management of resident species on federal lands. States should have been particularly heartened by the Court of Appeals statement that the United States has no ownership of wild animals within the various states.

Feral burros and horses are described as valued and natural components of the ecosystem under the Wild and Free-Roaming Horses and Burros Act of 1971 (United States). In 1974, a New Mexico rancher with a grazing permit on BLM land complained to the state Livestock Board about the presence of feral burros. The BLM made it clear that neither the Livestock Board nor the rancher could legally remove the burros. Nevertheless, the Livestock Board captured nineteen of the burros and sold them at public auction. In this case (Kleppe v. New Mexico, 1976), the Supreme Court ruled against the rancher and the Livestock Board and required restitution for the removal of the burros from federal land. In doing so, the Supreme Court emphasized that the Wild and Free-Roaming Horses and Burros Act and its decision in this case were legal exercises of congressional power under the Property Clause. Additional court cases and decisions are expected in the future relating to state and federal authority over wild animals on federal lands.

SUMMARY

A strong tradition of legislation relating directly to wildlife and fishery resources exists in the United States and Canada at the federal, state, provincial, and territorial levels. Wildlife and fishery legislation in North America reflects the traditions and democratic natures of the people that colonized the continent from Europe. In the United States, jurisdiction over wildlife and fisheries in the early years clearly fell to the states. Federal wildlife and fishery legislation in the United States began with the Lacey Act in 1900, in which the federal government exercised its constitutional power to regulate interstate commerce. Federal wildlife and fishery legislation in the United States has also resulted from the government's constitutional power to make treaties with foreign nations and its right to protect its own property (Property Clause).

In both Canada and the United States, the need to manage migratory birds, especially waterfowl, on a national and international basis was recognized in the early 1900s. The federal governments of both countries established such jurisdiction by signing a migratory bird treaty.

The Federal Aid in Wildlife Restoration Act of 1937 and the Federal Aid in Sport Fish Restoration Act of 1950 in the United States provided funding to states for many activities important to wildlife and fishery resources. Funds are collected by the federal government from taxes on firearms, ammunition, fishing gear, and other selected sporting goods used by hunters and anglers. A state can receive such funds only if state hunting and fishing license fees remain with the state natural resource agency for reinvestment in wildlife and fishery resources.

Environmental legislation in the United States, such as the National Environmental Policy Act, the Clean Water Act, the Federal Environmental Pesticide Control Act, and the Toxic Substances Control Act, has been important in terms of the welfare of wildlife and fishery resources. This legislation also greatly affects public health, safety, and welfare. Other legislation directed at U.S. federal lands, such as the two multiple use acts, the National Forest Management Act and the Federal Land Policy and Management Act, provides a legal basis for the conservation of wildlife and fishery resources on federal lands. Wildlife and fishery resources in the United States have also benefited from U.S. Department of Agriculture programs designed to retire cropland, reduce crop surpluses, and conserve soil. Programs such as the Soil Bank and Conservation Reserve Program provided for extensive habitat development on private lands.

Government treaties with indigenous peoples in North America often granted specific rights to harvest wildlife and fishery resources in exchange for advantages given to nonindigenous peoples. In the United States, these treaty-given rights generally supersede the right of a state to manage wildlife and fisheries within its borders. Many treaties grant rights that extend to usual and accustomed hunting and fishing areas outside reservation boundaries; these rights have been the subject of controversy. In Canada, indigenous people have broad treaty rights to gather food by hunting and fishing on public lands throughout the year.

Most Crown lands in Canada are under provincial control, while most federal lands in the United States are under federal jurisdiction. However, the states have traditionally been responsible for wildlife and fishery resources on most of those federal lands, while federal agencies manage the habitat on such lands.

Legislation, jurisdiction, and legal precedents have strong influences on wildlife and fishery resources. Many parties interested in wildlife and fisheries can affect the legislative process.

PRACTICE QUESTIONS

1. What is the "states' right doctrine" as it relates to wildlife and fishery jurisdiction?

2. Why is the Lacey Act important? What are its essential features?

3. How did the U.S. and Canadian federal governments successfully gain jurisdiction over migratory birds? Why might some states, provinces, and territories have opposed federal jurisdiction over migratory birds?

4. Why is the Migratory Bird Hunting Stamp Act (1934) so important to waterfowl conservation efforts in the United States?

5. What is meant by the "carrot-and-stick" approach as it relates to the Pittman-Robertson and Dingell-Johnson legislation?

6. Why is NEPA (1969) important to wildlife and fishery resources in the United States?

7. What is the legal basis for multiple use management on national forest lands? On BLM lands? How are the National Forest Management Act and the Federal Land Policy and Management Act related to earlier legislation on multiple use?

8. Describe the swampbuster and sodbuster provisions of the 1985 Food Security Act and their importance to wildlife and fishery resources.

9. What were the three primary goals of the initial Conservation Reserve Program?

10. What is the legal basis for allowing off-reservation hunting or fishing rights to some tribes?

11. Who has jurisdiction over most Crown lands in Canada? Compare this situation with jurisdiction over federal lands in the United States.

SELECTED READINGS

Bean, M. J. 1983. *The evolution of national wildlife law.* Praeger, New York.

Gilbert, F. F., and D. G. Dodds. 1992. *The philosophy and practice of wildlife management.* 2d ed. Krieger, Malabar, Fla.

Lund, T. A. 1980. *American wildlife law.* University of California Press, Berkeley.

Sigler, W. F. 1995. *Wildlife law enforcement.* 4th ed. Wm. C. Brown, Dubuque, Iowa.

LITERATURE CITED

Baldassarre, G. A., and E. G. Bolen. 1994. *Waterfowl ecology and management.* John Wiley & Sons, New York.

Bean, M. J. 1983. *The evolution of national wildlife law.* Praeger, New York.

Gilbert, F. F., and D. G. Dodds. 1992. *The philosophy and practice of wildlife management.* 2d ed. Krieger, Malabar, Fla.

Grue, C. E., M. W. Tome, G. A. Swanson, S. M. Borthwick, and L. R. DeWeese. 1988. Agricultural chemicals and the quality of prairie-pothole wetlands for adult and juvenile waterfowl: What are the concerns? Pages 55–64 *in* P. J. Stuber, coord. *Proceedings of the National Symposium on Protection of Wetlands from Agricultural Impacts.* Biological Report 88 (16). U.S. Fish and Wildlife Service, Washington, D.C.

Leopold, A. 1933. *Game management.* Scribners, New York.

Redmond, L. C. 1986. The history and development of warmwater fish harvest regulations. Pages 186–195 *in* G. E. Hall and M. J. Van Den Avyle, eds. *Reservoir fisheries management: Strategies for the 80s.* Reservoir Committee, Southern Division, American Fisheries Society, Bethesda, Md.

Sigler, W. F. 1995. *Wildlife law enforcement.* 4th ed. Wm. C. Brown, Dubuque, Iowa.

Thomas, J. W. 1984. Fee-hunting on the public's lands? An appraisal. *Transactions of the Forty-ninth North American Wildlife and Natural Resources Conference* 49:455–468.

19

WILDLIFE AND FISHERY LAW ENFORCEMENT

Laws and regulations, as discussed in chapters 11, 17, and 18, are important tools in wildlife and fisheries management; they are a facet of conservation of natural resources directed at human users. Such laws and regulations have little value, however, unless they are enforced by law enforcement personnel as well as in the court system. Some enforcement involves species taken for consumptive purposes, such as by hunters or anglers. Other enforcement involves species that are nonconsumptively important, as in the case of laws and regulations concerning nongame or endangered and threatened species. Enforcement may also involve laws and regulations relating to habitat, such as those associated with wetland conservation, clean water, or timber harvest. While a particular law or regulation may relate to any of the above areas, the procedures for its enforcement are usually quite similar.

A substantial number of wildlife and fishery professionals spend part or all of their time on law enforcement. The importance of law enforcement in the total picture of natural resource conservation is sometimes given insufficient attention by other wildlife and fishery professionals and by the public. This chapter is intended to place these activities in perspective with regard to their position in natural resource conservation. Other purposes of this chapter are to discuss some basic concepts of wildlife and fishery law enforcement and to provide insights into the nature of this work.

19.1 The Need for Law Enforcement in Wildlife and Fisheries

Wildlife and fishery professionals may be involved in different facets of research, management, education, administration, or any combination of these areas; they may also be involved in law enforcement. People with expertise in all of these areas are needed in the total effort to conserve wildlife and fishery resources.

Wildlife, fish, and other wild organisms are managed in part through **laws** and **regulations**. By dictionary definition, laws and regulations are essentially the same. However, in wildlife and fisheries, as in many other areas, the terms law and regulation often have different meanings. Laws can be defined as rules enacted by an elected governing authority, such as a federal, state, provincial, territorial, or local government. Regulations can be defined as rules made by nonelected government units, such as state or provincial wildlife and fishery commissions or boards, or by various units of a federal government, such as the Department of the Interior or the Department of Agriculture. Planning boards, park boards, and other local nonelected entities may also have regulatory responsibilities. The term **code** will not be used in this chapter, although it is often used interchangeably with the terms law and regulation. Both laws and regulations are enforceable under the court system.

While a law may be broad in scope, regulations are usually quite specific in their application. Regulations are often needed to effectively implement a law. For example, a federal law might require the U.S. Forest Service to implement a management plan in a particular area. The U.S. Forest Service would then need to promulgate regulations to optimize activities such as timber harvest, hiking, camping, hunting, rock climbing, or fishing in that area in order to implement the plan. Such regulations could relate to consumptive, nonconsumptive, direct, and indirect resource use activities.

It is appropriate here to provide some specific examples of the potential scope of law and regulation enforcement. In the case of golden and bald eagles, U.S. federal laws (Bald Eagle Protection Act and Migratory Bird Treaty Act) state that these species may not be taken by any means, with the possible exception of certain religious and ceremonial uses by indigenous peoples. Eagle feathers are

FIGURE 19.1   Eagle feathers, as in this Indian headdress and dance bustle (worn on the lower back), have a substantial value to collectors, but their possession is legal only for approved religious purposes by indigenous peoples in North America. (*Photograph courtesy of South Dakota Tourism.*)

illegal to possess, even if one is found lying on the ground as a result of molting. Because of their monetary value, eagle feathers and eagle talons, as well as headdresses, necklaces, and other items made from these parts, may be sold on the black market to collectors. Without law enforcement, the laws protecting eagles would be meaningless in the face of the market value of their feathers and other parts (fig. 19.1). In specific situations regulations may also control the distance from which people can view nesting bald or golden eagles at a particular time and place, or when and what types of motorized vehicles or boats can be used in an area where eagles are nesting. For any

of these rules to be effective in protecting the eagles, enforcement must occur.

In many trout streams in North America, Europe, New Zealand, and other countries, very specific regulations have been set to avoid depletion of fisheries in terms of both abundance and quality. For example, in a portion of the Snake River below Jackson Lake in Wyoming, there are specific regulations for cutthroat trout, allowing only catch-and-release fishing. The biological and social reasons for such regulations were discussed in chapter 17. Without enforcement efforts, however, some people fishing these waters would keep substantial numbers of cutthroat trout, and the social and biological benefits of these regulations would be negated. The same can be said for all regulations that help to conserve wild organisms.

Much of what administrators, research biologists, managers, and those involved in the rule-making process attempt to accomplish would be negated without law enforcement. Laws and regulations are ineffective unless a significant threat of enforcement exists. Thus, law enforcement is a vital and necessary part of natural resource conservation.

19.2 Types of Enforcement Officers and Their Training

Most of the discussion of the enforcement profession in this text is specific to wildlife and fishery personnel in the United States at the federal and state levels. However, most of the points discussed in this chapter also apply to wildlife and fishery enforcement personnel in other nations that have such enforcement programs. The enforcement of international agreements restricting trade in protected species or their parts requires a cooperative effort among nations.

A variety of personnel, such as park rangers, foresters, customs agents, and many others, can have enforcement responsibilities that involve fishery and wildlife resources; however, in this text we will limit our discussion to those people whose enforcement responsibilities are tied primarily to wildlife and fisheries. Wildlife and fishery enforcement personnel vary considerably in their education, training, interests, and responsibilities depending upon the natural resource agency for which they work. There is no standard educational requirement that fits all such agencies, especially with regard to state personnel. In most states, however, wildlife and fishery enforcement personnel have university educations with at least a four-year degree in wildlife and fisheries or a closely related area.

Personnel involved in wildlife and fishery law enforcement with state, provincial, or territorial natural resource agencies have a wide range of titles and responsibilities depending on the agency involved. **Conservation officer** is a title often used for personnel in these agencies who have wildlife and fishery law enforcement responsibilities in addition to many other duties. The title **game and fish warden** is sometimes used for resource personnel with strictly enforcement responsibilities. Other titles for positions with wildlife and fishery enforcement responsibilities include fish and wildlife officer, wildlife conservation officer, conservation enforcement officer, conservation ranger, natural resource officer, and renewable resource officer. For these positions, some natural resource agencies require only an interest in law enforcement with no specific academic training in the biological sciences; a bachelors' degree from a four-year college or university is, however, frequently required. In natural resource agencies where a college degree is not required, individuals with a college degree in wildlife and fisheries or a related area are generally more successful in the competition for available positions. Persons with little or no education in wildlife and fisheries or the biological sciences are less likely to understand the biological rationales for regulations and law enforcement efforts. Moreover, even if their responsibilities are strictly related to law enforcement, their ability to interact effectively with other wildlife and fishery professionals and

the public can be reduced if they possess little biological knowledge.

Wildlife and fishery enforcement personnel with the U.S. Fish and Wildlife Service are called **special agents** or, at a higher level of experience and responsibility, **senior resident agents**. Their education is highly variable, but a college degree is required. Degrees in criminology, criminal justice, police science, and wildlife and fisheries are all appropriate for these federal positions. Some education in wildlife and fisheries is highly recommended. Special agents and senior resident agents with degrees in subjects closely related to law enforcement, who also have a strong interest in wildlife and fishery work, have made highly effective professionals. Canada does not have a federal wildlife and fishery enforcement branch similar to that in the United States.

Increasingly, Indian reservations have their own natural resource agencies, which serve a broad public and provide employment opportunities for wildlife and fishery law enforcement personnel. Law enforcement personnel for natural resource agencies on reservations are usually titled conservation officers or fish and game wardens, and their responsibilities are similar to those of nontribal wildlife and fishery enforcement personnel.

Persons seeking careers in wildlife and fishery law enforcement need to develop effective communication skills and be able to interact with people positively. Conservation officers and other personnel involved in wildlife and fishery law enforcement are often the primary contact between the public and a natural resource agency (fig. 19.2). Thus, the demeanor and communication skills of these law enforcement professionals greatly influence the attitude of the public toward natural resources, and natural resource agencies and their personnel.

Individuals entering positions with wildlife and fishery law enforcement responsibilities are often required to obtain considerable additional training at special law enforcement schools. This training is generally broad, covering areas such as the rights of individuals, search and seizure, types of violations, interrogation, evidence, record

FIGURE 19.2 Contact with landowners, conservation groups, and other members of the public is an important part of the work of a conservation officer. The impression left by this conservation officer may influence the attitude of this landowner toward wildlife and fisheries professionals, the employing agency, and natural resources. (*Photograph courtesy of the Idaho Department of Fish and Game.*)

keeping, use of firearms, court procedures, and other legal and procedural matters that often confront law enforcement personnel. Such law enforcement training is extremely important. Many new officers have four or five years of college, but little law enforcement experience or knowledge. Enforcement is the area in which mistakes can be most costly to the individual and the natural resource agency. Poor preparation and training can lead to civil lawsuits against the agency and the officer or even to the officer's injury or death.

Most new nonfederal natural resource personnel with enforcement responsibilities attend a law enforcement training program along with new officers from other enforcement agencies, such as police and sheriff's departments. In addition, semiannual training and recertification in the use of firearms, defensive tactics, and legal issues is often required. Resource personnel with law enforcement responsibilities also receive training in cardiopulmonary resuscitation techniques and other emergency medical procedures because these people are often the first to respond to an emergency situation.

Enforcement personnel with the U.S. Fish and Wildlife Service attend a thirteen-week training course at the Federal Law Enforcement Training Center in Glynco, Georgia. After completion of this training, new officers are assigned to a senior resident agent for one year of on-the-job training. After that experience, these officers qualify for a single-person duty station as a special agent. Special agents and senior resident agents with the U.S. Fish and Wildlife Service are required to qualify twice annually in the use of firearms. Also, these agents annually receive forty-eight hours of in-service training that includes legal updates on search and seizure, evidence handling, forensics, intelligence on wildlife and fishery violation trends, physical training, firearms training, and policy.

Enforcement of wildlife and fishery laws and regulations can be dangerous for an officer; violators cause death or injury to wildlife and fishery officers in North America every year. U.S. Department of the Interior (1990) statistics for 1989 indicated that 1.7% of state conservation officers were assaulted during that year; in Canada, the assault rate on conservation officers was 3.6%. Federal personnel involved in wildlife and fishery law enforcement in the United States were assaulted at a 1.5% rate in 1989. Over half of the persons who assaulted wildlife and fishery officers in the United States and Canada in 1989 were under the influence of alcohol or a controlled substance. In the United States and Canada respectively, 42% and 62% of the people who assaulted these officers had a previous record of arrest and conviction.

For persons planning a career in wildlife and fishery law enforcement, additional education in criminal justice, criminology, and law enforcement are all desirable and can provide a competitive edge. Experience with law enforcement at a city, state, campus, or military level can also be valuable. For example, an undergraduate student in this field might want to complete a degree in wildlife and fishery sciences with a minor in criminal justice. While working on the degree, a few semesters of employment as a campus police or security officer is usually possible. During the summer the student might work as a park ranger in a state, provincial, territorial, or federal park where the duties involve some enforcement responsibilities. The primary ingredient in starting such a program is an interest in wildlife and fishery law enforcement as a career.

19.3 Types of Violators

Violators of wildlife and fishery laws and regulations vary greatly in terms of their intent and the degree to which they are conscientious and law-abiding. Three basic categories of violators are recognized by wildlife and fishery law enforcement personnel: the **accidental**, the **opportunist**, and the **intentional violator**.

The accidental violator is a person that law enforcement personnel generally dislike citing for a violation. For example, the person who attempts to fill an antlerless white-tailed deer license, but accidentally shoots a buck with very small antlers, is an accidental violator. Other examples might be an angler who misreads the regulations and is fishing with bait in an area restricted to fly fishing, or an out-of-state hiker who did not know that campfires are allowed only in specific locations. Officers must at times write such citations despite their great reluctance to do so. A law enforcement officer related the following example of a particularly embarrassing citation that had to be written. This incident occurred in a state where plumage or spurs were required to be left attached to male ring-necked pheasants while in transit. Many citations had been given at a particular roadside checking station for hunters that day, and several cited individuals were standing nearby when two Catholic nuns voluntarily stopped to show the officer two fully dressed ring-necked pheasants that had been given to them. The officer cited these accidental violators.

Opportunist violators are people who had no real intention of breaking a law or regulation,

but the opportunity to do so availed itself and they could not resist. Examples might be a hiker who comes across a fossil bone in a national park and tucks it into his or her backpack, or a wildflower enthusiast who happens to find a particularly beautiful example of a protected plant and collects it. Another example might be a quail hunter who is having too much success. This person shoots a bag limit of birds, but on the way back to his or her vehicle cannot resist shooting an additional bird when it flushes. These violations were not planned, but the opportunity was too much for them to resist. Perhaps another person has caught only a few largemouth bass in several outings and then really gets into good bass fishing. This angler replaces a smaller fish in the boat's live well with a newly caught, larger largemouth bass—a practice called highgrading, which is sometimes illegal. The smaller largemouth bass is alive but much weakened and when released has a poorer chance for survival than would the newly caught fish. Where highgrading is illegal, this angler has become an opportunist violator. Many people have at some time, either accidentally or because of opportunity, violated wildlife or fishery laws and regulations.

The intentional violator starts a particular activity with the intention of breaking the law. Someone who sets illegal gill nets for steelheads or paddlefish, someone who poisons golden eagles, someone who shoots white-tailed deer by night spotlighting for sale through a black market, or someone who captures a protected foreign species for import purposes is among the ranks of intentional violators. Intentional violators, especially those seeking economic returns from illegal natural resource exploitation, are a special problem for wildlife and fisheries law enforcement personnel. Such persons may be dangerous, and they are likely to be very difficult to reform. They are involved in serious criminal activity and should not be considered normal resource users; they are criminals in much the same way a person robbing a bank or burglarizing a home is a criminal. In addition, they are often involved in some other illegal activity. For example, one group involved in the smuggling of illegally taken walrus tusks in Alaska was found to be involved in drug smuggling as well; this group was considered to be very dangerous (see section 19.8). A little educational talk or even the embarrassment of apprehension is not likely to deter such people or permanently change their behavior.

A particularly common example of an intentional violator is the person who purchases a big game license for a nonhunting member of their family and then fills that license themselves. These people are not usually dangerous and may change their behavior after being cited for a violation or confronted by social pressure from friends that avoid such a practice. Such persons are often honest in their dealings with others and may have very high moral values, but for some reason believe that breaking certain wildlife and fishery laws is not morally wrong. This type of violation is planned, is often repeated, and occurs for a variety of harvested species. Section 16.4 contains a discussion of human ethics and values.

Some individuals intending to harvest, capture, or kill wildlife, fishes, or other wild organisms illegally for their own personal use can be dangerous and can represent a serious source of depletion of wildlife and fishery resources. Law enforcement officers must remain alert and cautious in approaching or apprehending any violator.

▼

19.4 Wildlife and Fishery Violations

Violations of wildlife and fishery laws and regulations can be placed in nine major categories, as illustrated in table 19.1. In addition to these violations, there other subcategories of violations with which natural resource law enforcement officers need to be familiar. The major classes of violations have generally been tested and upheld by the U.S. Supreme Court, as reviewed by Sigler (1995).

A person who shoots an eastern cottontail out of season in the United States has broken a state wildlife regulation. Likewise, a person hunting waterfowl in the United States who forgets to sign the "duck stamp" on their license has broken a federal wildlife regulation. Actions such as these are illegal because they violate laws and regulations meant to assist in the conservation of terrestrial and aquatic organisms. Other violators break these laws or regulations but also transgress moral principles. A person who legally shoots five snow geese and then decides to dispose of them in a road ditch because they are too difficult to dress has violated a law or regulation and has also transgressed standards of acceptable human conduct. Such acts of wanton waste are considered morally wrong in most societies regardless of their legal status in terms of actual laws and regulations. This distinction is important in considering the rehabilitation of violators because it is much more difficult to alter the behavior of those individuals or groups who violate the basic laws of human conduct.

Violations can be categorized as **felonies**, **high-grade misdemeanors** (sometimes called **class I misdemeanors**), and **low-grade misdemeanors** (sometimes called **class II misdemeanors**). In some situations additional classes of misdemeanors may be recognized; to avoid confusion, the terms high-grade misdemeanor and low-grade misdemeanor will be used in this book.

The seriousness of violations and their classification as felonies, high-grade misdemeanors, or low-grade misdemeanors varies greatly. A felony is a grave crime, ranging from burglary to robbery or even murder. An example of a felony natural resources violation in some states would be a second conviction for killing a deer during a closed season. Such a violation is not at all equivalent in seriousness to murder, but it is similar to a burglary or robbery in that these animals are being taken from the state and its people. Felony convictions are most common for violators breaking federal laws that protect wildlife and fishery resources. The smuggling of endangered species or their parts, for example, would generally be a

TABLE 19.1 Primary types of wildlife and fishery law and regulation violations.

Type of violation[a]	Examples or further description
Taking or attempting to take animals out of season	Hunting, fishing, or otherwise taking species of wild animals when the season is closed, including after-hours violations
Taking or attempting to take animals or plants in an illegal area	Includes any area closed to hunting, fishing, or taking of the species sought; posted private land for which the violator does not have permission is included here
Taking or attempting to take animals with an improper license or no license	Illegal purchase of resident license by nonresident; unsigned hunting stamp for waterfowl or other species requiring a signed stamp; no license
Taking or attempting to take animals by illegal methods	Use of a shotgun for waterfowl hunting that holds more than three shells; use of a small-caliber rifle (such as a .22-caliber) to hunt large ungulates; use of worms on a stream limited to artificial lures; use of gill nets to capture fish
Illegally possessing wild animals or plants or their parts	Exceeding possession limits on legally taken game or fish (a person with too many trout or ducks in the freezer would fit this description); possessing game, fish, or other species illegally even if it cannot be shown that the defendant actually killed or captured the animal
Following illegal procedures	Failure to properly tag a legally taken animal requiring tagging before transportation (most natural resource agencies require that a signed and dated tag be placed on large ungulates, usually around the leg, before placement in a vehicle or building or when the animal is brought into camp)
Illegally importing or exporting wild animals or plants or their parts	Many species are protected, but some, such as endangered or threatened species, are of particular concern; importing species that could cause agricultural or ecological damage without a permit is also restricted
Taking, possessing, or endangering nongame or endangered species	Killing, capturing, or possessing nongame vertebrates such as yellow warblers; many other organisms such as mollusks, butterflies, and plants are also included. Even harassment or destruction of habitat for endangered species could come under this category
Selling or attempting to sell wildlife, fishes, or other wild animals or plants in violation of state, provincial, territorial, or federal law	Sale of the parts or whole bodies of migratory bird species in North America is illegal with few exceptions; many other wild species are also protected from sale

Source: Adapted from Sigler 1995.
[a] A single violation may fit into more than one category.

felony; this is a federal offense. Felony convictions fall into various classes ranging from the most serious to those just somewhat more serious than a high-grade misdemeanor. Felony convictions for wildlife and fishery violations are relatively uncommon.

Misdemeanor means "misbehavior" or "misdeed" and refers to violations that are less serious than felonies. In some states a first offense for killing a deer or elk out of season would be a high-grade misdemeanor. Violations such as the lack of an appropriate hunting or fishing license, fishing with bait in an area open only to artificial lures, or illegally killing a female ring-necked pheasant during the pheasant season would be low-grade misdemeanors in most states. In wildlife and fishery law enforcement, most violations are low-grade misdemeanors.

19.5 Individual Rights

Law enforcement personnel in wildlife and fisheries must be aware of the legal rights of the individual. These fundamental rights are contained in the U.S. Constitution for specific reasons related to the democratic nature of our government. The courts have decided on the interpretation of the U.S. Constitution in various cases brought before them. A natural resource law enforcement officer must be aware of the rights of individuals as interpreted in certain important court cases. These cases are reviewed in many general law enforcement and criminology texts (Siegel 1992, Inciardi 1993, Sigler 1995). It is important for officers to remember that persons observed or suspected of committing a violation are not guilty of violating a law until convicted in court.

Every law enforcement officer in the United States, including those in wildlife and fisheries, should be aware of the protection provided to the individual by the Fourth Amendment to the U.S. Constitution. The Fourth Amendment protects individuals from search and seizure without probable cause. The specific wording of this amendment is as follows: "The right of the people to be secure in their persons, houses, papers, and effects, against unreasonable searches and seizures, shall not be violated, and no warrants shall issue, but upon probable cause, supported by oath or affirmation, and particularly describing the place to be searched, and the persons or things to be seized." Wildlife and fishery law enforcement officers must be aware of these rights as they have been defined by the courts and cognizant of exceptions to search and seizure protection. Violation of the Fourth Amendment rights of a person may result in loss of evidence because of the **exclusionary rule**, which states that evidence obtained through violation of the Fourth Amendment rights of a person cannot be used in court. In addition, the officer may be held liable and sued in civil court by the suspected offender for the violation of his or her rights. The exclusionary rule is important in deterring illegal conduct by law enforcement officers in the course of gathering evidence.

An **arrest** is generally considered to be the taking of a person into custody by a legal authority. During an arrest an officer can search the immediate person of the arrestee and the possessions within immediate control (**grabbing distance**) of that person without a search warrant. This ruling gives the officer access to items that might otherwise be destroyed or that might be a danger to the officer at the time of the arrest.

The search of houses, vehicles, and businesses is limited under the Fourth Amendment. Any searches of premises or belongings beyond the immediate reach of the arrestee should be made only with a search warrant. Search by consent is allowable in some situations, but the courts prefer the use of a search warrant. In many cases officers now tape conversations with suspected violators for possible use in court in case search by consent is allowed.

Vehicles present a special case in terms of search and seizure because of their mobility. In one important decision (Carroll v. United States,

1925), the U.S. Supreme Court upheld the actions of law enforcement officers who stopped and searched a vehicle based on its being heavily loaded and driven by two persons suspected (based on undercover work) of smuggling liquor. Liquor was also visible in the passenger compartment when the officers approached the stopped vehicle. Based on the Carroll decision, a vehicle can be stopped and searched without a search warrant if there is **probable cause** to believe that a crime has been committed. If enforcement personnel have facts and circumstances within their knowledge that would cause a person of reasonable caution to believe that wrongdoing has occurred or is occurring, then probable cause exists. A wildlife and fishery example of probable cause would be the stopping of a particular vehicle at 2:00 A.M. based on a vehicle description given by a landowner reporting a mule deer **poaching** (taking of wild animals illegally) incident earlier in the evening. Another example might be the stopping of a vehicle containing anglers who had bragged of overbagging American shad to a service station attendant, who then reported the vehicle's license number, description, and other information to appropriate authorities. In both cases, the officers involved would have probable cause to believe that a violation had occurred and that seizable evidence was in the vehicle. Upon stopping these vehicles, however, the officer might still need to observe additional visible evidence on or in the vehicle that fit the suspected violation. Such additional evidence might include visible rifles, shells, or a spotlight in the passenger compartment in the mule deer poaching incident, or fishing equipment in the American shad case. Probable cause must be based on more than just suspicion of the officer.

If an officer approached two deer hunters near their vehicle in an area closed to hunting and suspected that the trunk of the vehicle contained evidence of a violation, the officer should request permission, in a friendly and nonthreatening manner, to open and search the trunk space. If the evidence were limited to suspicion only, the officer might be restricted to checking licenses if the hunters would not voluntarily allow a vehicle search. If the officer

had no evidence at the start of the investigation, but observed blood or wild animal hair on the vehicle outside the trunk, that newly discovered evidence would serve as probable cause for a search. If the persons involved would not open the trunk even though evidence at the vehicle or testimony from other people provided probable cause, the vehicle could be impounded and a search warrant obtained, or the trunk could be forcibly opened.

Under the Fifth Amendment to the U.S. Constitution, a person has the right to remain silent if his or her testimony might be self-incriminating. The Fifth Amendment reads as follows: "No person shall be held to answer for a capital, or otherwise infamous crime, unless on a presentment or indictment of a Grand Jury, except in cases arising in the land or naval forces, or in the militia, when in actual service in time of war or public danger; nor shall any person be subject for the same offence to be twice put in jeopardy of life or limb; nor shall be compelled in any criminal case to be a witness against himself, nor be deprived of life, liberty, or property, without due process of law; nor shall private property be taken for public use, without just compensation." Because of a portion of the Fifth Amendment, a series of rights must be read by an arresting officer to the arrestee in custody. This **Miranda warning** is required whenever a person is taken into custody or is detained by an officer to a point at which the person may reasonably believe that he or she is in custody. Persons in custody, prior to questioning, must be informed that they have the right to remain silent, that any statement they make could be used as evidence against them in a court of law, and that they have the right to the presence of an attorney (the right to counsel is covered under the Sixth Amendment). The attorney can be retained by the defendant or appointed by the state. The defendant can waive these rights if desired. This procedure resulted from the case of Miranda v. Arizona (1966), in which the confession of Mr. Miranda could not be used as evidence because he was not aware of his right to avoid self-incrimination.

19.6 Arrests

Perhaps the first action that an officer should take when approaching a suspected violator is to clearly identify himself or herself as a law enforcement officer. Most violations in wildlife and fisheries are handled on a **citation** basis. The citation directs the violator to appear before a **magistrate** (a judge having jurisdiction over minor cases) at some later date or, in some cases, simply pay a set fine for the violation if such is agreeable. Many states consider a citation a type of arrest. After writing a citation, however, the officer can direct his or her attention to other responsibilities without having to take the violator directly to a magistrate.

If the officer observes a violation directly, he or she can arrest the violator without a warrant. Also, if there is probable cause to suspect that a felony or high-grade misdemeanor has been committed, the officer can arrest without a warrant.

In a warrant arrest, the officer obtains or is given a warrant from a legal magistrate. The warrant may be the result of a complaint filed by the officer or by other people who identified the individual and the violation. Thus, if the officer has good evidence that a person has committed a violation, an arrest can be accomplished through a warrant. The magistrate providing the warrant will require witnesses or other evidence to indicate probable cause. Evidence is usually provided in the form of an affidavit that lists probable cause. Once the violator is located, the officer must present the warrant and then make the arrest.

Good judgment is required with regard to the intentions and character of the individual to be arrested (including citations). Because officers must be acutely aware of dangers to their own lives, they will in most instances need to be able to make some type of character judgment. A night-time arrest of a person or persons poaching deer or elk is a dangerous situation that must be approached with particular care. Nighttime arrests of armed violators are particularly hazardous. If the officer is alone, it may even be necessary to hold the suspects under spotlight and loudspeaker while calling for backup support from other officers. At a minimum, the officer should provide information on the location, suspect vehicle license number, and other information associated with the arrest by radio prior to approaching the vehicle.

The arrest of someone approached for fishing without a license, cutting wood where they are not supposed to, or illegally digging wildflowers is normally a safe arrest. To begin with, no weapons are usually involved. However, even these situations can turn hostile. In a tragic situation that occurred recently in the southeastern United States, a conservation officer approached a man to check his fishing license and catch. The officer cited the angler for fishing without a license. As the officer turned to walk away after writing the citation, the man shot and killed the officer with a small-caliber rifle. The officer may not have even suspected a violation of fishing laws when he first approached this person; he certainly did not suspect any danger to his own life. Thus, even in a situation that appears safe, the officer must remain alert. The officer cannot use deadly force on a person resisting arrest for a misdemeanor unless the life of the officer is threatened.

In writing citations or making other arrests, the wildlife and fishery law enforcement officer needs to use good judgment. That good judgment comes from having common sense, being alert to all situations, knowing how to handle potentially dangerous situations in the safest manner possible, knowing the legal rights of individuals, and understanding the laws and regulations relating to wildlife and fishery resources. The wildlife and fishery law enforcement officer, in working with people who have violated the law, is repeatedly faced with decisions about whether or not to write a citation for various low-grade misdemeanors. The effective officer is able to make decisions regarding the issuance of citations with fairness and with a degree of leniency appropriate to the situation.

19.7 Evidence

A law enforcement officer investigating infractions or reports of infractions of wildlife and fishery laws or regulations faces a wide variety of situations in terms of gathering evidence. Direct encounters with most violators lead to the writing of a citation, which often requires no court appearance or additional evidence. Even for such encounters, however, the officer needs to maintain an accurate written record of the occurrence. Normally these records are adequately covered in the citation itself, but the officer may want to keep personal records containing other information about the citation. The date, time, nature of the infraction, and a description of vehicle and site, as well as other general information, should be recorded. Citation situations could still go to court, in which case accurate records will be needed. If a case reaches a judicial court, the officer may be carefully cross-examined as to the details of the incident; any written record would be valuable in this situation, especially because the time between the infraction and the court case can be lengthy.

In many cases, the infraction is not directly observed by an officer. Testimony from others, evidence from the site of the violation, or even information about a vehicle involved in transportation of the violators may be critical in these circumstances. The officer may visit the site of the observed violation to locate direct evidence. Photographs of the site should be taken for use in court if necessary. Animal hair, body parts, blood, spent shells, soil samples, prints from boots, tire tread prints, and other evidence may be collected. If possible, a cast or photograph of any tire tread and boot or shoe prints found at the site should be taken (fig. 19.3).

Even observations of vegetation may prove worthwhile. For example, suppose that law enforcement officers receive information that a trophy white-tailed deer buck that is on display as a

FIGURE 19.3 Photographs or casts of tire tracks and footprints from the site where a white-tailed deer was poached could serve as important evidence in court.

taxidermy head and shoulder mount was illegally taken in an area quite distant from the area for which the hunter held a permit. The vegetation in the suspected area of the violation and the area for which the hunter had a hunting permit may differ greatly. Conservation officers might discover seeds or other plant parts in the hair of the mounted head that could not have come from the claimed kill area; this evidence could result in arrest and conviction.

The officer, if successful in locating a suspect vehicle after it has been driven from the violation site, will first look for evidence on the outside of the vehicle or in the visible areas of the passenger compartment. Tire tracks from the suspect vehicle can be compared with those found at the violation site. If boot or shoe prints are found near the suspect vehicle, these too can be compared with those from the violation site. In all of these activities the officer must maintain thorough written records; photographs and physical evidence may also be useful.

If a case goes to court, the time lapse between the infraction and the court appearance can be many months. For this reason, wildlife and fishery law enforcement officers should be concerned about how they label and store evidence. Evidence must be stored in an organized fashion, and it must also be protected so that it will not be destroyed by accident or be accessible to unauthorized individuals. Evidence should be carefully marked by the officer at the time of receipt. It is recommended that specific marks be made, such as by removing a primary feather from a bird wing, clipping a fish fin, or making a file mark on a cartridge, to identify individual items of evidence in addition to a written tag. Small items of evidence can be sealed in an envelope and signed by a witness and the officer.

Physical evidence that can decompose should be placed in a locked freezer with access limited to authorized individuals. Freezers containing evidence should be checked periodically to guard against accidental defrosting and loss of evidence. Evidence that cannot be preserved should be carefully documented using photographs and written records. Any transfer of evidence should be carefully recorded on a receipt indicating the reason for the transfer as well as the date and time. If it can be shown that unauthorized individuals have had access to physical evidence in a case, the evidence may be challenged and dismissed in court.

The U.S. Fish and Wildlife Service Forensics Laboratory in Ashland, Oregon, provides assistance to wildlife and fishery agencies in the United States and throughout the world. The more than one hundred nations that have signed the Convention on International Trade in Endangered Species (CITES) can use this laboratory in endangered species investigations. Personnel at the laboratory have expertise in species identification, determination of the cause of death, and many other specialties. The laboratory has sections that deal specifically with morphology, criminalistics, serology, and evidence and property. The Morphology Section examines and identifies evidence such as claws, teeth, and fur, while the Criminalistics Section performs analyses on specimens such as powdered rhinoceros horn, sea turtle oil, or suspected poisons. The Serology Section uses procedures such as electrophoresis, immunoassays, and DNA sequencing to identify blood and tissue. Evidence and Property receives and catalogs evidence and distributes materials for scientific or educational purposes when they are no longer needed as evidence. For example, bald and golden eagle feathers are received in the Evidence and Property Section and are eventually distributed to indigenous peoples for religious purposes.

Some enforcement officers are highly trained in interrogation techniques. It is inappropriate here to provide in-depth coverage of these techniques; however, it should be evident that such training is a valuable tool for obtaining evidence and confessions from persons suspected of a violation. Officers trained in interrogation techniques know how to ask questions, how to read body language (eye contact, expressions of nervousness, and other actions), where and how far away to stand or sit in relation to the person being interrogated, and how to make it easier to obtain an admission of guilt. The skilled interrogator has both an innate potential for this work and the training and experience to develop that potential into actual skill.

19.8 Undercover Operations and Illegal Commercialization of Wildlife and Fishery Resources

Direct evidence of illegal activities is sometimes obtained by natural resource agencies through the use of undercover agents. These agents might go hunting with guides suspected of major infractions, purchase illegally taken plants, animals, or their parts from violators, or otherwise interact and work directly with those involved in illegal trafficking of wildlife and fishery resources. For example, special agents for the U.S. Fish and Wildlife Service posed as clientele to hunt waterfowl in Texas with hunting guides suspected of repeated **overkill** and illegal **baiting**. Overkill is killing more than the legal daily limit or accumulating more than the allowed possession limit for the season. Baiting involves hunting in an area where some type of material has been scattered to attract animals, such as grain to attract waterfowl. In baiting cases, prior knowledge of the baiting by those hunting the baited area does not need to be proved for conviction on a baiting violation. In the case described above, the agents observed severe overkilling as well as hunting over baited areas; several of these guides were later taken to court and convicted.

In another example of undercover work, a federal agent posed as a trophy hunter to uncover an aerial brown bear hunting operation in Alaska. The agent paid for the hunt and was able to directly observe the use of aircraft to locate and illegally drive brown bears to prospective hunters.

In a Midwestern state, state and federal officers worked cooperatively to obtain evidence leading to the conviction of a guide and out-of-state clients who were illegally hunting white-tailed deer and ring-necked pheasants. Two enforcement officers posed as out-of-state hunters and were able to hunt with this guide and other clients. The undercover officers observed multiple violations committed by the clients and the guide. After a few days of hunting, the undercover officers allowed the violating clients to leave the site and arrested them as they crossed the state border with illegal game, thus ensuring a Lacey Act conviction (see section 18.3). The guide was also arrested. During the undercover operation, the officers found that the guide was conducting illegal hunts rather openly during daylight hours. Some surrounding landowners were aware of many of the illegal activities, but had considerable fear of this guide and for this reason had not reported the violations.

Federal and sometimes state wildlife and fishery law enforcement officials may become involved in uncovering illegal international trade in wildlife and fishery resources. An extensive international market exists for illegal animal and plant parts, whole specimens, and live organisms. For example, the gall bladders of bears have a high economic value in some cultures in which they are used as medicine. The value of gall bladders on the black market leads to illegal bear killing in

North America and other areas of the world. Similar illegal traffic in animal parts such as walrus tusks, eagle feathers, and antlers also poses serious problems. Off the coast of Alaska, for instance, special agents uncovered a major walrus tusk operation in which the animals were being indiscriminately killed and only the tusks removed for sale on the black market. The chain of illegal activity stretched across the Pacific to Hawaii and to Asia and Europe; illegal drugs were also involved. Some of the people that were involved in the purchase and foreign trafficking of the walrus tusks were considered especially dangerous. Often the people involved in illegal international traffic in wildlife and fishery resources are also involved in other illegal activities. The fact that some persons are involved in both the illegal drug trade and illegal wildlife and fishery operations attests to the monetary returns possible from both of these activities.

Illegal trophy hunting within the borders of national parks in the United States has become a serious problem. Clients will pay large sums of money to hunt and kill trophy elk, mule deer, bighorn sheep, or other animals that are present in these national parks (fig. 19.4). In many cases trophy animals are shot and only selected parts, such as the antlers or canine teeth (bugler or whistler teeth) of elk, are taken. Guides involved in these activities are aware of the movements of park rangers and other park personnel and may even monitor their radio conversations. Most U. S. national parks are difficult to patrol adequately because of the extensive land areas involved and the inadequate numbers of enforcement personnel.

In fisheries, the illegal harvest of freshwater mussels for sale overseas is an area of concern; one reason that mussels are valuable is that the shells can be ground for use in pearl culture. At present, many states, especially inland states, have little knowledge concerning their shellfish resource. Pacific salmon, because of their high market values, have also been the target of poaching rings in several instances. In fact, many fishes, including walleyes and channel catfish, have

FIGURE 19.4 This mature male elk *(A)* was illegally killed and the trophy antlers taken *(B)* within the borders of Yellowstone National Park. *(Part A photograph courtesy of B. Lindholm and the National Park Service; part B photograph courtesy of the National Park Service.)*

market values that make them attractive to poachers; illegal nets are commonly used in these types of illicit fishing activities (fig. 19.5). The commercial value of sturgeon and paddlefish eggs on international markets has encouraged illegal fishing operations in various areas of North America.

The eggs are primarily destined for use as caviar in Europe. The eggs from a single fish can bring hundreds or even thousands of dollars.

Illegal imports or exports of live animals, including various parrot species and other birds, exotic fishes, butterflies, reptiles, amphibians, and mammals, are all within the enforcement responsibilities of wildlife and fishery officers. Parrots, for example, may be smuggled into the United States alive and then sold in legitimate pet stores as captive-reared animals. Illegally taken feathers, hides, ivory, internal organs, and other animal parts can also bring substantial income to violators, as described above. Not even plants are immune from illegal harvest. In the southwestern United States, for example, illegal trade in plants such as saguaro cacti can result in large monetary gains to people who traffic in these organisms. Such sales can constitute a particular threat to endangered and threatened organisms. International, federal, state, provincial, territorial, and local cooperation is often involved in enforcement efforts to control illegal imports and exports (see CITES and the Lacey Act, sections 11.2 and 18.3).

Raptors such as the gyrfalcon, especially in the white phase, can bring a substantial price through illegal markets. Single gyrfalcons have brought sellers tens of thousands in U.S. dollars. Illegal exploitation of the gyrfalcon, even though it is not currently listed as endangered or threatened by the United States, could negatively affect populations of this magnificent predator. Trafficking in falcons and hawks is an area of major concern in wildlife and fishery law enforcement and could pose problems for raptor populations that are already of special concern, threatened, or endangered. Raptors are sometimes illegally taken from the nest, reared, and sold on the black market; they may also be taken for personal use. Legitimate falconers abhor the illegal taking of raptors from nests and have themselves been active in restoration programs for species such as the peregrine falcon.

Undercover operations can be extremely dangerous if the identity of the undercover agent is

FIGURE 19.5 Gill nets may be set by poachers, thus robbing the public of a valuable resource. (*Photograph courtesy of J. Lott.*)

discovered. People involved in illegal activities often check the backgrounds of clients and test them in other ways to ensure they are not law enforcement personnel working undercover. Officers in undercover operations are often placed in compromising situations in which they themselves may be forced to break the law to maintain their cover. Officers in most undercover operations also must work with armed suspects. Nevertheless, undercover enforcement activities are critical to uncovering and eliminating illegal operations.

Unfortunately, the seriousness of infractions of wildlife and fishery laws and regulations is at times poorly recognized by the court system. For persons in illegal operations involving large cash benefits, the penalty for capture, prosecution, and conviction may not be sufficient to deter the activity. If a guide receives $10,000 per customer, which is a reasonable estimate for trophy elk taken in national parks in the United States, is a single fine of $10,000 likely to deter such an activity? Sometimes a few weeks of further illegal activity can easily pay the fines invoked by the court system. Fortunately, there is an increasing recognition by the general public and the court system of the seriousness of illegal activities involving a wide variety of natural resources.

SUMMARY

A variety of federal, state, provincial, territorial, and local laws and regulations exist to assist in the conservation of wildlife and fishery resources, but without adequate enforcement these laws and regulations cannot be effective. Enforcement personnel in wildlife and fisheries come from a variety of backgrounds, but a college degree in wildlife and fisheries or a related area is highly desirable. Education and training in criminology, criminal justice, and law enforcement can also be valuable. Most states as well as the federal government require special training in law enforcement for new officers as well as additional on-the-job and annual in-service training.

Violators of wildlife and fishery laws and regulations are often categorized as accidental, opportunist, or intentional violators. The accidental violator breaks laws or regulations unintentionally. The opportunist violator does not plan to break the rules but does so when an opportunity is too tempting to resist. Many people who use natural resources consumptively, nonconsumptively, directly, or indirectly have violated a law or regulation in one of these ways. The intentional violator starts an activity intending to break the rules.

In most cases, wildlife and fishery violations are low-grade misdemeanors. More serious violations can be high-grade misdemeanors or in a few cases, even felonies. An example of a possible felony would be the smuggling of endangered species or their parts.

The Fourth Amendment to the U.S. Constitution protects individuals from unreasonable search and seizure. Violation of the Fourth Amendment rights of a person by law enforcement officers can lead to the exclusion of evidence in later court proceedings. In general, the search of areas beyond the immediate reach of an arrested individual, including a house or place of business, is allowed only with a search warrant. A search warrant is granted based on probable cause that a serious violation has occurred and that evidence relating to that violation will be found through the search. Vehicles are an exception to the rule because of their mobility and in certain situations can be searched if probable cause is present. When persons are taken into custody, the officer must inform them of their rights through what is known as the Miranda warning.

Most violations in wildlife and fisheries are handled by issuing a citation rather than making an arrest. The cited individual often pays a set fine or may appear before a magistrate. Officers can make arrests or write citations upon directly observing a violation. If they have evidence of a violation but did not observe the incident, they can make an arrest with a warrant from a legal magistrate. Good judgment and common sense is required to be successful in wildlife and fishery law enforcement.

Wildlife and fishery personnel involved in law enforcement need to keep accurate written records of evidence and events. Photographs and other items of evidence are recommended in more serious cases. Court cases often occur months after the infraction, and recorded evidence can be especially valuable in the officers' testimony.

Undercover agents are sometimes used to obtain direct evidence of wildlife and fishery violations of the most serious nature. Undercover work can be dangerous to the officers involved but is sometimes needed to obtain evidence against violators. People involved in the smuggling of animals, plants, or their parts for commercial profit may also be associated with other illegal activities such as drug trafficking.

PRACTICE QUESTIONS

1. Give examples of four laws and four regulations pertaining to natural resource use.

2. What are the general in-service training requirements for wildlife and fishery personnel with law enforcement responsibilities?

3. List the three types of violators of wildlife and fishery laws and regulations and give an example of each.

4. List four examples of wildlife or fishery violations involving nonconsumptive users of resources.

5. What protection does the Fourth Amendment to the U.S. Constitution provide for individuals?

6. What is the exclusionary rule?

7. What is the Carroll decision, and how is it related to wildlife and fishery law enforcement?

8. What is the Miranda warning?

9. When does an officer need a warrant to make an arrest or issue a citation?

10. Assume that you are an officer called by a landowner to the site of a recent deer poaching. What types of evidence would you attempt to collect?

11. What kinds of illegal commercial operations relating to wildlife and fisheries might require undercover investigation? Try to use examples other than those provided in the text.

12. What are three positives and three negatives of a career with an emphasis in wildlife and fishery law enforcement? These may be very specific to your own life goals.

SELECTED READINGS

Goldstein, H. 1990. *Problem-oriented policing.* Temple University Press, Philadelphia, Pa.

Goode, E. 1994. *Deviant behavior.* 4th ed. Prentice Hall, Englewood Cliffs, N.J.

Inciardi, J. A. 1993. *Criminal justice.* 4th ed. Harcourt Brace Jovanovich College Publishers, Fort Worth, Tex.

Siegel, L. J. 1992. *Criminology.* 4th ed. West, St. Paul, Minn.

Sigler, W. F. 1995. *Wildlife law enforcement.* 4th ed. Wm. C. Brown, Dubuque, Iowa.

Zahn, M. C. 1993. *Law enforcement for natural resources managers.* Alaska Litho Incorporated and M. M. Zahn, Juneau, Ala.

LITERATURE CITED

Inciardi, J. A. 1993 *Criminal Justice.* 4th ed. Harcourt Brace Jovanovich College Publishers, Fort Worth, Tex.

Siegel, L. J. 1992. *Criminology.* 4th ed. West, St. Paul, Minn.

Sigler, W. F. 1995. *Wildlife law enforcement.* 4th ed. Wm. C. Brown, Dubuque, Iowa.

U.S. Department of the Interior. 1990. *Conservation officers killed and assaulted.* U.S. Fish and Wildlife Service, Division of Law Enforcement, Washington, D.C.

20

ORGANIZATIONS
AND AGENCIES,
BIOPOLITICS
AND CONFLICT
RESOLUTION, AND
PLANNING

The organizations and agencies that interact directly and indirectly with fishery and wildlife resources are numerous. Each year the National Wildlife Federation publishes the *Conservation Directory*, which lists most entities that affect wildlife and fishery resources. This publication, which is in excess of 450 pages, represents an important source of information. It contains material on international, national, and regional organizations; universities with wildlife and fishery academic programs; state, provincial, territorial, and federal agencies and commissions; professional societies and organizations; and a wide array of other natural resource-related entities. Even many indirect user groups, ranging from farm organizations to public health associations, are listed. The agencies, organizations, and other entities that are listed in the *Conservation Directory*, as well as others that are not listed, often interact; they are not just active as individual groups. Knowledge concerning the roles of these entities and how they interact is needed if one is to fully understand the fields of fisheries and wildlife.

The various entities listed in the *Conservation Directory*, others that are not listed, and individuals can interact in many ways in the biopolitical arena. **Biopolitics** involves the interaction of the biological aspects of natural resource conservation with a variety of political factors such as economic, legal, and social forces. Biopolitical activity can be divisive, or it can be positively channeled toward efforts at conflict resolution.

For a biologist or an entity to be effective and successful in conserving natural resources, planning is necessary. The planning process helps to ensure that vision and direction are maintained.

The intent of this concluding chapter is to describe a variety of agencies and organizations that are involved with natural resources, and also to provide examples of the ways in which they and individuals can interact in biopolitics and conflict resolution. The chapter concludes with a brief description of the planning process.

20.1 Professional Societies and Organizations

There are many professional societies and organizations devoted to various aspects of fisheries and wildlife. Some are international in scope but have a variety of subdivisions even at state, provincial, territorial, or local levels. Others operate primarily at the national or international levels. Most professional societies and organizations hold meetings at least once a year and produce scientific publications that contribute to knowledge about natural resources. The groups described in this chapter represent some of the larger professional societies and organizations involved specifically with North American wildlife and fishery resources; there are many other professional societies and organizations that are involved with these resources. Table 20.1 is a listing of the professional societies and organizations highlighted in this chapter.

The American Fisheries Society (AFS) is a professional scientific society formed in 1870 to promote the conservation, development, and wise use of fisheries, both recreational and commercial. AFS was originally formed by North American fish culturists. Those early culturists recognized that fishery resources were being depleted, and through their action federal fishery and wildlife resource agencies were initiated in the United States. AFS is international in scope (United States, Canada, Mexico, and other countries) and has numerous subdivisions. These subdivisions range from regional divisions to state and provincial chapters. There are also numerous special interest sections, such as those concerned with fish culture, fishery management, education, socioeconomics, and bioengineering. AFS has a professional certification program for biologists educated in fishery science as well as a code of ethics. AFS publishes several scientific journals and books, including *Transactions of the American*

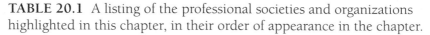

TABLE 20.1 A listing of the professional societies and organizations highlighted in this chapter, in their order of appearance in the chapter.

American Fisheries Society
American Society of Ichthyologists and Herpetologists
American Society of Limnology and Oceanography
North American Benthological Society
International Association of Fish and Wildlife Agencies
The Wildlife Society
American Society of Mammalogists
American Ornithologists' Union
Cooper Ornithological Society
Wilson Ornithological Society
Association of Field Ornithologists
Organization of Wildlife Planners
The Ecological Society of America
Society for Conservation Biology
Society for Range Management
Society of American Foresters
National Association of University Fisheries and Wildlife Programs
National Association of Professional Forestry Schools and Colleges
National Association of State Universities and Land-Grant Colleges
American Cetacean Society
Pacific Seabird Group
Southeastern Association of Fish and Wildlife Agencies

Fisheries Society, The Journal of Aquatic Animal Health, the *North American Journal of Fisheries Management, Fisheries,* and *The Progressive Fish-Culturist,* and conducts numerous meetings at the state, provincial, regional, and international levels.

The American Society of Ichthyologists and Herpetologists, established in 1913, promotes the scientific study of fishes, amphibians, and reptiles. Its primary publication is *Copeia.* The American Society of Limnology and Oceanography, established in 1936, promotes the advancement of various aquatic science disciplines through scientific and technical symposia, colloquia, and meetings, and promotes scientific research. Its major publication is *Limnology and Oceanography.* The North

American Benthological Society, founded in 1953, is an international scientific organization that promotes a better understanding of the biotic communities of lake and stream bottoms and their role in aquatic ecosystems. One of its publications is the *Journal of the North American Benthological Society.*

The International Association of Fish and Wildlife Agencies is an association of states, provinces, and territories of the United States and Canada, a variety of other Western Hemisphere government agencies, and individual associate members. Its mission is the conservation, protection, and management of wildlife and related natural resources.

The Wildlife Society (TWS) was founded in 1937. It is the wildlife counterpart to the American

Fisheries Society. TWS is an organization of professionals and students engaged in wildlife research, management, education, and administration that is dedicated to the sound stewardship of wildlife resources and the environments upon which wildlife and humans depend. It has special interest groups as well as regional and state subdivisions and is also international in scope. TWS has a certification program for biologists educated in the wildlife field and a code of ethics. TWS publishes *The Journal of Wildlife Management,* the *Wildlife Society Bulletin,* and *Wildlife Monographs,* along with several books, and conducts annual professional meetings at the state, provincial, regional, and international levels.

Professional mammalogists may belong to the American Society of Mammalogists, which was organized in 1919. The Society encourages research and learning in all phases of mammalogy, and publishes the *Journal of Mammalogy.*

There are four primary ornithological societies, all of which promote ornithological science and conservation through publications, annual meetings, and membership. These organizations and the research journals they publish are as follows: the American Ornithologists' Union (*The Auk*), the Cooper Ornithological Society (*The Condor*), the Wilson Ornithological Society (*The Wilson Bulletin*), and the Association of Field Ornithologists (*Journal of Field Ornithology*). The American Ornithologists' Union also publishes the *Checklist of North American Birds,* an important and much used listing of the common and scientific names of birds.

The Organization of Wildlife Planners was founded in 1978. This group consists of professional state and federal fishery and wildlife resource planners, natural resource educators, and professional conservationists who are dedicated to improving, through education and training, the quality of resource management and planning. It publishes *Tomorrow's Management.*

The Ecological Society of America, established in 1915, encourages the study of organisms in relation to their environment and promotes the exchange of ideas among those interested in ecology. The Society has a certification program for professional ecologists and a code of ethics, and publishes *Ecology* and *Ecological Monographs.* The Society for Conservation Biology was established in 1985; it is dedicated to providing the scientific information and expertise required to protect biological diversity in the world. Its publication is *Conservation Biology.*

Range scientists are represented professionally by the Society for Range Management, a group established in 1948 that promotes the understanding of rangeland ecosystems and their management and use for tangible products and intangible benefits. Its publications include the *Journal of Range Management* and *Rangelands.* The Society of American Foresters, founded in 1900, represents all segments of the forestry profession and is the accreditation authority for professional forestry education in the United States. Its mission is to advance the science, technology, education, and practice of professional forestry. Its publications include the *Journal of Forestry* and *Forest Science.*

There are three other professional organizations that should be mentioned because their activities can have substantial effects on both university students and natural resource issues. The National Association of University Fisheries and Wildlife Programs (NAUFWP) fosters improved communication among its members and other agencies, organizations, and the general public. The National Association of Professional Forestry Schools and Colleges serves the same function as NAUFWP, but for forestry programs. The National Association of State Universities and Land-Grant Colleges is an umbrella organization that involves many disciplines, including wildlife and fisheries.

There are many other organizations and societies that are concerned with scientific or professional activities in the fishery and wildlife fields. Some have very specific interests, such as the American Cetacean Society and the Pacific Seabird Group, while others are quite general, such as the Southeastern Association of Fish and

Wildlife Agencies. Most have specific missions and codes of ethics to which members must adhere. All are involved in some phase of professional or scientific enhancement of fishery or wildlife resources.

20.2 Advocacy Organizations

Advocacy organizations are groups that support particular views and function as advocates for those views. Their advocacy usually takes the form of legal, political, or public information activities. While professional societies and government agencies and commissions sometimes function in advocacy roles, that is not their primary purpose; the organizations described in this section have advocacy as their major role. There is a bewildering array of advocacy groups; only a few will be mentioned in this text.

It is difficult to categorize the many advocacy organizations that have wildlife and fishery resources as one of their major concerns, but some grouping is beneficial for discussion purposes. The advocacy organizations described in this chapter will be divided into six categories. The first category contains groups that are best described as general resource conservation organizations. They pursue a broad array of activities directed at natural resource advocacy. The second category contains general environmental action organizations. Their advocacy activities encompass areas that often go far beyond wildlife and fishery resources. The third grouping consists of animal rights and animal welfare organizations. These two types of organizations are quite different, and each has its own agenda for advocacy. The fourth grouping has as its major focal point the promotion of hunting and other outdoor activities. The fifth category contains organizations that support single species or groups of species. Some of the organizations included in this group go far beyond support of a single species or group of species,

but their primary activity is focused on the organism or organisms identified in the name of the group (for example, The Wolf Fund or Trout Unlimited). The last category consists of advocacy groups that support particular types of habitats or natural areas. Groups in all six categories interact with professional societies, government agencies and commissions, and other entities in the area of biopolitics.

Keep in mind that this is only a partial listing of natural resource advocacy organizations and that the classification scheme used is subjective. The description of what each group stands for is taken from its own description as provided in the *Conservation Directory*. In some cases, further information is provided, especially with regard to the group's position on the direct consumptive use of fishery and wildlife resources. Table 20.2 lists the advocacy organizations addressed in this chapter by their primary area of activity.

General Resource Conservation Organizations

The Wildlife Management Institute promotes the professional management of natural resources for the benefit of those resources. The Institute has been the longtime sponsor of an important annual meeting, The North American Wildlife and Natural Resources Conference. A large proportion of its activities are directed at the problem of habitat loss. The Institute considers activities such as sport hunting to be an essential element of wildlife conservation. It supports and encourages recreational hunting and harvest if they are within prescribed scientific guidelines, essential standards and traditions of fair chase, and laws and regulations established and enforced by state, provincial, and federal wildlife management agencies.

The newly restructured American Sportfishing Association (ASA) was created when the Sport Fishing Institute merged with the American Sportfishing Association in 1994. The primary mission of the ASA is insuring healthy and sustainable fish-

TABLE 20.2 A listing of the advocacy organizations highlighted in this chapter, in their order of appearance in the chapter.

General resource conservation organizations
    Wildlife Management Institute
    American Sportfishing Association
    National Wildlife Federation
    The Izaak Walton League of America

General environmental action organizations
    National Audubon Society
    The Nature Conservancy
    Sierra Club
    The Wilderness Society
    Worldwatch Institute
    Greenpeace
    World Wildlife Fund

Animal rights and animal welfare organizations
    The American Humane Association
    The Humane Society of the United States
    Defenders of Wildlife
    Friends of Animals
    The Fund for Animals
    People for the Ethical Treatment of Animals

Organizations that promote outdoor activities
    National Shooting Sports Foundation
    Boone and Crockett Club
    National Rifle Association of America
    The Wildlife Legislative Fund of America
    The Wildlife Conservation Fund of America

Organizations that support single species or groups of species
    Ducks Unlimited
    Foundation for North American Wild Sheep
    The National Wild Turkey Federation
    Pheasants Forever
    Quail Unlimited
    Rocky Mountain Elk Foundation
    Trout Unlimited
    Whitetails Unlimited
    Bass Anglers Sportsman Society
    The International Osprey Foundation
    North American Bluebird Society
    The Wolf Fund
    National Wildflower Research Center
    Desert Tortoise Council
    Save the Manatee Club

Organizations that support particular habitats or natural areas
    Chesapeake Bay Foundation
    Desert Fishes Council
    Grassland Heritage Foundation
    Wetlands for Wildlife
    Rainforest Alliance
    Grand Canyon Trust
    Save the Dunes Council

ery resources. It promotes an integrated program of ecological research, fish conservation education, increased profitability for the sport fishing industry, and aquatic science advisory services to help ensure optimum productivity for marine and freshwater ecosystems.

Many professional biologists and even greater numbers of nonprofessionals with an interest in natural resources belong to the National Wildlife Federation. This large organization, founded in 1936, has national, state, and local affiliates. Its mission is to educate, inspire, and assist individuals and organizations to conserve wildlife and other natural resources and to protect the environment. Among its other positions, the National Wildlife Federation supports hunting because, under professional regulation, wildlife populations are a renewable natural resource that can safely sustain harvest.

The Izaak Walton League of America promotes the education of the public to conserve, maintain, protect, and restore soil, forest, water, air, and other natural resources. The League also promotes the enjoyment and wholesome utilization of those resources. The League believes that hunting should be considered as a valuable management tool where it is compatible with other resource uses and purposes.

General Environmental Action Organizations

Many people, including numerous amateur and professional ornithologists, belong to the National Audubon Society, an organization that was founded in 1905 and uses solid science, policy research, forceful lobbying, litigation, citizen action, and education as tools to promote the conservation of air, water, land, and habitat. It has an extensive network of chapters and field offices in the United States and Latin America. The Society does not oppose hunting if it is done ethically and in accordance with laws and regulations designed to prevent depletion of wildlife resources.

The Nature Conservancy, founded in 1951, is an organization committed to preserving biological diversity by protecting natural lands and the life they harbor. The Conservancy works with states through its Natural Heritage Program to identify ecologically significant natural areas. As a portion of its advocacy program, the Conservancy directly purchases lands in need of protection using private, corporate, and other donations. It manages over 1,600 nature sanctuaries nationwide, but in many instances deeds purchased lands to governmental natural resource agencies.

The objectives of the Sierra Club, founded in 1892, are to explore, enjoy, and protect the wild places of the earth and to practice and promote the responsible use of ecosystems and resources. Sierra Club advocacy work includes lobbying, litigation, public information, and publishing. The Club has numerous field offices, chapters, and groups. The Club does not oppose hunting outside of appropriate sanctuaries, such as national parks, and believes that too much attention is paid to the hunting/antihunting issue, and that policy should not revolve around hunting.

Wilderness is synonymous with The Wilderness Society, an organization founded in 1935 that is devoted to preserving wilderness and wildlife, protecting prime forests, parks, rivers, and shorelines, and fostering an American land ethic. The Society has numerous field offices across the United States. The Wilderness Society subscribes to hunting as a legitimate use of wilderness areas, national forests, and certain wildlife areas, subject to appropriate regulation for species protection.

The Worldwatch Institute, founded in 1974, is quite different from the previously mentioned general environmental action organizations. Its interests are diverse, but include such areas as energy use, recycling, soil erosion, toxic wastes, natural resource conservation, and deforestation. The Institute stresses public education, and it produces an annual publication entitled *State of the World* and various other publications with which anyone interested in natural resources should be familiar.

Greenpeace, founded in 1971, is an organization dedicated to preserving the earth and the life it supports. Among other things, Greenpeace seeks to protect biodiversity, prevent pollution, and end all nuclear threats. This group has been involved in some highly confrontational and visible actions, such as attempts to stop whaling.

The World Wildlife Fund, founded in 1961, works to protect endangered species and wildlands, especially in tropical forests. The Fund supports scientific investigations, monitors international trade in wildlife, promotes ecologically sound development of natural resources, and seeks to influence public opinion and the policies of governments and private institutions. The Fund recognizes responsibly conducted hunting as a wildlife management tool for maintaining sustainable populations. It opposes hunting where it might adversely affect threatened or endangered species.

Animal Rights and Animal Welfare Organizations

A large number of local humane organizations are encompassed within The American Humane Association. Its nationwide programs deal with problems of child abuse and neglect and cruelty to animals. It seeks to protect small animals, livestock, and wildlife. This organization, founded in 1877, is opposed to the hunting of any living creature for fun, trophy, or sport.

The Humane Society of the United States, founded in 1954, is dedicated to the protection of animals, both domestic and wild. The Society is involved in humane and environmental education, federal and state legislative activities, laboratory animal welfare, wildlife and habitat protection, and a number of other areas. The Society is opposed to recreational hunting.

The Defenders of Wildlife, founded in 1947, advocates governmental, citizen, and legal action on behalf of endangered species, habitat conservation, predator protection, and wildlife appreciation. This group is neither an anti-hunting nor a pro-hunting organization. It advocates solutions to problems that will insure the best interests of wildlife.

Friends of Animals, established in 1957, is an international organization dedicated to the elimination of cruelty to animals. It strongly opposes hunting and the management of habitat to promote game animals. This group is primarily interested in animal rights.

Similarly, The Fund for Animals, founded in 1967, is adamantly opposed to hunting, and can best be described as an animal rights group. Its purposes are to preserve wildlife, save endangered species, and promote humane treatment of animals.

People for the Ethical Treatment of Animals (PETA) is also an animal rights organization. PETA opposes the use of animals for hunting or fishing, domestication, experimentation, or other purposes. According to its philosophy, keeping animals as pets is unethical.

Organizations that Promote Outdoor Activities

A wide variety of programs to create a better understanding of and more active participation in the shooting sports and practical conservation are sponsored by the National Shooting Sports Foundation, which was founded in 1960. One of the booklets that the Foundation produced, *What They Say About Hunting,* has been liberally used in this chapter as a source for the positions of various organizations on issues such as hunting, wildlife management, and consumptive use of wildlife and fishery resources. The position statements included in the booklet are those made by the organizations themselves. The Foundation itself supports hunting for a variety of reasons.

The Boone and Crockett Club was organized in 1887; it promotes hunting ethics and helps to establish wildlife conservation practices that will lead to the recovery of big game animals in North America. The Club also maintains size records of North American big game such as bighorn sheep and pronghorns.

The National Rifle Association of America, founded in 1871, has a variety of missions, including fostering and promoting the shooting sports; promoting hunter safety; and defending hunting as a valuable and necessary method of fostering the propagation, growth, conservation, and wise use of renewable wildlife resources. One high-profile activity of this organization is its support of the right to possess and use firearms.

The Wildlife Legislative Fund of America (WLFA) and the Wildlife Conservation Fund of America (WCFA) are companion organizations that were established to protect the heritage of the American sportsman and to promote scientific wildlife management principles. The WLFA is the legislative arm, and the WCFA is the legal defense, information, public education, and research arm. WLFA believes that hunting, fishing, and trapping are rights, not privileges, and supports such activities.

Organizations that Support Single Species or Groups of Species

Ducks Unlimited, founded in 1937, is involved with much more than just ducks. It has a large field staff and many chapters across the United States and Canada. The goal of Ducks Unlimited is to perpetuate waterfowl and other wildlife in North America, principally through the development, preservation, restoration, management, and maintenance of wetland areas. This large organization has been responsible for many important and beneficial wetland projects (fig. 20.1). Ducks Unlimited supports the concept of regulated sport hunting as an integral part of sound management and as a wise and prudent use of renewable natural resources.

The Foundation for North American Wild Sheep, founded in 1977, promotes the management of and safeguards against the extinction of wild sheep native to North America. The Foundation funds research on wild sheep, other wildlife studies, habitat improvement, and sheep transplants, and supports hunting and game management policies.

The National Wild Turkey Federation, founded in 1973, promotes the wise conservation of wild turkeys and assists in funding research projects. It has many state and local affiliates. This group also sponsors the National Wild Turkey Symposium every five years, at which research information on turkeys is reported.

Pheasants Forever, founded in 1982, is a conservation organization formed in response to the continued decline in ring-necked pheasant numbers. The mission of Pheasants Forever is to protect and enhance ring-necked pheasant and other upland wildlife populations through habitat improvement, public awareness and education programs, and land management policy refinement. This group has many chapters spread across North America and also has numerous field representatives.

A   B

FIGURE 20.1. Ducks Unlimited conducts many wetland restoration and conservation projects. The two photos show a wetland (*A*) before and (*B*) after a Ducks Unlimited-sponsored restoration project. By plugging the drainage ditch shown in (*A*), the wetland was restored. (*Photographs courtesy of R. Meeks.*)

Quail Unlimited, founded in 1981, is dedicated to improving quail and upland game bird populations through habitat management and research. This group also educates the public about the need for wildlife habitat management.

The Rocky Mountain Elk Foundation, founded in 1984, is dedicated to the conservation and management of elk and elk habitat. The Foundation funds projects in research, management, habitat improvement, and habitat acquisition. A field staff and local chapters characterize this group.

Trout Unlimited is an international conservation organization dedicated to the protection of clean water and the enhancement of trout, salmon, and steelhead fishery resources. This group, founded in 1959, has affiliates in countries such as New Zealand, Spain, and Japan.

Whitetails Unlimited, founded in 1982, supports programs that ensure the present and future well-being of the white-tailed deer and its habitat. It also educates the general public on the importance of sound conservation practices and supports research on white-tailed deer.

The Bass Anglers Sportsman Society was organized to fight pollution, assist state and national conservation agencies in their efforts, and teach young people good conservation practices. This organization, founded in 1968, has numerous chapters, including ones in Japan and Zimbabwe.

The International Osprey Foundation, founded in 1981, is dedicated to studying the problem of restoring osprey populations to stable numbers, making recommendations to enhance the continued survival of the osprey, and initiating educational programs. The North American Bluebird Society, founded in 1978, is concerned with increasing the populations of the three North American bluebird species and educating the public about the importance of these birds.

The Wolf Fund was founded in 1986. The sole mission of the Fund is to facilitate the reintroduction of the gray wolf to Yellowstone National Park. The National Wildflower Research Center, founded in 1982, is a nonprofit organization devoted to the reestablishment and conservation of native plants by promoting their use in public and private landscape designs. The Desert Tortoise Council, founded in 1975, was formed to assure the continued survival of viable populations of the desert tortoise. The Save the Manatee Club, founded in 1981, is concerned with public awareness and education, research, and lobbying for the manatee.

Organizations that Support Particular Habitats or Natural Areas

The Chesapeake Bay Foundation, established in 1966, promotes the environmental welfare and proper management of Chesapeake Bay, including its full watershed. The Foundation operates three programs: environmental defense, environmental education, and land management. The Desert Fishes Council, founded in 1969, provides for the exchange and transmittal of information on the status, protection, and management of the endemic fauna and flora of North American desert ecosystems.

The Grassland Heritage Foundation was organized to advance public understanding and appreciation of the cultural, historic, and scientific value of native American grassland. The Foundation, founded in 1976, also acquires and preserves representative tracts of native prairie. Wetlands for Wildlife was founded in 1960 and advocates and participates in the promotion, preservation, and acquisition of wetlands and wildlife habitat in the United States.

The Rainforest Alliance, founded in 1986, is dedicated to the conservation of tropical forests. Its primary mission is to develop and promote economically viable and socially desirable alternatives to deforestation. The Grand Canyon Trust, founded in 1986, is committed to the conservation of the natural and cultural resources of the Colorado Plateau. The Save the Dunes Council is dedicated to the preservation of the Indiana Dunes National Seashore for public use and enjoyment. The Council, founded in 1952, is concerned with protecting the ecological values of the dunes region, preserving Lake Michigan, and combating air, water, and hazardous waste pollution.

20.3 Federal Agencies and Commissions

The number of federal agencies and commissions that in some way affect fishery and wildlife resources is larger than most people assume. While the fishery and wildlife resource responsibilities of agencies such as the U.S. Fish and Wildlife Service and the National Marine Fisheries Service are obvious, many other agencies and commissions are also involved with these resources. It is not possible to describe all of them here, but most of the more important entities in the United States and Canada are listed and briefly described. Table 20.3 is a listing of the federal agencies and commissions highlighted in this chapter.

Within the U.S. Department of Agriculture (USDA), one natural resource agency is the U.S. Forest Service. It administers national forests, wilderness areas, and national grasslands and is responsible for the management of the forest and grassland resources in those areas. It also conducts research in forestry, fishery, and wildlife management through its experiment stations located throughout the United States. The U.S. Forest Service has many state and local offices and employs a large number of fishery and wildlife biologists.

USDA is also the home of the newly formed (1994) Natural Resources Conservation Service (NRCS). This government agency has assumed responsibility for all soil and water conservation programs that were previously carried out by the Soil Conservation Service, as well as for the Wetlands Reserve, Water Bank, Colorado River Basin Salinity Control, and Forestry Incentives programs previously carried out by the Agricultural Stabilization and Conservation Service. Many programs important to fishery and wildlife resources are administered by the NRCS. The Consolidated Farm Service Agency (CFSA) is also newly formed (1994); among its responsibilities is the continuance of the conservation reserve and agricultural conservation programs that were previously the responsibility of the Agricultural Stabilization and Conservation Service.

The Cooperative State Research, Education, and Extension Service (CSREES) is a portion of USDA that also has many responsibilities important to wildlife and fishery resources. Its extension activities include coordination of the fishery and wildlife extension specialists found at universities that receive federal extension funding. Its research responsibilities include agricultural research conducted at state agricultural experiment stations, of which wildlife and fisheries research represents a part. CSREES is also the federal entity that conducts wildlife and fishery academic program reviews at **land-grant universities**. Land-grant universities are certain institutions for higher education in the United States receiving federal aid under the Morrill Acts of 1862 and 1890. Examples are Oregon State University, South Dakota State University, Michigan State University, and the University of Georgia.

Another USDA agency that directly affects fishery and wildlife resources is the Animal and Plant Health Inspection Service (APHIS). One of its responsibilities is to resolve conflicts involving wildlife deemed injurious to agriculture, animal husbandry, forest and range resources, or other wildlife. A portion of what APHIS does is associated with control of damage to agricultural crops by wild animals. Included in APHIS are its Animal Damage Control and Biotechnology, Biologics, and Environmental Protection sections.

The National Oceanic and Atmospheric Administration (NOAA) is part of the Department of Commerce. A portion of its responsibilities involve oceanic research and management of the marine environment. Within NOAA is the National

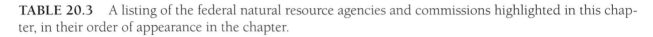

TABLE 20.3   A listing of the federal natural resource agencies and commissions highlighted in this chapter, in their order of appearance in the chapter.

U.S. federal government agencies and commissions	Canadian federal government agencies
U.S. Department of Agriculture	Forestry Canada
Forest Service	Environmental Conservation Service
Natural Resources Conservation Service	Canadian Wildlife Service
Consolidated Farm Service Agency	Ecosystem Conservation Directorate
Cooperative State Research, Education,	Biodiversity Convention Directorate
and Extension Service	State of the Environment Directorate
Animal and Plant Health Inspection Service	Environmental Protection Service
Animal Damage Control	Pollution Prevention Directorate
Biotechnology, Biologics, and	Response Assessment Directorate
Environmental Protection	Technology Development Directorate
U.S. Department of Commerce	Ecosystem Conservation Directorate
National Oceanic and Atmospheric Service	Department of Fisheries and Oceans
National Marine Fisheries Service	Wildlife Habitat Canada
National Ocean Service	Department of Canadian Heritage
U.S. Department of the Army Corps of Engineers	Parks Canada
U.S. Department of the Interior	
Bureau of Land Management	
Bureau of Reclamation	
National Park Service	
Fish and Wildlife Service	
U.S. Department of State	
Oceans and Fisheries Affairs	
Bureau of Oceans and International	
Environmental Scientific Affairs	
Independent U.S. agencies	
Environmental Protection Agency	
National Science Foundation	
Nuclear Regulatory Commission	
Tennessee Valley Authority	
U.S. commissions	
Great Lakes Fishery Commission	
International Whaling Commission	
Migratory Bird Conservation Commission	
Fish and Wildlife Information Exchange	
North-East Atlantic Fisheries Commission	
Marine Mammal Commission	
Pacific Salmon Commission	
Gulf States Marine Fisheries Commission	
International Pacific Halibut Commission	

Marine Fisheries Service, which provides management, research, and services for the protection and rational use of living marine resources. National marine sanctuaries and national estuarine research reserves are also administered by NOAA through its National Ocean Service branch.

Within the Department of the Army, the Corps of Engineers (COE) is involved with water management and waterway research. Many large reservoirs are managed by COE, and a part of its responsibilities lie in determining how natural resources are affected by its actions. COE also regulates the dumping of any type of fill into natural wetlands that are navigable or contribute water to navigable waterways through the Section 404 permit system derived from the Clean Water Act. The service branches, such as the Army, are also responsible for environmental concerns and natural resource management on military lands.

The Department of the Interior contains numerous important natural resource agencies. The Bureau of Land Management manages approximately 110 million hectares of land. These lands are managed under multiple-use principles and provide outdoor recreation, watershed protection, and fish and wildlife habitat. The Bureau of Reclamation is also part of the Department of the Interior and is involved in areas such as reservoir management. Recreational opportunities and fish and wildlife enhancement are included in its duties.

The National Park Service (NPS) is also part of the Department of the Interior. NPS administers national parks and seashores, national monuments, and other properties of national significance for their recreational, historic, and natural values. It also coordinates the wild and scenic river and national trail systems.

Another Interior Department agency is the U.S. Fish and Wildlife Service (USFWS). It is the lead federal agency in the preservation of migratory birds, threatened and endangered wildlife, certain mammals, and sport fishes. The USFWS provides technical assistance, manages refuges and fish hatcheries, conducts research, and educates the public about fish and wildlife resources.

Many students in university fishery and wildlife programs in the United States will become familiar with the Cooperative Fish and Wildlife Research Unit program in the Department of the Interior. The program, which is unique and successful, represents a cooperative effort among the Department of the Interior, universities, state natural resource agencies, and the Wildlife Management Institute. The missions of this program are graduate education, research, and service.

The Department of State is involved with foreign relations, and this at times encompasses some aspects of natural resource management. Two entities in the State Department concerned with fishery and wildlife resources are Oceans and Fisheries Affairs and the Bureau of Oceans and International Environmental Scientific Affairs. State Department responsibilities in the natural resource area include tropical forests, wildlife populations, and the environment.

There are also numerous independent agencies in the U.S. government that affect fishery and wildlife resources. The Environmental Protection Agency, the National Science Foundation, the Nuclear Regulatory Commission, and the Tennessee Valley Authority are examples.

The U.S. federal government is also a part of a variety of international, national, and regional commissions that are involved with fishery and wildlife resources. The responsibilities of these commissions are fairly specific. A good example is the Great Lakes Fishery Commission, which, among other activities, advises governments on fisheries management, including sea lamprey control. Some other examples of such commissions are the International Whaling Commission, the Migratory Bird Conservation Commission, the Fish and Wildlife Information Exchange, the Northeast Atlantic Fisheries Commission, the Marine Mammal Commission, the Pacific Salmon Commission, the Gulf States Marine Fisheries Commission, and the International Pacific Halibut Commission.

The Canadian federal government also has a variety of agencies that affect fishery and wildlife resources. Forestry Canada directs forestry policy in Canada. The Environmental Conservation Service includes the Canadian Wildlife Service, the Ecosystem Conservation Directorate, the Biodiversity Conservation Directorate, and the State of the Environment Directorate. The Canadian Environmental Protection Service includes the Pollution Prevention Directorate, the Response Assessment Directorate, the Technology Development Directorate, the Ecosystem Conservation Directorate, and other entities. The Department of Fisheries and Oceans is responsible for fishery development, management, and research. Wildlife Habitat Canada is dedicated to the conservation, restoration, and enhancement of habitat in order to maintain an abundance and diversity of wildlife in Canada. It is an independent, nonprofit foundation within the federal government. The Department of Canadian Heritage includes Parks Canada.

20.4 State, Provincial, and Territorial Agencies

Every U.S. state, Canadian province, and territory has at least one lead agency specifically responsible for wildlife and fishery resources. While these entities may have different designations, such as division, department, commission, or agency, and their titles may vary to include terms such as fish, wildlife, game, parks, natural resources, aquatic resources, conservation, lands, environment, forests, or renewable resources, all are involved with the conservation of wildlife and fishery resources. Examples are the Arizona Game and Fish Department, the Iowa Department of Natural Resources, the New Jersey Division of Fish, Game, and Wildlife, the Louisiana Department of Wildlife and Fisheries, and the Nova Scotia Department of Fisheries. Some of these agencies are headed by people

educated in wildlife and fisheries, while others are headed by political appointees.

Some unique features characterize most state fishery and wildlife agencies in the United States. In the majority of states, citizen boards or commissions oversee these agencies (fig. 20.2). In most cases, these boards or commissions make the final decisions about natural resource conservation; in some states, they function only in an advisory role. Where the board or commission is the final authority, professional biologists employed by the agency may recommend various actions, but decisions are made by the citizen board or commission. These boards or commissions are usually required by law to include a variety of types of citizen representation. Their composition may involve variation in such things as political party affiliation, regional representation, rural versus urban orientation, and occupation. Such variation in board or commission composition is mandated in an attempt to make these bodies as nonpartisan and representative as possible.

An additional unique characteristic of state fishery and wildlife agencies is that they are funded differently than most other state agencies. While most state agencies are primarily funded

FIGURE 20.2. Decisions affecting wildlife and fishery resources are made in most states by natural resource commissions or boards. (*Photograph courtesy of K. Moum, South Dakota Game, Fish and Parks.*)

through the direct appropriation of general funds (monies collected from all state citizens), most state fishery and wildlife agencies receive either none or a small percentage of their funding from general funds. Their primary sources of funds are license sales (hunting, fishing, trapping, and others) and excise taxes on fishing and hunting equipment (see section 18.5). Thus, these fishery and wildlife agencies are primarily funded only by those people in a state who are direct consumptive users of its fishery and wildlife resources (mainly anglers and hunters). Some states have unique funding sources that include nonconsumptive-user contributions. The state of Missouri, for example, has a one-eighth-cent sales tax that is specifically directed toward natural resources. Similarly, the state of Washington generates funds for nongame species through the sale of personalized license plates.

It is also important to remember that even though most state wildlife and fishery agencies are funded almost exclusively by direct consumptive users, the agencies are usually responsible for all wild organisms, not just those involved with sport hunting, fishing, or trapping. This can pose problems for fishery and wildlife agency commissions and boards, because they are answerable to the direct consumptive users for how funding is expended; some direct consumptive users are not always supportive of expenditures for nongame species.

Another unique feature of these state agencies is the resources that they control. In the United States, Supreme Court decisions have given control of fishery and wildlife resources to the citizens of each state (see section 18.1). Except for organisms such as migratory birds, endangered species, wild horses and burros, and marine mammals, which are under federal jurisdiction, each state has authority over and ownership of aquatic and terrestrial organisms within its borders. Even in the case of organisms such as migratory birds and endangered species, states may assume many conservation responsibilities within federal guidelines. On lands that are under the jurisdiction of a fed-

eral agency, such as the U.S. Forest Service or the Bureau of Land Management, the federal agency manages or controls the habitat, not the wild animals on those lands.

There are other state, provincial, and territorial agencies aside from those primarily responsible for fishery and wildlife resources that can have substantial effects on those resources. State, provincial, and territorial agencies involved with agriculture, environmental protection, water and natural resource development, and others can at times be in agreement with or in conflict with those state agencies responsible for fishery and wildlife resources.

20.5 Biopolitics and Conflict Resolution

An early lesson that any person being educated in wildlife and fisheries learns is that these natural resources are not necessarily managed based on the biology of the situation. Biopolitics influences all decisions, and it influences some decisions more than others. Biopolitical interactions ensure that economic, political, cultural, and other factors will be considered in the decision-making process. As the process of decision making becomes more difficult and complex, effective methods of conflict resolution become more important.

Depending on the scope and nature of a natural resource issue, a variety of federal agencies and commissions, state, provincial, or territorial agencies, directly or indirectly related advocacy groups, professional societies and organizations, and individual private citizens may become involved in biopolitical and conflict resolution activities. The term **stakeholder** is commonly applied to those individuals or groups that have something at stake or an interest in a particular decision or resource. Different stakeholders will have different degrees of interest in any biopolitical issue; that is, for some, the issue may be of paramount importance, while for others, interest may be only peripheral.

All stakeholders can be considered to function, at least to some degree, in an advocacy role, but motivations are quite variable from one stakeholder to the next. Most stakeholders, especially the nongovernmental and nonscientific ones, desire to appear representative of general public opinion. In an attempt to ensure that they are viewed as representing large numbers of people, they often use the terms "grassroots support" or "grassroots group." Some groups are truly "grassroots" in nature and represent many people, while others have few actual supporters and only cloak their activities in the guise of "grassroots" support. Many of these latter groups are motivated primarily by short-term economic gain. Their primary support generally comes in the form of monetary contributions from a few individuals, groups, or businesses that do not really represent the public sentiment. A good example of such a narrowly supported effort is the wise-use movement that was described in section 16.5.

Biopolitical activities can take many forms, but generally revolve around public opinion modification, political action, legal action, and conflict resolution. Depending on the power of each stakeholder to influence public opinion and political action, all will have some effect on the eventual outcome or decision. The more complex the interaction or problem, the more likely that a final decision will be unacceptable to one or more of the affected stakeholders. What should be obvious, however, is that the biology of the situation is just one factor considered in the decision-making process.

Biopolitics can involve attempts to sway public opinion through various means. Providing information to the public is usually an ongoing activity for all groups, including advocacy groups. This information can range from the totally factual to the untrue. At times, the discussion of an issue digresses to the point at which only emotionalism is used in an attempt to gain public support.

There are other situations in which actions taken by advocacy groups to sway public opinion may be or border on the illegal. Activities such as group demonstrations or harassment of various types of users are often directed at swaying public opinion, but can also be used in efforts to intimidate or ridicule others who hold different opinions. Such actions seldom result in solutions and often serve to further polarize the opposing sides on an issue.

Biopolitics can also involve political action by stakeholders as a means of attaining their goals. These attempts to influence the political process through elected officials can be aimed at issues ranging from the local to the national and even international level. One method by which stakeholders can influence the positions of elected officials is making campaign contributions to politicians who agree with their position. Another method is providing information to political decision makers concerning issues in which the stakeholders are interested.

If a solution to or compromise on a particular problem either is not reached or is reached and opposed by a particular advocacy group, legal action is often pursued. Legal action can be a time-consuming and costly procedure, but does represent an option that can be and often is used. For example, the Boldt Decision, described in section 18.10, was a result of numerous lawsuits.

Legal action can be taken in situations in which legitimate differences of opinion occur. Lawsuits intended to stop or modify a decision or to affect the decision-making process are common. While legal action is not necessarily negative, there can be problems associated with it. One disturbing legal method currently being pursued is represented by the increased number of lawsuits directed at stopping stakeholders, often individuals, from speaking out on controversial issues. These lawsuits are called **SLAPP suits**, which is an acronym for **Strategic Lawsuits Against Public Participation**. The intent of such lawsuits is not necessarily to win a court case against people or groups who speak out on an issue, but instead to stifle expression by costing those people or groups time, effort, and money to defend themselves. They are a form of intimidation. SLAPP suits are usually initiated by stakeholders who have large monetary resources and large economic interests at stake in the decision. For example, SLAPP suits are often directed at people and groups who speak out on issues such as the location of a local landfill or argue against building construction on lakeshores.

There are alternative conflict resolution techniques that can allow more reasonable solutions to biopolitical issues. One method of conflict resolution involves the **coordinated resource management** (CRM) process. This process has at times proven successful in solving conflicts that have arisen in natural resource conservation. It is primarily effective in solving problems related to publicly owned resources designed for multiple use, but other types of conflict can also be addressed with this process. An example might be a situation in which different individuals or groups have conflicting desires with regard to how a piece of public grassland should be managed. For example, some users might think that domesticated livestock grazing should be the primary consideration; others might think that soil conservation is being ignored; others might believe that wildlife and fishery resources are not being maintained correctly; others might think that their hiking and camping opportunities are being diminished. Such an issue has economic, social, biological, and ethical aspects. In the CRM process, the significant stakeholders involved are all represented in an effort to develop a compromise management plan that has consensus support. The process is monitored and led by a nonpartisan facilitator. Note that compromise is involved; not all stakeholders get everything they want. Also note the term consensus. Consensus means that all must agree, or at least agree not to disagree; this is not a majority decision process. If all cannot reach agreement, then the process has not worked. Even failure to reach a consensus is generally not a total loss, however, because each

stakeholder has the opportunity to learn the concerns of the others. To some people the CRM process makes sense; to others, it means that their ethics would have to be compromised and is therefore unacceptable.

While the CRM process is a method of solving an existing conflict, **integrated natural resource management** is a process designed to address areas of potential conflict before conflict occurs. It is a procedure for organizing the different human users of a natural resource to produce the greatest value of goods and services from that resource over a given period of time. For the process to be successful, "a comprehensive plan or accounting of the direct or intended resource effects plus all the indirect or unintended resource effects must be described, and where possible, quantified and put into common units that allow comparisons between alternative uses of natural resources" (Loomis 1993).

There are a multitude of specific examples that could demonstrate or depict wildlife- and fishery-related biopolitics and conflict resolution. Four examples are given here. The first is a hypothetical situation in which state biologists recommend an increase in the ring-necked pheasant bag limit; the second involves the setting of waterfowl seasons; the third involves damage to domesticated livestock by coyotes; and the fourth addresses marine fishery resources.

Ring-Necked Pheasant Bag Limits

Let us assume a hypothetical situation is which state natural resource agency biologists propose to their commission or board an increase in the daily bag limit of ring-necked pheasants from two birds to three birds. What stakeholders might become involved in such a decision? Obviously the biologists are involved. They have used the biological information available to them, which indicated that the resource could accommodate the additional harvest. The state citizen commission or board is involved because, in most states, it will make the final decision. Numerous other stakeholders may take an interest in the decision and attempt to influence its outcome.

Tourism or economic development entities in the state may support an increased bag limit so that they can use it as a selling point in their promotional activities. People or organizations representing service industries such as motels, service stations, restaurants, and sporting goods stores may think the increase is a good idea and be supportive of it. Organizations that represent hunters may be supportive because their members like the idea.

It is probable that there would be others who would not be supportive of the increase. Organizations representing landowners may oppose an increased bag limit because of the added bother imposed by more hunters. They may also want a greater voice in future decisions on issues such as this one. Wildlife—the ring-necked pheasant in this example—is often in the unique situation of being a publicly owned resource that can live on privately owned land. Landowners do not own the birds, but they control access to many of them.

An increase in bag limits on ring-necked pheasants may be opposed by animal rights groups. They may argue that a zero-bird limit should be considered. Some groups that represent hunters may be opposed because of the possible reduction in hunting opportunities for residents if the number of nonresident hunters increases due to the increased limits. Hunting preserve operators may be opposed if they perceive an increased bag limit as having a negative effect on their business; they may also be supportive if they believe such a change is economically beneficial to them. Some people may oppose the increase because they believe that the added hunting pressure would harm the resource. Some may be opposed or in favor just because some other stakeholder they oppose or support is involved in the process. Depending on the situation, there would probably be numerous other people and groups with an interest in this issue. However, it should be appar-

ent that much more is involved than just the issue of whether the daily bag limit should be two birds or three birds.

A decision would still have to be made, and all stakeholders would work to have that decision made in a manner that they favor. While the decision is in the process of being made, political pressure can be brought to bear, either for or against the increase, by state legislators, county commissioners, and others. Even after the decision is made, attempts to change the decision, regardless of its outcome, can be made through legislative or legal means. This complex of interactions can make what initially appears to be a simple situation into a complicated one. To further exacerbate the situation, the final decision can affect future related or even unrelated situations. If some stakeholders perceive the decision as a loss or a win, it could affect their future actions. In the end, the issue of how the ring-necked pheasant resource might be affected by the increase may have little influence on the outcome.

Waterfowl Regulation Setting

Another example that demonstrates the intricacies of biopolitics and conflict resolution is the system for setting regulations on the harvest of North American waterfowl. These migratory birds are managed by international treaty. Initially it might seem like a simple procedure to determine when and how long waterfowl seasons will be, how many birds can be harvested, and which species can be harvested, but the determination procedure is more complicated than most people realize. Figure 20.3 depicts the steps in this process.

The initial step involves the collection of biological data on waterfowl population trends. This information is then assessed by a Flyway Technical Committee. There are four of these committees, one for each waterfowl flyway in North America (Pacific, Central, Mississippi, and Atlantic) (fig. 20.4). The committees are composed of waterfowl biologists from each state, province, or territory in the fly-

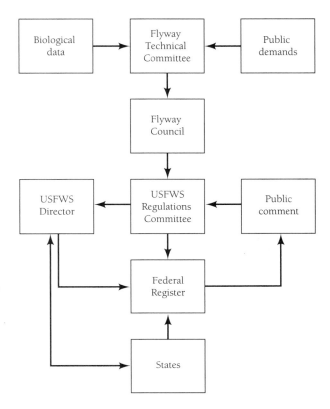

FIGURE 20.3. The regulation development system for the harvest of migratory waterfowl in the United States represents a good study in biopolitics and conflict resolution. *(U.S.F.W.S. is the U.S. Fish and Wildlife Service.)*

way. This phase of the process addresses the biology of the situation.

Each Flyway Technical Committee submits its recommendations to its respective Flyway Council. The Council is composed of one person from each state, province, or territory in the flyway, usually the chief natural resource agency administrator. This administrator may or may not be educated specifically in natural resource conservation. Based on input from the Flyway Technical Committee and other information received, the Council makes a recommendation.

In the United States, each Council's recommendations go to the U.S. Fish and Wildlife Service (USFWS) Regulations Committee. This Committee publishes the proposals in the ***Federal***

Register (a publication that contains in printed form all activities of the U.S. federal government) and solicits public comment. The public comment portion of the process involves public meetings at which any and all stakeholders can comment about the proposed regulations. Testimony can be verbal or written; all input is entered into a subsequent issue of the *Federal Register.*

A review of the *Federal Register* of August 23, 1993 revealed that the following stakeholders were just a few of those who provided testimony with respect to proposed waterfowl regulations for the next season: the Central Flyway Council, Texas Parks and Wildlife Department, The Humane Society of the United States, Alabama Department of Conservation and Natural Resources, Louisiana Department of Wildlife and Fisheries, Wisconsin Department of Natural Resources, National Audubon Society, The Fund for Animals, Minnesota Department of Natural Resources, California Department of Fish and Game, Oregon Department of Fish and Wildlife, Cedar Gun Club, Massachusetts Beach Buggy Association, Parker River Refuge Access Committee, Delta Waterfowl Foundation, Concerned Coastal Sportsmen's Association, Andover Sportsman Club, Essex County League of Sportsmen, an individual from Washington, an individual from Oklahoma, and a college student from Virginia.

After all testimony is received and considered, the USFWS Regulations Committee sets a framework for what the states can do with regard to seasons, bag limits, and other aspects of waterfowl harvest regulation. Each state then can begin its decision-making process to set regulations in that individual state. These decisions are usually made by the citizen commission or board in the state, again with biological and public input. After the state decisions are made, the state must notify the federal government of how it has decided to work within the guidelines provided, and a final total season framework, by state and flyway, is published in the *Federal Register.*

Such decisions do not preclude dissatisfied stakeholders from making legal challenges if they

FIGURE 20.4. In North America, migratory waterfowl are managed by flyway. There are four such flyways.

believe legal recourse is an option. More than just waterfowl biology is involved in this lengthy process. Biopolitics can become quite convoluted in situations in which a large number of stakeholders have varying opinions.

Coyote Damage

Another example of biopolitical and conflict resolution activity in natural resource conservation that can involve a variety of government agencies and commissions, professional organizations, ad-vocacy groups, and individual citizens occurs in the area of animal damage. There are a wide variety of animals, ranging from insects to elk, that can cause damage to agricultural crops, habitat, and other organisms. Coyotes will be the focus of this discussion.

Coyotes are wild animals over which states have jurisdiction, and they can cause damage to domesticated livestock. One aspect of this issue is the question, "Who is responsible for damage caused by a publicly owned resource (the coyote) when it damages a privately owned resource (do-mesticated livestock)?" To further confuse the issue, the domesticated livestock may be on either privately or publicly owned land. A wide variety of other issues are also involved.

The array of potential stakeholders spans a wide spectrum of state and federal agencies and an even wider spectrum of advocacy groups and individuals. The state agencies that can become involved in the coyote damage issue include natural resource and agriculture departments as well as environmental and health departments. The natural resource agency is usually responsible for managing coyotes because they are wild animals. Agriculture departments are involved because the coyotes damage an agricultural commodity (domesticated livestock). State environmental and public health departments may become involved if coyote control efforts include chemicals, drugs, or some other agent that may pose environmental or human health risks. Public health agencies may also become involved if coyotes are perceived to be reservoirs or intermediate hosts of parasites that might pose a risk to humans or other animals.

There are also a number of federal agencies that can be involved in the coyote damage issue. The U.S. Fish and Wildlife Service can be a participant, as can the U.S. Department of Agriculture (USDA). Within USDA, the Animal Damage Control program of the Animal and Plant Health Inspection Service can be involved, as can other USDA branches. If the damage occurs near or on federally owned or controlled land, then the federal agency with jurisdiction over the land, such as the Bureau of Land Management, the U.S. Forest Service, the military, or the National Park Service, may be involved. In addition, if control measures that include chemicals or drugs are an option, the Environmental Protection Agency or the Food and Drug Administration may be involved.

There are innumerable other stakeholders who may interact in the biopolitical arena around the coyote issue at the local, state, regional, national, or international levels. Agricultural advocacy groups, such as the American Farm Bureau Federation, and a variety of cattle and sheep grower associations can become involved. If the damage problem is occurring on publicly owned land that is leased to domesticated livestock grazers, those grazers may act individually on the issue. There are also organized advocacy groups that specifically represent public-land leaseholders.

Animal rights advocacy groups usually become involved in the coyote damage issue because any measure to control coyotes is against their philosophy. In addition, some animal welfare proponents may be against some aspects of control. Other advocacy groups, ranging from general resource conservation supporters, such as the Wildlife Management Institute, to environmental groups, such as The Wilderness Society, to outdoor ethics groups, such as the National Shooting Sports Foundation, can become involved. Even advocacy groups that support single species or groups of species, such as Pheasants Forever or Ducks Unlimited, may become involved in this issue. Professional societies, such as The Wildlife Society, may also become involved.

The particular permutation of agencies, organizations, groups, and individuals that interacts around a particular aspect of the coyote damage issue is to a large extent dependent on the particular situation. Is the problem on public land, private land, or both? Is the problem caused by coyotes on public land going onto private land? What type of control is proposed—trapping, aerial shooting, poisons, or other methods? What are the current legal methods of removing coyotes? Are bounties proposed? Who will be doing the control? Who will be paying for the control? Are agencies responsible for damage? How will other wild animals be affected? Are these affected organisms game or nongame animals? Are endangered or threatened species involved? Are there habitat problems involved? Will the control be targeted at specific coyotes or coyotes in general? Will public health be involved? Will human food supplies be involved? The list could continue, but even these few questions demonstrate the complexity of this biopolitical issue. It is so complex that, depending on the specific situation, stakeholders that normally find themselves on opposite sides of other issues may find themselves in agreement in a particular situation. If the coyote damage issue was simple to solve, then solutions that satisfy everyone would have been found long ago.

Marine Fisheries

Fishery resources are not immune from biopolitical influences; conflict resolution is also needed to solve problems in this area. For example, marine fishery stocks can be a point of contention. These fishery stocks may be found in the open ocean or in coastal waters. They may therefore exist in waters controlled by one or more countries, or they may exist in international waters well beyond political boundaries (commons of the world). In the United States, one of the obvious indicators that marine fishery resources represent a biopolitical issue is the placement of primary governmental responsibility for those resources. Rather than being controlled by a natural resource agency, most marine fishery resources are under the control of the Department of Commerce and the Department of State.

Special commissions, sometimes representing many countries, have been established to help in the conservation of marine resources. These commissions often act to resolve conflicts concerning the use of a particular marine resource. A few examples can best demonstrate the magnitude and scope of such commissions. The International Pacific Halibut Commission was organized in 1923 to investigate and manage the Pacific halibut resource in North America. It was established by a convention between Canada and the United States. The International Whaling Commission was established in 1946 under international convention to regulate whaling, to provide for the conservation of whale stocks, and to provide for the orderly development of the whaling industry. Over forty countries are members of this international convention. One problem associated with such conventions is that membership is voluntary. Countries do not have to belong, and if they do

not belong, they are not bound by commission decisions.

Not all such commissions are international. For example, the Atlantic States Marine Fisheries Commission, organized in 1942, is the result of a compact among states to promote the better utilization of the fishery resource of the fifteen Atlantic seaboard states (Maine to Florida).

There are many functions for which such commissions may be responsible. They may have research responsibilities, or they may be responsible for determining the condition or viability of a particular fishery stock. Often a substantial portion of their efforts is directed toward conflict resolution. One example of such an activity is **resource allocation**. If a particular marine fishery resource cannot be sustained under the pressure of harvest by competing resource users, such a commission may need to allocate portions of that resource to particular users. It may allocate resources between sport and commercial fishing interests, among different countries, between indigenous people and other users, or among any other number and kind of users. Whenever a fishery resource is in high demand conflict will almost invariably result from resource allocation.

The habitat that supports a fishery resource may also support human economic interests other than fisheries; this situation results in a whole new set of problems and conflict. What happens if a particular shellfish fishery is negatively affected by electric power generation, shipping corridor activities, logging, or a host of other alternative uses? To the extent that these alternative uses adversely affect the production of the shellfish fishery, an allocation of a formerly harvestable portion of the shellfish stock must be made to cover these non-fishery-induced losses. Only a certain proportion of the stock can be harvested while still maintaining an optimum sustained yield.

There are other issues that marine fishery commissions may need to address. One of these is **incidental catches (bycatches)**, in which the utilization of one fishery resource affects another through unintentional capture. For example, when tuna are purse-seined, incidental catches of porpoises and dolphins may occur. Even seabirds, such as horned puffins and marbled murrelets, can become a bycatch when caught in gill nets. Such conflicts must be resolved. Another issue that often needs resolution by such commissions is **ghost fishing**. Ghost fishing refers to the capacity of a variety of different fishing gears, such as gill nets and traps, to continue to capture organisms even after the gear is lost or abandoned.

The above are only a few of the issues that may cause conflict leading to biopolitical action in marine fishery situations. It should be obvious that conflict resolution is essential if these fishery resources are to be maintained. It should also be obvious that different stakeholders have different agendas with regard to how a resource is managed.

The biopolitical and conflict resolution activities described in the ring-necked pheasant, waterfowl, coyote, and marine fishery examples are not necessarily the most complex examples that could have been provided. When a bill such as the Food Security Act (commonly referred to as the Farm Bill), which has many effects on wildlife and fishery resources, is introduced in the U.S. Congress, an extremely complex interaction is set in motion. Other legislative actions, such as those dealing with endangered species or clean water, similarly increase the complexity of biopolitical interactions.

There are many issues and problem areas other than the ones mentioned in this chapter that involve, to varying degrees, fishery and wildlife biopolitics and conflict resolution. Some contemporary issues facing wildlife and fisheries include restoration efforts for species such as the gray wolf and black-footed ferret, genetic engineering, the multiple use principle on federal lands, the effectiveness and ecological appropriateness of many stocking programs, the use of electricity and piscicides for sampling fishes, who is responsible for nongame organisms, ecosystem management as it affects target species management, preservation versus conservation, maintenance of genetically unique populations such as some Pacific salmon

stocks, the effects of climate change and habitat fragmentation, animal damage concerns, the negative effects of commercial aquaculture on the environment, limitation of some nonconsumptive uses such as national park visitation, game ranching, maintenance of minimum instream flows, biodiversity efforts and their effects on economically important species, consumptive versus nonconsumptive use of some species, and the ecological appropriateness of exotic and transplanted organisms. Most of these issues have been mentioned in other portions of this text; others have not and are worthy of further inquiry.

Biopolitics and conflict resolution can be an exasperating area for biologists, but biologists must be involved in these processes. In fact, biologists are becoming much more adept at successful and positive interaction in the biopolitical and conflict resolution arenas. A good deal of their success can be attributed to the increased education and training they are receiving in human dimension aspects. Success comes from learning the "art of the possible," that is, knowing that any problem is solvable if correctly approached. Biologists must continue to improve their ability to communicate with individuals and groups in a positive way to resolve differences. With this skill, biologists can serve as facilitators in bringing people, knowledge, and money together for the benefit of natural resources. As human populations increase and natural resources come under more stress, biologists will need to become even more involved and proficient in the biopolitical and conflict resolution arenas, or they will fail to attain their biological goals.

20.6 Planning

It may seem incongruous to place planning at the end of this text, given that planning is usually one of the first undertakings of any successful action.

However, good planning is not possible unless a previous knowledge base has been obtained. The entire text before this section and the information you already possessed represent a start toward that knowledge base.

The eventual success or failure of any undertaking is usually a result of the effectiveness of its planning. Those responsible for a wildlife or fishery resource, or for that matter, any public or private undertaking, must develop plans to determine what results they want to achieve and what methods they will use to attain those results. Such plans must then be implemented and evaluated.

There are a wide variety of different planning modes and management systems, such as **Management by Objective (MBO)** and **Total Quality Management (TQM)**. While planning and management methods often have substantive differences in their approaches, most have some basic elements of similarity. Following are some basic elements that are present in most planning methods. Figure 20.5 depicts the steps in the planning process.

Any agency, department, or individual should first be aware of what their mission entails. This can be accomplished by developing a single mission statement or, if an entity has numerous missions, a set of mission statements. A **mission statement** identifies the purpose or reason for the existence of an entity; that is, why the entity exists, whom it serves, and its ultimate aims. Correct determination of a mission can be much more difficult than one would initially think. There are a multitude of examples to illustrate how an incorrect perception of mission can lead to problems. Railroad companies in North America represent one such example. What was once a preeminent mode of transportation has fallen on rather hard times, as have many railroad companies. Many of their problems stem from a time when they planned and managed for being in the "railroad business" while in reality their mission should have been defined as the "transportation business." Similarly, ranchers who raise domesticated livestock by grazing are not just in the "cattle

FIGURE 20.5. Planning is a continuous process; it really has no beginning or end, except to specific aspects that are completed during the process.

or sheep business;" they are also, and maybe more importantly, in the "grass business." In any situation it is necessary to have or develop a mission statement because without an accurate and clearly defined mission, planning is unlikely to be effective.

The following is an example of a mission statement for a state natural resource agency:

The Division of Wildlife and Fisheries will manage the wildlife and fishery resources of the state and their associated habitats for their sustained and equitable use, and for the benefit, welfare, and enjoyment of the citizens of this state and its visitors.

Such a mission statement sets the stage for the next step in the planning process.

The next step is the development of goals and objectives. The number and direction of these goals and objectives will vary depending upon what one expects or is expected to accomplish toward the mission. The mission statement is usually relatively static and unchanging compared with the goals and objectives.

Goals are broad categories that are specific, but do not identify measurable outcomes. Goals represent outcomes that the entity desires to reach in relation to its mission. For example, two goals (of many) for the natural resource agency described by the previously defined mission statement might be as follows:

To conserve and enhance the stream resources of the state and to increase public knowledge about them.

To manage wild turkeys in the state so as to maximize user opportunity while maintaining populations consistent with the ecological, social, aesthetic, and economic values of the people of the state.

Objectives need to be specific, measurable, output- or benefit-focused, realistic, and have a particular time frame. They are directed at attaining a specific goal; they are what one specifically wants to accomplish. For example, some objectives directed at the previously described goal of conserving and enhancing stream resources might be as follows:

Develop a watershed-based aquatic resource management policy for streams by 1996

Provide an annual minimum of 800,000 angling days of sustainable fishing on state streams by 1996.

Conduct stream preservation and restoration projects at the rate of at least 10 kilometers of stream annually through 1996.

The next planning step is the development of **strategies** to meet particular objectives. Strategies are actions that can be undertaken to insure that objectives will be met; they represent how one will attempt to accomplish something. To accomplish the above-described objective of 800,000 angling days, some strategies might be as follows:

Inventory existing aquatic uses and preferences through appropriate survey mechanisms.

Inventory existing aquatic resources of state streams.

Develop and initiate watershed-scale management plans for sustainable recreational fishing.

Develop and initiate a stream and stream corridor access program to enhance stream user opportunities.

The development of missions, goals, objectives, and strategies represents the **strategic planning** phase of planning. After that phase is completed, the plan must be put into operation. This next phase is the **operational planning** phase. The operational planning process has a number of key components. Goals, objectives, or strategies require monetary resources; thus, planning for expenditures (budget development) is required. This step may involve the allocation of available funding or requests for additional funding. Goals, objectives, or strategies require human resources, and planning with regard to who will do what is also needed. All persons involved with the plan may not contribute to each strategy or even to each goal or objective. Operational planning requires a delineation of what responsibilities will be addressed by particular persons.

The development of missions, goals, objectives, and strategies and the implementation of an operational plan have little utility if results are not evaluated. Not only should end results be evaluated, but progress should also be periodically assessed. The results of periodic evaluations may dictate the development of new or revised strategies to meet the objectives.

Planning involves many aspects and is much more complex than indicated in this brief section. One of the primary purposes of planning is to provide direction to an activity that will result in the successful completion of that activity. Planning is a continuous process; it really has no beginning and no end, except in specific aspects that are completed during the process.

▼ SUMMARY

The Conservation Directory, published annually by the National Wildlife Federation, provides information on a wide array of entities that affect wildlife and fishery resources. Among the entities listed and described in this chapter are various professional societies and organizations, advocacy organizations, federal agencies and commissions, and state, provincial, and territorial agencies. Included among advocacy organizations are groups concerned with general resource conservation, general environmental action, animal rights and welfare, outdoor activities, single species or groups of species, and particular habitats or natural areas.

An early lesson for any person being educated in wildlife and fisheries is that natural resources are not necessarily managed based on the biology of a situation. Biopolitics is a major factor in conservation decisions and ensures that economic, political, cultural, and other factors will be considered in the decision-making process. Conflict resolution must be undertaken before many biological actions are implemented. Two methods of conflict resolution are coordinated resource management and integrated natural resource management. There are a wide array of issues around which various stakeholders such as professional societies, advocacy groups, federal agencies and commissions, state, provincial, and territorial agencies, local entities, and individuals can interact. These interactions can be complex and contentious. Biologists need to continue to improve their biopolitical and conflict resolution skills. These skills will become even more important in the future.

The planning process is a continuing activity. A mission statement, goals, objectives, and strategies must first be developed; this strategic planning phase is followed by operational planning. Periodic and final evaluations of the results of the plan are also necessary. If a natural resource–oriented entity is to be successful, effective planning is essential.

PRACTICE QUESTIONS

1. Which of the described professional societies were you aware of before reading this text? Do you belong to any? Which ones?

2. Name two advocacy groups with which you are familiar? Are they among the ones mentioned in this text? If they were not described, what are their missions?

3. List eight different federal agencies in the United States that affect wildlife and fishery resources. Describe their responsibilities.

4. What are the unique characteristics of most state wildlife and fishery agencies with regard to how they are funded, who usually makes the final decisions on natural resource questions, and what resources they are responsible for managing?

5. Biopolitics often involves public opinion modification, political action, legal action, and conflict resolution. Provide an example of a resource conservation issue in the area where you live, and describe how each of these facets of biopolitics interacts with that issue.

6. What is a SLAPP suit?

7. With regard to your own future, develop a mission statement, goals, objectives, and strategies.

SELECTED READINGS

Anderson, S. H. 1991. *Managing our wildlife resources.* 2d ed. Prentice Hall, Englewood Cliffs, N.J.

Crowe, D. M. 1983. *Comprehensive planning for wildlife resources.* Wyoming Game and Fish Department, Cheyenne.

Decker, D. J., and C. C. Krueger. 1993. Communication: Catalyst for effective fisheries management. Pages 55–75 *in* C. C. Kohler and W. A. Hubert, eds. *Inland fisheries management in North America.* American Fisheries Society, Bethesda, Md.

Fazio, J. R., and D. L. Gilbert. 1981. *Public relations and communications for natural resource managers.* Kendal/Hunt, Dubuque, Iowa.

Hutchings, J. A., and R. A. Myers. 1994. What can be learned from the collapse of a renewable resource?

Atlantic Cod, *Gadus morhua*, of Newfoundland and Labrador. *Canadian Journal of Fisheries and Aquatic Sciences* 51:2126–2146.

Krueger, C. C., and D. J. Decker. 1993. The process of fisheries management. Pages 33–54 *in* C. C. Kohler and W. A. Hubert, eds. *Inland fisheries management in North America*. American Fisheries Society, Bethesda, Md.

Loomis, J. B. 1993. *Integrated public lands management*. Columbia University Press, New York.

National Wildlife Federation. 1995. *Conservation directory*. 40th ed. R. E. Gordon, ed. National Wildlife Federation, Washington, D.C. (New edition each year.)

Reed, D. F. 1981. Conflicts with civilization. Pages 509–535 *in* O. C. Wallmo, ed. *Mule and black-tailed deer of North America*. University of Nebraska Press, Lincoln.

Stickney, R. R., and F. G. Johnson. 1989. Resource uses in conflict. Pages 297–311 *in* F. G. Johnson and R. R. Stickney, eds. *Fisheries: Harvesting life from water*. Kendall/Hunt, Dubuque, Iowa.

Towell, W. E. 1979. The role of policy making boards and commissions. Pages 49–54 *in* R. D. Teague and E. Decker, eds. *Wildlife conservation: Practices and principles*. The Wildlife Society, Washington, D.C.

LITERATURE CITED

Loomis, J. B. 1993. *Integrated public lands management*. Columbia University Press, New York.

GLOSSARY

abiotic. Nonliving, as the abiotic portion of an ecosystem.

abomasum. The fourth and last chamber of a ruminant stomach.

abundance. A quantitative measure of a resource, usually expressed in biomass or numbers per unit of area.

accidental violator. A person who accidentally violates a wildlife or fishery law or regulation.

accuracy. The closeness of a measure to the true value.

acid deposition. The deposition of rain, snow, fog, dew, or solid airborne particles having high levels of acidity. This phenomenon is often referred to as acid rain or acid precipitation, but acid deposition is a more accurate term.

acidity. The capacity to donate hydrogen ions.

acid precipitation. See **acid deposition.**

acid rain. See **acid deposition.**

active capture device. A capture device that requires human or mechanical action to move the device in order to capture the target organism.

active transport. The use of body energy to transport salts or other ions across a semipermeable membrane against a concentration gradient.

adaptation. Changes in structure, physiology, behavior, or mode of life that allow species to adjust to their environment.

additive mortality. Mortality that exceeds the total mortality rate that would have been expected due to natural mortality alone. This term is typically used to refer to the possible added mortality resulting from excessive harvest by humans.

adult. An animal that has reached normal breeding age or size.

advocacy organization. A group that supports particular views and functions as an advocate for those views. Its advocacy usually takes the form of legal, political, or public information activities.

aerial. Having flight capabilities; occurring in flight.

aerobic. Living in the presence of free molecular oxygen.

aerobic decomposer. A decomposer that uses oxygen during its activities.

aestivation. Hypothermy during the summer.

after-hatching year (AHY). An age classification used to denote a bird that is one year of age or older.

age class. A classification for wildlife species by chronological age in years or by some other age or life stage grouping. This term is increasingly being used by fishery biologists.

age group. A classification for fishes based on the number of years a fish has lived. Fish in their first year of life are denoted as age group 0, fish in their second year as age group 1, and so on. Often the word group is deleted, for example, age 0. Age class is increasingly being used as a synonym for age group by fishery biologists.

age pyramid. A graphic representation of the age structure of a population, beginning with the youngest age as the pyramid base.

age structure. The age distribution of a population.

agonistic behavior. Behavior involving aggression or threat displays.

algal bloom. A rapid increase in algal populations.

alidade. A straight-edged device that is used for sighting along a straight line; used to map surface area.

alkalinity. A measure of the carbonates and bicarbonates present in a solution.

alleles. The alternative forms of a gene.

Allen's rule. The tendency of ears, bills, legs, tails, and other extremities to be shorter in endotherms found in the colder portions of the range of a species.

allopatric. Having nonoverlapping ranges of distribution.

alpha cell. A specialized cell in the pancreas that produces glucagon.

alpine tundra. An altitudinal biome similar in its characteristics to the tundra biome.

altitudinal biome. A biome resulting from changes in altitude rather than changes in latitude.

altricial. Poorly developed and relatively helpless at hatching or birth. This term is not commonly applied to fishes; however, a fish such as the northern pike that is poorly developed at the time of hatching may be considered altricial.

altruism. Behavior involving aid-giving that has a cost to the aid-giving individual.

alveoli (singular: alveolus). Small dead-end sacs in mammalian lungs where gas exchange with the blood occurs.

ambient temperature. The temperature of the environment.

ampullary organ. A specialized structure found in some fish species that can function as an electroreceptor, can be sensitive to pressure, touch, and temperature, or can function as a chemical stimulus receptor.

anabolic process. A metabolic process that builds up or maintains body tissue and stores energy.

anabolize. To assemble proteins or store energy by anabolic processes.

anadromous. Reproducing in fresh water but living most of the life cycle in a marine environment.

anaerobic. Living in the absence of free molecular oxygen.

anaerobic decomposer. A decomposer that does not use oxygen during its activities.

anchor ice. Ice formed on rocks and other streambed substrates resulting from ice crystals falling into the water from above the stream.

androgenesis. Reproduction without fertilization that produces viable young that have only the genetic material of the father.

anemometer. A mechanical device used to measure wind velocity.

animal rights. A philosophy based on the belief that other individual animals have rights equal to those of humans.

animal welfare. A philosophy advocating the correct handling and humane treatment of nonhuman animals.

anion. A negatively charged ion.

annulus (plural: annuli). An annual mark found on fish scales and bones, mammal teeth, and other hard structures.

anthropocentric. Believing that humans are the center of the universe.

anthropomorphism. The attribution of human shape or characters to, among other things, other animals.

antimycin. An antibiotic that is used as a general-purpose piscicide.

aphotic zone. The portion of a freshwater ecosystem that is unlighted, and in which photosynthesis does not occur. This zone is generally considered to be the portion that receives less than 1% of the surface light intensity.

aquaculture. The culture and husbandry of aquatic organisms.

aquifer. An underground water body in a permeable formation of sand, gravel, or rock.

arboreal. Living in trees.

Arctic tundra. See **tundra.**

arithmetic mean. A statistical parameter obtained by summing all observations, then dividing by the number of observations.

arrest. The taking of a person into custody by a legal authority.

artificial selection. Selection by humans of the parents of each generation based on desired traits.

Australian region. The zoogeographic region comprising Australia, New Zealand, and New Guinea.

autonomic response. An innate behavior that acts independently of volition.

autotroph. An organism, such as a plant, that can produce its own food.

backfire. A fire purposefully set upwind from a barrier such as a road or a cleared area. Backfires are sometimes set to stop future fires from passing the backfired area. They move slowly and burn thoroughly.

baculum (plural: bacula). The bony inclusion found in the penis of some mammals.

bag census. A complete count of animals taken by hunters. This term is also used occasionally in relation to fisheries.

bag limit. The number of a wildlife species that a person may legally harvest in a day. This term is sometimes used in reference to fishes.

bag seine. A seine with an added bag of mesh that increases capture efficiency. The bag is usually located midway between the two ends of the seine.

bag survey. A partial count of animals taken by hunters. Such a survey is often expanded to an estimate of the total harvest. This term is also used occasionally in relation to fisheries.

baiting. Hunting over an area where some type of material has been placed to attract animals.

balancing selection. Selection that occurs when the heterozygote has superior fitness and the two homozygotes have equal fitness.

bal-chatri trap. A passive capture device, usually used for raptors, consisting of a number of monofilament loops that ensnare the feet of a bird landing on the trap.

bank cover. An artificial instream device used to provide overhead cover.

basal area. The cross-sectional area of standing tree trunks at breast height; often determined on a per hectare basis.

basicity. See **alkalinity.**

Bayer 73. A selective piscicide used primarily on sea lampreys.

beach seine. See **seine.**

beam. The main trunk of an animal antler.

beam trawl. An active capture device for aquatic organisms consisting of a net whose mouth is held open by a rigid beam.

beard. A specialized group of hairlike feathers such as those found on the breast of a male wild turkey.

benthic. Associated with the bottom of an aquatic ecosystem.

benthic zone. The zone in aquatic ecosystems associated with the bottom of a water body.

Bergmann's rule. The tendency of endotherms living in colder climates to have larger body sizes than subspecies or closely related species living in warmer regions.

beta cell. A specialized cell in the pancreas that produces insulin.

bias. The distance of an estimated value from the actual value or true parameter.

big game. Large animals, primarily mammals, that are hunted for sport purposes.

biocenosis. The European term for biotic community.

biocentric. A philosophy in which all creatures in an ecosystem are viewed as being equally important. A more "biology-centered" philosophy than an anthropocentric viewpoint.

biodiversity. The variety and variability among living organisms and the ecological complexes in which they occur.

biogeocenosis. The European term for ecosystem.

biogeochemical cycle. The cycling through an ecosystem of an element that is essential for growth of living cells.

biological amplification. See **food chain biomagnification.**

biological diversity. See **biodiversity.**

biology. The study of living things.

biomass. The weight or mass of a particular species in some specific unit of area; for example, the total weight of all the American robins in a square kilometer area. This term can also refer to the total weight or mass of all organisms in a specific unit of area.

biome. A major terrestrial community or major ecosystem that recurs on the various continents, classified by the dominant vegetation present.

biopolitics. The interaction of the biological aspects of natural resource conservation with a variety of political factors such as economic, legal, and social forces.

biosphere. The layer around the earth in which all living things occur.

biota. All of the animals and some of the plants involved in a fishery or a wildlife system. Biota is most often defined as all living organisms.

biotelemetry. The use of devices placed on or in animals to relay information about the animals to a biologist; these devices usually produce radio or ultrasonic signals.

biotic. Living, as the biotic portion of an ecosystem.

biotic community. The interacting assemblage of living organisms in an area; at times, this assemblage is subdivided into plant and animal communities.

biotic potential. The maximum rate of population increase under ideal conditions; reproductive potential.

black basses. Largemouth bass, smallmouth bass, spotted bass, and other members of the genus *Micropterus*.

block cutting. A forestry practice that involves the clear cutting of small areas of timber.

bog. A wet area characterized by floating spongy mats of vegetation often composed of sphagnum, sedges, and heaths.

bole. The trunk of a tree.

booming ground. A lek used by greater or lesser prairie-chickens.

boreal forest. See **northern coniferous forest.**

box trap. A passive capture device used for both birds and mammals in which the entry of an animal triggers a mechanism that closes the trap entrance.

brail. A pole or rod placed on each end of a seine that allows a worker to hold onto and pull the seine.

branchiostegal ray. A bony portion of the hyoid (throat) support system in fishes that is sometimes used in age determination.

breeding tubercle. An epidermally derived skin structure on fishes that is usually used during reproductive behavior.

browse. The leaves, shoots, and stems of trees and shrubs. This term is generally used in reference to food production for terrestrial animals.

browsing. The act of feeding on the leaves, shoots, and stems of trees and shrubs.

brush bundle. An artificial instream device that collects silt and is most often used to narrow a streambed.

buffering capacity. The ability of water or soil to react with and neutralize acids.

Burlese-Tullgren funnel. A sampling device primarily used for terrestrial invertebrates. The medium containing the animals is placed in the funnel, and the animals are forced out of the medium into a collecting jar.

bursa of Fabricius. A dorsal pocket in the cloaca that is present in young birds, but disappears or becomes shallow in adult birds.

bycatch. See **incidental catch.**

calamus. The bare shaft of a feather below the feathering.

cannon net. An active capture device primarily used on mammals and birds, consisting of a lightweight net that is carried over a group of animals by mortar-type projectiles.

canopy. See **overstory.**

canopy closure. See **canopy coverage.**

canopy coverage. The percentage of forest floor with tree cover directly overhead; also used in relation to shrubs.

canopy volume. The space occupied by tree crowns per unit of area or per tree.

carnivore. An animal that primarily feeds on heterotrophs.

carrion. Dead animal flesh; the primary food source for animals such as California condors and black vultures.

carrying capacity. The maximum biomass of a population that can be sustained within a defined area throughout a specified period of time. In wildlife, the definition usually includes an added factor such as "without causing damage such as from overbrowsing."

catabolic process. A metabolic process that breaks down substances and releases energy.

catabolize. To release energy through the chemical breakdown of substances.

catadromous. Reproducing in the ocean, but living most of the life cycle in a freshwater environment.

catch-and-release fishery. A fishery in which regulations require the live release of all fishes that are captured.

catch-curve method. A method of determining the mortality rate in a fish population based on the declining abundance of successive age groups in a single sample.

cation. A positively charged ion.

cementum annuli (singular: cementum annulus). Annual deposits on the teeth of some mammals that can be used for age determination.

census. A complete count of organisms present or harvested.

chaparral. A biome characterized by a mix of trees and shrubs, often with hard, thick evergreen leaves. The coastal area of southern California is representative of this biome.

checklist scale method. A survey or census method used to measure human attitudes in which the respondent is provided with a list of items and is asked to indicate whether each applies in a certain context.

chemical imprinting. Imprinting of a young animal on the particular chemical characteristics of its natal area.

chloride cell. A specialized cell in fishes used to obtain or excrete salts against an osmotic gradient.

chromosome. A structure composed of DNA that carries genes, and is found in the nuclei of cells.

chromosome aberration. Genetic alteration resulting from changes in chromosomes, such as deletion, duplication, inversion, and translocation.

chute. An intermediate area located between the pools and riffles of a headwater lotic freshwater system.

cichlid. A member of the fish family Cichlidae, which includes fishes such as the Nile tilapia, the blue tilapia, and the Jack Dempsey.

circadian behavior. Activity patterns or behaviors that change over the course of the day.

circannual behavior. Activity patterns or behaviors that change over the course of the year.

circuli (singular: circulus). Concentric markings around the focus of a fish scale that result from the fish's growth.

citation. A document that directs the violator of a law or regulation to appear at some later date in court or, in some cases, simply pay a set fine for the violation.

Clarke-Bumpus sampler. A specialized capture net used primarily to sample plankton.

class I misdemeanor. See **high-grade misdemeanor.**

class II misdemeanor. See **low-grade misdemeanor.**

clear cutting. A forestry practice that involves the removal of all trees from a large area of forest in one cutting operation.

cleithra (singular: cleithrum). Paired bones that provide support to the pectoral fin of a fish and can be used to determine fish age.

climax community. The final stable stage in ecological succession.

clinometer. A device used to estimate angles; commonly used to determine tree height or the angle of a cable leading to a trawl.

cloaca. The common chamber in birds and some other animals into which the intestinal, kidney, and reproductive canals empty.

Clover trap. A passive capture device used primarily to capture deer in which the entry of an animal activates a device that closes the trap entrance (and exit).

clumped dispersion. A dispersion pattern in which members of a species are found in aggregations; clumped dispersions usually reflect the distribution of habitats as well as social tendencies.

clutch. A group of eggs.

coastal wetland. A transitional area between land and ocean.

code. A term that is sometimes used interchangeably with law or regulation.

codominance. The genetic condition in which both alleles at a locus are expressed phenotypically to the same degree.

cohort. A group of individuals in a population born or hatched during a particular time period, such as a year. The term also represents a rank of taxonomic hierarchy below class and above order.

coldwater species. Freshwater fish species that survive, grow, and reproduce best in cold water. They generally spawn at water temperatures below 13°C (55°F).

commons. Those resources such as air, water, wildlife, fish, and publicly owned land that are owned by everyone but by no individual in particular.

community. See **biotic community.**

compensatory mortality. An increase or decrease in one form of mortality that compensates for an increase or decrease in another form of mortality. Compensatory mortality occurs, for example, if increased mortality from predation results in decreased mortality from disease, and total mortality remains stable.

competition. The relationship between two or more organisms living in the same area that have overlapping niche requirements for a resource that is in limited supply.

competitive exclusion principle. The principle that no two species can simultaneously and completely occupy the same niche for an indefinite period of time; also called Gause's principle.

concentrate selector. A ruminant that is capable of digesting only vegetation with low fiber concentrations.

condition. The relative plumpness of a fish; various indices are used to quantify fish condition.

conductivity. A measure of the ability of water to convey electrical current. In general, the greater the concentration of ions, the greater the conductivity.

confidence interval (CI). A measure of precision associated with statistical methods; typically, the interval within which a parameter has a certain probability of occurring.

conservation. The wise management and use of natural resources.

conservation biology. A multidisciplinary science seeking to investigate human effects on biological diversity and to develop practical approaches to prevent the extinction of species.

conservation officer. A title often used for a state, provincial, or territorial natural resource worker who has both law enforcement and other responsibilities.

Conservation Reserve Program. A provision of the 1985 and 1990 U.S. Food Security Acts (Farm Bills) that idled some highly erodible cropland.

conspecific. A member of the same species.

consumers' surplus. The theoretical value that would be collectable if the user could be induced to pay in excess of actual expenditures; the value of a resource beyond what is actually expended on its use.

consumptive user. A person who harvests a resource, such as a deer, a trout, a tree, or a flower.

contest competition. Competition in which some individuals compete better for a resource at the expense of other competing individuals.

controlled burn. See **prescribed burn.**

convergent evolution. A process by which species with different origins and histories develop similar traits and characteristics and fill similar biological niches.

coolwater species. Freshwater fish species that survive, grow, and reproduce best in cool water. They generally spawn at water temperatures between 4 and 16°C (40 and 60°F).

coordinated resource management. A conflict resolution technique that attempts to get all stakeholders interested in a resource conservation issue to agree on a particular management plan or course of action; often referred to as CRM.

coprophagia. Eating feces to recover nutrients that escaped initial digestion.

corpus luteum (plural: corpora lutea). A hormone-secreting body in the ovary that develops during pregnancy in mammals.

corral trap. A passive capture device used for a variety of terrestrial animals in which the animals are usually driven into the trap and the entrance (exit) is closed behind them.

corridor. A narrow strip of land that differs, usually in terms of dominant vegetation, from the surrounding areas. This term can also refer to aquatic corridors such as rivers or streams.

countercurrent exchange. The exchange of gases or heat by means of an arrangement of closely associated blood vessels that facilitates more effective transfer.

cover. Habitat that is specifically used by animals for activities such as shelter, concealment, or escape.

coverage. The percentage of the ground obscured by plants when viewed from above. This term is usually used in relation to grasses and forbs.

coverts. Small overlying feathers on the wings and tails of birds.

creel census. A complete count of fishes caught by anglers.

creel limit. The number of a fish species that a person may legally catch and harvest in a day.

creel survey. A partial count of fishes caught by anglers. Such a survey is typically expanded into an estimate of the total harvest.

crepuscular. Of or pertaining to the twilight hours at dawn and dusk.

critical habitat. The geographic area or ecosystem essential for the survival of a threatened or endangered species. This term is specifically related to the U.S. Endangered Species Act.

crop. An expanded area in the esophagus of some birds where food can be stored until space is available in the stomach.

current deflector. An instream device most often used to narrow a stream by redirecting water flow.

cursorial. Having a highly developed ability to run.

cyclic population. A population in which year-to-year abundance increases and declines on a regular basis.

dancing ground. A lek used by sharp-tailed grouse.

dart gun. An air or powder-charged gun that shoots a syringe-like projectile carrying an anesthetic, used primarily to capture large mammals.

Daubenmire quadrat. A 20 × 50 centimeter plot used to estimate plant coverage; usually used in the assessment of grasses and forbs.

deamination. Removal of an amine group from an amino acid.

decomposers. Organisms, such as bacteria and fungi, that reduce dead organic matter into elemental components.

decreaser species. A plant species, usually a grass of high forage quality, that decreases in number or biomass with increased grazing intensity.

deletion. A form of genetic change that results when a chromosome breaks, then reunites, but loses a segment in the process.

density-dependent factor. An environmental factor that affects a population in proportion to its density; that is, whose effect increases or decreases as population size increases or decreases.

density-independent factor. An environmental factor that acts independently of population density; that is, whose effect is constant regardless of population size.

dentine lake. An area in a mammal tooth where surface enamel has been worn away, exposing the dentine layer; dentine lakes can be used to estimate age.

depositional zone. The inner portion of a curve in a bending or turning lotic system. Current flow and water depth are less here than in the erosional zone on the outside bend.

desert. An arid biome usually having less than 25 centimeters of precipitation in a year. In North America this biome is primarily found in the western and southwestern regions.

determinate growth. A growth pattern in which an animal reaches some general adult size and essentially stops growing. Mammals and birds exhibit this growth pattern.

detritivore. An organism that feeds primarily on detritus.

detritus. Dead organic matter.

dewlap. A fold of skin on the neck of an animal such as that found on some lizards.

diadromous. Migrating between freshwater and marine environments. Anadromous and catadromous fishes are diadromous.

diameter at breast height. The diameter of a tree trunk approximately 1.4 meters above the ground; usually referred to as DBH.

diaphysis. The shaft of a long bone.

diel behavior. See **circadian behavior.**

diploid. Having two complete sets of chromosomes.

direct user. A person who contributes to, uses, or directly benefits from a wildlife system or a fishery.

disclimax. An ecosystem maintained by continual disturbances, such as fires, such that the climax community or later seral stages of ecological succession are not allowed to occur.

disease. Any departure from health; a destructive process in an organ or organism with a specific cause and symptoms.

disease agent. A physical, chemical, or biological factor that causes a disease.

dispersal. The outward movement of individuals away from their established areas of activity.

dispersion. The spatial pattern of individuals in a population; pairs or groups of animals can also demonstrate spatial patterning.

dispersive process. A genetic process that causes random change in allelic frequencies and a loss of genetic diversity in populations over time.

diurnal. Of or pertaining to the daytime; occurring during daylight hours.

DJ funds (Dingell-Johnson). Funds derived from a U.S. federal excise tax on a variety of fishing gears and equipment. These funds are available to states for the acquisition, restoration, rehabilitation, and improvement of fish habitat and for fishery conservation and research.

DNA (deoxyribonucleic acid). The chemical used to store genetic information in most organisms.

dominant allele. An allele that is typically expressed over a recessive allele.

drag line. A rope or cable that is pulled by two people or two vehicles across grassland cover to flush nesting birds.

drawdown. A water level management practice involving the removal of water from a water body. In relation to wildlife, it can also refer to reduced water levels resulting from evaporation.

dredge. An active capture device used in fishery work that is dragged over the bottom of a water body to capture benthic organisms. A dredge usually has blades, rakelike teeth, or other structures that enhance capture efficiency.

drive net. A passive capture device used for some mammals and birds that entangles the target organisms in a single panel of netting.

drop net. An active capture device primarily used for birds and mammals consisting of a net that is suspended above the ground; when target animals move under the net, it is dropped on them.

duplication. A form of genetic change that results when a segment of a chromosome is duplicated; that section appears twice in the genetic composition of the organism.

D-vac sampler. A large field vacuum cleaner used for capturing terrestrial invertebrates.

ecological pyramid. A graphic representation of the relationships among the various trophic levels in terms of numbers, biomass, or energy flow, in which the greatest quantity is represented by the base of the pyramid.

ecological succession. The process that occurs as an ecosystem changes from an early, immature stage to a stable final stage.

ecology. The study of the interrelationships of organisms with other organisms and the environment.

economics. The study of the allocation of resources.

ecoregion. A category in a system for the classification of the biotic world based on regional similarities associated with soils, land use, geology, land surface form, climate, and potential natural vegetation.

ecosystem. An interacting system of biotic and abiotic components in a particular area or place.

ecosystem function. The way an ecosystem works, particularly the way in which nutrients and energy move through and within the system.

ecosystem structure. The composition of an ecosystem with regard to the species of plants, animals, and microbes present, the nutrients available, and the availability of water, or, more generally, a "grocery list" of what is present.

ecotone. The transitional area between two different ecosystems.

ecotourism. Tourism associated primarily with the nonconsumptive use of natural resources.

ectothermic. Having a body temperature that is generally the same as the ambient temperature. Fishes, reptiles, and amphibians are generally ectotherms; their primary source of heat comes from outside their bodies.

edge. See ecotone.

edge species. A species (or subspecies) that requires a variety of habitats in close proximity to one another. This term is most commonly used for birds, less often used for mammals, and is not used for fishes.

Ekman grab. A sampling device used to obtain both organisms and substrate from the benthic zone of an aquatic habitat.

emergence trap. A device used to capture aquatic invertebrates as they emerge from their aquatic life stages into their terrestrial life forms.

emergent vegetation. Rooted aquatic plants that grow through the water column and above its surface.

emigration. Migration of an animal out of an area.

emulsify. To finely divide globules of fat.

endangered species. A species that is in danger of becoming extinct throughout all or a portion of its range. This term commonly pertains to the U.S. Endangered Species Act and to similar acts in some other countries, although other entities, such as states, provinces, and territories can also develop endangered species lists.

endothermic. Maintaining body temperature at some specific level regardless of ambient temperature. Birds and mammals are endotherms; their primary source of heat comes from within their bodies.

energy flow. The movement of energy through an ecosystem.

environment. The total surroundings of an organism, including both biotic and abiotic components.

environmental impact statement. A legally required assessment of the effects of federally funded projects or activities in the United States that can be expected to have a substantial effect on the quality of human life. Under the National Environmental Policy Act, the effects of such a project or activity on fishery and wildlife resources must be considered.

environmental resistance. The environmental factors that interact to limit the abundance of a population; examples include predation and competition.

enzootic. A low-level chronic disease problem.

epilimnion. The uppermost water layer in a freshwater lentic system that is thermally stratified.

epiphyseal cartilage. The cartilage separating the epiphysis and the diaphysis of a long bone.

epiphysis. The end of a long bone.

epizootic. A large-scale eruptive disease outbreak.

erosional zone. The outer portion of a curve in a bending or turning lotic system. Current flow and water depth are greater here than in the depositional zone.

esophagus. That portion of the digestive tract of an animal located between the pharynx and the stomach.

essential amino acid. An amino acid that an organism must obtain in its diet, either because it cannot manufacture that amino acid or because the quantities it can manufacture are insufficient to provide for its needs.

estuarine. Of or pertaining to an estuary.

estuary. A transitional area where a freshwater system enters the ocean.

ethics. Standards of conduct and moral philosophy.

Ethiopian region. The zoogeographic region comprising Africa south of the Sahara, southern Arabia, and Madagascar.

euphotic zone. See **photic zone**.

eutrophic. Rich in nutrients. A eutrophic water body is generally characterized by reduced water clarity.

evolution. The theory pertaining to the process of continuous and gradual transformation of lines of descent from a common ancestor; the changes in the gene pool of a population over time.

exclusionary rule. The rule that evidence obtained through the violation of the Fourth Amendment rights of a suspect cannot be used in court.

exotic. An organism introduced from another zoogeographic region.

experimental gill net. A gill net with numerous sections having different-sized meshes.

exploitation. Harvest of a resource by humans; usually conveyed as a rate or a percentage.

exploitative competition. Competition in which competing individuals divide the resource somewhat equally and all suffer similar effects from resource shortages.

extirpate. To eliminate an organism from a portion or all of its native range.

fall overturn. An event that occurs during fall in a thermally stratified lentic system in which water temperature throughout the water body is equalized and the entire water body mixes.

false annuli (singular: false annulus). Marks formed on hard body parts, such as scales or bones, which appear to be annuli but are not; these marks can be caused by growth variation during periods of stress.

fecal pellets. The droppings of some animals; also called pellets or scats. These terms are usually used in reference to mammals.

fecundity. The potential number of reproductive products or units that can be produced by an organism.

federal aid funds. See **PR funds; DJ funds**.

Federal Register. A publication that contains in printed form all activities of the U.S. federal government.

feeding habits. Feeding behavior; when, where, and how an organism obtains food.

fee fishing. The payment of fees for the right to trespass and take either publicly or privately owned fishery resources on private land.

fee hunting. The payment of fees for the right to trespass and take either publicly or privately owned wildlife resources on private land.

felony. A grave crime, such as burglary or murder. Some serious violations of wildlife and fishery laws and regulations fall into in this category.

feral animal. An animal that has reverted to the wild state after having been domesticated.

filial imprinting. The process by which a young animal becomes behaviorally attached to a parent.

fingerling. The life stage of a young fish between 25 millimeters in length and the length at age 1.

first law of thermodynamics. In any ordinary chemical or physical change, energy is neither created nor destroyed, but merely changed from one form to another.

fish. A single fish, or more than one representative of the same species; one largemouth bass or 200 largemouth bass are fish.

fisheries. More than one fishery.

fishery. A system composed of three interacting components: the biota, the habitat, and the human users. The term fishery can relate to a single target species or multiple target species, their habitat, and human effects on both organisms and the habitat.

fishery biology. The biological study of a fishery.

fishery conservation. The wise management and use of fishery resources.

fishery ecology. The study of the interrelationships of organisms that are considered to be a part of the fishery realm with other organisms and the environment.

fishery management. The art and science of manipulating the biota, habitat, or human users of a fishery to produce some desired end result.

fishery science. The process of obtaining knowledge about and studying fishery resources.

fishes. One or more representatives from at least two fish species; one northern pike and one muskellunge are fishes.

fishway. A structure designed to allow fish passage over or around a barrier.

fitness. A measure of the reproductive success of individuals or populations; the product of genes and the environment interacting throughout the lifetime of an organism.

fixed radius plot method. A method used to estimate tree density and basal area in which the plot radius is a specific measured distance from the plot center.

fledge. To grow feathers necessary for flight; to reach the life stage at which flight is possible.

flight-intercept trap. A device used to capture and hold both terrestrial and emerged aquatic flying invertebrates.

float line. A buoyant line attached to the top of a net used to capture aquatic organisms.

focus. The center or origin of a fish scale.

follicle. A structure in the ovary that may develop into an ovum.

food chain. A simple, linear, graphic representation of energy flow through various trophic levels; for example, from the sun to clover to a while-tailed deer to a gray wolf.

food chain biomagnification. The concentration of a substance, such as a pesticide, at increasing levels in animals higher in a food web.

food habits. What an organism eats.

food web. A graphic representation of the complex interrelationships that occur as energy flows through the various trophic levels of an ecosystem.

forage fish. See **prey.**

forb. A nonwoody plant that is not a grass or grasslike.

foregut fermenter. A mammal with a ruminant stomach.

fossorial. Burrowing into the earth.

founder effect. The effect of inbreeding that occurs when a few individuals leave a larger population to establish a new population.

frequency scale method. A survey or census method used to measure human attitudes in which the respondent indicates the likelihood of a particular event or action.

fry. The life stage of a fish from the time it hatches until it reaches 25 millimeters in length.

fundamental niche. The niche occupied by an organism when there is no competition from other species.

funnel trap. A passive capture device, used for birds and mammals, that involves an arrangement of netting or wire mesh such that animals can readily find their way into the trap, but not out of it.

furbearer. A mammal commonly harvested for its hide.

fyke net. A passive capture device in which a lead net directs fishes into a trap that contains funnels. A fish can readily find its way into the trap through the funnels, but not out of it.

gabion. An instream device consisting of a rock-filled wire basket that can be used as a current deflector or to stabilize banks.

gallinaceous. Of or pertaining to the order Galliformes, which includes birds such as gray partridge and spruce grouse.

game animal. An animal species that is harvested for recreational purposes; these species are usually birds, mammals, or fishes. At times commercially harvested animals are also placed in this category.

game farming. The equivalent of game ranching (both definitions), except that smaller organisms, such as game birds, are used. See **game ranching.**

game ranching. The husbandry of native animals, usually large ungulates, for the production of meat and other products. Also, a pay-for-hunting enterprise in which animals, primarily exotic ungulates, are harvested for sport purposes.

game and fish warden. A title often used for a state, provincial, or territorial natural resource worker who has only law enforcement responsibilities.

gap analysis. A landscape-level analysis of gaps in the protection of species and species diversity.

gas bladder. An organ found in fishes in which a fish can increase or decrease internal gas content and thus its body density.

gastric caeca (singular: gastric caecum; alternative spellings: ceca/cecum). Outpocketings from the intestine containing microbial populations that break down complex plant matter and improve digestive efficiency; found in birds and mammals.

Gause's principle. See **competitive exclusion principle.**

gear. A fishery term for equipment used to capture or sample aquatic organisms.

gene pool. All of the alleles present within the individuals constituting a population.

generalist. An organism that can utilize a wide variety of resources or habitats.

genes. Segments of hereditary material that are positioned on chromosomes within cells, are heritable, and determine the characteristics of an organism. A segment of DNA that codes formation of a particular protein.

genetic drift. Random changes in allelic frequencies due to natural sampling errors that occur in each generation.

genetic integrity. The genetic characteristics of a group of organisms that distinguish it from other groups or populations of that species.

gene transfer. The artificial insertion of genetic information from one species into another.

genital papillae (singular: genital papilla). External fleshy protuberances through which fish eggs and sperm are discharged; can be used to differentiate the sexes in some species.

genotype. The specific genetic information possessed by an organism.

geographic information system (GIS). A system for the computerized mapping and analysis of geographic features using digitized data.

geographic range. The geographic area within which a species occurs.

geophagia. The ingestion of soil by animals to obtain minerals.

gestation. The growth of an embryo in the uterus. This term is primarily applied to mammals.

gestation period. The period of time during which gestation occurs.

ghost fishing. The capacity of some fishing gears, such as gill nets, to continue capturing organisms even after they are lost or abandoned.

gill net. A passive capture device used in fishery work in which fishes are captured by becoming wedged or tangled in a single panel of netting.

gizzard. A portion of the digestive tract of some animals that is muscularized for food-grinding purposes.

glandular stomach. The anterior stomach in birds, the simple stomach in nonruminant mammals, and the abomasum in ruminant mammals.

global positioning system (GPS). A device that uses earth-orbiting satellites to locate a specific position on the earth's surface.

glomeruli (singular: glomerulus). Conglomerations of blood vessels associated with the functional units (nephrons) of a kidney.

glucagon. A hormone that causes stored polysaccharides and amino acids in the liver to be converted to glucose.

gluconeogenesis. The process of glucose synthesis from noncarbohydrates such as fats and proteins.

goal. A broad, specific, but nonmeasurable outcome desired by an entity in relation to its mission.

gonadosomatic index (GSI). An index used to determine the reproductive status of a fish; obtained by dividing gonadal weight by body weight and multiplying by 100.

grabbing distance. The area within the immediate control of a suspect; this area is generally searchable for evidence without a search warrant.

grass/roughage eater. A ruminant that is capable of digesting coarse and fibrous foods.

grazing. The act of feeding on grasses and forbs.

greater secondary coverts. The largest covert feathers overlying the secondary feathers on a bird wing.

greenhouse gas. A gas, such as carbon dioxide, water vapor, ozone, methane, nitrous oxide, or chlorofluorocarbons, that lets visible light from the sun reach the surface of the earth, but inhibits infrared radiation from escaping.

gross economic value. The total economic value or benefits of a fishery or a wildlife system.

guild. A group of organisms that uses environmental resources in a similar manner.

gular sac. A distended membrane in the throat area of some animals.

guzzler. An artificially constructed watering area for terrestrial animals. Guzzlers are often constructed for wild animals in arid regions.

gynogenesis. Reproduction without fertilization that produces viable young that have only the genetic material of the mother.

habitat. The specific set of environmental conditions under which an individual, species, or community exists.

habitat fragmentation. The breaking of larger patches of habitat into smaller patches.

habitat suitability index model. A model based on suitability indices formulated from variables that affect the life cycle and survival of a species; usually referred to as an HSI model.

habituation learning. Learning not to respond to a meaningless stimulus.

hacking. A wildlife term usually used to denote the methods involved in raising captive raptors and releasing them into the wild.

half-log. An instream device consisting of a split portion of a log that is staked in a streambed to provide overhead cover.

hardness. A water quality parameter that is governed by the content of calcium and magnesium salts in water, largely combined with bicarbonate and carbonate and with sulfates, chlorides, and other anions of mineral acids.

harvestable surplus. The surplus production that can be removed without adversely affecting a population.

harvest mortality. Mortality resulting from human activities directed at taking organisms for food or sport purposes.

harvest quota. The specific number or total weight of a species that may be legally harvested; for example, the biomass of black basses from a lake or the number of Canada geese from a population that may be taken during a given time period.

hatching curve. A plot of the number of bird clutches hatched in relation to the progression of the reproductive season.

hatching year (HY). An age classification used to denote a bird that is less than one year old.

headfire. A fire purposefully set to move downwind. Headfires burn more rapidly but less thoroughly than backfires.

herbage. Vegetation consisting of grasses and forbs.

herbivore. An organism that eats autotrophs; also called a primary consumer.

heterosis. See **hybrid vigor.**

heterotherm. An ectothermic animal that maintains a body temperature above the ambient temperature.

heterotroph. An organism that feeds on autotrophs or other heterotrophs.

heterozygosity. The condition of having one or more pairs of dissimilar alleles at one or more loci in homologous chromosome segments.

heterozygous. Having two dissimilar alleles at a particular locus.

hibernation. Hypothermy during the winter, during which the body temperature drops to approximately the ambient temperature.

hierarchial order. The order of dominance in an animal social group.

high-grade misdemeanor. A serious violation of a law or regulation, but one less serious than a felony.

hindgut fermenter. A species in which fermentation occurs primarily in the gastric caecum and intestine.

homeostasis. The tendency to maintain relatively stable conditions within the body.

homeothermic. See **endothermic.**

home range. The area within which an individual animal normally travels in the course of its daily activities.

homing. The capacity of an animal to return to a familiar site that is outside the range of the direct senses such as sight and hearing.

homologous chromosomes. The two chromosomes that carry the same set of genes.

homozygosity. The condition of having identical alleles at one or more loci in homologous chromosome segments.

homozygous. Having two copies of the same allele at a particular locus.

hoop net. A passive captive device consisting of mesh-covered hoops and funnel-shaped throats for capturing fishes. A fish can easily find its way into the net through the funnels, but not out of it.

human dimension. See **human users.**

human users. Direct, indirect, consumptive, and nonconsumptive users of fishery or wildlife resources; the human component of a fishery or a wildlife system.

hybrid. An offspring resulting from the mating of parents that are genetically unlike. In most cases hybrids represent crosses between different species, subspecies, or strains.

hybrid vigor. An increase in some phenotypic value of a hybrid relative to its parents; also called heterosis.

hydric plant. A plant associated with a wetland.

hydric soil. A soil developed in a wetland.

hydroacoustics. A technique involving the use of sonar that is used in fishery work to locate, count, or study fishes.

hydrologic cycle. The cycling of water through the biosphere.

hydroxyapatite matrix. A mineral matrix of calcium phosphate of which vertebrate bones and teeth are primarily composed.

hyperosmotic. Having a greater ion concentration than its surroundings; freshwater bony fishes are hyperosmotic to their environment.

hyperphagy. Rapid and lengthy feeding.

hypolimnion. The lowermost water layer in a freshwater lentic system that is thermally stratified.

hyposmotic. Having a lower ion concentration than its surroundings; marine bony fishes are hyposmotic to their environment.

hypothermy. The lowering of body temperature in endotherms to save energy during periods of stress.

hypoxic. Having a deficiency or lack of oxygen in body tissues.

ichthyocide. See **piscicide.**

immature. An organism that is too young to breed and can be distinguished from an adult based on external characteristics; also called a juvenile. The term immature can also be used to denote fishes that are juveniles and have not reached sexual maturity.

immigration. Migration of an animal into an area.

impoundment. A water body formed by blocking a natural watercourse such as a river or stream.

inborn behavior. See **innate behavior.**

inbreeding. Mating between closely related individuals.

inbreeding depression. A reduction in fitness or vigor due to increased homozygosity resulting from inbreeding.

incidental catch. Those organisms unintentionally captured in addition to the target organisms or species.

incomplete dominance. Partial dominance by one allele such that both alleles are expressed phenotypically, but to different degrees.

increaser species. A plant species, usually a grass of lower forage quality than decreaser species, that increases in number or biomass in an area with increased grazing intensity.

indeterminate growth. A growth pattern in which an animal essentially continues to grow throughout its life; fishes generally exhibit this growth pattern.

index. A value, often a ratio, used in lieu of the actual number; for example, the number of fish caught per hour of electrofishing is sometimes used as an index to population density.

indicator species. A key organism that serves as an indicator of the status of ecological conditions in an area.

indigenous. Originally found in or native to an area.

indirect user. A person who uses or manages some aspect of a natural resource and in doing so affects some other aspect of the resource. For example, someone who harvests trees and thus has an effect on wildlife and fishery resources.

individual distance. The minimum distance maintained by an animal between it and another animal, usually in group situations.

inguinal canal. The canal in some mammals through which the testes descend into the scrotum.

innate behavior. A behavior pattern with which an animal is born.

inorganic. Of or pertaining to minerals or other materials made of noncarbon compounds.

insectivore. An organism that eats insects.

insight learning. Learning that requires an animal to have the ability to perceive relationships.

instinct. An innate stereotyped behavior that is characteristic of a given species.

insulin. A hormone produced by beta cells in the pancreas that stimulates liver, muscle, and fat tissues to take up glucose.

integrated natural resource management. A conflict resolution technique designed to address areas of potential conflict in natural resource conservation before conflict occurs.

intentional violator. A person who starts a particular activity with the intention of violating a wildlife or fishery law or regulation.

interference competition. See **contest competition.**

interior species. A species (or subspecies) that requires an extensive area of relatively homogenous habitat. This term is commonly used for birds, less often used for mammals, and is not used for fishes.

interspecific. Between or among species.

interspersion. A measure of the intermixing of different habitat types. It is also a measure of the horizontal diversity of habitats.

intraspecific. Within a species.

introduced organism. An exotic or transplanted organism.

introduction. See **introduced organism.**

introductory stocking. The release of a fish species into a new or renovated body of water, or the introduction of a new fish species (either native or introduced) into a water body with an existing fish community. This term can also be used in relation to wildlife species.

introgression. The process by which the phenotype of one group of organisms becomes dominant after several generations of interbreeding.

invader species. A plant species that invades from other areas into heavily grazed areas. These plants often reduce the carrying capacity of an area for domesticated livestock and wildlife.

inverse stratification. Stratification that occurs in freshwater lentic systems under ice cover in which the warmest (and most dense) water (4°C) is at the bottom of the water column.

inversion. A form of genetic change that results when the sequence of genes on a chromosome becomes inverted.

inversity. An inverse (negative) relationship between population density and natality or recruitment.

irruptive population. A population that increases or decreases dramatically at irregular intervals.

isolating mechanism. A mechanism that tends to keep organisms from interbreeding.

isosmotic. Having an osmotic pressure similar to that of the surrounding environment.

J-curve. A curve that describes population growth when a population in a new and favorable environment increases at an exponential rate.

Juday net. A specialized capture net used primarily to sample plankton.

juvenal plumage. The plumage following natal down in birds.

juvenile. See **immature.**

juxtaposition. A measure of the proximity of habitats needed by an animal. It is also a measure of the horizontal diversity of habitats.

Kemmerer water sampler. A metal device that can be lowered into a water body, remotely triggered to close, and brought to the surface with a water sample from a particular depth.

ketogenesis. A catabolic process involving the hydrolysis of fat.

kin selection. Natural selection operating on the interactions between closely related cooperating individuals.

K-selected species. A species with a reproductive strategy that involves the production of small numbers of offspring and high levels of parental care.

kype. The hooked jaws found on some male salmonids during the reproductive season.

lactation. Milk secretion in mammals.

lacustrine. Of or pertaining to lentic systems.

lampricide. A selective piscicide that primarily affects or kills lampreys.

land-grant university. An institution of higher education in the United States receiving federal aid under the Morrill Acts of 1862 and 1890.

landscape. A heterogeneous land area composed of interacting ecosystems that are repeated in similar patterns throughout an area or region. A landscape can include aquatic systems.

landscape ecology. The study of structure, function, and change in landscapes.

larval fish. A newly hatched fish.

lateral line system. A sound reception system found in fishes and some amphibians that is used to perceive mechanical sound.

law. A rule enacted by an elected governing authority, such as a federal, state, provincial, territorial, or local government.

law of tolerance. The principle that the absence or failure of an organism in a particular habitat is controlled by a qualitative or quantitative deficiency or excess with respect to any one of several factors that may approach the limits of tolerance for that organism.

lead line. A weighted line attached to the bottom of a net used to capture aquatic organisms. (In this term the pronunciation of lead is like that of "pencil lead.")

lead net. A net radiating from a passive capture device that directs animals into the device. (In this term the pronunciation of lead is like that of "lead dog.")

learned behavior. Behavior that is not innate but instead results from remembered experience.

leghold trap. A passive capture device, primarily used on mammals, in which an animal trips a device that then captures the animal by its leg.

lek. A term that is sometimes used to describe an area where animals gather to perform courtship displays.

length-frequency analysis. A method used to age a fish population sample, based on the principle that fish of similar age, especially younger fish, in a particular body of water tend to be of approximately the same length.

lentic. Of or pertaining to standing water, as in a pond or lake.

lesser secondary coverts. Covert feathers located immediately above the middle secondary coverts on a bird wing.

level ditching. Earth-moving activity done to modify a wetland so that deeper water areas are present; often done to ensure that some areas retain water even during low precipitation periods.

life table. In wildlife, a summary of survivorship or mortality in a population from one age class to the next.

Likert scale method. A survey or census method used to measure human attitudes by which both direction of affect (positive or negative) and intensity of affect can be measured.

limiting factor. A single habitat component that limits population growth.

limiting resource. See **limiting factor.**

limnetic zone. The lighted portion of the deeper water zone beyond the littoral zone in a freshwater lentic water body.

Lincoln method. In wildlife, a method of estimating population number by means of a single mark and recapture period.

line-intercept method. A technique used to assess coverage or species composition of plants in which a tape is stretched above the vegetation and the percentage of the tape overlying plants is determined.

littoral zone. The shallow water zone in freshwater lentic systems, generally characterized by rooted aquatic vegetation.

live trap. See **box trap.**

locus (plural: loci). The location of a particular gene on a chromosome.

logistic growth curve. See **S-curve.**

long-range navigation-C (LORAN-C) system. A device that uses a system of radio stations to locate a specific position.

lotic. Of or pertaining to flowing water, as in a river or stream.

lower critical temperature. The ambient temperature below the thermoneutral zone of an animal at which increased energy consumption must occur for the animal to maintain temperature homeostasis.

lower zone of metabolic compensation. The temperature zone below the lower critical temperature within which an animal can maintain temperature homeostasis by expending additional energy.

low-grade misdemeanor. A low-level violation of a law or regulation.

macroelement. An inorganic element required in relatively large amounts by an organism.

magistrate. An official having jurisdiction over minor court cases.

magnitude scale method. A survey or census method used to measure human attitudes in which respondents are given a list of items and asked to allocate points based on importance or value to them.

maintenance burn. One of a series of prescribed burns set regularly to maintain conditions resulting from previous burns.

maintenance stocking. The release of a fish species into a population that has no or very limited natural reproduction. This term can also be used in relation to wildlife species.

Management by Objective (MBO). A planning or management process that involves a systems approach to managing an organization.

marginal coverts. Covert feathers above the lesser secondary coverts on a bird wing (toward the leading edge of the wing).

marine. Of or pertaining to the ocean.

marsh. A low, treeless wet area characterized by plants such as sedges, rushes, and cattails.

mast. Fruit produced by trees and shrubs, such as acorns and wild plums. This term is generally used in reference to food production for terrestrial animals.

maximum length limit. A regulation under which all fish of a particular species that reach or exceed a specified length must be released.

maximum sustained yield (MSY). A traditional management philosophy with the goal of achieving the maximum yield that can be sustained over a series of years from a population; the maximum biomass of a population that can theoretically be harvested without affecting future harvest.

megaherbivore. An herbivore that weighs over 1,000 kilograms.

metabolic rate. The rate at which chemical reactions occur in an organism.

metalimnion. See **thermocline.**

metapopulation. A population existing as a set of geographically separate subpopulations, but with some genetic interchange.

meter net. A device commonly used to capture fish eggs and larvae. The opening of the net is 1 meter in diameter.

microbivore. An organism that feeds on the nutrients and energy in bacteria and fungi.

microclimate. The climate on a very local scale. Microclimate differs from the general climate of an area.

microelement. An inorganic element required in trace amounts by an organism.

microflora. Small organisms such as bacteria or small fungi.

microhabitat. A small area of intensive use within the overall habitat of an organism.

middle secondary coverts. Covert feathers between the greater secondary coverts and lesser secondary coverts on a bird wing.

migration. A two-way movement of an animal to and from an area with characteristic regularity or with changes in life history stage.

milk teeth. The first set of teeth in mammals.

milt. The sperm-bearing fluid of fishes.

mineralization. The release of inorganic nutrients that have been immobilized or kept out of nutrient cycles because of their inclusion in something such as the body tissues of an organism.

minimum instream flow. The minimum flow needed to ensure sufficient water for aquatic organisms at all times of the year.

minimum length limit. A regulation under which all fish of a particular species that are less than a specified length must be released while those exceeding the length may be kept.

Miranda warning. A procedure for informing a suspect of his or her constitutional rights.

mission statement. A statement identifying the purpose or reason for the existence of an entity, who it serves, and its ultimate aims.

mist net. A passive capture device used on birds that entangles them in a single panel of netting.

mitigation. The legally required replacement of certain habitats, such as wetlands or riparian areas, that are lost due to projects that receive federal funds. Mitigation may involve the purchase of private lands in some other area of a state to replace habitat that is lost to project construction. Mitigation can also be required for some nonfederally-funded activities.

mixed-grass prairie. That portion of the temperate grassland biome characterized by grasses that are intermediate in stature and by precipitation levels that are intermediate between those of short-grass and tall-grass prairies.

model. A conceptual or mathematical description used to portray a complex system.

modified-fyke net. A passive capture device in which a lead net directs fishes into a trap containing funnels. A fish can readily find its way into the trap through the funnels, but not out of it.

molt. In birds, the loss of feathers in the process of renewal. Many other animals also molt various body parts.

monoculture. The growing or raising of a single species over a large area.

monogamous. Of or pertaining to a mating system characterized by pair bonding between a single male and a single female.

monosaccharide. A simple sugar such as glucose.

monosex population. A population that consists of only one sex.

montane coniferous forest. An altitudinal biome similar in characteristics to the northern coniferous forest biome.

mortality. The death of an individual. Mortality rate is commonly expressed as the percentage of the population that dies in a year. In addition, mortality rates can be determined for age groups, year classes, or age classes.

movement. A change in the position or posture of an animal; these changes are not necessarily regular in nature as would occur in migrations.

multidimensional niche. The variety of niche components that constitute the total niche of an organism.

mutation. A change in the DNA of a gene.

natal area. An animal's place of birth or hatching.

natal down. The short, fluffy feathering on newly hatched birds.

natality. The birth or hatching of an animal. Natality rate is commonly expressed as the number of individuals born or hatched within a specified time period.

native range. The geographic area in which a species is indigenous.

naturalized organism. A naturally occurring offspring produced by exotic or transplanted organisms.

natural lake. A naturally formed lentic water body.

natural mortality. Mortality caused by predation, starvation, disease, accidents, or other natural causes.

natural resource. Something such as land, water, or wildlife that is useful and valuable in the condition in which it is found.

natural resource economics. The study of the allocation of natural resources.

natural selection. A theory that attempts to account for the adaptation of organisms to their environment, which states that such adaptation results from organisms of different genotypes in a population contributing differently to the gene pool of succeeding generations.

Nearctic region. The zoogeographic region comprising North America and Greenland.

Neotropical migrant. A bird that migrates between the Nearctic and Neotropical zoogeographic regions. These birds would more appropriately be called Nearctic-Neotropical migrants.

Neotropical region. The zoogeographic region comprising South America, Central America, and the West Indies.

net gun. A gun that fires projectiles that propel a small net away from the gun. This device is primarily used to capture flying birds.

neuromast. A specialized sensory structure associated with the lateral line system of fishes that can detect the displacement of water.

niche. The role or function of an organism in a biotic community.

niche segregation. The tendency for two species that live in the same area and require similar resources to have niche requirements that differ in one or more dimensions.

night-vision scope. See **starlight scope.**

nitrogen fixation. Conversion of inert, gaseous nitrogen to ammonia, nitrites, or nitrates.

nocturnal. Of or pertaining to the night; active during periods of darkness.

nonconsumptive user. A person who does not harvest a resource, but does use it for some purpose.

nonessential amino acid. An amino acid that an organism can manufacture in sufficient quantities within its own body.

nongame animal. An animal species that is not harvested for recreational or commercial purposes; an animal that is legally described as a nongame species.

nonhomologous chromosomes. Chromosomes that carry different sets of genes.

nonpoint source pollution. Pollution emanating from diffuse or dispersed sources.

northern coniferous forest. A biome characterized by evergreen trees and cool climate. In North America this biome is located south of and adjacent to the tundra biome.

Nudds board. A 2.5 meter long, 30 centimeter wide, marked board used to determine the concealment or cover value of vegetation.

nutrient cycle. See **biogeochemical cycle.**

nutrient immobilization. The keeping of nutrients out of circulation, as in the case of those in the body tissues of an organism.

objective. A specific, measurable, output- or benefit-focused, and realistic outcome desired by an entity within a certain time frame in relation to attaining a specific goal.

olfactory. Of or pertaining to the sense of smell.

oligosaccharide. A short chain of two or more simple sugars.

oligotrophic. Low in nutrients. An oligotrophic water body is generally characterized by high water clarity.

omasum. The third chamber of a ruminant stomach.

omnivore. An animal that eats both autotrophs and heterotrophs.

operational planning. Planning for the implementation of a strategic plan.

opportunist violator. A person who had no intention of violating a wildlife or fishery law or regulation, but did so when the opportunity occurred.

optimal foraging theory. A theory based on the premise that an animal should not expend more energy than necessary to obtain food.

optimum sustained yield (OSY). A management philosophy that considers ecological and socioeconomic factors; the optimum biomass of a population that could be harvested without negatively affecting population quality.

organic. Of or pertaining to carbon-containing compounds, including living or dead organisms.

Oriental region. The zoogeographic region comprising India, Burma, Indochina, Malaysia, Sumatra, Java, Borneo, and the Philippines.

osmotic balance. Water and mineral balance in an organism.

osmotic pressure. The pressure difference resulting from differing water and ionic concentrations on two sides of a semipermeable membrane.

otolith. An earstone found in the inner ear of animals. In fishes, otoliths are commonly used in age determination.

otter trawl. An active capture device for aquatic organisms in which the mouth of the trawl is maintained in an open position by weighted boards called otter boards.

outbreeding. Mating between distantly related individuals.

outer coastal plain forest. A portion of the temperate forest biome that ranges from Florida to Louisiana, characterized by mangroves and live oaks.

overexploitation. Variously defined as the point at which harvest mortality causes undesirable reductions in fish, wildlife, or other populations; the level of harvest at which adult organisms cannot replace themselves; or the level of harvest at which population quality is impaired.

overharvest. See **overexploitation.**

overkill. The harvesting of more than the legal daily limit or the accumulation of more than the possession limit for the season.

overstory. The topmost vegetation in an area. This term is most often used in reference to tree crowns in forests, which are also known as the canopy.

oxygen-demanding wastes. Pollutants such as domestic sewage, domesticated livestock wastes, and other biodegradable organic wastes that cause water to be depleted of dissolved oxygen.

Pacific coast coniferous forest. A small North American biome extending from Alaska to northern California that is characterized by a cool climate and abundant rainfall. Douglas firs, giant redwoods, and Sitka spruce are the dominant trees in this biome.

Pacific salmon. A group of salmonid fishes that includes pink, chum, coho, sockeye, and chinook salmon.

Palearctic region. The zoogeographic region comprising Europe and Asia north of the Himalayas.

palustrine. Of or pertaining to bogs, marshes, and swamps.

panfish. A small game fish such as a bluegill, white or black crappie, white perch, or yellow perch.

parabronchi. The site of gas exchange in bird lungs.

parasite. An organism that obtains energy by living in or on another organism.

parr (plural: parr). The life stage of a salmonid fish from the time of initial feeding until the fish develops sufficient pigmentation to obliterate the parr marks.

parr mark. A vertical pigmentation band on the sides of young salmonids and certain other fishes. Numerous parr marks are usually present.

partial rehabilitation. See **partial renovation.**

partial renovation. A term used in fisheries to denote the removal of a portion of a population or community. This term can also be used in relation to wildlife species.

passerine bird. A member of the largest order of birds, the Passeriformes.

passive capture device. A device that is put in place, essentially remaining stationary, such that the movement or behavior of the target animal results in its capture.

passive entanglement device. A device that is put in place, essentially remaining stationary, and the target animal when encountering the device becomes ensnarled.

passive entrapment device. A device that captures animals that move into an enclosed area and are retained in that area.

patch. A continuous, nonlinear (not long and narrow) surface area within a landscape that differs from surrounding areas.

patch size. The size of a particular patch within a landscape; for example, the surface area of a forest surrounded by urban habitat.

pathogen. A disease-causing biological agent.

pecking order. See **hierarchial order.**

pellets. See **fecal pellets.**

peptide group. A combination of two or more amino acids.

permafrost. The presence of frozen ground, usually a short distance below the soil surface, during the entire year.

Petersen grab. A sampling device used to obtain organisms and substrate from the benthic zone of an aquatic habitat.

Petersen method. In fisheries, a method of estimating population number from a single mark and recapture period.

pH. A measure of the hydrogen ion concentration in a solution.

pharynx. That portion of the digestive tract of an animal between the oral cavity and the esophagus.

phenotype. The actual physical expression of genetic traits. The phenotype is a product of the interaction of an individual's genotype with its environment.

pheromone. A chemical cue used in animal communication.

photic zone. The portion of a freshwater ecosystem that is lighted and in which photosynthesis occurs. The photic zone is generally considered to be the portion that receives 1% or more of the surface light intensity.

photoperiod. The proportion of light and dark in a 24-hour period.

photosynthesis. Energy fixation by autotrophs using sunlight as a source of energy.

physoclistous. Fishes that as adults do not have a duct connecting the digestive tract and the gas bladder.

physostomous. Fishes that as adults have a duct connecting the digestive tract and the gas bladder.

phytoplankton. Tiny aquatic plants that drift or float in water.

pigeon milk. A rich milklike material produced in the crops of reproductively active doves and pigeons.

pineal gland. A structure located in the brain that can function in light sensitivity. Other potential functions are also attributed to this endocrine gland and structures derived from it.

pineal organ. See **pineal gland.**

pinna (plural: pinnae). The external ear structure in mammals that directs sound into the middle ear.

pinnae feathers (singular: pinna feather). Elongated feathers on the necks of some birds, such as greater prairie-chickens.

pipping. Breaking of the eggshell during hatching in birds.

piscicide. A toxicant or poison that affects fishes.

piscivorous. Feeding on fishes.

pit trap. A passive capture device used on a wide variety of animals in which the target animal falls into a pit or hole from which it cannot escape.

poaching. The illegal taking of a wild organism.

poikilothermic. See **ectothermic.**

point-centered-quarter method. A method used to determine tree density in which the distance from a sampling point to the nearest tree is measured in each of four quarters.

point frame. A metal frame into which metal rods are inserted that is used to determine basal and canopy coverage of plants, usually grasses and forbs.

point source pollution. Pollution emanating from a specific, identifiable location.

polar planimeter. A manual or electronic device used to trace the outline of a feature on a map to determine its area. Using the planimeter reading and the map scale then allows for actual area determination.

polyandrous. Of or pertaining to a polygamous mating system in which one female mates with multiple males.

polygamous. Of or pertaining to a mating system in which a single individual of one sex mates with multiple individuals of the opposite sex.

polygenic trait. A genetic trait controlled by multiple genes.

polygynous. Of or relating to a polygamous mating system in which one male mates with multiple females.

polyploidy. The genetic condition of having more than two sets of chromosomes.

polysaccharide. A straight or branched chain of hundreds or thousands of sugar molecules.

pond. A lentic water body that is less than 4 hectares in surface area. A pond is usually constructed by blocking a small intermittent watercourse.

pool. That area of a headwater freshwater lotic system characterized by being deep, wide, and soft-bottomed. Current flow in pools is less than that in riffles because of the greater volume of space present.

population. All the individuals of one species within a specified area at a given time.

population density. The number of organisms of one particular species present in a specific unit of area at a given time.

population dynamics. The study of changes in numbers or biomass of organisms in populations and the factors (natality or recruitment, growth, and mortality) that influence those changes.

population estimate. The expansion of a sample into an estimate of the total population number.

population form. See **population structure.**

population function. See **population dynamics.**

population structure. The size structure, age structure, or sex ratio of a population. The three dynamic rate functions (natality or recruitment, growth, and mortality) interact and result in population structure.

possession limit. The total number of a species that one person may legally possess at any one time. Possession limits are commonly two to five times the daily creel or bag limit for a species.

post-breeding. An age classification term used in wildlife to denote animals that have bred.

pre-breeding. An age classification term used in wildlife to denote animals that have not bred.

precision. The closeness of repeated measures to one another.

precocial. Well developed, alert, and mobile soon after birth or hatching. This term is not generally applied to fishes; however, a fish such as the walleye that is well developed at the time of hatching may be considered precocial.

predator. An animal that attacks, kills, and eats other animals.

prescribed burn. An intentional and controlled burning of an area for habitat management purposes.

preservation. A conservation philosophy that involves leaving natural systems as they are.

prey. An animal that a predator eats.

PR funds (Pittman-Robertson). Funds derived from a U.S. federal excise tax on such items as arms and ammunition. These funds are available to states for the acquisition, restoration, rehabilitation, and improvement of wildlife habitat and for wildlife conservation and research.

primaries. See **primary feathers.**

primary consumer. An organism that primarily eats autotrophs; also called an herbivore.

primary coverts. Covert feathers overlying the primary feathers on a bird wing.

primary feathers. Large feathers arising on the posterior portion of a bird wing beyond the "wrist."

primary producer. An autotroph or plant; an organism that can produce its own food.

primary succession. Ecological succession occurring on a site unchanged by living organisms, such as a new volcanic island that appears above the surface of the ocean.

probable cause. The situation in which enforcement personnel have facts and circumstances within their knowledge that would cause a person of reasonable caution to believe that wrongdoing has occurred or is occurring.

production. The biomass accumulated by a population during a year or some other specific period, including both living organisms and those that died during that period.

productivity. The rate at which energy is captured and stored in an ecosystem.

profundal zone. The unlighted portion of a freshwater ecosystem located below the limnetic zone.

promiscuous. Of or pertaining to a mating system that is not characterized by male-to-female pair bonding.

propagation. The planting of seeds or transplanting of trees and shrubs.

Property Clause. A clause in the U.S. Constitution that allows the federal government jurisdiction over its property.

proportional stock density (PSD). The percentage of stock-length fish in a sample that are also of quality length.

protected slot length limit. A regulation under which all fish of a particular species that are within a specified length range must be released, while those outside that length range may be kept.

protection. The prevention of damage to plants until they mature.

proventriculus. The glandular stomach in birds.

proximal. Near the body, as opposed to distal (away from the body).

pseudoreplication. Treating replicate samples as if they represent replicate experimental units.

purse seine. An active capture device used in fishery work in which a purse line on the net bottom, when drawn together, prevents fishes from escaping.

put and take stocking. The stocking of fish of catchable size for immediate angler use. This term could also be used in relation to wildlife species.

put, grow, and take stocking. The stocking of small fish into a water body where they will grow to catchable size and become available to anglers. This term could also be used in relation to wildlife species.

pygmy water meter. A meter used to determine water current velocity.

pyloric caeca (singular: pyloric caecum; alternative spellings: ceca/cecum). In fishes, structures found between the stomach and intestine that probably function in both digestion and absorption.

race. See **subspecies.**

raceway. An intermediate area located between the pools and riffles of a headwater freshwater lotic system. This term is also applied to rectangular rearing units, often made of cement, used in aquaculture.

radiometer. An electronic device that can be used to estimate the biomass of grasses and forbs.

raghorn. A male elk approximately 2.5 years of age.

random dispersion. A dispersion pattern in which the position of an organism is independent of the positions of its conspecifics; this pattern occurs when a species lacks social tendencies toward clumping and where there is little variation in the environment.

random-stratified sample. A survey in which random samples are collected from predetermined categories (stratifications). A random-stratified creel survey might use boat anglers and shore anglers as stratifications, but would require random samples of each angler type.

range. See **geographic range** in reference to organism distribution. In reference to statistics, range is the difference between the largest and smallest observations in a sample.

ranging optometer. An optical device that provides distance readings, usually in meters.

ranking scale method. A survey or census method used to measure human attitudes in which the respondent is asked to order a list of items according to personal preference or some other standard.

raptor. A predatory bird that has feet with sharp talons or claws adapted for seizing prey and a hooked beak for tearing flesh.

realized niche. The niche occupied by an organism when some competition with other species is occurring.

recessive allele. An allele that is typically not expressed in the presence of a dominant allele.

reclamation burn. A prescribed burn conducted on mistreated or unmanaged land.

recovery plan. A plan developed to bring an endangered or threatened species back to heathy population levels. This term is specifically related to the U.S. Endangered Species Act.

recruitment. A term denoting the number of individuals hatched or born in any year that survive to reproductive size. Some alternative definitions include that number of individuals that reach harvestable size, a particular size or age, or a size captured by a particular sampling gear.

rectrices (singular: rectrix). The relatively large, conspicuous tail feathers on birds.

reflex. A simple innate behavior exhibited by an animal that is an autonomic response to a stimulus.

regulation. A rule enacted by a nonelected government entity such as a state fishery and wildlife commission, a federal agency, or a local board.

regulatory gene. A gene that controls the expression of a structural gene during development or among different tissues.

relative weight. An index of condition in fishes in which the actual weight of a fish is divided by a standard weight selected to represent an "optimal" weight for that species at that length, then multiplied by 100.

release. The result of reduction of competition from other, less desired plant species. Release can also occur after a reduction in the numbers of the target species.

remote sensing. The measurement of or acquisition of information about some property of an object or phenomenon by means of a recording device that is not in physical or intimate contact with the object or phenomenon under study. Aerial photography, for example, is a form of remote sensing.

reservoir. An impoundment that exceeds 40 hectares in surface area. Some reservoirs can be quite lotic in nature, while others are primarily lentic.

resource allocation. The allocation of portions of a particular resource, such as a particular fishery, to various resource users.

restoration. The act of reestablishing a native population that has been extirpated from a portion or all of its native range.

restored species. A species that has been extirpated from a portion or all of its native range and then successfully reestablished in the vacated area.

rete mirabile. A specialized netlike assemblage of blood vessels that brings large numbers of small arteries and veins into close contact to increase countercurrent exchange.

reticulum. The second chamber of a ruminant stomach.

reverse slot length limit. A regulation under which all fish of a particular species that are within a specified length range may be kept, while all those outside that length range must be released.

rhythmic behavior. An activity pattern or behavior of an animal that is associated with daily or seasonal changes.

riffle. That area of a headwater freshwater lotic system characterized by being shallow, narrow, and rock-bottomed. Current flow in riffles is stronger than that in pools because of the reduced volume of space present.

riparian zone. The habitat associated with the edges of rivers and streams.

riverine. Of or pertaining to lotic systems.

r-K **continuum.** The range of reproductive strategies exhibited by animals, from producing large numbers of offspring and providing no parental care (*r*) to producing small numbers of offspring and providing substantial parental care (*K*).

Robel pole. A marked pole used to determine the concealment or cover value of vegetation.

rocket net. A lightweight net that is carried over a group of animals by small rocket-type projectiles.

rookery. A nesting colony, usually of birds.

rotenone. A type of general-use piscicide.

r-**selected species.** A species with a reproductive strategy that involves production of large numbers of offspring and little parental care.

rumen. The first chamber of a ruminant stomach.

ruminant. An ungulate with an even number of toes that regurgitates and chews its food repeatedly.

run. An intermediate area located between the pools and riffles of a headwater lotic freshwater system. This term can also be used in reference to a particular stock or stocks of fish.

rut. The period associated with male breeding activity in some ungulates.

sagebrush rebellion. A movement on the part of some people, primarily those holding leases on U.S. federal grazing lands, in opposition to the Public Rangeland Improvement Act of 1978.

salmonid. A member of the fish family Salmonidae.

salt gland. A gland used for the active transport of salts out of the bloodstream.

salt wedge. Water with a higher salt content that, because of its higher density, is located under water with less salinity. Salt wedges are often associated with estuarine areas.

sample. In reference to animal populations, a sample is a subset or portion of the total number of organisms in a population. In a broader sense, a sample represents any subset or portion of the whole.

sample replication. Repeating a sample multiple times.

saprophyte. A nonanimal, such as a fungus, that feeds mainly on dead plant material.

scale annuli (singular: scale annulus). Annual growth rings found on fish scales.

scats. See **fecal pellets.**

scavenger. An animal that eats dead animals.

S-curve. A J-curve that has flattened at the upper end because the population is being restricted by one or more limiting factors.

seabirds. Birds that are associated with oceans and, at times, their shorelines.

Secchi disk. A metal, wooden, or plastic disk 20 centimeters in diameter that is used to estimate water clarity. Opposite quarters of the disk are usually colored white and black.

secondaries. See **secondary feathers.**

secondary consumer. An organism that feeds on primary consumers.

secondary feathers. Large flight feathers on the posterior margin of a bird wing arising between the "wrist" and "elbow."

secondary succession. Ecological succession occurring on a site that was previously occupied by other living organisms, such as a burned forest.

secondary tree cavity. A cavity that has already been constructed by another animal.

second law of thermodynamics. When energy is transferred or transformed, much is lost or dispersed in unusable forms, such as heat.

seine. An active capture device used in fishery work consisting of a panel of netting that is pulled through the water to capture the target animals.

selective cutting. A forestry practice that involves cutting only specific kinds and sizes of trees in an area so that some trees remain in the cut area.

semantic differential method. A survey or census method used to measure human attitudes in which an object or concept must be rated by the respondent on a scale with each endpoint anchored by an adjective or phrase.

semicircular canals. A portion of the inner ear of vertebrates that contributes to the sense of balance and equilibrium.

senior resident agent. A title used for a law enforcement officer employed by an agency such as the U.S. Fish and Wildlife Service who has more experience and responsibility than a special agent.

seral stage. One of the transient but recognizable stages in an ecological succession.

sere. The entire sequence of seral stages in an ecological succession.

sexual dichromatism. Differences in the color, pattern, hue, or shade of animals due to sex.

sexual dimorphism. Differences in the shape or form of animals due to sex.

shorebirds. A group consisting primarily of small wading birds, most of which feed along shores and mudflats.

shoreline development. The ratio of shoreline length to the circumference of a circle with the same area as the water body. Shoreline development is basically a measure of shoreline irregularity.

short-grass prairie. That portion of the temperate grassland biome characterized by grasses that are short in stature; the driest of the prairie types.

size limit. The size of an animal that can be legally harvested.

size structure. The size distribution of a population; for example, the relative proportions of small, medium-sized, and large individuals in a population.

SLAPP suit. A lawsuit intended to stifle expression by costing people or groups time, effort, or money to defend themselves. The intent of such a lawsuit is not to win a legal decision, but instead to intimidate someone into silence.

small impoundment. A constructed lentic water body that is between 4 and 40 hectares in surface area and intermediate in size between a pond and a reservoir.

smolt. The life stage of a salmonid fish at the time of physiological adaptation to life in the marine environment. The process is termed smoltification and occurs even in forms that spend their entire lives in fresh water.

snap trap. A passive capture device, usually used on mammals, such as the familiar "mousetrap."

snare. A passive capture device, most often used to capture mammals, that consists of a wire or rope noose that captures the animal by the leg or neck.

sociality. The tendency to gather in organized groups.

sociobiology. The study of the biological basis of social behavior and the organization of societies.

sodbuster provision. A provision in the 1985 and 1990 U.S. Food Security Acts (Farm Bills) that requires landowners who convert grasslands to croplands and who receive federal funding to comply with certain regulations.

songbird. A bird species that produces long vocal displays consisting of various series of complex notes and repeated patterns.

southeastern mixed forest. A portion of the temperate forest biome that ranges from Virginia through the Carolinas and Alabama and contains numerous evergreen trees.

spaced dispersion. See **uniform dispersion.**

spawn. In fishes, to release the male and female reproductive products. The term can also be used as a noun to denote either the male or female reproductive products.

spawning check. A false annulus on a fish scale (or other hard body part) caused by growth pattern anomalies at the time of spawning. See **false annuli.**

special agent. A title used for a law enforcement officer employed by an agency such as the U.S. Fish and Wildlife Service.

specialist. An organism that can utilize only a narrow range of resources or habitats.

speciation. The process of species formation.

species (plural: species). A naturally occurring group of organisms that can successfully interbreed and produce fertile offspring.

species diversity. A measure that includes both species richness and species equity or evenness. Most often, only certain portions of a community are assessed at one time, such as the trees in a forest.

species equity. A measure of the evenness of abundance of the various species in a community or portion of a community. Maximum equity occurs when all species in a portion of a community—birds, for example—are equally abundant.

species evenness. See **species equity.**

species richness. The number of species in a community or portion of a community.

specific conductance. See **conductivity.**

speculum. A colored, sometimes iridescent, patch on the secondary feathers of some bird wings.

spherical densiometer. A device used to determine tree canopy coverage that consists of a concave, mirrored surface marked with a grid.

sport animal. See **game animal.**

spot-mapping. A noncapture population estimation technique commonly used for songbirds that sing or call within established territories that involves plotting the locations of individual birds on a gridded map.

spring overturn. An event that occurs during spring in a thermally stratified lentic system in which water temperature throughout the water body is equalized and the entire water body mixes.

stakeholder. An individual or group that has something at stake in or an interest in a particular decision or resource.

standard deviation (SD). A statistical term denoting the average deviation between multiple observations and the mean.

standard error (SE). A statistical measure of variation; the standard error of the mean is obtained by dividing the standard deviation by the square root of the sample size.

standardized sampling. Sampling with effective devices using the same device at the same sites at the same time of year and from year to year. Standardized sampling is a common procedure when the purpose of sampling is to monitor long-term trends in a population.

standing crop. See **standing stock.**

standing stock. The abundance (number or biomass) of organisms present at a given time in a given area.

starlight scope. A type of optical equipment that gathers light and can be used to observe animals at night.

states' rights doctrine. In the United States, the power of individual states to control the wildlife and fishery resources within their borders, derived from the Tenth Amendment to the Constitution.

steelhead. A rainbow trout form that normally spends a portion of its life in marine waters.

stock. A group of animals with a common ancestry or parentage that is adapted to a particular environment. When used as a verb, the term denotes the release of organisms, as in fish stocking.

stock-recruitment curve. A curve that demonstrates the influence of parental abundance on subsequent population recruitment. Theoretically, the production of recruits is low when parental abundance is very low or very high and is generally high at moderate levels of parental abundance.

strain. See **stock.**

strategic planning. The development of missions, goals, objectives, and strategies.

Strategic Lawsuits Against Public Participation. See **SLAPP suits.**

strategy. An action undertaken to ensure that an objective will be met.

stress. The result of biotic and abiotic forces that extend the internal stabilizing mechanisms of an organism beyond its ability to maintain homeostasis.

structural gene. A gene that encodes an amino acid sequence or the structure of a protein.

strutting grounds. A lek used by sage grouse.

stunted population. A fishery term denoting a population characterized by high density and slow growth.

subadult. An age classification used in wildlife to denote an individual that is too young to breed but is externally indistinguishable from an adult.

submergent vegetation. Aquatic vegetation that grows in the water column but generally does not extend above the water surface.

subspecies. A formally recognized subdivision of a species. The members of a subspecies resemble one another to some extent and differ from other subdivisions in some recognizable way. Where ranges of subspecies overlap, they regularly interbreed to produce fertile offspring.

subyearling. See **young-of-the-year.**

suction trap. A trap that draws in air; used primarily to capture aerial insects.

summerkill. The death of aquatic organisms during summer resulting from dissolved oxygen depletion.

sunfish. A member of the fish family Centrarchidae. Most often the term is used in reference to smaller members of the family, such as bluegills and rock bass, but even largemouth bass are members of the sunfish family.

supplemental stocking. The release of fish into a population to augment natural reproduction. The term can also be used in reference to wildlife species.

Surber stream sampler. A device used to capture larger, primarily benthic, aquatic organisms, such as insects, in lotic habitats.

surplus production. That portion of production that can be removed from a population by natural causes or human harvest without adversely affecting future population levels.

survey. A partial count or sample of organisms present or harvested.

swamp. A wetland that usually contains standing trees.

swampbuster provision. A provision in the 1985 and 1990 U.S. Food Security Acts (Farm Bills) that requires landowners who receive federal funding to follow certain regulations with regard to wetland conservation.

sweep net. The familiar "butterfly net" used primarily to capture terrestrial invertebrates.

symbiotic. Of or pertaining to a relationship in which interacting species are closely and permanently dependent on each other.

sympatric. Having overlapping ranges of distribution.

systematic process. A genetic process that changes the allelic frequencies of a population in some predictable, nonrandom fashion.

taiga. See **northern coniferous forest.**

tailrace. The area immediately below a dam that receives water released from an impoundment.

tall-grass prairie. That portion of the temperate grassland biome characterized by grasses that are tall in stature; the wettest of the prairie types.

taxis. The oriented movement of an organism toward (positive taxis) or away from (negative taxis) a stimulus.

taxon (plural: taxa). The general term for a taxonomic group, whatever its rank.

temperate forest. A biome generally characterized by deciduous hardwood trees and moderate precipitation. In eastern North America it is found south of the northern coniferous forest biome.

temperate grassland. A biome characterized by the dominance of grass as a vegetation type. In North America it is primarily located west of the temperate forest biome and south of the northern coniferous forest biome.

temperate rain forest. A portion of the Pacific coast coniferous forest biome characterized by exceptionally high precipitation levels.

terminal gear. See **terminal tackle.**

terminal tackle. The fishing equipment located at the end of a fishing line.

territory. The portion of the home range of an animal that it defends against others of the same or sometimes closely related species.

tertiary consumer. An organism that primarily feeds on secondary consumers.

TFM (3-trifluoromethyl-4-nitrophenol). A selective piscicide used primarily on sea lampreys.

thermal stratification. The seasonal presence of layers with different temperatures and dissolved oxygen levels in freshwater lentic water bodies.

thermocline. The middle temperature zone in a lentic water body that is thermally stratified; the thermocline is the zone of greatest temperature variation.

thermoneutral zone. The range of ambient temperatures within which an animal requires no additional energy beyond that needed for normal body metabolism to maintain temperature homeostasis.

threatened species. A species that is likely to become endangered throughout all or a portion of its range. This term commonly relates to the U.S. Endangered Species Act, although other entities, such as states, provinces, and territories can also develop threatened species lists.

timberline. The highest altitude at which trees can grow in a particular area.

tines. The points on animal antlers.

titration. The measurement of the quantity of a given constituent of a solution through the addition of reagents of known concentration until a particular chemical reaction is completed. Water quality characteristics, such as dissolved oxygen content, can be determined by titration.

torpor. A period of lethargy or sluggishness. See **hypothermy.**

total mortality. The sum of natural and harvest mortality. Total mortality is commonly expressed as the overall death rate of a population in a one-year period.

total rehabilitation. See **total renovation.**

total renovation. A term used in fisheries to denote the complete removal of a fish community. This term can also be used in relation to wildlife species.

Total Quality Management (TQM). A planning or management process developed by Dr. W. Edwards Deming, who stressed employee empowerment to promote product quality.

trace element. See **microelement.**

trammel net. A passive capture device used in fisheries that entangles organisms in two or three panels of netting.

transect. A straight line along which samples are taken. Transects can be used in terrestrial or aquatic situations and are variable in length, depending on what is being sampled.

translocation. A form of genetic change that results when a segment from one chromosome becomes attached to a homologous chromosome.

transplant. An organism moved outside its native range but within the same zoogeographic region.

trap net. A passive capture device used in fisheries that contains either rectangular frames that stabilize the net or floats and stakes that hold the net open. The netting is arranged so that a fish can readily find its way into the trap, but not out of it. Lead nets are usually present.

trial-and-error learning. Learning by means of sorting through several nonrewarding actions to find an action that will provide a reward or avoid a punishment.

triploidy. The genetic condition of having three sets of chromosomes.

trophic level. The position of an organism in a food web or food chain; for example, the first trophic level consists of primary producers.

trophogenic zone. See **photic zone.**

tropholytic zone. See **aphotic zone.**

tropical rain forest. A biome characterized by high precipitation levels, hot climatic conditions, and lush vegetation. Large portions of equatorial South America, Southeast Asia, and the Malayan archipelago, and to a lesser extent Africa, consist of this biome.

tropical seasonal forest. A biome characterized by wet and dry seasons and hot temperatures throughout the year. The southern tip of Florida is representative of this biome.

tundra. A biome characterized by short, cool summers, long, cold winters, and permafrost. Large areas of northern Canada and Alaska consist of this biome.

turbidity. Water opaqueness due to suspended particulate matter.

two-story fishery. A fishery in which warmwater and/or coolwater fishes are present in one temperature stratum and coldwater fishes are present in another temperature stratum.

ultratrace element. An inorganic element required in extremely small amounts by an organism, and generally only lacking in some sterile captive situations.

ultraviolet survey trap. A device used to capture night-flying insects, both terrestrial and aquatic.

understory. Vegetation that lies beneath and is shaded by a canopy or overstory.

ungulate. A hoofed animal.

uniform dispersion. A dispersion pattern in which organisms are regularly or systematically spaced; often related to territoriality or strong intraspecific competition.

upland game. A category that includes most birds and small mammals that are used for sport hunting and are found in terrestrial systems.

upper critical temperature. The ambient temperature above the thermoneutral zone of an animal at which increased energy consumption must occur for the animal to maintain temperature homeostasis.

upper zone of metabolic compensation. The temperature zone above the upper critical temperature within which an animal can maintain temperature homeostasis by expending additional energy.

values. The qualities of a thing according to which it is thought of as being more or less desirable, useful, or important.

van Dorn water sampler. A plastic device that can be lowered into a water body, remotely triggered to close, and brought to the surface with a water sample from a particular depth.

variable radius method. A method of estimating tree density and basal area in which a prism or angle gauge is used to determine which trees are included in the inventory.

velvet. The soft, highly vascularized tissues that cover developing antlers.

visual obscurity measurement device. See **visual obstruction measurement device.**

visual obstruction measurement device. A marked board or rod used to determine the concealment or cover value of vegetation.

volatilize. To readily vaporize at a relatively low temperature.

Wallop-Breaux Amendment. An amendment to the Dingell-Johnson Act that places an excise tax on a variety of fishing gears and equipment not covered in the initial Act.

warmwater species. Freshwater fish species that survive, grow, and reproduce best in warm water. They generally spawn at water temperatures above 16°C (60°F).

waterfowl. A bird group consisting of ducks, geese, and swans.

Waterfowl Production Areas. Federally owned areas in the United States purchased with funding provided by the Migratory Bird Hunting Stamp Act (1934) and other sources to provide habitat primarily for migratory bird species. Many other wild organisms are benefited by these areas.

wattle. A fleshy lobe on the throat of a bird, such as that found on the male wild turkey.

wetland. An area where the water table is near ground level or just shallowly covers the land surface, such as a bog, marsh, swamp, or coastal wetland. This term can also be used to denote any area with surface water regardless of water depth or size.

wetland easement. A federally owned easement on private land in the United States purchased with funding provided by the Migratory Bird Hunting Stamp Act (1934) and other sources to provide habitat primarily for migratory bird species.

wild animals. Animals that are legally defined as wild by state, provincial, territorial, or federal governments.

wildlife. Wild terrestrial and partly terrestrial vertebrate animals. In a broader sense, this term can refer to all wild terrestrial, partly terrestrial, and aquatic organisms.

wildlife biology. The biological study of wildlife.

wildlife conservation. The wise management and use of wildlife resources.

wildlife ecology. The study of the interrelationships of organisms that are considered to be a part of the wildlife realm with other organisms and the environment.

wildlife management. The art and science of manipulating the biota, habitat, or human users of a wildlife system to produce some desired end result.

wildlife science. The process of obtaining knowledge about and studying wildlife resources.

wing dike. A bank stabilization structure that is used to concentrate current into the middle of a river channel.

winterkill. The death of aquatic organisms primarily resulting from dissolved oxygen depletion and increased toxic gases under ice cover.

Wisconsin plankton net. A specialized capture net used primarily to sample plankton.

wise-use movement. A group of people who advocate resource management that primarily favors grazing, logging, and mineral extraction on federally owned lands over other uses.

year class. The year in which a fish was hatched or born; for example, a fish hatched in 1992 will always be a member of the 1992 year class. This term also has some wildlife usage.

year-class strength. The abundance of a particular year class. Year-class strength is typically high for years when environmental conditions are appropriate for reproduction and recruitment and poor for years when they are not. This term is most commonly used in fisheries.

yearling. A term commonly used for an ungulate in its second summer and fall after birth. In fisheries, this term refers to an age-1 fish.

yield. That portion of a population harvested by humans.

young-of-the-year. A fish at age 0. This term is also used to identify the young of wildlife species.

zooplankton. Tiny aquatic animals that primarily float or drift in water.

APPENDIX

Vertebrates Used as Examples in this Book

Scientific names for North American fishes were obtained from C. R. Robins, R. M. Bailey, C. E. Bond, J. R. Brooker, E. A. Lachner, R. N. Lea, and W. B. Scott. 1991. *Common and scientific names of fishes from the United States and Canada.* 5th ed. American Fisheries Society, Special Publication 20, Bethesda, Md. Scientific names for North American amphibians, reptiles, birds, and mammals were obtained from R. C. Banks, R. W. McDiarmid, and A. L. Gardner. 1987. *Checklist of vertebrates of the United States, the U.S. Territories, and Canada.* U.S. Department of the Interior, Fish and Wildlife Service, Resource Publication 166, Washington, D.C. Species and subspecies not found in the above publications were obtained from a variety of sources.

FISHES

albacore	*Thunnus alalunga*
alligator gar	*Lepisosteus spatula*
American eel	*Anguilla rostrata*
American shad	*Alosa sapidissima*
Apache trout	*Oncorhynchus apache*
Arctic char	*Salvelinus alpinus*
Arctic grayling	*Thymallus arcticus*
Atlantic cod	*Gadus morhua*
Atlantic mackerel	*Scomber scombrus*
Atlantic salmon	*Salmo salar*
Atlantic sturgeon	*Acipenser oxyrhynchus*
banded pygmy sunfish	*Elassoma zonatum*
bigmouth buffalo	*Ictiobus cyprinellus*
black bullhead	*Ameiurus melas*
black crappie	*Pomoxis nigromaculatus*
black hagfish	*Eptatretus deani*
blacknose dace	*Rhinichthys atratulus*
blacktail shiner	*Cyprinella venusta*
bleeding shiner	*Luxilus zonatus*
blue catfish	*Ictalurus furcatus*
bluegill	*Lepomis macrochirus*
blue pike	see walleye
blue sucker	*Cycleptus elongatus*
blue tilapia	*Tilapia aurea*
bluntnose minnow	*Pimephales notatus*
bowfin	*Amia calva*

brassy minnow — *Hybognathus hankinsoni*
bridgelip sucker — *Catostomus columbianus*
brindled madtom — *Noturus miurus*
brook silverside — *Labidesthes sicculus*
brook stickleback — *Culaea inconstans*
brook trout — *Salvelinus fontinalis*
brown bullhead — *Ameiurus nebulosus*
brown trout — *Salmo trutta*
bull trout — *Salvelinus confluentus*

Cape Fear shiner — *Notropis mekistocholas*
central stoneroller — *Campostoma anomalum*
chain pickerel — *Esox niger*
channel catfish — *Ictalurus punctatus*
chinook salmon — *Oncorhynchus tshawytscha*
chum salmon — *Oncorhynchus keta*
coho salmon — *Oncorhynchus kisutch*
Colorado squawfish — *Ptychocheilus lucius*
common carp — *Cyprinus carpio*
creek chub — *Semotilus atromaculatus*
cutthroat trout — *Oncorhynchus clarki*
   greenback cutthroat trout — *O. c. stomias*
   Lahontan cutthroat trout — *O. c. henshawi*
   Yellowstone cutthroat trout — *O. c. bouvieri*

desert sucker — *Catostomus clarki*
Devils Hole pupfish — *Cyprinodon diabolis*
dollar sunfish — *Lepomis marginatus*

emerald shiner — *Notropis atherinoides*

fantail darter — *Etheostoma flabellare*
fathead minnow — *Pimephales promelas*
firemouth cichlid — *Cichlasoma meeki*
flat bullhead — *Ameiurus platycephalus*
flathead catfish — *Pylodictis olivaris*
freshwater drum — *Aplodinotus grunniens*

gizzard shad — *Dorosoma cepedianum*
golden shiner — *Notemigonus crysoleucas*
golden trout — *Oncorhynchus aquabonita*
grass carp — *Ctenopharyngodon idella*

great barracuda — *Sphyraena barracuda*
greenback cutthroat trout — see cutthroat trout
green sunfish — *Lepomis cyanellus*
guppy — *Poecilia reticulata*

haddock — *Melanogrammus aeglefinus*
hornyhead chub — *Nocomis biguttatus*

Jack Dempsey — *Cichlasoma octofasciatum*
johnny darter — *Etheostoma nigrum*

Lahontan cutthroat trout — see cutthroat trout
lake chub — *Couesius plumbeus*
lake sturgeon — *Acipenser fulvescens*
lake trout — *Salvelinus namaycush*
largemouth bass — *Micropterus salmoides*
longnose dace — *Rhinichthys cataractae*
longnose gar — *Lepisosteus osseus*
longnose sucker — *Catostomus catostomus*

Mexican tetra — *Astyanax mexicanus*
mountain madtom — *Noturus eleutherus*
mountain sucker — *Catostomus platyrhynchus*
mountain whitefish — *Prosopium williamsoni*
muskellunge — *Esox masquinongy*

Nile tilapia — *Tilapia nilotica*
northern hog sucker — *Hypentelium nigricans*
northern pike — *Esox lucius*

orangebelly darter — *Etheostoma radiosum*
orangespotted sunfish — *Lepomis humilis*
orangethroat darter — *Etheostoma spectabile*
Ozark cavefish — *Amblyopsis rosae*

Pacific barracuda — *Sphyraena argentea*
Pacific hagfish — *Eptatretus stouti*
Pacific halibut — *Hippoglossus stenolepis*
Pacific sardine — *Sardinops sagax*
paddlefish — *Polyodon spathula*
Pahrump poolfish — *Empetrichthys latos*
pallid sturgeon — *Scaphirhynchus albus*
peacock cichlid — *Cichla ocellaris*

pink salmon	*Oncorhynchus gorbuscha*	Utah chub	*Gila atraria*
plains minnow	*Hybognathus placitus*		
pollock	*Pollachius virens*	walleye	*Stizostedion vitreum*
pumpkinseed	*Lepomis gibbosus*	blue pike	variously regarded as a color phase of the walleye, a subspecies *S. v. glaucum,* or a species *S. glaucum*
rainbow darter	*Etheostoma caeruleum*		
rainbow smelt	*Osmerus mordax*		
rainbow trout	*Oncorhynchus mykiss*	wedgespot shiner	*Notropis greenei*
razorback sucker	*Xyrauchen texanus*	western mosquitofish	*Gambusia affinis*
redbreast sunfish	*Lepomis auritus*	white bass	*Morone chrysops*
redear sunfish	*Lepomis microlophus*	white catfish	*Ameiurus catus*
Red River pupfish	*Cyprinodon rubrofluviatilis*	white crappie	*Pomoxis annularis*
red shiner	*Cyprinella lutrensis*	white perch	*Morone americana*
riffle sculpin	*Cottus gulosus*	white shark	*Carcharodon carcharias*
river carpsucker	*Carpiodes carpio*	white sturgeon	*Acipenser transmontanus*
rock bass	*Ambloplites rupestris*	white sucker	*Catostomus commersoni*
		woundfin	*Plagopterus argentissimus*
sand shiner	*Notropis stramineus*		
sauger	*Stizostedion canadense*	yellow bullhead	*Ameiurus natalis*
sea lamprey	*Petromyzon marinus*	yellow perch	*Perca flavescens*
shorthead redhorse	*Moxostoma macrolepidotum*	Yellowstone cutthroat trout	see cutthroat trout
shortnose sturgeon	*Acipenser brevirostrum*		
shovelnose sturgeon	*Scaphirhynchus platorynchus*		
sicklefin chub	*Macrhybopsis meeki*		
silver chub	*Macrhybopsis storeriana*		

AMPHIBIANS AND REPTILES

silver lamprey	*Ichthyomyzon unicuspis*	American alligator	*Alligator mississippiensis*
skipjack herring	*Alosa chrysochloris*		
smallmouth bass	*Micropterus dolomieu*	bullfrog	*Rana catesbeiana*
smallmouth buffalo	*Ictiobus bubalus*		
snail darter	*Percina tanasi*	common kingsnake	*Lampropeltis getulus*
sockeye salmon	*Oncorhynchus nerka*		
southern cavefish	*Typhlichthys subterraneus*	desert tortoise	*Gopherus agassizii*
spoonhead sculpin	*Cottus ricei*	diamondback water snake	*Nerodia rhombifera*
spottail shiner	*Notropis hudsonius*		
spotted bass	*Micropterus punctulatus*		
spotted gar	*Lepisosteus oculatus*	eastern diamondback rattlesnake	*Crotalus adamanteus*
spotted sucker	*Minytrema melanops*		
starhead topminnow	*Fundulus dispar*		
steelhead	see rainbow trout	Kemp's ridley	*Lepidochelys kempii*
striped bass	*Morone saxatilis*		
sturgeon chub	*Macrhybopsis gelida*	long-nosed leopard lizard	*Gambelia wislizenii*
suckermouth minnow	*Phenacobius mirabilis*	long-toed salamander	*Ambystoma macrodactylum*
		Santa Cruz long-toed salamander	*A. m. croceum*
threadfin shad	*Dorosoma petenense*		
thresher shark	*Alopias vulpinus*		

marine iguana	*Amblyrhynchus cristatus*
smooth green snake	*Opheodrys vernalis*
snapping turtle	*Chelydra serpentina*
southern leopard frog	*Rana sphenocephala*
southern water snake	*Nerodia fasciata*
tiger salamander	*Ambystoma tigrinum*
western fence lizard	*Sceloporus occidentalis*
western hog-nosed snake	*Heterodon nasicus*

BIRDS

Adelie penguin	*Pygoscelis adeliae*
Aleutian Canada goose	see Canada goose
American avocet	*Recurvirostra americana*
American black duck	*Anas rubripes*
American coot	*Fulica americana*
American crow	*Corvus brachyrhynchos*
American dipper	*Cinclus mexicanus*
American goldfinch	*Carduelis tristis*
American kestrel	*Falco sparverius*
American redstart	*Setophaga ruticilla*
American robin	*Turdus migratorius*
American white pelican	*Pelecanus erythrorhynchos*
American wigeon	*Anas americana*
American woodcock	*Scolopax minor*
Andean condor	*Vultur gryphus*
Arctic tern	*Sterna paradisaea*
Attwater's greater prairie-chicken	see greater prairie-chicken
Baird's sandpiper	*Calidris bairdii*
Baird's sparrow	*Ammodramus bairdii*
bald eagle	*Haliaeetus leucocephalus*
bank swallow	*Riparia riparia*
barn swallow	*Hirundo rustica*
belted kingfisher	*Ceryle alcyon*
black-and-white warbler	*Mniotilta varia*
Blackburnian warbler	*Dendroica fusca*
black phoebe	*Sayornis nigricans*
black vulture	*Coragyps atratus*
blue grouse	*Dendragapus obscurus*
blue jay	*Cyanocitta cristata*

blue-winged teal	*Anas discors*
bobolink	*Dolichonyx oryzivorus*
Brandt's cormorant	*Phalacrocorax penicillatus*
brant	*Branta bernicla*
Brewer's sparrow	*Spizella breweri*
brown-headed cowbird	*Molothrus ater*
brown kiwi	*Apteryx australis*
brown pelican	*Pelecanus occidentalis*
brown towhee	*Pipilo fuscus*
burrowing owl	*Athene cunicularia*
California clapper rail	see clapper rail
California condor	*Gymnogyps californianus*
California quail	*Callipepla californica*
Canada goose	*Branta canadensis*
Aleutian Canada goose	*B. c. leucopareia*
giant Canada goose	*B. c. maxima*
canvasback	*Aythya valisineria*
Carolina parakeet	*Conuropsis carolinensis*
chestnut-collared longspur	*Calcarius ornatus*
chicken (domesticated)	*Gallus domesticus*
chimney swift	*Chaetura pelagica*
chipping sparrow	*Spizella passerina*
chukar	*Alectoris chukar*
clapper rail	*Rallus longirostris*
California clapper rail	*R. l. obsoletus*
cliff swallow	*Hirundo pyrrhonota*
common barn-owl	*Tyto alba*
common grackle	*Quiscalus quiscula*
common loon	*Gavia immer*
common merganser	*Mergus merganser*
common murre	*Uria aalge*
common nighthawk	*Chordeiles minor*
common poorwill	*Phalaenoptilus nuttallii*
common snipe	*Gallinago gallinago*
common tern	*Sterna hirundo*
Cooper's hawk	*Accipiter cooperii*
dodo	*Raphus cucullatus*
double-crested cormorant	*Phalacrocorax auritus*
downy woodpecker	*Picoides pubescens*
dusky seaside sparrow	see seaside sparrow
dwarf cassowary	*Casuarius bennetti*

eared grebe	*Podiceps nigricollis*	Iceland gull	*Larus glaucoides*
eastern bluebird	*Sialia sialis*	ivory gull	*Pagophila eburnea*
eastern kingbird	*Tyrannus tyrannus*		
eastern meadowlark	*Sturnella magna*	Japanese quail	*Coturnix japonica*
emperor penguin	*Aptenodytes forsteri*		
emu	*Dromaius novaehollandiae*	killdeer	*Charadrius vociferus*
European starling	*Sturnus vulgaris*	king eider	*Somateria spectabilis*
		king penguin	*Aptenodytes patagonicus*
ferruginous hawk	*Buteo regalis*	Kirtland's warbler	*Dendroica kirtlandii*
field sparrow	*Spizella pusilla*		
Florida scrub jay	see scrub jay	least tern	*Sterna antillarum*
		lesser golden-plover	*Pluvialis dominica*
gadwall	*Anas strepera*	lesser megalapteryx	*Megalapteryx didinus*
Gambel's quail	*Callipepla gambelii*	lesser prairie-chicken	*Tympanuchus pallidicinctus*
giant Canada goose	see Canada goose	lesser rhea	*Pterocnemia pennata*
gila woodpecker	*Melanerpes uropygialis*	lesser snow goose	see snow goose
glaucous gull	*Larus hyperboreus*	little blue heron	*Egretta caerulea*
golden eagle	*Aquila chrysaetos*	long-billed dowitcher	*Limnodromus scolopaceus*
grasshopper sparrow	*Ammodramus savannarum*		
gray-cheeked thrush	*Catharus minimus*	mallard	*Anas platyrhynchos*
gray duck	*Anas superciliosa*	marbled murrelet	*Brachyramphus marmoratus*
gray partridge	*Perdix perdix*	Merriam's wild turkey	see wild turkey
great auk	*Penguinus impennis*	mountain bluebird	*Sialia currucoides*
great blue heron	*Ardea herodias*	mourning dove	*Zenaida macroura*
great egret	*Casmerodius albus*		
great horned owl	*Bubo virginianus*	northern bobwhite	*Colinus virginianus*
greater prairie-chicken	*Tympanuchus cupido*	northern flicker	*Colaptes auratus*
Attwater's greater		red-shafted flicker	a race of northern flicker
prairie-chicken	*T. c. attwateri*	yellow-shafted flicker	a race of northern flicker
heath hen	*T. c. cupido*	northern goshawk	*Accipiter gentilis*
greater rhea	*Rhea americana*	northern harrier	*Circus cyaneus*
greater white-fronted		northern mockingbird	*Mimus polyglottos*
goose	*Anser albifrons*	northern oriole	*Icterus galbula*
greater yellowlegs	*Tringa melanoleuca*	northern pintail	*Anas acuta*
green kingfisher	*Chloroceryle americana*	northern spotted owl	see spotted owl
green-winged teal	*Anas crecca*		
gyrfalcon	*Falco rusticolus*	oldsquaw	*Clangula hyemalis*
		osprey	*Pandion haliaetus*
Hawaiian goose	*Nesochen sandvicensis*	ostrich	*Sturthio camelus*
heath hen	see greater prairie-chicken	ovenbird	*Seiurus aurocapillus*
horned grebe	*Podiceps auritus*		
horned lark	*Eremophila alpestris*	painted bunting	*Passerina ciris*
horned puffin	*Fratercula corniculata*	passenger pigeon	*Ectopistes migratorius*
house finch	*Carpodacus mexicanus*	pectoral sandpiper	*Calidris melanotos*
house sparrow	*Passer domesticus*		

peregrine falcon	*Falco peregrinus*
pied-billed grebe	*Podilymbus podiceps*
pigeon	see rock dove
pileated woodpecker	*Dryocopus pileatus*
piping plover	*Charadrius melodus*
purple martin	*Progne subis*
razorbill	*Alca torda*
red-eyed vireo	*Vireo olivaceus*
redhead	*Aythya americana*
red-shafted flicker	see northern flicker
red-tailed hawk	*Buteo jamaicensis*
red-winged blackbird	*Agelaius phoeniceus*
ring-billed gull	*Larus delawarensis*
ring-necked duck	*Aythya collaris*
ring-necked pheasant	*Phasianus colchicus*
rock dove (pigeon)	*Columba livia*
rock ptarmigan	*Lagopus mutus*
rock sandpiper	*Calidris ptilocnemis*
rose-breasted grosbeak	*Pheucticus ludovicianus*
Ross' goose	*Chen rossii*
ruby-throated hummingbird	*Archilochus colubris*
ruddy duck	*Oxyura jamaicensis*
ruffed grouse	*Bonasa umbellus*
rufous-crowned sparrow	*Aimophila ruficeps*
sage grouse	*Centrocercus urophasianus*
sage thrasher	*Oreoscoptes montanus*
sandhill crane	*Grus canadensis*
scaled quail	*Callipepla squamata*
scarlet tanager	*Piranga olivacea*
scissor-tailed flycatcher	*Tyrannus forficatus*
scrub jay	*Aphelocoma coerulescens*
Florida scrub jay	*A. c. coerulescens*
seaside sparrow	*Ammodramus maritimus*
dusky seaside sparrow	*A. m. nigrescens*
sharp-shinned hawk	*Accipiter striatus*
sharp-tailed grouse	*Tympanuchus phasianellus*
slender moa	*Dinornis torosus*
snail kite	*Rostrhamus sociabilis*
snow goose	*Chen caerulescens*
lesser snow goose	*C. c. caerulescens*
snowy egret	*Egretta thula*
snowy owl	*Nyctea scandiaca*

spotted owl	*Strix occidentalis*
northern spotted owl	*S. o. caurina*
spruce grouse	*Dendragapus canadensis*
Swainson's hawk	*Buteo swainsoni*
three-toed woodpecker	*Picoides tridactylus*
trumpeter swan	*Cygnus buccinator*
turkey (domesticated)	see wild turkey
turkey vulture	*Cathartes aura*
upland sandpiper	*Bartramia longicauda*
vesper sparrow	*Pooecetes gramineus*
western grebe	*Aechmophorus occidentalis*
western kingbird	*Tyrannus verticalis*
western meadowlark	*Sturnella neglecta*
western tanager	*Piranga ludoviciana*
white-breasted nuthatch	*Sitta carolinensis*
white-crowned sparrow	*Zonotrichia leucophrys*
white-winged dove	*Zenaida asiatica*
white-winged scoter	*Melanitta fusca*
whooping crane	*Grus americana*
wild turkey	*Meleagris gallopavo*
Merriam's wild turkey	*M. g. merriami*
Wilson's phalarope	*Phalaropus tricolor*
wood duck	*Aix sponsa*
wood thrush	*Hylocichla mustelina*
worm-eating warbler	*Helmitheros vermivorus*
yellow-headed blackbird	*Xanthocephalus xanthocephalus*
yellow-shafted flicker	see northern flicker
yellow warbler	*Dendroica petechia*

MAMMALS

African elephant	*Loxodonta africana*
Arctic fox	*Alopex lagopus*
Arctic ground squirrel	*Spermophilus parryii*
Arizona pocket mouse	*Perognathus amplus*
Asiatic elephant	*Elephas maximus*
Audubon bighorn sheep	see bighorn sheep
aurochs	*Bos primigenius*

badger	*Taxidea taxus*	eastern cottontail	*Sylvilagus floridanus*
barbary sheep	*Ammotragus lervia*	eastern mole	*Scalopus aquaticus*
beaver	*Castor canadensis*	elk	see wapiti
big brown bat	*Eptesicus fuscus*		
bighorn sheep	*Ovis canadensis*	fallow deer	*Dama dama*
Audubon bighorn		feral pig	see pig
sheep	*O. c. auduboni*	Florida panther	see mountain lion
bison	*Bison bison*	fox squirrel	*Sciurus niger*
black bear	*Ursus americanus*	Franklin's ground squirrel	*Spermophilus franklinii*
black-eared mouse	*Peromyscus melanotis*		
black-footed ferret	*Mustela nigripes*	giraffe	*Giraffa camelopardalis*
black rat	*Rattus rattus*	goat (domesticated)	*Hircus aegagrus*
black rhinoceros	*Diceros bicornis*	gorilla	*Gorilla gorilla*
black-tailed deer	see mule deer	mountain gorilla	*G. g. beringei*
black-tailed jack rabbit	*Lepus californicus*	gray squirrel	*Sciurus carolinensis*
black-tailed prairie dog	*Cynomys ludovicianus*	gray wolf	*Canis lupus*
blue wildebeest	*Connochaetes taurinus*	grizzly bear	*Ursus arctos*
bobcat	*Lynx rufus*	guinea pig	*Cavia porcellus*
bottle-nosed dolphin	*Tursiops truncatus*	Gunnison's prairie dog	*Cynomys gunnisoni*
brown bear	see grizzly bear		
brown lemming	*Lemmus sibiricus*	hamadryas	*Papio hamadryas*
burro (domesticated)	*Equus assinus*	harbor seal	*Phoca vitulina*
		hooded seal	*Cystophora cristata*
California sea lion	*Zalophus californianus*	horse (domesticated)	*Equus caballus*
caribou	*Rangifer tarandus*	house cat (domesticated)	*Felis silvestris*
cat	see house cat	house mouse	*Mus musculus*
cattle (domesticated)	*Bos taurus* and *Bos indicus*	human	*Homo sapiens*
chamois	*Rupicapra rupicapra*	humpback whale	*Megaptera novaeangliae*
cheetah	*Acinonyx jubatus*		
chimpanzee	*Pan troglodytes*	Indian mongoose	*Herpestes auropunctatus*
collared peccary	*Tayassu tajacu*		
Columbian white-		killer whale	*Orcinus orca*
tailed deer	see white-tailed deer	kit fox	*Vulpes macrotis*
coyote	*Canis latrans*	San Joaquin kit fox	*V. m. mutica*
Dall sheep	*Ovis dalli*	least shrew	*Cryptotis parva*
deer mouse	*Peromyscus maniculatus*	little brown bat	*Myotis lucifugus*
desert cottontail	*Sylvilagus audubonii*	little pocket mouse	*Perognathus longimembris*
dog (domesticated)	*Canis familiaris* (probably	lodgepole chipmunk	*Tamias speciosus*
	not a distinct species but	long-tailed weasel	*Mustela frenata*
	instead descended from	lynx	*Lynx canadensis*
	various species of *Canis*)		
duck-billed platypus	*Ornithorhynchus anatinus*	manatee	*Trichechus manatus* (also
			called West Indian
eastern chipmunk	*Tamias striatus*		manatee)

mandrill	*Papio sphinx*
marsh rabbit	*Sylvilagus palustris*
marten	*Martes americana*
meadow vole	*Microtus pennsylvanicus*
Mexican long-nosed bat	*Leptonycteris nivalis*
Mexican vole	*Microtus mexicanus*
mink	*Mustela vison*
montane vole	*Microtus montanus*
moose	*Alces alces*
mouflon	*Ovis musimon*
mountain goat	*Oreamnos americanus*
mountain gorilla	see gorilla
mountain lion	*Felis concolor*
Florida panther	*F. c. coryi*
mule deer	*Odocoileus hemionus*
Sitka black-tailed deer	*O. h. sitkensis*
muskox	*Ovibos moschatus*
muskrat	*Ondatra zibethicus*
northern flying squirrel	*Glaucomys sabrinus*
northern fur seal	*Callorhinus ursinus*
northern grasshopper mouse	*Onychomys leucogaster*
northern sea lion	*Eumetopias jubatus*
Norway rat	*Rattus norvegicus*
nutria	*Myocastor coypus*
Nuttall's cottontail	*Sylvilagus nuttallii*
Ord's kangaroo rat	*Dipodomys ordii*
Pacific jumping mouse	*Zapus trinotatus*
pallid bat	*Antrozous pallidus*
pig (domesticated and feral)	*Sus scrofa*
polar bear	*Ursus maritimus*
porcupine	*Erethizon dorsatum*
pronghorn	*Antilocapra americana*
pygmy rabbit	*Sylvilagus idahoensis*
raccoon	*Procyon lotor*
red deer	see wapiti

red fox	*Vulpes vulpes*
river otter	*Lutra canadensis*
roe deer	*Capreolus capreolus*
sacred baboon	see hamadryas
San Joaquin kit fox	see kit fox
sea otter	*Enhydra lutris*
sheep (domesticated)	*Ovis aries*
short-finned pilot whale	*Globicephala macrorhynchus*
Sitka black-tailed deer	see mule deer
snowshoe hare	*Lepus americanus*
southern flying squirrel	*Glaucomys volans*
sperm whale	*Physeter catodon*
spiny anteater	*Tachyglossus aculeatus*
spotted skunk	*Spilogale putorius*
Steller's sea cow	*Hydrodamalis gigas*
striped skunk	*Mephitis mephitis*
swift fox	*Vulpes velox*
thirteen-lined ground squirrel	*Spermophilus tridecemlineatus*
tiger	*Panthera tigris*
walrus	*Odobenus rosmarus*
wapiti	*Cervus elaphus*
red deer	*C. e. elaphus*
warthog	*Phacochoerus aethiopicus*
West Indian manatee	see manatee
white-footed mouse	*Peromyscus leucopus*
white rhinoceros	*Ceratotherium simum*
white-tailed deer	*Odocoileus virginianus*
Columbian white-tailed deer	*O. v. leucurus*
white-tailed jack rabbit	*Lepus townsendii*
white-tailed prairie dog	*Cynomys leucurus*
wisent	*Bison bonasus*
wolverine	*Gulo gulo*
yellow-bellied marmot	*Marmota flaviventris*
yellow-nosed cotton rat	*Sigmodon ochrognathus*

INDEX

Note: Page numbers in *italics* indicate illustrations; those followed by *t* indicate tables.